By Christopher Dow

Fiction
Effigy
    Book I: Stroud
    Book II: Oakdale
The Books of Bob
    Devil of a Time
    Jumping Jehovah
The Clay Guthrie Mysteries
    The Dead Detective
    Landscape with Beast
    The Texas Troll Unlimited
    Darkness Insatiable
Roadkill
The Werewolf and Tide, and Other Compulsions

Nonfiction
Lord of the Loincloth (nonfiction novel)
Book of Curiosities: Adventures in the Paranormal
Occasional Pilgrimage: Essays on Film, Literature, and Other Matters
Living the Story: The Meandering, True, and Sometimes Strange
    Adventures of an Unknown Writer
    Vol.I: Growing Up Takes a Long Time
    Vol. II: Growing Old Takes Longer

Martial Arts
The Wellspring: An Inquiry into the Nature of Chi
Circling the Square: Observations on the Dynamics of Tai Chi Chuan
Elements of Power: Essays on the Art and Practice of Tai Chi Chuan
Alchemy of Breath: An Introduction to Chi Kung
Leaves on the Wind: A Survey of Martial Arts Literature (Vol. I–VI)

Poetry
City of Dreams
The Trip Out
Texas White Line Fever
Networks
A Dilapidation of Machinery
Puzzle Pieces: Selected Poems

Editor
The Abby Stone: The Poetry of Bartholo Dias
The Best of Phosphene
The Best of Dialog

Lord of the Loincloth

Lord of the Loincloth

Christopher Dow

Phosphene Publishing Company
Temple, Texas

Lord of the Loincloth
© 2007 by Christopher Dow
ISBN 13: 9780985147709

Published by:
Phosphene Publishing Company
Temple, Texas, U.S.A.
phosphenepublishing.com

All rights reserved. No part of this work may be copied or otherwise produced or reproduced in any form, printed or electronic, without express permission of the author, except for brief excerpts used in reviews, articles, and critical works.

This book is dedicated to those who dare to go on despite travail and seemingly insurmountable obstacles. Most of all, it is dedicated to my wife, Julie, who did the same with me.

Contents

Prologue —— 13

Book I: On Land
Chapter I: The War in Africa —— 19
Chapter II: Preparations —— 37
Chapter III: England to Cape Town —— 69
Chapter IV: Cape Town to Fungurume —— 105
Chapter V: Fungurume —— 141
Chapter VI: On the Road —— 185
Chapter VII: The End of Bad Luck —— 203
Chapter VIII: Back on Track —— 235
Chapter IX: The Mitumba Mountains —— 251

Book II: On Water
Chapter X: The Lualaba River —— 281
Chapter XI: The Belgians —— 321
Chapter XII: Kalemie —— 353
Chapter XIII: *Kingani* —— 389
Chapter XIV: January —— 413
Chapter XV: *Hedwig von Wissmann* —— 443
Chapter XVI: Bwana Chifunga Tumbo —— 459
Chapter XVII: Searching for *St. George* —— 481
Chapter XVIII: Bismarckburg —— 503
Chapter XIX: The Duration —— 517

Acknowledgements/Sources/Bibliography —— 535

Lord of the Loincloth

Prologue

LIEUTENANT COMMANDER GEOFFREY BASIL SPICER-Simson was in high spirits as he stepped onto the deck of his command ship, the H.M.S. *Niger*. His swagger stick briskly tapping a counterpoint on the plating, he glanced neither left nor right as he strode imperiously to the rail. He stood there for a moment, stroking his graying goatee, then he lit a cigarette, inserted it into a foot-long ebony holder, stuck the holder between his teeth, and took a few puffs. The crisp, early morning March air made his pulse quicken as his piercing, frosty blue eyes surveyed the watery domain over which he had such important charge.

Although he'd been in command of the Downs Boarding Flotilla for a little less than two weeks, already the doldrums and failures of his past were fading and the mantle of leadership settling easily upon his shoulders. True, a couple of elderly gunboats and half a dozen tugs wasn't a fleet to complement the *Niger*, but damn it, at least he was no longer beached! And in all truth, the appointment wasn't half-bad. Ramsgate, at the mouth of the Stour River, was about as close to France as one could get and still be in England, and the Great War was young, yet. If German U-boats came this way, surely his flotilla would see action.

Spicer-Simson nodded, filled with a sense of inevitability. He was more than a mere a naval man, more than a simple warrior. He knew that now, knew it in the marrow of his bones. He was born to lead, and his rightful place was on the bridge of a warship in times of war. His destiny was on him, and one good engagement was all he asked. Then he'd show the Admiralty of what stuff he was made.

Not today, though. Today was such a fine day. Plenty of time later for battle. Today was a day for celebration. What a pity Amy was eighty miles away in London. He would have liked celebrating with his wife, but he wouldn't be seeing her for weeks longer. Ah, well. He knew a couple of interesting young ladies in town. Perhaps they would care to join him for a drink. Surely they'd like to hear of that time he'd faced down a tiger in Borneo, armed only with a Webley & Scott self-loading pistol.

Spicer-Simson had his lieutenant anchor the *Niger* in the outer reaches of the harbor, near the Deal Bank buoy, and drop a launch to take him in to the docks. He found that the young ladies he'd been thinking of were unoccupied, and he took a sea-view room in a quayside hotel and ordered several bottles of wine to be brought up. More than a couple of pleasant hours passed, and the young ladies consumed several of his more extravagant tales of derring-do as readily as they did the wine. Perhaps they were ripe for his current achievements.

"As you can see," Spicer-Simson said, gesturing out the window, "the Admiralty has great faith in my abilities. The boats under my command are U-boat hunters, instrumental in protecting the Thames, and London itself, from German attack. The U-boats are especially dangerous in this area. Do you see that ship anchored out there, near the mouth of the harbor? The larger navy one? That's my flagship, the *Niger*. An appropriate name considering my African exploits, don't you think?"

The ladies could not help but see the *Niger*. The bright day was completely devoid of mists, and visibility was clear enough that, from their third-story vantage, they could just make out the dark line of the French coast.

"What are those other boats next to it?" asked the pretty little brunette.

"The two close by are French and Dutch steamers under my protection." Spicer-Simson quaffed off the rest of his glass of wine and reached for the bottle of burgundy. "The six tugs are used for boarding inspections."

"What about that little boat?" the redhead asked. "The one that just came up?"

"Just came up?" Spicer-Simson turned and squinted out across the bright harbor. A little beyond his flotilla, a small, dark shape floated. He did not immediately realize that it was the conning tower of a German U-boat, but when a silvery underwater streak sped from the U-boat toward his flotilla, he was shocked into recognition.

"Look out, you fools!" he screamed through the window, but the men on the flotilla were more than a mile away and paying no more attention to him than they were to the torpedo speeding toward them.

In seconds, the underwater missile struck the *Niger*, and a great fountain of water and debris flung itself into the air. Another second passed before the sound of the explosion booming across the quay reached Spicer-Simson's ears. He dropped the bottle of wine, and it fell onto its side, gurgling its contents over the floor.

"Shit!" he bellowed. "Shit!" Jerking on his tunic, he raced for the door.

Speed was futile. In the twenty minutes it took his launch to reach the location where the *Niger* had been anchored, all that remained was an oil slick, some debris, and floating sailors, some swimming toward the two nearby steamers, hauling wounded companions, some floating face down.

The German U-boat was long gone.

When Spicer-Simson was called before the naval review board convened to crucify him, he was ready for the worst they could hand him. Willing. What more could they do, anyway? He'd lost his second command in a way more ridiculous, if that was possible, than his first. His career was scuttled, destroyed beyond belief, beyond repair.

"You have failed miserably in your duty," the chairman stated bluntly.

Spicer-Simson needed no assistance, at the moment, in feeling guilty. The sight of his dead crewmen floating pointlessly in pools of blood and oil had been enough to unseat him thoroughly.

"What explanation have you for this unlikely incident?"

"No explanation, sir."

A second review board member snorted derisively and said, "You had your command shot out from beneath your arse, and you have no explanation?"

"It was most unfortunate, sir," Spicer-Simson said quietly.

A third board member held up a report that Spicer-Simson recognized too well, and his heart sank.

"It says here," the man said, "that your first ship, the *Alcott*, rammed and sank a liberty boat in the English Channel."

"Yes, sir," Spicer-Simson admitted. "But that was nearly ten years ago. If you'll look at my record since, you'll see that I have served faithfully and successfully."

"I wouldn't call a survey of the Gambia River a significant achievement."

You would if it had been you going up that muddy slough, broiling under the African sun for four years, Spicer-Simson thought. Aloud, all he said was, "No, sir."

"If the war effort didn't demand every available hand, we would cashier you immediately," the chairman said. "But regular career naval officers are in high demand, and the Admiralty is willing to keep you on."

"Thank you, sir."

"Don't thank me," the chairman snapped. "We are not overlooking your idiotic blunders. You will remain in the Royal Navy, at your present rank, but in view of your apparently total incompetence as a field commander, you are being reassigned to a post where, we pray, you will not be able to do more damage to us than to the enemy."

Spicer-Simson was dismissed with a contemptuous wave, and he slinked over to his new post in the Admiralty's personnel department. Here he was to sit, a glorified clerk, and review application forms from merchant seamen who wished to join the navy. That was the explicit order. The implicit one was that he was to cause the Admiralty as little trouble as possible.

Once again, Spicer-Simson was beached, as he had been many times in his life, but this time, it promised to be permanent.

Book One: On Land

Chapter I: The War in Africa

ON APRIL 21, 1915, SIR Henry Jackson, the British First Sea Lord, sat in his office, brooding over strategic planning and the indigestion that had been plaguing him since he'd eaten that curried beef for lunch. There was a knock on his door, and his adjutant entered.

"Mr. Lee to see you, sir."

"Lee?" Sir Henry had to suppress a belch.

"The big game hunter. You remember. He's here straight away from Cape Town. It's about...."

"Yes, yes. Something about Africa. Well, show him in."

The adjutant disappeared then reappeared with a big, sun-browned man dressed in the khaki appropriate for his profession. A wide-brimmed Boer hat was on his head.

"Mr. John R. Lee," the adjutant said. "Mr. Lee, this is Lord Admiral Sir Henry Jackson."

"Pleased to meet you, Sir Henry," Lee said, doffing his hat.

Sir Henry rose, came around his desk, and shook Lee's hand. It was rough, strong, and reassuring.

"A drink, Mr. Lee?"

"Scotch and soda, if you please."

"I'd have thought gin and quinine water, considering where you're from."

"To tell the truth, sir, I've had my fill of quinine, and when I get back, I'll have my fill again. But here, I'd like something a little more civilized."

Sir Henry laughed and turned to his adjutant. "You heard the man, Whitley. And my usual for me."

As the adjutant left, Sir Henry gestured toward a pair of leather-covered easy chairs near the fireplace, and the two men sat. Lee placed his hat on the floor next to his chair, brim up. He didn't seem too awed to be in the presence of the British First Sea Lord, but then he'd spent the better part of his adult life hobnobbing with kings and chiefs. That they'd been black and their domains in central and southern Africa rather than Europe made no difference to Lee. They were as regal and personally impressive as anything Europe had produced for the last two hundred years, and if they were less powerful, that was only due to the accidents of technology that allowed their European counterparts to wage wars with gunpowder and machines.

"Well, Lee, you're a big game hunter, I hear."

"A bit of that," Lee admitted. Lee was not just a big game hunter; he was a prospector, and he'd served with Rimington's Scouts during the Boer War and later as an officer of the Canadian Scouts. His life in Africa had taught him the wisdom of practicality, and his intelligence and ingenuity had carried him through situations where practicality could not prevail.

"You look the part," Sir Henry said. "Know the territories well?"

"Well enough."

"I've always had a fascination for the African interior—you know, Burton, Livingstone, and whatnot. I was a young man when all that was going on, and I even was stationed there for a time. I actually met Stanley once. Tough as nails. A real man of iron. But frankly, I can't imagine why you've come to me. Africa, and all. I'm a naval man, not the army."

"That's precisely why I've come to you, Sir Henry. His Majesty's boats are required."

The adjutant entered with the drinks.

"Set those on the table, Whitley," Sir Henry said, "And send for Gamble and Duggan. Might as well get Corley in, as well."

As the adjutant left, Sir Henry turned back to his visitor. "So, you have some information on the Hun?"

"I have," Lee said. "You know, of course, of von Lettow-Vorbeck."

Sir Henry knew the name well. Colonel Paul Emil von Lettow-Vorbeck was the commanding officer in charge of German East Africa, the most strongly held German territory in Africa. His twenty-five years of military service in conflicts from Europe to China to Africa not only had hardened him but had given his natural military genius considerable practical experience. A sharp observer of African warfare and guerrilla tactics, von Lettow-Vorbeck had had great success in convincing several central African tribes—in particular the Ba-HoloHolo—to act as spies and guerrillas on his behalf. His method to control the Ba-HoloHolo was alternately sly, nasty, and brutish. He easily took advantage of the Ba-HoloHolo's overpowering hatred for the Belgians, whose cruelty toward them and their tribal neighbors was well known, and at the same time, he made sure that they kept a healthy respect for him by occasionally shelling their villages. He did appease them on a gut level, however, by encouraging their proclivity for cannibalism.

"Von Lettow-Vorbeck has amassed a considerable number of troops near Taveta and Mombassa on the north side of the colony," Lee said, "but his real advantage is a pair of armed steamers on Lake Tanganyika. Using those, he can deploy nearly a thousand fully equipped askaris from one end of the colony to the other in days. Their mobility has virtually paralyzed the Belgians, forcing them to hole up in their several forts along the western shore. But even that situation is deteriorating as the Germans are now using the lake as a means to conduct raids against British as well as Belgian posts. In short, Sir Henry, German supremacy on the lake has to be broken if the Allies ever hope to launch a successful invasion against German East Africa."

Sir Henry's attention was immediately piqued. Just the previous week, he'd attended a strategic planning committee meeting at which a major point of discussion was a joint operation against German East Africa, with British troops squeezing in from Northern Rhodesia to the south and Kenya to the north and the Belgians rounding the northern end of Lake Tanganyika from the Congo. Sir Henry hadn't paid much attention since the operation would take place on land, but if the truth were told, he was rather irritated to be left out of such a major undertaking. About all he could do was provide coastal support, and even that wouldn't be much since he could do little more than restore the blockade off the Rufiji delta, just south of Dar-es-Salaam. And he'd planned to do that, anyway. The German cruiser *Königsberg* was a fly in the ointment along the coast there, running out of the Rufiji to attack supply and troop vessels, then running back in to hide.

But he knew that restoring the blockade might be a daunting task with the resources at his command in the area. The fleet out of the Cape, under Vice Admiral Herbert G. King-Hall, was down to minimal size after most of its ships had been detached for duties elsewhere. King-Hall had only two decent ships left—the *Challenger* and the *Laconia*—one floating wreck—the *Hyacinth*—and a handful of smaller vessels that could in no way be termed warships. Thank God King-Hall already had cornered the *Moewe* at Dar-es-Salaam and destroyed the German supply ship *Präsident* in the Lindi River. But the formidable *Königsberg* remained to be dealt with. That didn't leave Sir Henry in a good mood, especially with talk that von Lettow-Vorbeck might soon move against Kenya.

"Excuse me, Mr. Lee." Sir Henry pressed a button on a box on his desk, and his adjutant entered.

"Sir David and Admirals Duggan and Corley are on their way, sir."

"Excellent. Bring me a chart of Lake Tanganyika."

"Lake Tanganyika, sir? In Africa?"

"That's right, Whitley. Off you go."

When the door shut behind the adjutant, Sir Henry turned back to Lee.

"I am aware of the difficulties in central Africa, Mr. Lee," Sir Henry said, taking a sip of his drink. "If von Lettow-Vorbeck concentrates his forces at just a handful of critical points, he could sufficiently puncture the British and Belgian defenses to cause a definite breach. Easy troop movements on Tanganyika would certainly give him all the mobility he needs. But undoubtedly our own vessels on the lake and those of the Belgians will inhibit him."

"I wish that were so, Sir Henry. Unfortunately, I bring you fresh intelligence since I've just now arrived from the region. The Germans now have total control of the lake."

"What did you say, sir?" Sir Henry stiffened.

"The Germans have seized control of Tanganyika."

"Be damned! How did that happen?"

"Von Lettow-Vorbeck assigned command of the lake to a Captain Gustav Zimmer, and Zimmer set up in Kigoma, a port town about one quarter of the way down the east coast."

"Why there?" Sir Henry asked.

"It's the lakeside terminus of the rail line that runs directly from Dar-es-Salaam," Lee explained, "so supplying his base is easy. When the *Moewe* was scuttled, all her crew, armaments, and supplies were in Kigoma within a couple of weeks."

"What kind of guns?"

"Three 57mm rapid-firing six-pounders with a range of about two miles, a 37mm one-pounder with five revolving barrels that operates like a Gatling gun, and a pair of 88s with a range of several miles. They mounted two 57s on the bow of their larger steamer, the *Hedwig von Wissmann*, and the 37 on her stern along with a Maxim gun. The other 57 went onto their smaller ship, the *Kingani*. They don't have a ship large enough for the 88s, so they installed them on a large raft that they can tow around the lake with them. The 88s have more range than anything in any of the shore fortifications."

"Damn inconvenient."

"It gets worse, Sir Henry. One of the *Moewe* crew was a Lieutenant Horn. Quite a capable officer. Zimmer assigned him command of the *Hedwig*, and he's used her and the raft to sink every Allied ship on the lake and bombard fortifications along the shore—ours as well as the Belgians'."

"Nothing is left of the Allied fleet?"

"The Belgians have managed to salvage the hulk of their largest ship, a ninety-ton steamer named the *Alexandre Delcommune*, but they can't repair it without protection from the Germans. The Belgians are maintaining tentative control of Tanganyika's western shore, but the Germans own the lake and the eastern shore. With the only armed craft on the lake, they will soon own it all."

And, Lee went on, there were two other important aspects to consider, both matters of alliance. First, German hostilities against the Belgians on the lake had done much to convince the Belgians to side with the British and take an active role in the war, which could have a decisive effect if the second matter did not too quickly come to a head. This second matter was von Lettow-Vorbeck's powerful influence over his Ba-HoloHolo spies and guerrilla warriors.

"It's partially the Belgians' own fault," Lee said, his mouth twisting in disgust. "They think that terrorizing the natives will bring control. Just last month, they hanged a whole crowd of them all at once."

"I heard about that," Sir Henry said. "It did quiet the rest of them."

"And make them even angrier." Lee responded. "If terror were the answer, Sir Henry, the Congo would have been made as docile as the Cotswold long ago. Besides, while the Belgians hang a few dozen, the Germans shell their villages for sport. Who wins the war of terror under those circumstances?"

"Point taken," Sir Henry said.

"It's this simple," Lee went on. "As long as the Germans are allowed to remain in control of Lake Tanganyika, a major tribal uprising against the Belgians—and the British—is a distinct possibility. If that happens, the Germans will become even more heavily entrenched in central Africa and will be defeated only at great cost of men and arms. In short, the Belgians can't do much to aid the Allied war effort against von Lettow-Vorbeck as long as the German boats command the lake. Nor can Britain."

"Africa's always been a problem since Gordon botched things at Khartoum," Sir Henry frowned. "Thank God there was Lord Kitchener to bail us out at Omdurman." He drained his glass. "These are the very acts of aggression that showed our Belgian friends that neutrality in this conflict is a presumption."

"If we help the Belgians, it will secure their alliance both in Africa and in Europe," Lee said. "And an Allied victory in Africa will assure us unquestioned rule there after the war."

"Sensible," Sir Henry nodded. "And I take it you have a plan to effect control and allegiance, all in one swoop?"

"I believe I have," Lee said. "I want to use von Lettow-Vorbeck's own guerrilla tactics against him."

2

At that moment, Sir Henry's adjutant entered with three other men. Sir Henry and Lee stood as they came over.

"Mr. Lee," said Sir Henry, "these are Admiral Sir David Gamble, Vice Admiral Duggan, and Commodore Corley. Have a seat, gentlemen. Mr. Lee is here straight from Africa with a proposition." He turned to the adjutant. "Whitley, where is that chart of Tanganyika."

"I've been looking, sir, but I don't think there is one in the whole of the Admiralty."

"Preposterous! That blasted lake is an inland sea. We bloody well better have a chart!"

"I'll see what I can do, sir." Whitley hurried for the door.

"Drinks, Whitley," Sir Henry reminded him in a mellower tone.

"Sir."

The adjutant disappeared, and while he was gone, Lee repeated his background information for the benefit of the newly arrived officers. The adjutant returned with a tray of drinks and left again, but he was barely noticed.

When Lee finished, Sir David said, "We know the obvious: Neither the Royal Navy nor the Belgians have a presence on the lake. But what about the German force? What is this *Hedwig von Wissmann* like?"

"She's a seventy-foot steamer with a steel hull and wooden decks and cabins, displaces one hundred and fifty tons, and has a top speed of ten knots. She is armed with two six-pounder Hotchkiss guns in the bows with a range of two miles and a revolving one-pounder Hotchkiss with a range of about a mile and a half mounted aft. There's a Maxim gun next to the one-pounder. Her main weakness is that the two 57s only have a one-hundred-twenty-degree area of fire."

"That's all?" Corley asked. "Doesn't sound very formidable."

"Formidable enough if there's nothing to stand up to her," Sir Henry replied. "And their second ship?"

"The *Kingani*, a fifty-five-foot steam yacht with a steel hull and wooden decks and cabins. She displaces fifty-three tons, has a top speed of seven knots, and is armed with a single six-pounder on the foredeck and a Maxim gun aft."

Just then, the adjutant returned, somewhat timidly.

"Yes, Whitley?" Sir Henry demanded.

"I'm afraid we haven't a chart, sir. None at all for Lake Tanganyika. But I did find this." He hopefully held out a *Bartholomew's Atlas* opened to a map of Africa.

"It will do until you come up with the real thing."

Whitley laid the open book on a table, and Lee and the officers rose and gathered around as he hurried from the room to find the real thing.

"Gentlemen, Mr. Lee tells me he has a plan."

"What I propose is this," Lee said. "I want to ship an armed motor launch from England to Cape Town and, from there, take it overland to Fort Lukuga. That's here," he pointed. "It's the Belgians' principal fort on the western shore, only about fifty miles south of Zimmer's base at Kigoma. Once on the lake, we can use it to attack and destroy the *Hedwig von Wissmann* and the *Kingani*."

"You're mad, man," Duggan sputtered. "There's more than six thousand miles of sea between here and Cape Town, and what...?" He quickly consulted the map. "Another three thousand miles of African bush, narrow hill tracks, and primeval forest!"

"More than three thousand, Vice Admiral," Lee replied with aplomb. "I know the terrain well."

"Why Lukuga?" Corley wanted to know. "Why not simply launch from Northern Rhodesia?"

"Too far," Lee answered. "Tanganyika is four hundred and fifty miles long, and its shores are alive with enemy eyes. The Germans would know we were coming well in advance, and we'd lose the element of surprise. Besides, it would take a gunboat two or three days to travel to the north end of the lake where it's needed. Even if the crew managed to arrive undetected, they'd be in no condition to fight after that long in an open cabin. Remember, we will have only the single boat against two Germans, and we can't afford to have them ganging up on her. Consider also the problem of supplying them with petrol, ammunition, and food at such a distance. We need the harbor at Lukuga both for its proximity to German territory and because Lukuga is a ready base for supplies."

"Can't we approach the lake from any other direction?" Corley asked.

"Not safely," Lee replied. "Kenya is the only other possibility, but the Germans constantly threaten the Mombassa rail line."

"Why not just sail them up the Congo?" Corley asked. "The Lualaba River, here," he pointed to a crooked blue line wavering across a few inches of map, "is one of the Congo's main upper tributaries, and it leads to the Lukuga, which goes right to the fort."

"We will use the Lualaba," Lee said. "But we can't sail up the Congo to get to it because of these." He indicated a darkened area of the map through which the Congo flowed. "The Crystal Mountains. The cataracts there preclude using the river for any great distance, and the portages around the cataracts are by means of a narrow-gauge railway that winds through a great number of covered bridges and tunnels, none of which are much more than three feet wide."

"Clearly not wide enough to pass any boat large enough to do sufficient damage to the Germans," Sir Henry said.

"And using the Lukuga River for the entire distance is out of the question because its own cataracts are interspersed with very shallow stretches. But there is a rail line between the Lualaba and Fort Lukuga."

"I see the wisdom of the route," Gamble said, "But one motor launch against two German warships?"

"As Commodore Corley pointed out, their ships are not large," Lee said. "They are really little more than oversized steam yachts. And again, we will have the element of surprise. And speed. The *Hedwig* and the

Kingani are both slow and not especially maneuverable, and surprise will give our boat a large advantage, no matter how small she may be."

"Impossible, I say," Duggan said firmly. "It is impos...." Sir Henry cleared his throat, and Duggan was instantly quiet.

"Perhaps we should consider Mr. Lee's idea, Vice Admiral," Sir Henry said, a musing tone in his voice.

"But we are the Royal Navy, sir," Duggan argued. "Not some bloody marines. Tanganyika is a lake, a thousand miles at least from any real body of water."

"From the looks of it," Sir Henry said, running a finger down the image of Tanganyika on the map, "it is something more than a lake." He looked at Duggan. "We defeated the enemy on Lake Victoria and Lake Nyasa, did we not?"

"Yes, sir," Duggan admitted.

"And it was the Royal Navy that did the job?"

"It was, sir."

"And they are equally far from, as you put it, any real body of water?"

This time Duggan merely nodded silent assent.

"I believe that dispenses with your objections," Sir Henry smiled. "In any case, it is both the duty and the tradition of the Royal Navy to engage the enemy wherever there is enough water to float a ship. Or," he nodded toward Lee, "a boat. I want to hear more of Mr. Lee's plan."

"It won't be simple, I agree," Lee admitted. "There isn't a direct rail link between Cape Town and Tanganyika, but I have surveyed a likely route. It will be difficult, not impossible. What makes it a likely success is that a rail line already exists from Cape Town to Fungurume, here." He touched the map. "That takes care of about twenty-five hundred miles. That leaves a single overland stretch without rail, which is only 150 miles. Then there is about 350 miles by river— the Lualaba, as Commodore Corley noted—to that last short rail ride of 180 miles. Once the expedition reaches the lake, the boat can harbor under the guns of the Belgian shore batteries at Fort Lukuga."

"The rail and river journeys make sense," Sir David put in. "But how do you propose to cart the boat overland?"

"We can bring in a traction engine," Lee replied. "And teams of oxen. We'll probably need them."

"How is the road?" Duggan asked.

"I'm afraid there isn't a road. We'll have to make one."

"What's this?" Corley asked, pointing to a patch of brown that lay directly across the overland route.

"The Mitumba Mountains," Lee answered.

"Mountains, you say?" Duggan asked. "No road? Over mountains? Really, sir!"

"The Mitumba are only six thousand feet," Lee replied smoothly. "And I know a pass. Dragging the boat over shouldn't be too difficult."

"Will you obtain this traction engine locally?" Corley asked

"There aren't any locally, as far as I know," Lee admitted. "I don't think they've been used before in that part of the continent. They're hauling howitzers with them in Tanga, but that's on the far side of enemy territory. I propose that we ship one up by rail from Northern Rhodesia."

"You seem certain that it will perform satisfactorily. After all, central Africa is notoriously wet."

"Much of the terrain we will have to cross is open plain crossed by rivers and streams that only flow in the rainy season. If we make the crossing some time between April and October, before the rainy season begins, I am confident we will succeed."

"I am curious why you propose to take the boat fully assembled," Sir David said. "Why not carry it in pieces and assemble it on the lake?"

"That was my first thought," Lee admitted. "But it would never survive the assembly," Lee said. "Look what's happened to the *Alexandre Delcommune*. The Germans are always afraid that the Belgians are going to begin building another steamer, and they patrol the area frequently. And even if they didn't spot our boat immediately, the Ba-HoloHolo will inform Captain Zimmer of its location posthaste."

"Won't they inform him of a fully assembled craft as well?" Sir Henry asked.

"Quite possibly," Lee said. "But a fully assembled boat could be launched much more rapidly and engage the enemy immediately rather than remaining a static target on the shore. And another factor is the lake's geography. It's long and narrow and six-thousand-foot mountains hem in much of the shore. Because of these features, typhoon-force storms frequently whip up at less than a moment's notice. That also makes a rapid launch imperative."

"It strikes me, gentlemen, that Mr. Lee has given much thought to his plan," Sir Henry said. "It also strikes me as a plan worthy of further consideration." He looked around at all of them.

Even Duggan nodded.

Sir Henry turned to Lee. "I have some inquiries to make. Will you please return here tomorrow at 3 PM for further discussion?"

"Certainly, Sir Henry." Lee tossed off the remainder of his drink and stood. "Gentlemen." He nodded to them all then left the room.

"Remarkable," Sir Henry said, watching the door close behind Lee. Then he turned to the others. "I like the idea, but with one change. As ingenious and loyal as Mr. Lee is, he's not a military man. We can't have a civilian leading a military expedition. It just isn't good form, don't you agree?"

The others gave their assent.

"Certainly we can find someone appropriate for the mission," the Lord Admiral smiled. "It should be someone with both command and battle experience. Preferably someone who's been stationed overseas in the tropics."

"That will be a tall order, sir," Duggan frowned. "Available men like that are few and far between. Most of them already are stationed in command posts."

"There must be someone," the Lord Admiral said.

"We can't force a command like this on anyone," said Gamble.

"Perhaps not, but we can entice."

"Yes, sir," Gamble said. "That's about all we have to offer. Enticement. But I'm not sure a mad expedition to tropical Africa can be much of an enticement to men already in cozy commands much closer to home."

"Someone will do it," Sir Henry said, assurance in his voice. "They always do."

3

That very afternoon, Geoffrey Basil Spicer-Simson sat in his office in the Admiralty feeling quite despondent and tired. His assignment in the personnel department was even more tedious than the three years he'd spent on dockside watch-keeping duties after the *Alcott* incident. Supervising the lengthy refitting of a couple of old ships had been uninteresting, unrewarding work and a bitter blow to an ambitious man who had traveled the world only to be relegated to observing other travelers as they came and went on important tasks. But at least he'd still been dealing with ships. Now all he dealt with were applications filled out almost uniformly by common sailors unfit to polish his boots.

Even worse, his new quarters were a far cry from the remarkable places he'd visited or the open deck of his own command. His realm was a small dusty office on a floor crowded with other small dusty offices, all crammed with ancient, scarred desks and drab men in musty uniforms. This was the end of the line for the old, the weak, the stupid, and the incompetent. Which was he? He didn't know—perhaps all of them, except old, and that would come soon enough in a place like this. What else but stupid incompetence could account for the sunken liberty boat and this recent fiasco? If he were being honest with himself—which he rarely was—he could not imagine that he even belonged here, or *anywhere* in the Royal Navy. Yet here he was, and here he was certain to spend the remainder of his career. Hidden and hiding.

But Spicer-Simson was not a man to be kept down for long, at least in his own mind. His indomitable spirit bubbled once more, this time with indignation at his superior officers and with animosity for a history that had failed him. He was here only because of unlucky flukes and the blind obstinacy of the Admiralty, who chose to weigh a couple of completely unavoidable and relatively minor accidents more heavily than his many years of dedication and service and his numerous achievements. True, the sight of the floating dead he had commanded and led, ultimately, to futile deaths, haunted him, but would things have been different had he been aboard the *Niger*? The U-boat still would have spit its torpedo, and the men still would have died.

But the truth of it—the hell of it—was that he hadn't been there.

Now, he alternately rationalized and pondered his plight as he sat in the dusty office at the drab desk, sandwiched between a has-been major of the Royal Marines and an insipid sublieutenant who obviously had won his commission solely because of family connections. The fellow was insufferably dull and had the temerity to doubt some of Spicer-Simson's tales of adventure, saying that only an idiot would face down a tiger armed only with a pistol. What did the young fool know? He'd never set foot outside of London in his entire life. And as for the major, all that old blunderbuss did was wax and trim his handlebar mustache, polish his buttons, and count his days to retirement. The fact that he only spoke to Spicer-Simson when absolutely necessary was fine by the lieutenant commander. He had no use for the old windbag, anyway. He could only thank God that Amy had stood by him, soothing him, telling him that all the bad luck had to lead to something good, that he was just banking all his good luck for a time when he would need it most.

That evening, after an especially trying day, Spicer-Simson, his muscular, round shoulders slumping, stepped off the tram at the edge of Russell Square. He crossed to the front door of his hotel and trudged up the stairs to his rooms. In his present state of mind, even the hotel was an insult. When he'd married Amy, he'd promised her a fine house in the city and, perhaps, a summer home in some quiet spot not too distant from London. Instead, they were constrained in what Spicer-Simson deemed little more than shabby temporary quarters.

In truth, though the hotel where they lived may have been unpretentious, it was well kept and certainly decent, and Amy had done wonders in transforming the simple rooms into what amounted to an exotic paradise. She reveled in the artifacts, such as animal skin rugs, barbaric weaponry, and a real elephant's foot umbrella stand, that her husband had brought back from far-flung ports of call almost as much as she adored the equally far-flung tales he told of his exploits.

Inside the blessedly cozy flat, he dropped his walking stick into the elephant's foot, hung his hat on a peg behind the door, and went into the drawing room to fix himself a drink. But before he could reach the liquor cabinet, a little gray terrier bounded in, and Spicer-Simson bent and scooped it up.

"Little Toutou," he crooned as the dog's tongue flicked out and licked the tip of his aquiline nose. "At least you love your master, don't you?"

Toutou licked him some more but was interrupted when Amy bustled in from the bedroom, took the dog, and set it on the floor. The expression on her handsome face seemed faintly alarmed, and Spicer-Simson noticed that she was wearing one of her better dresses.

"Oh, dear," she said, bracing him by the shoulders and beginning to unbutton his jacket. "You're late. You must hurry."

"Late for what, if I may ask?" His tone was peevish, but she ignored it.

"Why the Hanschells, of course. Didn't I tell you we were joining them for dinner at Mitchell's?"

"Of course." Spicer-Simson shook his head tiredly. "I simply forgot. Work at the office is piling up, and I'm drowning in merchant seaman applications."

"Oh, Geoffrey," Amy said soothingly as she finished undoing his jacket. "I know you'd rather be commanding a ship than a desk. Perhaps your chance will come again."

"That will be highly unlikely." He took off the jacket and hung it up, then paused to stroke the cat, Mimi, who sat imperiously on an ottoman.

"It will, dear," his wife said, hoping fervently that it would. She'd never seen her husband this despondent. "I know it will. But come along, now. We're due to meet the Hanschells in the lobby in three quarters of an hour."

As he changed into something more suitable for the evening, Spicer-Simson tried to relax, but it wasn't easy. Everything seemed so bleak and his job such a dead end that he found it difficult to work up any excitement for an evening out, even if it was with the Hanschells.

Amy and Mrs. Hanschell had been schoolgirls together but had lost touch until, by chance, they'd found each other several years before the war began. Their old friendship had re-flowered, forcing their husbands into close acquaintance. It was not something that made Spicer-Simson comfortable, at least not initially. It was one thing for schoolgirls to chum around, but youngsters had little regard for social proprieties, and he dreaded having to associate with someone not quite on his level.

He was relieved, then, to discover that Hother M. Hanschell was a physician, albeit just then embarking on his career. Even better, the two men hit it off splendidly, fueled in part by the fact that, like Spicer-Simson, Hanschell had spent some time overseas—the Gold Coast of Africa, to be specific, where he'd participated in a search for the source of yellow fever. Hanschell hadn't found yellow fever, but he had met with a case of amoebic dysentery severe enough to send him back to England.

The two men developed a friendship of their own, and if Spicer-Simson had been as worldly in his personal perceptions as he was in his travels, he might have noted that Hanschell was, indeed, his only friend. Hanschell accepted Spicer-Simson's supercilious arrogance, more amused than grated by it and the lieutenant commander's affected nasal drawl. Obviously, Spicer-Simson thought the drawl patrician, though nearly everyone else found it snobbish and irritating. It did bother Hanschell, though, to hear Spicer-Simson use it to scathingly deprecate individuals he felt occupied social and professional levels lower than his own, which was just about everybody.

Spicer-Simson's condescending rhinolalia may have been disagreeable at times, but it was balanced out by his tales of derring-do, which Hanschell loved to listen to. These stories were incredible enough to make Spicer-Simson a serious rival of Baron Munchausen, and in them, he invariably emerged as a hero. The stories were so good, in fact, that Hanschell thought Spicer-Simson tended to believe them himself.

One of Hanschell's favorites concerned an experience Spicer-Simson claimed to have had while conducting a triangulated survey of the Yangtze River. He said he'd discovered an unknown channel—unknown to outsiders, that is. The locals knew it quite well since a great number of their kin had drowned in its rapids and whirlpools. "The assent was so perilous," Spicer-Simson related, "that the Chinese built a temple beside it in a grove of fir trees, and the pilots and junk men go there to pray before attempting it. I like to think that I was able to help the poor devils. I expect they still remember me for surveying it for them."

Hanschell didn't necessarily entirely believe Spicer-Simson's tales, but he suspected that each contained a grain of truth beneath the embellishments. In any case, the man was far more interesting and entertaining than Hanschell's colleagues at the hospital. For her part, Mrs. Hanschell thought the odd naval officer who had married her friend a bit absurd but relatively harmless.

"I'm beached," Spicer-Simson confessed, or complained, to his wife and the Hanschells later that evening as they relaxed over after-dinner drinks. It wasn't the first time he'd made the statement. "I suppose I'll have to resign myself to being the oldest lieutenant commander in the Royal Navy."

"Oh, I shouldn't worry, old chap," Hanschell said. "There's plenty of time. You'll find your spot yet."

4

"Welcome, Mr. Lee," Sir Henry said as soon as his adjutant brought the big game hunter into his office. Admiral Gamble also was present. "I have interesting news." He waved to the chairs, and they sat.

"After we spoke yesterday, your plan so intrigued me that I immediately conferred with my colleagues at the War Office and the Colonial Office," Sir Henry began. "As you mentioned, the native situation around Tanganyika is extremely unstable. We have confirmation from General Tombeur, the Belgian's chief commander in the region. The hell of it is that we haven't enough of our own troops to fight the Hun and his askaris all at once, but we daren't arm any natives to help us for fear they will turn our own weapons against us."

The First Sea Lord rose and stepped up to the fireplace, where he turned to warm his backside.

"Interestingly enough," he continued, "the War Office has just received a request very like yours from General Tombeur himself through Brigadier General Malleson. Frankly, just between the three of us, Malleson isn't particularly well regarded, but in this case, he is simply a messenger. Tombeur is the source. He is presently in Ruchuru, organizing three Belgian brigades to aid in the united assault against von Lettow-Vorbeck. He tells us that he is constantly plagued by the German boats, which shell his camps and in two days can transport German troops the distance it takes his troops two weeks to march. The short of it is that he, too, has requested that we put a gunboat on the lake."

"He's even asked for airplanes and a miniature submarine equipped with torpedoes," Gamble put in with a chuckle.

"Clearly the submarine is out of the question," Sir Henry snorted an echo to Gamble's laugh, "though we may be able to accommodate him with an airplane or two."

"However we resolve the matter," Gamble said, "we cannot allow the Germans to control Lake Tanganyika. Not only would taking it from them make possible a series of joint operations by Nyasaland, Rhodesian, and Belgian forces that is already in the planning stages, but it poses a challenge which the Royal Navy cannot, with clear conscience, ignore."

"There you have it, Mr. Lee," Sir Henry said, returning to his chair. "We are determined to treat these waters as an outlying sea within the sphere of British naval power. So, with rather more haste than is my custom, I am going to approve your plan. It will be put into action immediately. We will call it the Royal Naval African Expedition, and I've asked Sir David here to take charge. I will let him fill you in on a few of the details we have."

"Thank you, Sir Henry." Gamble nodded at his superior, then he turned to Lee. "We have approved your overall plan with one minor alteration. Yesterday, you made the point that our single armed launch on the lake would face two opponents. With that in mind, Sir Henry has insisted that we send two armed launches rather than one. If you can drag one boat across the continent, Mr. Lee, certainly you can drag a second."

"I always carry a brace of pistols into the bush, Sir David."

"Excellent." Then Gamble's brow clouded. "Your proposal has poured oil on the coals, so to speak, and Sir Henry wanted to give recognition where it is due by placing command of the Royal Naval African Expedition in your hands. He has been advised, however, that it is only fitting and proper for a regular naval officer to be appointed commander. Our proposal is to appoint you second in command with the formal rank of lieutenant commander in the Royal Navy Volunteer Reserve. What do you say?"

"I accept," Lee said, "if you will allow me to choose several of the men."

"Who are they?"

"Officers I know and trust," Lee said. "A couple are from the 25th Royal Fusiliers, and a couple of others will be useful where we're going."

"Done," Sir Henry said. "Give a list to my adjutant, and we will have them released from their current duties and commissioned in the navy. Sir David will see to it that qualified petty officers and technical ratings are assigned."

"The Belgians have agreed to provide a safe harbor at Fort Lukuga. Since their guns were knocked out by the Germans, they are bringing in several four-inch guns from Shinsakasa on the coast. That will take a month or more, but they should be in place by the time the expedition arrives. With those guarding the harbor, the German boats won't dare come close enough to trouble our launches."

"All that remains," said Sir Henry, "is to find the right man to take command of the expedition. Any ideas, Gamble?"

"The task may be difficult, Sir Henry. I've made preliminary inquiries, but all our seasoned officers are presently assigned to sea duty, and with our shortage of men at this early stage, I hesitate to pull any of them for what may become— forgive the term, Lee—a futile exercise."

"No offense taken," Lee said. "I know it will be a long shot."

"You're right in saying that all our best men are already at sea and that we should leave them there," Sir Henry said. "We are the Royal Navy, after all. No volunteers, I take it."

"I've approached several men, but most thought me daft, though they didn't say it in so many words. I have a few more on my list, and I'm sure one will accept."

"What about the other branches?"

"There's an idea," Gamble said musingly. "Yes. I know a fellow who might just do the trick. He's a major in the Royal Marines. Considering the amphibious nature of the operation, he will have the experience on both land and sea."

"Where is he, now?"

"The last I heard, he was in the personnel department. Semi-retired. I'll track him down immediately. I'll also investigate the sort of launches we'll need. In the meantime, Mr. Lee, perhaps you'd better make plans to return to Africa to chart our route and prepare the road."

After Lee had gone, Gamble raised an eyebrow at the First British Sea Lord. "Do you think it will work, sir?"

"Not a chance in hell," Sir Henry shook his head. "If they do manage to make the lake, they'll probably be sunk in the first engagement. But it will be only a small expeditionary force, and if they fail, it will be on their heads and Lee's, not ours."

"And if they succeed?"

"Well," Sir Henry said with a smile, "If our expedition manages to destroy the *Hedwig von Wissmann* and the *Kingani*, I suspect that the War Office will be drinking a toast to us, won't they?"

5

The next morning, Spicer-Simson was in the midst of telling the sublieutenant the story of how he had single-handedly fought off five members of the secret Chinese Boxer society who had survived the initial rebellion.

"It was while I was on the Yangtze," Spicer-Simson said, thoughtfully stroking his goatee. "This was only a few years after the rebellion, and there were still plenty of those Boxers about. You probably don't know that the actual term the Chinese used was *I Ho Chuan*, which means 'righteous harmony fist.' It was no mean feat that I fought off the five of them; they are quite adept at pugilism...."

"Shhh." The junior officer hissed. "Isn't that Admiral Gamble?"

Spicer-Simson glanced around, and sure enough, Admiral Sir David Gamble was standing in the doorway, looking around. A lieutenant accompanied him.

Good God! Spicer-Simson thought. He's coming in!

He grasped the paper nearest at hand—a midshipman's application—and began poring over it with a diligence that did not include attention. He was wondering what Sir David was doing here in the office of flotsam and jetsam.

He was approaching the major at the next desk!

"Attention!" cried the sublieutenant, and everyone shoved back their chairs and started to rise.

"That's all right, men," Gamble said. "As you were. Hello, John." Gamble's lieutenant hurried over with a chair, and Gamble sat. "It's been a long time."

"Right you are, Sir David. Not for a couple of years."

"Ah, for the good old days," the admiral said. "Remember that supporting action against the Boers?"

"Like it was yesterday, though it's been more than ten years ago."

"Too long for a warhorse like you, that's for certain." Gamble glanced around the room. His eyes lighted briefly on Spicer-Simson, who still pretended to read the application, and passed on without taking note of anything more than Spicer-Simson's rank. "How do you like it here?"

"It's quiet, sir," the major replied.

"Perhaps just a little too quiet?"

"What did you have in mind, sir?"

"I've got something important. Something I need someone to take charge of. Someone I can trust to carry it through."

"I'm flattered that you'd look me up for old times' sake, Admiral. But may I respectfully point out the old part? I haven't seen action since that time you so kindly remember."

Spicer-Simson's disparaging opinion of the major's expertise died. This was Admiral Sir David Gamble, himself, recalling old campaigns and asking for help in the new. Grudgingly, if not any more charitably, Spicer-Simson altered his estimation of the major's abilities.

"I admit it'll be a tough task," Gamble said, "but I'm sure you're up to it. You've got at least one more good fight left in you."

"That I might," the major laughed.

Gamble spoke quickly and quietly, outlining Lee's plan and the subsequent interest of the Admiralty.

"We need someone to command the expedition," he concluded. "Because it is, in essence, an amphibious assault, we thought a man from the Royal Marines would be ideal. I immediately thought of you."

"I appreciate your faith in me, Admiral," the major said, "but I'm afraid, in all good conscience, I cannot accept your offer."

"You won't even consider?" Gamble asked, obviously disappointed.

"With all due respect, sir, it is an impossible adventure." The major's voice rang with a tone of finality.

And that tone was a bell ringing in Geoffrey Spicer-Simson's head. He dropped the application he'd been holding and abruptly stood up, nearly knocking his chair over backward.

"I'll take the assignment, Admiral Gamble!"

Chapter II: Preparations

"I THINK I'VE FOUND OUR man, Sir Henry. He's a career officer, and he's had experience commanding small warships."

Gamble bore a serious expression, but Sir Henry, glaring over Spicer-Simson's military record, had serious doubts.

"You call this commanding small warships? On his first command, he rammed a liberty boat and sank it."

"Actually, sir, that was his second command. His first was the H.M.S. *Widgeon* on the Yangtze."

"Don't split hairs with me, David. That was a river steamer, not a warship. And look at this: His second command was a U-boat hunter torpedoed at anchor in broad daylight. In the mouth of the Thames, for God's sake." Sir Henry snorted and rubbed his chin. "Strikes me that I remember the incident. He was reportedly carousing with a pair of young floozies in some cheap hostel. The joint chiefs nearly laughed me out of our meeting that week. Damn me if I didn't blush like a schoolgirl. I'm not sure what the hell you see in this fellow, Gamble."

"Aside from the fact that he volunteered for an assignment that no one else is willing to touch?" Gamble asked, and he saw that he'd gotten Sir Henry's attention. "I do admit he doesn't seem much like leadership material, but there are several factors in his favor, not the least of which is his African experience."

Sir Henry nodded. "Yes, I see that." He looked at the report. "Four years, eh? Director of the Gambia Survey."

"And he's participated in other surveys in out-of-the-way places where travel is difficult—the Yangtze and northern Borneo. That means he has the skills not only to navigate the lake but to survey the terrain during the overland portion of the journey."

"Lieutenant commander," Sir Henry mused. "He's got the rank, or close enough."

"He's an expert linguist," Gamble pressed on. "Fluent in several languages, including French. He'll be able to deal with the Belgians, and even the Germans if it comes to that, without an intermediary."

"Another strong point," Sir Henry conceded, though he still looked concerned. "I hope you're right in this, David, but look at the man's age. He's thirty-nine and still only a lieutenant commander, and he's sunk two commands through incompetence."

"Perhaps his past losses will make him more cautious," Gamble ventured, but Sir Henry waved off the notion.

"Once burned, twice shy, but twice burned, and you're a bloody fool." Sir Henry sat for several long moments, his head bowed, lost in thought. At last he looked up at Gamble. "But perhaps this is a fool's errand, eh, David? All right. We'll give him the chance. If he succeeds, all to our good, and if he fails...." Sir Henry shrugged. "He will have to accept it as one more personal failure, as will we." He glanced at the door. "Is he outside?"

Gamble nodded, and Sir Henry buzzed for his adjutant to bring in Spicer-Simson. When he did, Spicer-Simson stood stiffly at attention before the First British Sea Lord.

"At ease," Sir Henry said, perfectly aware than no one of lesser rank could be completely at ease in his presence. Spicer-Simson relaxed somewhat, though tension still laced the tendons of his neck, and his cool blue-gray eyes were bright. "You are aware of the outline of the proposed expedition?" Sir Henry asked.

"I am, sir."

"There will be hardship. Danger."

"I am prepared for them, sir."

"You've already lost two commands," Sir Henry reminded the lieutenant commander, his tone lending no charity to his voice. "Quite frankly, I'm not certain that an action station is right for you. By all reports, your organizational skills are without reproach, but you realize that command is another thing entirely."

"Battle is in my blood, sir."

"What?"

"I said, battle is in...."

"I heard you," the Lord Admiral said, waving his hand. He paused for a moment, his eyes piercing Spicer-Simson, who stood it as well as any man could be expected. That, as much as anything, weighed in his favor. "Very well. Under Admiral Gamble's recommendation, I hereby place you in command of the Royal Naval African Expedition. For the time being, you are to be given the temporary rank of full commander. What say you?"

"Thank you for your faith, sir."

"Faith?" Sir Henry almost looked genuinely amazed, though he hadn't felt that emotion in years. "I have no faith, only belief. And there is nothing that so engenders firm belief as results. Follow your orders and produce results, and I'll have no need of faith."

"I'll remember that, sir."

"Report to Admiral Gamble's office tomorrow. That is all."

"Sir!" Spicer-Simson came to attention, slammed his heel against the floor, pivoted smartly, and marched out of the room. As the door closed behind him, Sir Henry shook his head and glanced at Gamble.

"By God, David, he's a queer duck. I hope you're right about him."

"I hope so, too, Sir Henry."

Outside, Spicer-Simson continued his stiff march across the anteroom, conscious that all eyes were on him. Out in the hallway, however, he loosened visibly and sauntered cockily out of the building.

Full commander, he thought. An African adventure. A chance at redemption. It'll carry me far. I'll make it the crowning point of my career. I'll....

Just outside the doors, he pivoted and glanced at his reflection in the narrow glass edging the frame. The slightly distorted image looked natty enough. He drew his foot-long ebony cigarette holder from his tunic, inserted a cigarette, stuck it between his teeth, and lit the cigarette with a flourish. The next minutes saw him strolling down the sidewalk, the cigarette holder jutting at a jaunty angle.

The day was glorious. To hell with his dusty desk in that musty office back at the Admiralty. He had a command; he had a promotion; he wasn't old or a failure or ineffectual.

He was going to Africa! This time to fight!

Maybe dear Amy was right. Maybe all the bad luck was just a saving up of the good. She had faith in him, and now, if ever, was the time for him to have faith in himself. The good luck would follow.

Damn, but it was a glorious day!

He was still feeling great when he got home. Mimi was on a chair, sleeping in the sun. She yawned and rolled over as Spicer-Simson came in.

"Good cat," he said, rubbing her gray tabby belly. At the sound of his voice, Toutou rushed in, yipping excitedly, a little gray and white blur as he bounced around his master's feet. Spicer-Simson scooped him up, scratched behind the perky little ears, and headed toward the kitchen.

"Amy," he called out.

"I'm in here, lambkin." Her voice came from the sitting room, and he changed direction.

"You're home early. Aren't you feeling well?"

"I'm feeling fantastic." He set Toutou on the floor.

"Look," she said and held up the garment she'd been sewing. It was a skirt done in tasteful tweed, although its cut was a trifle shorter and broader than the fashion of the day. "I've just finished it. Want to try it on?"

"Not now, dear," Spicer-Simson said, pulling her to her feet. "I have wonderful news, but it's very secret. You mustn't tell a soul." His eyes twinkled. "In fact, it's so secret that I might not tell even you."

"Oh, Geoffrey. You're teasing. Come on and tell me."

"Oh, all right. But only if you say the magic word."

"And what word is that, lambkin?"

"Commander," he said, relishing the way it came off his tongue, taking pleasure in the authority of its sound. "Commander, my dear."

2

A few days later, Spicer-Simson sat in Sir David's office, listening to the admiral, nodding and attentive, but never obsequious. He didn't think he needed to be. Sir David had let on the names of a few of the men who'd been approached before the old major, and the list was, if not stellar, then certainly impressive. The expedition was being offered only to the best, and he'd been the one chosen. He had to admit that he was a little peeved that they hadn't actually offered it to him outright—even before the old major—since he had experience at command and in Africa, but he'd stood up for himself and taken matters into his own hands. Nothing ventured, nothing gained, his mother used to say, and she was usually right.

"We have the blessings of the War Office, the Colonial Office, and the Belgian government," Gamble was saying. "And just this morning a dispatch came in from General Tighe in Kenya. Bloody hell, the fellow didn't know the first thing about the situation on the lake."

"Bloody unbelievable, sir," Spicer-Simson said, just as a knock sounded on the door.

"Enter," Gamble ordered, and a young, plain-looking woman with bobbed brown hair and the rank of petty officer stuck her head in the door.

"Lieutenant Commander Lee and Petty Officer Magee to see you, sir."

"Send them in, Smith."

Spicer-Simson threw her a sly wink behind the admiral's back.

"Sir." The petty officer ducked out of the room without so much as a blush.

She was back a few seconds later with Lee and a stocky, moon-faced man of about thirty. Both wore uniforms that looked out of place on them.

"Lieutenant Commander Lee and Petty Officer Magee," PO Smith announced, pointedly ignoring Spicer-Simson. The two men came into the room, and Smith left, shutting the door.

"Good to see you again, Lee," Gamble said as the two men snapped awkward salutes. "Oh, don't bother with that. I know you're not used to the formalities. Let me introduce you to Commander Geoffrey Spicer-Simson. He's to lead the expedition. Commander, this is John Lee, your second in command. He is the chap who came up with this idea."

"Commander," Lee said, stepping forward and sticking out his hand. "It's a pleasure."

Spicer-Simson took Lee's hand and gave it a hearty shake. "Fresh out of Africa, are you, Lee?"

"Less than a month."

"Wonderful place. Can't wait to get back."

"You've been to Africa?" Lee smiled companionably.

"Up along the Gambia, primarily," Spicer-Simson grinned back. He stroked his mustache. "Was there for a number of years before the war. Remind me to tell you of the time I faced down an angry rhino armed only with a rifle."

"I never heard of rhino that far northeast," Lee said. "But I commend your courage. They're tough, stubborn-headed beasts and hard to take down. I prefer to leave them alone, myself."

"Probably wisest," Spicer-Simson said, nodding. Then he cast an eye on Lee's companion. "And you are?"

"Frank J. Magee," the stocky one said. "Petty Officer Magee, that is. Sir."

"Magee was a journalist before the war," Gamble said. "Spent some time in Tripoli and South Africa. He's to be the recorder and clerk for the expedition."

"A journalist," Spicer-Simson mused. "An interesting profession. I'm sure you're always looking for a good story."

"That I am, sir."

"Remind me to tell you some of the tales of my years in the Far East. Positively hair raising."

"I'm sure you will have all the time you need for stories on the trek," Gamble said. "Right now, we have an expedition to organize."

"Quite right, sir," Spicer-Simson said smoothly. "Duty calls."

"We've approved four of your five recommendations, Lee," Gamble said. "In addition to Magee, there are the three from the 25th Royal Fusiliers."

"I don't believe I've heard of that branch," Spicer-Simson said. "Some new regiment for far-flung locales?"

"Rhodesian." Lee responded. "They also call themselves the Legion of Frontiersmen. Commanded by Colonel Driscoll."

"Wouldn't be the same Driscoll that led Driscoll's Scouts during the Boer War, would it? Fine chap."

"You know him, then?"

"Not personally," Spicer-Simson said. "But I know something of his exploits. One hears much on the African continent if one keeps his ears open, don't you think?"

"Quite right," Lee said in a slightly amused tone. "African drumbeats, and all that."

"Yes, well...." Spicer-Simson let it hang, but something flashed in his eyes.

"Lee has been making some preliminary requisitions for the expedition," Gamble said.

"I hope you don't mind, Commander, but I've taken the liberty of ordering up some of the arms we'll need."

"Indeed?" said Spicer-Simson, a slight chill in his voice. "I had assumed that would be my prerogative."

"No sense in wasting time," Lee said. "If the expedition has any hope of reaching Tanganyika this year, we must get there before the rains begin."

"I fully understand the need for haste, Lee. I can only hope the arms you've found are adequate."

"There are two Hotchkiss guns for the boats. Both three-pounders. And three Maxims. We can put one on each of the boats, with the third held as a reserve. As for small arms, I've collected thirty-three Lee-Enfield .303s and an equal number of Webley revolvers."

"I prefer the Webley & Scott self-loaders myself," Spicer-Simson sniffed.

"A fine weapon with a lot of stopping power," Lee agreed. "But I always found them uncomfortable to fire."

"They can be for some," Spicer-Simson drawled. "They are very accurate, though, and seldom foul, even under the worst conditions."

"Well, bring yours along, by all means," Lee said. "They chamber the same .455 rounds as the revolvers, so there'll be plenty of ammunition to go around."

"Commander," Gamble said, breaking in. "Lee also has suggested you take a supply of mines with you."

"Will they be useful?"

"Very possibly," Lee said. "There is no coal on the lake, so the Germans burn wood, which they stockpile at several fueling stations along the shore. We can perhaps impede them if we mine the approaches to the fuel depots."

"We're giving you three Vernon Booms," Gamble said. "Thirty small percussion mines to each."

"We'd better bring along an experienced demolition man," Spicer-Simson said with a dry chuckle. "I wouldn't trust our ordinary seamen to handle mines."

"Right you are," Gamble replied. "We'll add one to the roster. Well, gentlemen, matters are moving along nicely, don't you think?"

The others nodded or voiced assent, but as they bent to look at *Bartholomew's Atlas*, opened to its map of Africa, Spicer-Simson shot a look of veiled hostility in Lee's direction.

3

By mid May, Spicer-Simson still occupied the same office, but the air of dusty futility that had pervaded the room just weeks earlier had vanished. People were coming and going on important errands, and the callow young sublieutenant now hopped to Spicer-Simson's every command. Even the old major showed signs of rejuvenation as he took a hand in the preparations.

A knock on the door did not interrupt the commander's concentration as he studied *Bartholomew's Atlas*, which still was the best map that could be found of the area through which he was to lead the Royal Naval African Expedition. His expedition. But he looked up when the sublieutenant came over to his desk.

"Captain Hope to see you, sir."

"Bring him over," Spicer-Simson said. "And a chair."

Douglas Edward Hope and Spicer-Simson had known each other for seventeen years. Hope's long career stretched back to a youth in the Merchant Service, where he eventually earned a mate's ticket. Between then and now, he had served in the Bechuanaland Border Patrol and the British South African Police, and during the Boer War, he'd held the rank of lieutenant in the North Staffordshire Regiment.

He was here, in part, because of his seaman's skills and his great familiarity with Africa but also because, in his most recent position, he had been a staff captain in the Railway Transport Division of the War Office. Even more to the point, Spicer-Simson trusted him, and the commander knew that the long road ahead required the backing of men he could lean on for support. After all, Lee had handpicked several men, and that needed to be balanced, just so everyone knew who really was in charge.

"Congratulations on your promotion, Commander," Hope said as Spicer-Simson waved him to the chair. "I know you couldn't be more pleased."

"It's been a long time in coming," Spicer-Simson said. "Now that it's here, I plan to make the most of it. I suppose you wonder why I've called you in."

"I've heard that something hush-hush is up, but no one can say just what."

"Information known only to a few," Spicer-Simson affirmed. "An expedition, and I want you in on it."

"You're to lead?"

"Into the heart of Africa. But say yes or no before I go further since this is completely confidential, and you must be sworn to secrecy."

"Africa," Hope mused. "Count me in. What's the assignment, and what can I do?"

Hope interjected amazed grunts as Spicer-Simson explained.

"Your expertise with rail transport will be invaluable during the first part of the journey," the commander finished. "And once we reach the lake, I'll want to rely on your seamanship."

"You'll have everything I can offer," Hope said.

"I know that." Spicer-Simson leaned forward and lowered his voice. "That's why I've called you. We've known each other a long time, and I need someone dependable for a rather delicate assignment."

"Go on, Commander."

"It's regarding Lieutenant Commander Lee. He isn't really a military man—he's been commissioned for this expedition, but he's just some kind of rough bush guide. The Admiralty seems to think highly of him and made him second in command, but I don't trust the man. He's going to Cape Town to make arrangements for the rail journey and then on up to the railhead at Fungurume to begin surveying and building the road. I want you to go with him to keep an eye on his movements. I may even want you to send a few messages here and there regarding his behavior and competence. As I say, he's not really military, and as commander, I can't have an amateur bollixing up matters before we've even begun."

"I understand."

"One last thing. I'm afraid that I can only give you a sub-lieutenancy. I know that it's a drop in rank, but the navy is loath to give equivalency in rank to army personnel. My apologies, but it can't be helped."

"No matter," Hope replied. "It's worth it to get the chance to go back to Africa and fight the Hun all at the same time. Anyway, it sounds like jolly good fun."

After Hope left, Spicer-Simson decided he'd had enough of *Bartholomew's Atlas* for the morning. He left the Admiralty and caught a cab for home, where he ate a light lunch and dawdled with Amy for a couple of pleasant hours. When he returned to the office, he found two extremely large men nearly blocking the corridor. Both wore kilt uniforms of the London Scottish Regiment bearing lance corporal insignia.

"Out of the way," Spicer-Simson snapped at the two, who had come to rigid attention as he approached, making entry to his office completely impossible. He rapped his cane impatiently on the floor. "Out of the way, I say."

"You wouldn't be the commander, would you, sir?" one of the giants asked in a heavy London accent.

"If you haven't learned to read rank properly," Spicer-Simson sniffed, "I believe that it is too late for me to instruct you."

"He means Commander Spicer-Simson, sir," said the other man. He wasn't quite as tall as his companion but was somewhat bulkier.

"I am Commander Spicer-Simson."

"Well, sir," said the second man, "we've come to volunteer."

"Volunteer?"

"For Africa, sir. We want to go fight with you on that Lake Tanganyiker." Stunned, Spicer-Simson could only stare up at the two giants, then he hurriedly glanced up and down the hall.

"Come into the office," he ordered, and inside, he learned that the taller man was Mollison and the bulkier was Tait. Both were serving in an officer's training camp after having been champion forwards in the London Scottish Regiment's famous rugby team.

"Now what's this you say about Africa?" the commander demanded.

"We heard you was working up an expedition to Africa," Tait amplified. "We want to go along."

"And just where did you hear this? In the officer's training camp?"

"No, sir. We was in a bar off Piccadilly," Mollison responded. "Some fellers was talking about it."

A Piccadilly bar? Spicer-Simson's eyes widened as the shock that his top secret expedition had become the talk of the streets hit him.

"Just who were these 'fellers?'"

"We don't know, sir," Mollison said. "Some navy fellers."

Good God, Spicer-Simson thought. The Germans are going to learn about the expedition even before we completely form it.

"Look, you two," he said to Mollison and Tait. "This mission is very confidential. You're not to speak of this to anyone under any circumstances. The penalty could be severe. Do you understand?"

"Yes, sir," Tait said, holding a hand over his heart. Spicer-Simson noticed he was missing a finger.

"Where did you lose that?" he asked.

"Ypres," Tait said. "We was both there. Mollison was wounded too. Took some shrapnel in the leg."

"Want to see the scar, sir?" Mollison made as if to hoist his kilt, but Spicer-Simson halted him.

"That will do, Mollison." He looked the two over. "Ever been to sea?"

"No, sir," Tait replied.

"But you've seen battle?"

"Yes, sir. We was both wounded at Ypres."

"Very well," Spicer-Simson said. "Considering that you already know about the expedition, I might was well take you along. Talk to the sublieutenant here. He will see to it that you are transferred. After that, you may join your mates at the Crystal Palace."

"The Crystal Palace, sir? The one where they had the Festival of Empire a few years ago?"

"That's the place."

"You'll report to the former monkey house," the sublieutenant put in a bit caustically.

"An appropriate billet for this lot," the old major muttered.

"Thank you, sir," Tait said. "You won't regret it."

"I had better not," Spicer-Simson glowered then, after dismissing the two, he went to his desk to finish memorizing the atlas.

4

It was May 21, and an air of expectation permeated Admiral Gamble's office as Spicer-Simson met with Lee and Magee for the last time in England. The two were leaving for Cape Town in the morning aboard the *Llandovery Castle*.

"When you arrive, you will first make arrangements for rail transport from Cape Town to Fungurume and for the traction engines to rendezvous with the expedition there," Gamble said. "Then you will proceed into the bush to prepare the first twenty miles of road and scout out the route. We've sent orders ahead by wireless, so you will have everything you need at your disposal. Understood?"

"There should be no difficulties, sir," Lee replied. "I've already been over much of the route, and I'm familiar with many of the people with whom we will be dealing."

"Excellent. We are organizing a crew of Rhodesians to help supervise your advance road-building party. We expect them to be ready in about a fortnight, so they should be on hand in Cape Town when you arrive. Survey the route as rapidly as possible and build the initial stretch of road so that it will be prepared when Commander Spicer-Simson arrives in Fungurume with the main party. He will leave England as soon as we find suitable boats and lay in sufficient supplies, say a couple of weeks after your departure."

"We'll be ready, Admiral," Lee promised.

"I'm sure you will. By the way, you'll be taking a man Commander Spicer-Simson has suggested who has served in the Railway Transport Division of the War Office. His name is Sublieutenant Hope, and he has extensive African experience. He will help ascertain the suitability of the rail line to Fungurume."

Gamble turned to Spicer-Simson. "As for the remainder of the crew, Commander, aside from the few you and Lee have chosen, we've brought in several of our most-qualified junior officers and technical ratings. Some have experience in the bush, and the gunners have exceptional records."

"What about a surgeon, sir?"

"We're pulling one from hospital duty here in London."

"Might I make a suggestion, sir?"

"By all means, speak up."

"I am personally acquainted with Hother Hanschell, the assistant director of the London School of Tropical Medicine, and since the war, he's been acting medical superintendent at the Seaman's Hospital in the Royal Albert Dock. And he's served in Africa. Considering where we're going, he's just the man to take along."

"He sounds like a first-rate choice. Have you approached him?"

"Not yet, but I have no doubts he will accept. He's often spoken to me about his desire to continue to study tropical diseases at their source, and after all, that's where we are headed."

"Tell him there's a commission in it for him if he volunteers," Gamble said, then he looked around. "Is that everything?"

It seemed to be.

"Well, gentlemen, there you have it," Gamble said. "The Royal Naval African Expedition is formed. I suggest a toast. Magee, would you pour the drinks? I'll have brandy, and I believe the Commander prefers vermouth and Lee Scotch and soda."

When the drinks had been poured and distributed, Gamble raised his glass. "To success."

After the meeting, Lee went off to his lodgings, but Spicer-Simson cornered Magee."

"A word, PO," he said.

"Certainly, sir."

"How long have you known Lieutenant Commander Lee?"

"Lee?" Magee's brow wrinkled. "I first met him a few years back, but I can't say I really know him. I was surprised, actually, when he requested me for the expedition. He knows the bush and natives pretty well, but I guess he thought I'd come in handy dealing with the colonial authorities. I know them better than he does, and I've got friends in Cape Town."

"Friends are admirable assets." A note of barely concealed warning lurked in Spicer-Simson's voice.

"I think you're right, sir," Magee said, eyes narrowing thoughtfully as he looked at his new commander.

"As a journalist, you are, I'm sure, aware of the power of words. A few of the right sorts of words in appropriate places can accomplish what daring and courage cannot. Don't you agree?"

"Yes, Commander. Words are powerful."

"Yes," Spicer-Simson said musingly. "Words can carry a man far, or they can destroy him. The right sorts of words."

"If I may be so bold, Commander, may I ask your meaning?"

"Oh, no meaning, Magee. Just curiosity. Have you ever thought about your standing in the Royal Navy?"

"My standing, sir? You mean my rank?"

"We are in a time of war. Many things are possible in war that would not be probable otherwise. Field promotions, for example. Did you know a commander has the power to grant field promotions to men under his command who prove themselves worthy of special consideration?"

"I am aware of that, sir. Yes."

"If I were a young fellow like yourself, a young fellow who will probably not remain in the Royal Navy after the war is over, I think I would try to rise as far as possible in the ranks before I left. Greater prestige, and all that, not to mention a better pension."

"I believe I see your point, Commander."

"I'm glad you do." Spicer-Simson stroked his beard. "And this fellow, Lee. You say you know him but that you do not consider him a friend?"

"Not a friend. No."

"Ah. Another interesting point, don't you think, Magee?"

"Yes, sir. Another interesting point."

"Well, Magee, thank you for your time. I doubt if I will see you again until we meet in Africa, but I'm sure you will do the right thing by God and your country. And for the good of the expedition."

"I'll do right, Commander. You have my word on it."

"Good, Magee. Very good."

Spicer-Simson patted Magee on the shoulder then strode off. Magee watched his retreating back, eyes pensive, then he, too, turned and went his own way.

5

Following his rounds on the morning of June 2, Hother Hanschell returned to his office at the Seaman's Hospital to find a note waiting for him.

"Urgent you see me in my office at the Admiralty," the message read. "Come immediately." It was signed Spicer-Simson, but Hanschell didn't need the signature to recognize the familiar scrawl.

Hanschell thought about ignoring the note. It wasn't that he didn't want to see his friend—in fact, he'd seen very little of Spicer-Simson for the past month, ever since he'd been promoted to commander. At the time, Hanschell had congratulated Spicer-Simson and asked what his new posting was, but Spicer-Simson had begged off, saying it was a military secret. Hanschell knew it must be something serious because Spicer-Simson, who was normally prone to brag about the slightest matter, had remained extremely tight-lipped.

So Hanschell was curious, but with the war on and his new position at the hospital, he also was overworked, and the matter had retreated from the forefront of his consciousness. And now that he was reminded of it, his innate curiosity remained buried beneath a weight of duty. Already, today's schedule was full, and he really didn't have the time to travel all the way to Spicer-Simson's office at the Admiralty just to have a chat.

He looked again at the note and the words "urgent" and "immediately" and took in the terse tone. Whatever the note was about, it must be important for his usually verbose friend to pen such a brief message. For a moment, he wondered if he ought to take along his medical bag.

Telling the duty nurse that he'd be back in an hour, he left the Seaman's Hospital and caught a cab for the Admiralty.

He found Spicer-Simson in his usual basement office, but a great change had been wrought over both the office and his friend.

"You wanted to see me, Geoffrey?"

"I do, Doctor. Come in."

Hanschell couldn't help but wonder what this was about since Spicer-Simson always called him Hanschell, never Doctor. There were two other desks in the office—a young sublieutenant sat at one, and the other was unoccupied.

"Run on up to the quartermaster's office," Spicer-Simson said to the sublieutenant, "and see if those petrol requisitions have gone through. And bring over your chair for the doctor. There's a good lad."

"You're in a chipper mood," Hanschell said as he sat and the sublieutenant vanished through the door. "And I must say that you're looking splendid, too. Younger. Have you dyed your hair or found some healing waters to bathe in?"

"Even better than that," Spicer-Simson said, leaning forward. His whole demeanor was afire, and Hanschell could practically feel the heat. "You know of my promotion, though until now, I have been unable to divulge the reasons for it. But before I can tell you anything, you must be sworn to the same secrecy that has prevented me from saying anything to you."

"My dear chap, you have my Hippocratic Oath."

"I have been selected to lead an expedition." Spicer-Simson spoke rapidly and with great excitement. "A most extraordinary expedition. Or rather, I've volunteered and they've given me the chance. And how do you think I got the command?"

He sat back, his eyes gleaming with mischievous humor.

"I couldn't begin to guess," Hanschell replied, finding himself infected with his friend's excitement even if he didn't yet know the cause.

"Simply by eavesdropping!" Spicer-Simson let out a laugh. "Admiral Gamble came in here to offer it to my colleague, a major from the Royal Marines. I listened on tenterhooks, for I saw at once what a chance it would be for me. When, much to my relief, the major turned it down, I stepped up and offered to go in his place!"

"What wonderful brass, old chap," the doctor said. "But I always knew you had yet to meet your destiny."

"I have two gunboats under my command and nearly thirty men. You'll never guess where I'm taking them. Africa!"

"Africa?" Hanschell was astounded.

"Central Africa, Hanschell. Lake Tanganyika, to be exact."

"But how…?"

"Quite simple, old boy, but quite complex, too. Let me tell you."

He succinctly outlined the entire enterprise.

"An amazing situation, isn't it?" he finished.

"But can it be done?" Hanschell wanted to know.

"This Lee fellow thinks it can, and so does the Admiralty, or they'd never have gone this far with it."

"But what about you?"

"I'm certain of it. I've studied the maps and the route. If anyone can do it, I'm the man."

"God, what an adventure it will be."

"Yes, old man, and I want you to come with me."

"What?"

"We'll be traveling through some of the thickest, most disease-infested bush in the world," Spicer-Simson said. "Who knows the dangers better than you? Think of it: malaria, sleeping sickness, dysentery, and God knows what else. It would be your responsibility to get every man out to the lake alive and well and safely back again."

"Yes, I'll go," said Hanschell without hesitation. He heard the words emerge as if from the voice of another. He hadn't thought but simply reacted and spoken and sealed his fate. "I'll go with you."

He felt numb, disoriented, like one of his shell-shocked patients, but as the implications of his assent began to work their way through his mind, the numbness vanished, replaced by a strain of his friend's infectious fever.

"Splendid!" Spicer-Simson exclaimed with a grin. "And there is a commission in it for you, too." Then he leaned back in his chair and said more seriously, "But there is one favor I must ask."

"Anything."

"As I am now officially your commanding officer, you must not address me as Geoffrey, but as Commander or sir. After all, military decorum must be maintained. Understood?"

"Yes..., sir." Hanschell expected the word to sound foreign on his lips, but it didn't. It felt right in a way that vindicated all his faith in his unlucky friend. "Yes," he repeated, standing and saluting. "Commander, sir."

Spicer-Simson rose, too, and beaming, returned the salute.

As soon as he arrived home that night, Hanschell told his wife that he would be going away—possibly for the duration of the war.

"But I'm afraid I can't tell you anything more about it," he said. "It's all terribly hush-hush."

"Oh, but I already know everything," she said. "You're going to Lake Tanganyika."

"But how did you know that?" asked an astonished Hanschell.

"Why, Amy Spicer-Simson phoned me this afternoon, and we had a long talk about it. It sounds very exciting. I just hope you don't catch your death like you did the last time you went to Africa."

Hanschell hoped not, too.

6

It was June 4, and John Lee was long gone. And good riddance, thought Spicer-Simson. The temerity of the man! He spends a few months in the bush and thinks that qualifies him to command a naval expedition. Why, he wasn't even military, much less navy!

But Spicer-Simson had faith in Hope and Magee. Had faith that they would do the right thing and advance not only the cause of the expedition but of their commander—if not for the sake of the expedition and Spicer-Simson, then for the sake of themselves.

The cab he was riding in pulled to a stop, interrupting his thoughts. "Thornycroft Shipyards," the driver said.

Spicer-Simson stared at the man.

"I say, Thornycroft Shipyards," the man repeated.

"Well don't just sit there like a bloody fool," Spicer-Simson barked. "Perform the duties for which you are paid."

"Sir?" The man bristled at Spicer-Simson's tone.

"Get out and open the bloody door!"

"Oh, yeah. Right, sir." With exaggerated haste and subservience, the driver hopped out of the cab, hustled to the other side, and opened the door.

"Your destination, sir."

"Don't be impertinent," Spicer-Simson grated. "Don't you know there's a war on? Some of us have important duties that mean the life and death of thousands. We hold the sway of whole continents in our hands."

"Right you are, sir. And I suppose you are one of those favored few. Perhaps you might tell me a little about...."

"Silence!" Spicer-Simson barked. "I am sworn to silence, my entire crew is sworn to silence, and you, you impertinent dog, you should swear yourself to silence before you find yourself in water deeper than you care to be."

"I'm sure you naval chaps have a lot of experience with deep water," the cabbie said. "That'll be one pound ten."

Gritting his teeth at the cabby's impervious insolence, Spicer-Simson dug out his wallet, produced a note, then added a ten-cent piece to it.

"Shrink your mouth, and you'll see your gratuity grow," he said down his nose, then he spun and stalked off toward the guardhouse at Thornycroft's main gate. By the time he reached it, however, the anger that had arisen due to the insufferable behavior of that coarse cab driver was fading, and he regained his good humor. After all, such people didn't rate much attention.

"Commander Spicer-Simson," he told the sublieutenant at the guardhouse. "Admiral Gamble is expecting me."

"Quite right, sir," said the sublieutenant, saluting. "Mr. Evans," he said to the petty officer in the booth with him. "Man the gate while I take the commander to the admiral." He stepped out of the booth, gave Spicer-Simson a sharp nod, said, "If you will please follow me, sir," and strode off down a roadway that cut between two large buildings.

Now, that's the proper way to treat the commander of an important expedition, Spicer-Simson thought, and all memory of the cabbie vanished from his mind.

As the sublieutenant led the way through the Thornycroft yards, Spicer-Simson was impressed at the size of the operation. Warehouses and manufactories lined the many lanes and alleys, and everywhere was the bustle of workers preparing the armaments of war for the seafaring warriors who would employ them and champion England's cause. The sublieutenant's course through the labyrinthine yards turned this way and that, but at no time did Spicer-Simson lose his direction. In about four minutes, as the sublieutenant took a left-hand turn, Spicer-Simson thought, "I should see the Thames just ahead," and he was right. Around the corner, less than fifty yards distant, the waters of England's legendary artery gleamed in the morning light. The sublieutenant did not take Spicer-Simson much closer to the river, however, but stopped by a door in the building to their left, opened the door, and ushered Spicer-Simson inside.

The interior was not particularly spacious, and most of it was occupied by two large boats, propped upright in wooden cradles. Admiral Gamble stood nearby, talking to a barrel-chested man with a full gray beard.

"I see the admiral, Sublieutenant. Dismissed."

The sublieutenant snapped a smart salute, stomped, turned, and headed off the way he'd come. Spicer-Simson, making his way around piles of fabrication materials and tools, approached the admiral.

"There you are, Commander," Gamble said as Spicer-Simson came up to him and saluted. "At ease. Let me introduce you to Mr. Meredith, one of the master shipbuilders here at Thornycroft. Meredith, this is Commander Spicer-Simson. He will be taking your little boats to a most extraordinary launching."

"Commander," Meredith said, shaking Spicer-Simson's hand. "So you've been telling me, Admiral, but I have yet to learn the exact nature of this extraordinary launching."

"Ah, Meredith, alas, you know I cannot reveal that. You I trust implicitly, and I'm sure you can vouch for your men, but the Germans have ears everywhere. If they were to get wind of our enterprise, Commander Spicer-Simson might have a devil of a time."

Spicer-Simson winced, thinking of the way that Tait and Mollison had learned of the expedition, but he kept that information to himself.

"I understand. But I need not know where my little darlings are going to make them seaworthy."

"They appear quite fit," Spicer-Simson said, running his eyes over the two craft.

"They are that," Meredith said. "Tough little boats, if I do say so myself. They were originally built as tenders for the Greek Air Force. There were eight of them, and six had already gone on to Greece. We were just about to ship out these last two when the Sir David asked if we had anything like them on hand." Meredith waved Spicer-Simson and the admiral over to one of the boats.

"Each of these little beauties is forty feet in length and eight abeam, with hulls of three-eighths-inch mahogany." He pounded the hull with the flat of his hand, and the small warehouse resounded with a dull boom. Then he led them astern and slapped the hull again. "There's a one-hundred-horsepower petrol engine in here. You can see she's fitted with twin-screws. She'll do nearly nineteen knots."

"You've had them in the water, then?" Spicer-Simson asked.

"Oh, yes. We test-run all our small craft before they leave the yard. We brought them back in to fit them with guns. Here, look." He led them to a ladder, and they climbed into the cockpit. In the middle of the foredeck, which occupied nearly half the boat's total length, a hole had been cut in the planking, and two men labored to install a steel reinforcement around the hole. "We're mounting a three-pound Hotchkiss here."

The gun was suspended by chains and pulleys above the working men.

"That gun appears practically antique," Spicer-Simson commented dryly. "One of the arms selected by Lee, no doubt."

"It's seen some use aboard an armed trawler," Meredith admitted. "But it's in excellent condition, and its four-foot barrel will give you plenty of range for a gun this size. It's the best we could do given the short timing. There were a pair of new six-pounders available, but they were too big for these craft. As it is, we've had a little trouble mounting these since the decks on the boats were not originally built to take guns. We had to cut down the forecastles to make it possible. And we had to shorten the gun mounts."

"Won't that make firing the guns difficult for the gun-layer?" Gamble asked. "A Hotchkiss is normally fired from a standing position."

"We had no choice, Admiral. The stock Hotchkiss mount was too high for the size of the craft. Threw off the center of gravity. The gun layer will have to kneel down to fire it. I know that'll be damn hard, but there was no choice if we're going to mount a gun with enough firepower to engage in combat."

"You're the shipbuilder, Meredith. We'll have to leave it to your good judgment and make do with the results. But I do have one question about the weight of the gun. Won't that unbalance the boat?"

"We thought of that, Admiral, and solved it in a practical way." Meredith hopped off the foredeck, into the cockpit, and led the way to the stern, where he opened an access hatch. "We put the petrol tanks underneath the stern sheets and surrounded them with steel plating to protect them from small arms and machine gun fire. The weight of the plating counterbalances the weight of the Hotchkiss. And we're going to mount a Maxim gun on the aft deck. That'll help, too. Of course, the guns and the plating will make the boats draw a deeper draft, maybe an additional four inches, and you'll have a reduced maximum speed— somewhere in the range of fifteen knots."

Gamble looked at Spicer-Simson. "That's still better than anything you're likely to face."

"I'm satisfied," Spicer-Simson said, affectionately running his palm along the mahogany rear deck. "I can hardly wait to get them into the water."

"First, you've got to get them wherever it is you're going," Meredith said. "Let me show you the trailers."

The trailers sat in an adjacent area. Workers were still constructing the frame of the second trailer, but the first was complete. It had small, light wheels fore and aft.

"We understand that you will need to transport the boats some distance by ship and rail," Meredith said. "With that in mind, we've built these so that the boats' cradles can be unbolted from their trucks and lifted as a single unit onto the deck of a ship or a railway platform car." He pointed out the appropriate fittings.

"The basic design looks sound enough," Spicer-Simson said, then he walked over to one of the smallish wheels at the front of the trailer and poked at the tire with his walking stick. "Pneumatic," he commented.

"The best we could find," Meredith said.

"If I may make a suggestion, Admiral?"

"By all means, Commander."

"Considering the terrain we will be crossing, I would like to see solid rubber tires rather than pneumatic. Except for a single horse, the battle was lost, and, to paraphrase, except for a single tire, this battle might be lost. I don't want flats to bog us down. And might they be larger? Larger wheels traverse rough terrain better than small."

"Excellent suggestion. Meredith, see to it."

"Yes, sir."

"Any other suggestions, Commander?"

"Again thinking of the terrain, Admiral. I'm concerned about the center supporting beams of these trailers. I'm afraid they just won't do."

"They're six inches by two," Meredith said. "Sound enough to take any strain they're likely to see. You could transport the boats the length of Great Britain on these."

"Perhaps you could take them that distance over macadam," Spicer-Simson said. "But we won't be seeing an inch of paving. I must insist, if I may," he glanced at Gamble, "that those center beams be doubled in size."

"We'll have to rebuild everything from the ground up," Meredith said.

"How long will that take?" asked Gamble

"Several weeks at the least. Possibly a month or more. Timber large enough to make beams the size the commander suggests is difficult to obtain at the moment."

"Not enough time," Gamble said. "What do you say, Spicer-Simson? Can you make them do?"

"I suppose I'll have to, Admiral. Where there's a will, there's a way."

"What about the lorry?" Gamble asked Meredith.

"We have a three-ton Dennis with a caravan-style body that should do nicely.

If you don't mind, sir, we'd like to send one of our mechanics along, just to make sure all goes well, if you know what I mean, sir."

"What's his name?"

"Reggie Mullens, sir. Reginald, that is. And a first-rate mechanic he is, too. He'll come in handy maintaining the boats' engines."

55

"I'm sure he will." Gamble turned to Spicer-Simson. "We'll have to give this mechanic some rank."

"Simple enough," Spicer-Simson replied. "We can make him a petty officer mechanic."

"Have him report to the commander," Gamble told Meredith. "What kind of cargo capacity does the lorry have? The commander will be transporting a considerable quantity of supplies."

"What kind of quantity, if I may ask?"

"The petrol alone amounts to seven thousand gallons," Gamble said. Meredith whistled and looked at Spicer-Simson.

"I don't know where you're taking my little darlings, but you might as well be going to the middle of Africa for all the preparations you're making. The lorry is rated for two tons, but it sounds like you'll be hauling a lot more than that."

"We'll take it in relays if we have to," Spicer-Simson said. "Drop it off and come back for more. But the more we can take in each trip, the better."

"Right you are," Meredith said. "Well, we'll see if we can reinforce the lorry's frame and suspension. Now if you gentlemen will excuse me, I'll have my men get to work on those new wheels for the trailers. They'll be ready when the boats are." He left the naval officers and headed toward his workers.

Spicer-Simson rode back to the Admiralty with Gamble, and when they arrived, they went straight to Sir Henry's office.

"Seaworthy little craft, from the look of it," Sir Henry commented after he had glanced at the sheets bearing the boats' specifications. "Small enough to transport but large enough to do some damage, eh?"

"They're quite satisfactory, sir," Spicer-Simson said.

"One thing, Spicer-Simson," Sir Henry said, tapping the sheets. "I see that they still bear numbers, not names. As they are not convicts, you had better find proper names for them."

"Do you have any suggestions, Admiral?"

"You will be their commander," Sir Henry said. "I'll leave their naming to you. I simply suggest that the names be short and easily distinguishable from each other."

"As a matter of fact, sir, I've been thinking on it and have picked out a couple of names that are just as you suggest. Short and distinguishable."

"What are they?"

"*Dog* and *Cat*, sir."

Sir Henry looked as if something had caught in his throat.

"*Dog* and *Cat!*" he snorted. "I sincerely hope you do not wish to pursue such absurd names. I am aware that we have the H.M.S. *Lion* and *Tiger* as well as *Wolf, Beagle, Bulldog,* and *Harrier,* but the animals they refer to are fighters. The names you have suggested seemed rather bald and unpoetical."

Spicer-Simson, who also knew of the *Salamander* and *Pickle*, not to mention his own first command, the *Widgeon*, was a little surprised that Sir Henry had reacted so strongly to the names he'd suggested.

"I'm sorry you don't like them, Lord Admiral. I'll think of something else."

"You do that," Sir Henry said, then he turned to Admiral Gamble. "Will the boats be ready on schedule?"

"Meredith says that Commander Spicer-Simson will be able to take them out on a test run on the Thames in four days."

"Excellent," Sir Henry said. "Let me know what time the test run will take place. I want to be on hand."

7

At the beginning of the war, Hanschell had tried to volunteer for the Royal Navy, but his request had been turned down on the grounds that his duties at the London School of Tropical Medicine and the Seaman's Hospital were too important. But with Spicer-Simson's influence, he was now in the service of his country. And he was going to Africa!

But he didn't have to wait that long to encounter his first patient. It seemed that Lieutenant Higgins, one of the men selected by Lieutenant Commander Lee, had just returned to England from Africa, but he failed to show up at the Crystal Palace where the men were being billeted. Instead, he sent a message to Spicer-Simson's office saying that he was unable to report for duty due to illness. That had been several days ago, and since then, no one had heard a word from him.

On the cab ride to the rooming house where Higgins was staying, Hanschell reflected on the expedition's second in command. Hanschell hadn't had a chance to meet Lee before the big game hunter had departed for Cape Town, but he'd heard that Lee had references from several influential administrators and army officers and had good connections, especially through his wife, who was the daughter of the late president of the Conway Training Ship Committee in Liverpool. Thus, he wasn't sure what Spicer-Simson's attitude of aloof suspicion and resentment toward the man meant, but he didn't think it boded well. What made him even more concerned was that he'd learned that Douglas Hope, one of Spicer-Simson's old cronies, who Hanschell had met on a couple of occasions, had been sent ahead with Lee. Hanschell, who hadn't particularly liked Hope, wasn't sure exactly what *that* meant either.

Hanschell found Higgins's room easily enough, and when a faint acknowledgement came to his knock, he let himself in.

The room stank—not with the odor of uncleanliness but with the fetor of disease. Higgins lay prostrate on a simple bed against one wall. He lifted a weak hand to acknowledge Hanschell's presence, but the hand fell back to the cover and limply remained there.

"I'm Doctor Hanschell," the doctor said, hurrying over to the bed. "With the Royal Naval African Expedition. Commander Spicer-Simson sent me over to see about you." He bent over and looked carefully at Higgins.

The man's complexion was a gray-green pallor, his lips blanched, and his eyes yellow with jaundice. Hanschell picked up the limp hand from the blanket and felt for a pulse. Rapid and erratic, it wasn't hard to find. The hand was hot, and Hanschell estimated that Higgins must be running a fever as high as 105 degrees.

"How long have you been this way?" he asked, opening his bag.

"A week," came the faint reply. Maybe longer. Don't really know. It's been worse the past couple of days."

"Why didn't you go to a doctor?"

"I did. The one just down the street. He said I was having another bout of malaria."

Blackwater fever, more like it, Hanschell thought, and he crouched and peered under the bed. There, he saw a large bowl that Higgins had been using as a bedpan, and when he pulled it out, it sloshed with a reeking, ugly, reddish-black liquid.

"We have to get you to the hospital right away," Hanschell said, taking the bowl of bloody urine to the toilet, dumping it, and flushing.

Within the hour, he had Higgins transported to the hospital attached to the School of Tropical Medicine, but it was no use. Higgins had gone too long without proper medical care, and even if he had been admitted days earlier, his chances only would have been fifty-fifty. He died the next morning.

"Dead?" Spicer-Simson seemed more baffled than shocked. "Blackwater fever, you say? Here in London?"

"An extreme complication of malaria," Hanschell told him. "It's common in the Congo and Northern Rhodesia, but it generally doesn't occur until a person has had four or more malarial infections. Higgins undoubtedly was infected before he returned to England. It's a shame that the doctor who first treated him didn't recognize the symptoms; we might have been able to save his life."

"Thank God I asked you along, then," Spicer-Simson said. "We'll be going right through those areas. You may become our most valuable man. After all, the Germans must sleep sometime, but tropical diseases know no rest."

Higgins had been the expedition's transport officer, and his loss left a gap that had to be filled. Spicer-Simson, after looking over the roster, selected Sublieutenant A. E. Wainwright. Despite being one of Lee's choices, Wainwright not only had experience in Africa but had been a railroad construction foreman and a locomotive engineer. Wainwright was promoted to lieutenant.

Hanschell could not so easily fill in the vacant spot left in himself by Higgins's death. He'd been asked to join because of his expertise, but he'd lost his first patient to a tropical illness even before the expedition had begun. No matter what, he vowed, I'll not lose another. Not to illness or snakebite or injury or to any other cause.

He knew he had a tremendous amount of new medical information to learn before the expedition left, and the departure was only two weeks off, so he went back to his office and rang up the Admiralty and was given the name of a naval surgeon captain who would take care of his requisitions. After that, he began collecting as many of the latest medical journals and texts on the diagnosis and treatment of tropical diseases that he could lay his hands on and packing them in a trunk.

8

On the morning of June 8, Sir Henry Jackson, Admiral David Gamble, and a small entourage that included Amy Spicer-Simson gathered on a dock at Thornycroft Shipyard to observe the launching and test run of the boats. Amy was wearing a light-colored dress and a little, dark hat with a round crown and brim. Pride gleamed in her eyes as she watched her husband's every move.

Mr. Meredith supervised as his workmen handled the actual launching, while the two expedition members Spicer-Simson had inexplicably selected as pilots—the giants, Tait and Mollison—climbed aboard. Spicer-Simson had the rest of his crew standing at full attention to observe the proceedings. He strode up and down in front of his men—gunners, machinists, engine room artificers, engineers, and supply clerks—looking them over.

Most notable among them was Lieutenant A. E. Wainwright, the man Spicer-Simson chose to replace the dead transportation officer, Higgins, and who was to serve as third in command. At forty-five, Wainwright was older than everybody else in the expedition save one, but his sandy blond hair, sharp features, lively eyes, and compact, tough body made him seem considerably younger. Modest, unpretentious, and possessed of a quick and ready wit fueled by a well-read, alert, and inquisitive mind, he seemed at polar opposites to the expedition's commander.

Wainwright's Irish brogue testified to his Belfast upbringing, but after serving as a lance corporal machine gunner in the Middlesex Regiment, he'd moved to Africa, where, from Mozambique to Rhodesia, he'd done everything from construction work to driving locomotives. He'd finally settled in Rhodesia, becoming a moderately prosperous rancher in the process. He'd volunteered for the 25th Fusiliers at the outbreak of the war largely to help protect his holdings from German encroachment. Although he'd never been to sea, his knowledge of Africa and railways would be of vital importance to the expedition.

The other officers were Hanschell; Lieutenant Rickson, the bridge engineer; "Tubby" Eastwood, a paymaster with a profoundly religious bent; Engineer Lieutenant Cross, who was a master mechanic and had twice driven to a winning finish at the Grand Prix; and Sublieutenant Tyrer, a warrant officer in the intelligence division of the Royal Naval Air Service. Tyrer was fifty—the one man older than Wainwright—but his handsome face looked ten years younger, abetted by the fact that he dyed his hair a canary yellow. He wore an affected air that went well with the monocle in his eye, and he prefaced nearly every statement with, "Dear boy." Tait, however, let every "dear boy" in the expedition know that Tyrer was better known to the barmaids of London's West End as "the Piccadilly Johnny with the glass eye." Maybe it was the Coronation Medal ribbon he kept pinned to his chest, or maybe it was his supercilious attitude, but Spicer-Simson took to him enough to name him his personal aide-de-camp.

The noncommissioned officers were Chief Petty Officer James Waterhouse, the chief gunner; PO William Cobb, a skilled motor mechanic; PO Flynn, another gunner; and PO Murphy, also a gunner; as well as Mullen, Tait, and Mollison. Ten ratings rounded out the company—except for Lee, Magee, and Hope, who would meet them in Africa.

And a checkered lot they were, Spicer-Simson thought. Tait and Mollison wore the kilts of the London Scots, and the attire of the rest varied from petty officer uniforms and seaman's bellbottoms to khakis and mufti. New uniforms obviously would be necessary. But to a man, they had come highly recommended and were considered experts in their duties. He would take that on faith for the time being, but should any of them fail in his duty....

The commander went over to Sir Henry and saluted. "Nearly ready, sir."

"Very good, Commander."

"Ah, sir, if I might...."

"Yes?"

"The names. For the boats."

"You've come up with something more poetic than *Dog* and *Cat*, I presume."

"I believe so, sir. Something with a French flair. I want to call them *Mimi* and *Toutou*."

Sir Henry's lips pursed as if he'd swallowed something bitter.

"*Mimi* and *Toutou*?"

"Yes, sir."

"I must say, they do have a bit of the French sound to them." There was a slight edge of sarcasm in the First British Sea Lord's voice that Spicer-Simson missed.

"So you approve, sir?"

Meredith gave the signal that his workmen were ready to launch, and Spicer-Simson glanced expectantly at Sir Henry.

"They're your boats, Commander," Sir Henry waved tiredly. "Give your first order."

Puffing up and strutting to the edge of the dock, Spicer-Simson called down to Meredith, "Launch *Mimi* and *Toutou*!"

There was a titter from the ranks, while over in the admiral's group, Gamble turned to Sir Henry.

"You wouldn't let him name them *Dog* and *Cat*, but you let him name them *Mimi* and *Toutou*?" Gamble asked, amusement in his voice.

Sir Henry shrugged and gave a resigned grimace.

"I told him to be more poetic. If he wasn't that, at least he was inventive. That's in his favor, even if the names are a touch eccentric."

The boats splashed into the water, and their engines roared to life. *Mimi* approached the dock first, inexpertly piloted by Tait. The huge, taciturn ex-lance corporal may have done his duty at Ypres, but unfortunately, he'd never been to sea and even more obviously knew nothing about boats. Under his guidance, *Mimi* scraped the dock and nearly rebounded into the Thames before the mooring lines were made fast. Mollison—who also had absolutely no experience with boats but was nonetheless piloting *Toutou*, proved to have a gentler hand.

"Pardon me, sir."

Spicer-Simson looked at the ranks, and the chief petty officer stepped forward.

"Waterhouse, isn't it?"

"Yes, sir. The chief gun-layer."

"What do you want, Mr. Waterhouse?"

"It's about the boats, sir. Their names, I mean." "What about them?"

"Did you say they're to be called *Mimi* and *Toutou*?" This time, with Spicer-Simson staring in their general direction, the men in the ranks kept silent.

"Do you have a difficulty understanding English, Mr. Waterhouse?"

"No, sir. I understood what you said. It's just...."

"Well, spit it out, man."

"Is that dignified, sir, to call warships names like that?"

"Just what do you find undignified, CPO? Do you find the French language undignified?"

"No, sir. I...."

"For your information, the words *mimi* and *toutou* are French childish expressions for cat and dog and stand for 'meow' and 'bow-wow.'"

Gamble glanced at Sir Henry, but the First British Sea Lord merely arched an eyebrow, and Gamble knew when to keep quiet.

"Now, stand back, Mr. Waterhouse." Spicer-Simson glared at the rest of the ranks. "Do I hear any further comments on our worthy craft? Good."

He looked at Sir Henry, who thought he detected a strange light in the commander's eyes that he first defined as villainous. But he immediately realized that Spicer-Simson's expression was much more complex, compounded of pride, a desire for fealty, and a burning need for acceptance born of success. Those things Sir Henry, an experienced ruler of men and judge of character, understood and prized in leaders under his command, but there was something else there that gave extra brightness to Spicer-Simson's piercing, frosty blue eyes—something that Sir Henry, for all his experience, knowledge, and understanding, felt he could never comprehend. Something tightly leashed but a touch too maniacal for any man to fully tame.

The First British Sea Lord felt a momentary, almost overpowering urge to call a halt to the entire proceedings right then and there, but the look vanished from the commander's eyes as if it never had existed, and he was speaking to Sir Henry.

"With your permission, Lord Admiral, we will proceed with the test run."

Rather than trust his voice—and it had been many a long year since that had been the case—Sir Henry merely nodded, and Spicer-Simson turned back to his men.

"Boat crews, man your stations," Spicer-Simson called out, and as the crews clambered into the boats, he went over to his wife and held out his hand.

"Would you join us, my dear?"

Amy put one hand in her husband's, and with the other at her throat, followed him to the edge of the dock.

"Oh, Geoffrey, this is so exciting," she said as he guided her down into *Mimi*'s cockpit.

"You'll be most comfortable in the stern sheets," he told her, and when she looked confused, he smiled tolerantly and led her to the back of the boat and sat her in the seat just beneath the Maxim gun. "Do shut your ears when we test the gun," he said. "The report will be quite loud."

She nodded and tied her round hat beneath her chin. Meanwhile, the commander went forward to the pilot's station, just behind the foredeck. Chief Waterhouse stood on the foredeck, bracing himself against the Hotchkiss.

"Take her out, Mr. Tait," Spicer-Simson ordered. "Hold offshore until *Toutou* joins us."

"Sir," Tait said, and he pulled the boat away from the dock. Soon, *Toutou*, with Wainwright in command, joined them on the water.

As the two boats moved onto the river, Sir Henry pondered the look he'd seen in Spicer-Simson's eyes and felt a flood of relief that he had said nothing. Perhaps an insane venture such as this, traveling to the very borders of the possible, might require a man such as Spicer-Simson to bring it off. He relaxed and watched the test run with professional interest.

Out on the water, Spicer-Simson was feeling pretty good. He put the boats through several maneuvers, and they obeyed splendidly. They were sound craft, even if they *were* the smallest gunboats in the Royal Navy, and he was proud to command them.

His men weren't feeling quite so proud. In *Toutou*, PO Flynn, crouching beside the Hotchkiss, muttered against the stiff wind, "Bloody hell if I'm going to serve aboard a boat with a pansy name. *Mimi* and *Toutou*. Can you believe it?"

"It's scandalous," agreed Lamont from the Maxim, but Lieutenant Wainwright broke in.

"You'll serve aboard 'em and you'll like it, the both of you," he said. "This is still the Royal Navy, and you're sailors, and that's the end of it."

"Well, I may have to serve aboard 'em," Flynn said, "but damn me if I'll call them by those names."

"How about *Mimmie* and *Toto*?" Lamont piped in. "I've got an Aunt Mimmie in Brighton, and we can think of the other like Dorothy's little pup."

"Call them what you like among yourselves," Wainwright said. "But don't let *him* hear you." He nodded his head across the water toward Spicer-Simson, who was steadying himself against the edge of the foredeck as *Mimi* bounded across the Thames.

"No, sir," the men agreed.

At last, after several turns up and down the river, it came time to test the guns. As *Mimi*, going full speed, passed the dock where Sir Henry and Sir David stood, patiently observing, Spicer-Simson leaned over the foredeck and called to Waterhouse.

"We have obtained permission to fire a round into that old dock over there." He pointed to a half-collapsed wharf clinging to the shore just upstream from their position. "Fire on it as we come abreast, and God help you if you hit anything else."

"I won't," Waterhouse promised, and he meant it. He'd already taken a taste of his new commander's temper, and he wasn't about to let himself in for another browbeating if he could help it. He kneeled beside the Hotchkiss, loaded a round, then aimed the barrel over the starboard rail.

"Ready, sir."

Spicer-Simson glanced back to make certain that Sir Henry and Sir David were watching then chopped his hand down dramatically and cried out, "Fire!"

Waterhouse pulled the trigger, and the sound of the Hotchkiss boomed across the water. Simultaneously, *Mimi*'s deck lurched with the recoil, and with a splintering sound, the Hotchkiss, carrying Waterhouse with it, flew backwards over the port side and disappeared with a great splash into the water.

"Bloody hell!" Spicer-Simson exclaimed, feeling his gut lurch in unison with the deck. He did note, though, that Waterhouse's shot exploded on target, ripping the crumbling dock to pieces.

A second later, Waterhouse, sputtering, bobbed to the surface. Seeing him, several of the crew members laughed, but Tait wasn't one of them. He was having a difficult enough time maneuvering *Mimi* back across her own wake, and besides, he wasn't feeling much like laughing. There was a nasty hole in his deck.

"What the hell happened, man?" Spicer-Simson spat after the crew had hauled the chief into the boat and set him, dripping and chagrined, in front of his commander.

"It would seem, sir, that the gun mount wasn't strong enough for the recoil," Waterhouse said.

Spicer-Simson's brow darkened, and he said, "Not strong enough? Well, we'd better get it shipshape, otherwise our battles will be short-lived."

"Yes, sir," the chief agreed dourly. A drop of water hung ridiculously from his nose.

Suddenly Spicer-Simson grinned. "Damn fine shot, though, considering. Eh, Mr. Waterhouse?" He waved to the pilot. "Take her in, Mr. Tait."

When Tait and Mollison had jostled *Mimi* and *Toutou* up to the dock with a minimal loss of varnish, and the men again stood at attention on the planks, Sir Henry looked them over with an eye half hopeful, half jaundiced. There were twenty-four of them—seven officers, seven noncommissioned officers and ten able seamen. Twenty-seven if you counted Lee, Hope, and Magee. Twenty-seven men on whom the fate of the African theater of war might well depend. Twenty-seven men for a fool's errand.

9

Divers recovered the lost Hotchkiss off the bottom of the Thames, and *Mimi*'s foredeck was quickly repaired. Extra steel reinforcements were added to both boats to accommodate the stress of the guns' recoils, but despite that, the guns could only be fired straight ahead—even the extra bracing would not keep them from being torn from their mountings and thrown overboard if they were fired abeam. Admiral Gamble expressed some concern that the lack of sweep for the guns would hamper the strategic effectiveness of the craft, but Spicer-Simson was in no mood for bleak outlooks.

"If I may say, sir, *Mimi* and *Toutou* are quite quick in the water and have excellent maneuverability. The best we will be going up against will be only half as fast, and if I may be so bold to point out, *Hedwig von Wissmann* and *Kingani* have similar constraints in terms of the sweep of their main guns, all of which reportedly can fire only forward."

Gamble saw the truth of that, and Sir Henry agreed, and plans proceeded as petrol, equipment, and supplies were rounded up.

Meanwhile, Hanschell discovered that laying in proper medical supplies was going to be more difficult than he imagined. The doctor had spent his entire professional career studying the many diseases that the expedition would face except one—the disease of ignorance that ran like the plague through the Royal Navy's medical supply system. Its chief symptoms were purblindness, rigor mortis of the imagination, and delusions of competence.

Through the School of Tropical Medicine, Hanschell had known several members of the Royal Army Medical Corps, and he knew that the army, with a century of practical experience in equatorial climes, provided its expeditions into those regions with an adequate array of medical supplies packed into insect-proof, zinc-lined boxes that could be carried easily by bearers. That arrangement seemed perfect to him, so within days of being commissioned for the expedition, he paid a visit to the naval surgeon captain he had been told would process his requisitions.

"Doctor Thomlinson," he said after he had introduced himself to the gray-headed medical officer. "I've learned that the Royal Army Medical Corp has equipment on hand that is just what we will need. Perhaps they would be willing...."

"Young man," the aging naval surgeon captain interrupted, a sharp look trying to pierce through the glaze of his rheumy eyes. "I understand that you are joining the *navy*."

"Yes, I am, sir," Hanschell answered. "Now, these army supplies...."

"Doctor Hanschell," the old man interrupted. "The navy has no need of assistance from the army."

"But Doctor, they have just what we need. The matter is very urgent."

"I am aware that it is urgent, thank you," Thomlinson said testily. "Now go do as I say. Sit on a park bench and draw up a list of your requirements. And remember, you are in the *navy*." He shook Hanschell's hand. "Good luck."

As Hanschell left the surgeon captain's office, his modern medical spirit bloomed with indignation. He went straight to Spicer-Simson's office, where he found his commander pouring over the most detailed map of Central Africa that the Admiralty had been able to find. *Bartholomew's Atlas*, at last, was lying discarded on a shelf.

"It's intolerable," Hanschell said after he'd explained the surgeon captain's case of myopically rigid ignorance. "God knows what we'll have to face—malaria, cholera, diphtheria, sleeping...."

He stopped as Spicer-Simson laughed abruptly, though not unkindly.

"Don't worry, Doctor," the commander said. "Everything will turn out all right. Just do the best you can."

"But I must have medicines and equipment," Hanschell insisted. "I can't cure men with my bare hands." He was a trained scientist and medical man, and he had no great faith in great faith.

"I'll see what I can do," Spicer-Simson replied then turned back to his map. "Now go make up your list."

Hanschell returned to his office to brood on tropical diseases and visions of the men of the expedition littering its trail like cast-off supply crates. Things changed the next day, but not necessarily for the better, when a midshipman entered and handed him a pamphlet titled *Medical Stores as Supplied to Gunboats on the West African Station*.

The date on the thing was 1898.

Damn fools, Hanschell thought. *This* is all they have to offer?

In the end, he could only shrug. If the list in the pamphlet was what he had to work with, he'd do his best. But he refused to be complacent. After he read the pamphlet, he shook his head and reached for a pen to bring the list up to date, adding quinine, morphine, chloral, chloroform, aconite, pyretic salts, and a dozen other modern drugs that he would need to keep his charges alive and well. He complemented the drugs with a complete surgical kit and a microscope.

When he was done, he compiled a memorandum to be distributed to the officers and men of the expedition. The memorandum recommended that the men wear clothing made of heavy twill, with long sleeves and trouser legs to protect them from the hostile African environment, which would bedevil them with everything from a merciless sun to insect bites to superficial wounds caused by the thorny bush. Most of the men would contract exotic tropical sicknesses, and half probably would come down with fungal infections and veldt sores, as well. He took his revised supply list and the memorandum to Spicer-Simson for his approval.

"Good work, Hanschell," Spicer-Simson said after he'd reviewed the list. "I'll pass this on to Admiral Gamble." Then he turned his attention to the memorandum. He stroked his mustache with a deliberate forefinger then looked up at Hanschell.

"Very good, very good. I agree with your recommendation most heartedly. It is an excellent rule. But we must keep in mind that the exception is what makes the rule."

"I don't understand. Tait and Mollison will have to forego their kilts, but standard Royal Navy uniforms will comply in all respects with my recommendations."

"Doctor, Doctor. What you have here," Spicer-Simson tapped the memorandum, "is all well and good for everyday military purposes, but we cannot limit ourselves to anything so prosaic for our extraordinary endeavor."

The commander rose and strode over to a wardrobe that stood against the wall to the left of his desk. "I have here a special uniform for our expedition's officers and men." He opened the wardrobe and took from it a suit of clothes that he proudly displayed for Hanschell.

"This is mine, of course. I designed it myself, and Monsieurs Gieves of Bond Street sewed it up for me. I've contracted with them to outfit the whole expedition. Do you like it?"

Hanschell looked at the uniform and didn't know what to say. The trousers and shirt were made of a light-weight, light blue-gray material, finished with a blue necktie. The peaked cap, trimmed with gold braid, bore badges of rank for both navy and army.

"The officers shall have the cap," Spicer-Simson went on, "but the petty officers and ratings will wear pith helmets. Appropriate for Africa, don't you think? Of course, as commander, I must have some distinguishing accessories." Spicer-Simson reached into the wardrobe once more. "I am adding these to my own uniform." He withdrew a curved naval cutlass and a silver-headed ebony walking stick.

"Are you quite sure that sword is right?" Hanschell asked. "Wouldn't it be better suited for the cavalry?"

The look that the commander shot him almost made him step back a pace. He'd often seen Spicer-Simson bully waiters and cab drivers, but this was the first time that he'd felt the effect himself.

"Do not presume, Doctor, to tell a career military man what is or isn't a proper uniform. And in particular, please refrain from contradicting your superior officer in matters where you lack expertise. I will accept your diagnosis for blackwater fever and the like, but you will remember that in all other matters I am in charge."

"Yes, sir. I'll remember."

"See that you do. I'm relying on you to help set a good example for the rest of the men. A successful outcome to this expedition will carry me far in the navy, and I intend to be successful. You, of all people, should know how important this is to me."

Spicer-Simson then turned his attention back to the cutlass and cane, holding them up to better admire them.

"Marvelous, don't you think?" he asked, giving a grin.

Marvelous, Hanschell thought, not quite knowing what to think. Simply marvelous.

Chapter III: England to Cape Town

EARLY ON THE MORNING OF June 15, Spicer-Simson, his officers and seamen, and Dr. Hanschell assembled at St. Pancras Rail Station. In the dawn light, the gothic windows and the numerous slender brown columns of the as-yet unfinished station looking like so many petrified tree trunks embedded in the red brick walls made for a foreboding atmosphere beneath the vault of crisscrossed iron beams. But Spicer-Simson was oblivious to the setting. For nearly half an hour, he made a great show of parading the men across the gray flagstones before their gathered families and a contingent from the Admiralty.

"Damn odd looking lot," Sir Henry commented to Gamble, referring to the mixed and sundry uniforms worn by both officers and men. "I don't believe I recognize the commander's uniform."

Spicer-Simson, dressed in his new blue uniform, swung about, and his long cutlass swayed behind him like a wagging tail.

"I am unfamiliar with it, sir," Gamble admitted, then he was surprised to hear his superior chuckle. "Do you find it amusing, Sir Henry?"

"I find it absurd," the Sea Lord said, "and thus completely in keeping with the tenor of this entire enterprise."

Behind them stood Amy Spicer-Simson. She hadn't heard their exchange, but her husband's attire was on her mind, too—what he was wearing now and what was in the small parcel under her arm, wrapped in brown paper and tied with twine. How dashing he looked in his blue uniform, how completely in command of matters. She only regretted that she wouldn't see him wearing the garment in the parcel. But she knew he would love it.

At last, Spicer-Simson had enough parading, and he gave the men a short break to have a final few minutes with their families. While they did, he went over to the admirals and saluted.

"Very nicely done, Commander," Sir Henry said.

"Thank you, sir. I think the men are spot on considering the short notice."

"Perhaps not short enough," Gamble said. "We've only this morning received word from the Colonial Office that von Lettow-Vorbeck is on the move. He's using his Tanganyika steamers to ferry troops down as far as Northern Rhodesia. Our forces in the area are pretty sketchy right now, and the Germans would have prevailed if it weren't for the arrival of a contingent of Belgian officers and five hundred askaris. But now the Belgians want their troops back up in the Congo. Understandable, perhaps, but pointless since no real fighting is going on there. Only our promise that you'll have your boats on the lake by August prevented them from pulling out."

"August, sir?" Spicer-Simson's complexion turned a little ashen. "But we won't even arrive in Cape Town until the beginning of July."

"Don't fret, Commander," Sir Henry said with a smile. "We know it will take you a little longer than that, but we'll let the Belgians find that out in their own good time. Before they do, however, we have the use of their askaris, and if we can keep them another two or three months, perhaps von Lettow-Vorbeck will think them permanent and give up on Northern Rhodesia."

"One more thing, Commander—your orders," Gamble said, handing over a thick, sealed packet of papers. "You'll note one stipulation that we haven't discussed previously. It seems that you will technically outrank the commander of Fort Lukuga, and the Belgian government is concerned about their autonomy. They have made it clear that you will be operating within their jurisdiction, and while you may command on the lake, their commander has sole authority ashore."

"Does that include my harbor encampment, sir?" Spicer-Simson asked. His color had returned, and now his eyes narrowed.

"I think not," Sir Henry said. "We will consider the naval installation under your purview."

"Report to Vice Admiral King-Hall in Cape Town," Gamble said. "But once you travel inland, your immediate superior will be General Edwards. He's the commandant general of Rhodesia, stationed in Salisbury. He'll give you any help you need."

"Thank you, sir."

"Good luck with your mission, Commander," Sir Henry told him. "We know you have a great deal of hardship ahead, but we are confident of your success. Keep us apprised of your progress."

"I will, sir. Thank you, Sir Henry. I won't fail you." Spicer-Simson tucked the bundle of orders under his left arm and snapped a sharp salute. Then the admirals stepped back to let Amy through.

She embraced her husband and clung to him for a moment before she pulled away and said, "Oh, Geoffrey, shall I ever see you again?"

"Don't be ridiculous, my dear. Of course you shall."

"I wish you weren't going. Not without me."

"Have no fear. I will be perfectly safe."

"But this is war, not surveying. Terrible things can happen in war. Here," she said, pressing the paper-wrapped parcel into his hands. "The fliers' wives are giving their husbands scarves for good luck, but I thought you'd prefer this more. I made it of leather, so it will be very durable."

"Thank you, my dear. I'm sure it is up to your usual high standards. Now, I must go and do my duty."

"I'm so wonderfully proud of you, Geoffrey." Tears gleamed in her eyes. "Keep safe."

He gave her one final peck, then he turned to his crew.

"Fall in, men, and prepare to board the train. We're off to Tilbury Docks!" At Tilbury, their ship, the Union Castle passenger liner *Llanstephen Castle*, was nearly ready to get under way. *Mimi* and *Toutou* had been put aboard the previous day, their cradles chained to the aft deck. The boats were fitted with tarpaulin covers, though those did little to disguise the fact that what they covered were far too large to serve as emergency lifeboats. Soon the men had their gear stowed away, but activity on the deck did not completely subside.

"I say, officer." The speaker was Joseph Hyde, who had spent the better part of the past ten years in South Africa and was now returning to Cape Town with his wife and three children after their annual visit to England. With him were four other men, all in a similar condition. The officer he addressed was the second mate of the *Llanstephen Castle*. "Are those boats under the tarpaulin on the rear deck gunboats? I've seen a large number of military men aboard."

"I believe they are gunboats, sir. Bound for Cape Town."

An angry murmur arose among the passengers.

"Are we to understand," Hyde pressed, "that the British government condones sending war materiels on a civilian passenger liner during a time of war?"

"What about our safety?" demanded one of Hyde's companions. "There are women and children aboard."

"I'm not in charge of what the government does, sir," the second mate said. "I just know what's aboard."

"Would you mind taking us to the captain, please."

"Very well, sir."

"I can't believe that the Union Castle lines would deliberately place civilians in mortal danger during a time of war," Hyde said to the captain of the *Llanstephen Castle* when the second mate brought the civilian contingent onto the bridge. "If the Germans discover us, we're likely to be torpedoed."

"Yes," agreed one of the other civilians. "Look what happened when they sank that liner in Cardigan Bay."

"Why weren't we informed of this ahead of time?" Hyde went on. "Many of us may not have boarded at all. It is positively unconscionable."

"Let me apologize for the line, Mr. Hyde," the captain replied. "Unfortunately, the operative word here is *war*, and the Royal Navy has requisitioned space aboard this vessel."

"But surely the navy has ships of its own," said one of Hyde's companions. "Why not use one of them?"

"Precisely because they *are* navy," spoke a new voice. The civilian contingent looked at the speaker, a largish man in an odd blue uniform with the insignia of a naval commander on his gold-braided cap. He sported a goatee, and clenched between his teeth was a foot-long cigarette holder.

"You are...?"

"Commander Geoffrey Spicer-Simson," Spicer-Simson said, removing the cigarette holder. "I am the commander of the two gunboats and the military men aboard this vessel."

"Perhaps you can explain...."

"I am afraid, sir," Spicer-Simson said abruptly, looking down his nose at Hyde, "that I simply cannot. We are on a secret mission with orders directly from the First British Sea Lord Sir Henry Jackson himself not to divulge the nature of our venture to anyone. If I can't tell the good captain, here, there's certainly no reason to say anything to you lot."

Hyde looked shocked at Spicer-Simson's haughty attitude, but he drew himself up and said, "I can't imagine what could be so secret that it is necessary to place women and children in harm's way without the courtesy...."

"Are we to let the Hun overrun the British Isles as they've overrun France simply because a few civilians are too fainthearted to accept that they, too, have a role in the war?" Spicer-Simson snapped.

"I'm not sure I care for your tone, sir," Hyde said.

"If I seem abrupt, it is because I have little time for extraneous matters. You mentioned the naval men aboard this ship, and you will do well to remember that they are going into combat, perhaps to face death. I, personally, am responsible not only for their welfare and very lives but for seeing to it that they accomplish the mission upon which they are being sent. That mission will involve more danger and hardship than you can imagine, but they are all, to a man, volunteers who are gladly undertaking it for the sake of you and your families."

"But *we* are *not* military men. The safety of our families...."

"If you are so concerned for the safety of your families," Spicer-Simson said caustically, "you shouldn't be shipping them out during a state of war."

"You, sir," said Hyde, "are insufferable."

"And you, sir," Spicer-Simson replied, "are a tempest in a teapot. The fate of the world rests in the hands of a few, yet you would tie those hands least you receive an accidental buffet. Good, God, man, have the courage of necessity even if you haven't the courage of your convictions."

Hyde, turning livid, sputtered but was at a loss for words.

"Gentlemen, gentlemen," the captain temporized. "I'm certain we can come to some kind of mutual understanding. Please understand that the Union Castle Lines does not actively seek to carry military men and munitions, but we *are* at war, and when the Crown asks our help, we are obliged to give it. I realize that when you purchased your tickets, many of you were not aware that this trip would include a naval force. In light of that, before we depart, I will ask my employers to refund in full any ticket should the holder not wish to travel at this time. I will point out, however, that nearly every Union Castle liner, and vessels from all of the other major lines, as well, frequently carry war materiels. It is a fact of life which I'm afraid we will have to live with for the duration." He looked around at the civilian contingent. "Now, are there any of you who wish a refund?"

None of them, not even Hyde, whose color gradually was returning to normal, took the offer.

"Very well, then," the captain said. "Then may I suggest that you return to your families? We are about to sail." He blew the ship's horn.

2

Out on deck, Doctor Hanschell was dealing with his own crisis. He'd watched anxiously as pile after pile of crates, boxes, tins, drums, and bundles were lifted to the deck, but none contained his precious medical supplies. As the final crane-load was hoisted and deposited on the deck, he rushed over, only to suffer further disappointment. Not a single parcel, package, or crate of his medical supplies had arrived.

A loud horn blew, and he watched in horror while the gangplank was swung aboard. With a dreadful sinking sensation seizing hold of his stomach, he realized that he was heading off to Africa almost completely unprepared to deal with the health of the nearly thirty men in his charge.

As tugs hauled the *Llanstephen Castle* away from the pier, Hanschell, envisioning men suffering and dying for lack of treatment, hurried in panic to find Spicer-Simson. The last place he knew the commander to be was on the bridge with the captain, so he headed that direction. He had to pause in a narrow corridor to let several civilians pass. They were grumbling, and the one in front looked positively furious, but Hanschell couldn't tell what had upset them. Nor did he particularly care. He had far more profound problems. After the civilians passed, he rushed on. Spicer-Simson was just coming down the passageway from the bridge when Hanschell nearly ran into him.

"We're leaving port, Commander," Hanschell panted.

"Quite right, Doctor. Our wonderful journey has just now begun."

"You don't understand, Commander. None of it arrived."

"None of what?"

"The supplies."

"Nonsense. I personally observed them all being brought aboard. Everything is in order."

"But the medical supplies, Commander. Not a single bit. No equipment or instruments. No drugs. I have nothing."

"My good Doctor," Spicer-Simson said, patting Hanschell on the shoulder. "I wouldn't fret. We don't need more than a little quinine, do we? I never had more than that on the Yangtze or in Gambia. We can buy some in Cape Town."

"Quinine?" Hanschell gaped. "In Cape Town?"

"My good fellow, you look as if you need a doctor yourself. Follow me. The bar's open—come and have a drink!"

Crestfallen, Hanschell obediently followed Spicer-Simson to the bar. He ordered a whiskey while Spicer-Simson had his usual vermouth.

"Here's a bit of medical knowledge I'm sure you never found in your medical books," the commander said.

"What's that, sir?" Hanschell asked, and was surprised to see Spicer-Simson hold up his glass of liquor. "Vermouth? I suppose you could use it in a pinch as an antiseptic."

"Much more than that, Doctor. At least in the tropics. It's the best thing for those sorts of climes. Whiskey, too, for that matter." The commander indicated Hanschell's own glass. "A lot of men will drink beer and wine there, but those are headachy drinks. Vermouth is the best, and really such a pleasant, wholesome drink, too. In fact, the idea of vermouth alone is attractive, for it is made from the dried flowers of chamomile, to which the later pressings of the grape have been added. One has only to smell dried chamomile flowers to find that their fragrance is that of hay meadows in an English June."

"Perhaps you're right, Commander," Hanschell said. "Chamomile preparations are quite common in medicine."

Even if Spicer-Simson was right about vermouth, the doctor didn't think that liquor was going to solve his lack of real medicines, and he had no idea of how he was going to protect the safety and health of the crew on the trek across Africa. Spicer-Simson's faith in quinine would do little to combat sleeping sickness, diphtheria, and the host of other ailments Hanschell knew lurked in the Dark Continent. Ailments that were acting especially strongly on his imagination thanks to all the extra reading he'd done since watching Higgins succumb to blackwater fever. A least he'd had the foresight to raid the supply cabinet of the School of Tropical Medicine for smallpox vaccine and typhoid serum. He'd even managed to come away with a set of basic surgical tools, but there would be trouble if he had to treat any serious conditions or, God forbid, battle wounds.

He was so lost in his own thoughts that he barely noticed the half-dozen civilians grouped around the far end of the bar. He did notice that the one who seemed to be their spokesman shot an occasional glare at Spicer-Simson, but he soon forgot them. He was a medical man without the tools of his art but in great need of them. He downed the last of his drink, excused himself to Spicer-Simson, and went outside. What more was there for him to do but lounge on deck and take in the sea airs? It was a holiday. Nothing but a bloody damn holiday!

He moved to the side rail and, soon afterwards, saw the tugs let loose and head back toward Tilbury. He stood at the rail for several long minutes watching the land slip past as the ship slowly gained headway and made for land's end. Even a cry from the watch barely roused him from his despondency. He looked in the direction the watch was pointing and, in the near distance, observed a small motor launch bounding over the increasingly heavy waves in the wake of the *Llanstephen Castle*.

Moving closer to one of the sailors on deck, he asked, "What is it?"

"Don't know, sir. They're sending out something, but if they don't get to us by the sea reach, they never will."

Curiosity almost always is more attractive than lethargy, and Hanschell let himself get caught in its net. He drifted toward the stern to watch the chase, and gradually the pursuing launch drew alongside. Land's end was only minutes away. Sailors cast mooring lines down to the launch, then lowered small nets as the tiny craft bucked dangerously in the ship's wash. In minutes, half a dozen crates, boxes, and packages were hauled to the deck, and the launch cast off. The relief was obvious in the faces of the men on the tiny boat as their craft diminished with distance.

The excitement over, Hanschell turned to head back to his former position by the rail, when he heard a voice say, "Atkins, go find Doctor Hanschell."

"I say, Eastwood" Hanschell called out. "Here I am."

The man who'd spoken was Tubby Eastwood, the expedition's quartermaster. One of the men recruited by Lee, he had seen action in Elizabethville in Thomas Cook's regiment, but like most of Lee's men, he had never been to sea.

"There you are, Doctor. This lot's for you." Eastwood waved over the pile of cardboard boxes, badly tied packages, and mismatched crates that lay where the sailors who had hauled them aboard had haphazardly deposited them. As Hanschell moved closer, his eyes lighted on the symbol stenciled on one of the crates, and he gasped then whooped. Pictured there was a staff wrapped with twin serpents. His medical supplies had arrive after all!

"Get these down to my cabin," he ordered, and he grabbed one of the crates himself and led the way. When everything was safely in his cabin, he discovered that there was little room left for himself, but he didn't care. He'd sleep in the corridor if he had to, as long as he had his supplies.

With Eastwood's help, he began an inventory. Prying open a likely looking crate, he found to his great joy that it contained a first-rate microscope. Further examination, though, did not bear out the good tidings of this first discovery. Only a small fraction of his requisition was there, and worst of all, there were no surgical instruments, so he'd have to make do with what he'd brought. But all was not bleak. At least, there were enough drugs for him to begin a program of inoculation.

A knock on the door brought him back to the moment. He called out, "Hello," and Spicer-Simson squeezed into the cramped cabin with the two men. The commander glanced at the opened crates and boxes and grinned one of his bright grins.

"Supplies, eh, Doctor? What did I tell you? Have faith. Everything will be all right."

3

Time was something that Lee, Magee, and Hope had on the rest of the expedition, and they were making good use of it. They'd left England on May 22 and were in Cape Town a couple of weeks later. During a little more than a week in the city, they arranged for supplies, visited some of Lee's friends, and met up with the ten Rhodesians who were to serve as supervisors for the advance road building crews. The Rhodesians were led by a man named Locke. Locke had been an officer in the Boer War, and he still dressed like one in jodhpurs and a dark, military-style shirt. With his party in tow, Lee boarded the train for Elizabethville. It was June 15, the same day that the *Llanstephen Castle* sailed from Tilbury.

The journey to Elizabethville was quite long, and Lee spent part of it briefing the road building supervisors on the purpose for the road, the sorts of terrain they might expect to find, and possible hazards along the way. After his briefing, though, there was little to do but relax and enjoy the scenery as the train chugged north.

"It's a lot different from the English countryside," Magee commented.

"England's all well and good for those back in England," Lee said. "Give me Africa any day. Raw and untamed it may be, but that gives a man a chance to make his mark." He laughed. "Back in England, I'd be a shopkeeper or some such nonsense. Can you imagine?"

Magee couldn't. Before the expedition, he'd only known Lee casually, but Magee was a journalist, and his quick insights into character had found ample opportunity during the past six weeks to assess the man who had originated this wild scheme. Magee liked and admired what he saw, and he couldn't see Lee anywhere but Africa. At least not for long. Lee even tended to use the African names for places rather than European equivalents.

"Do you suppose we'll have any difficulty with the natives?" Hope asked. He had some experience in Africa, but not in this region.

"I doubt it," Lee said, a trace of bitterness in his voice. "The Belgians have pretty well destroyed anyone who might be a danger. That Yank taught them well." He was referring, the other men knew, to Henry Morton Stanley, though they also knew that Stanley was not really an American but a transplanted Englishman. "The terrain will be more of a problem," Lee went on. "If we have trouble with the natives, it'll be getting enough of them to work for the wages we can offer."

At last, the train arrived in Elizabethville, and as it pulled into the tiny station, Magee stared with astonishment out the window. There was no platform, just a dusty margin between the tracks and the small, spare station building, and this area was completely filled with people, most of whom were white. Off to the side clustered a small brass band, and arrayed along the front of the station building were Belgian askari troops in full-dress uniform. A great many local Africans in colorful civilian dress crowded at either end of the station.

"What do you suppose they're all here for?" Magee asked following Lee out of the coach.

If Lee had an answer, it was drowned in the noise from the brass band, which struck up "God Save the Queen."

A sweating middle-aged man wearing a suit and a broad smile came up, accompanied by an older man in a Royal Army uniform with a colonel's insignia. In turn, they shook Lee's hand.

"Welcome to Elizabethville," the man in the suit shouted happily over the music in a voice strongly tinted by a French accent. "I am the vice governor general of the Katanga region. I believe you know Colonel Sawyers."

"The colonel and I are old friends," Lee affirmed, warmly clasping Sawyers's hand. Sawyers was the town's leading British citizen.

"It is a great honor for us to welcome the Royal Navy to our modest town," the vice governor general continued.

"Thank you, sir," Lee said, looking somewhat embarrassed at the attention. "I must admit, though, I don't know quite what to make of all this."

"What?" the vice governor asked loudly.

"I said, I don't know what to make...."

"I suggest we speak later," Sawyers called out. "When we have some quiet."

"Yes, Colonel Sawyers," the vice governor agreed. "We will speak later. Now is the time for celebration." Taking Lee by the arm, he led the way out of the station, followed by Sawyers and the other Europeans, who took the rest of the advance party into their midst. After them came the band, then the honor guard, then a long train of curious Africans. The whole parade made its way down a dusty street to a simple two-story wooden building whose sign proclaimed it to be a hotel.

There, the Europeans all jammed into the lobby and bar, leaving the band, the honor guard, and the crowd of Africans outside.

Elizabethville had approximately fifty European inhabitants—about fifteen British and thirty-five Belgians. Most of the Belgians were support staff for the vice governor's office, and the British were loosely associated with the British Foreign Office. As Magee mingled with them, he quickly realized that they had long since become inured to their own company and were hungry to meet fresh faces from back home.

About halfway through the evening, while Lee, Colonel Sawyers, and the colonel's wife caught up on each other's lives since they'd last met, Magee spied Hope edging through the door. Probably stepping out for some air, the journalist thought, or to get away from the noisy party. The latter seemed likely, since Hope was a quiet sort who kept his thoughts to himself. Magee didn't really know the first thing about Hope, but there was something about the man that made him a bit uneasy. It wasn't that Hope was unfriendly as much as uncommunicative and a bit cold, and his sharp, constantly roving eyes didn't seem to miss a thing, especially where Lee was concerned. It was the scrutiny, Magee finally decided, of a predator stalking its prey.

Oh, well, Magee shrugged. Sometimes there was no accounting for people. So, instead of worrying over it, he broke away from the rest of the advance party, who were eating and drinking and having a good time, and went to the table where Lee sat with the vice governor and the Sawyers.

"This is the first visit the Royal Navy has ever made to our humble city," the vice governor was saying.

"I'm not surprised," Lee replied. "You're not on any of the usual shipping lanes."

"Ah, yes," the vice governor laughed. "But you are not a sailor, yourself, Mr. Lee. Colonel Sawyers speaks highly of you."

"The colonel and I go back a long way," Lee said. "And, no, I'm no sailor. I was specially commissioned for this expedition. My party is here to survey and begin opening a route for the main expedition when it arrives."

"I am pleased to say that my government has ordered me to give you whatever assistance you require," the vice governor told Lee. "Your undertaking is remarkable. To take boats to Tanganyika...."

"You know of our purpose and destination?"

"But where else would the Royal Navy take boats at this time?" Sawyers asked genially.

"At this rate," Magee put in, "the Germans will know all about us."

"And what if they do?" the vice governor shrugged and smiled. "They can do nothing about you."

Another man joined them at the table.

"Ah, Mr. Lee, let me introduce Monsieur La Plae. He is one of the engineers who helped build the Katanga Railway."

"Monsieurs," La Plae nodded to Lee and Magee. "I understand you have a serious undertaking ahead of you."

"We are not at liberty to discuss it," Lee said, laughing, "though it seems common knowledge."

"You will be going overland north of Fungurume?"

"Roughly."

"Then you must cross the Mitumba Mountains."

When Lee nodded, La Plae shook his head.

"Impossible, monsieur," he said. "The Mitumba are 6,000 feet high and much too steep for a road. Or at least such a road as you may build given your timetable. They will be a wall beyond your capabilities."

"I've been that way more than once," Lee said confidently. "I know a route that will make it possible."

"I do not doubt your sincerity, monsieur," La Plae said. "Nor do I underestimate the viciousness of the terrain you propose to conquer. I salute your attempt and will weep at your inevitable defeat."

He raised his glass in a toast, but Lee said, "I cannot drink to my own defeat, but I will accept your challenge."

La Plae nodded, a faint smile on his face, and everyone drank.

"Now, Vice Governor," Lee said. "You offered your aid, and I'll need it. I know the countryside around here fairly well, but not the local chiefs. Is there someone who can provide me with introductions to the tribal leaders along our route? I'll need to hire as many of their people as possible for work crews."

"That should be easy enough," the vice governor replied. "And you can arrange for bearers here, too. Anything else?"

"Oxen," Lee said. "We'll need forty-five to fifty."

"That, too, can be arranged," the governor said. "I'll see to it in the morning. But now," he poured another round of drinks. "Now we celebrate."

4

Hanschell was sitting in the lounge—at the end of the bar, precisely—sipping on a whiskey. The date was June 23, and the *Llanstephen Castle* was about halfway to Cape Town—if the weather held, and Hanschell hoped to hell it did. Just a few days out of Tilbury, they'd hit rough weather, and for the next seventy-two hours, Hanschell wished he was dead. The few times he'd staggered out on deck to empty his bucket over the rail, he'd had just enough presence of mind to be vexed by the nonchalance of the seasoned naval personnel, who walked the pitching deck as if they were on a Sunday stroll.

But as interminable as the bad weather had seemed, at last the skies cleared, the wind softened, and the ship quit its nauseating combination of roll, pitch, and yaw. Hanschell emerged from his cabin with a ravenous hunger that was matched only by a tremendous thirst. Or overmatched. He headed straight for the bar, and he'd been there practically ever since.

He wasn't the only one. Commander Spicer-Simson had become a regular, too, but unlike Hanschell, who carefully nursed his drink, the commander sometimes slung back the vermouth with careless abandon. Amazingly, his tolerance for liquor seemed as mighty as his sea legs were worthy. The two or three occasions that Hanschell had seen Spicer-Simson during the gut-wrenching storm, he had been swaggering down the deck or a corridor, jauntily swinging his cane, that infernal grin plastered on his face. It had been positively sickening, and Hanschell had promptly complied.

Now, however, the commander was sitting in the center of the lounge, relating one of his adventures on the Gambia. He was surrounded by passengers, most of whom had grown somewhat easier about the naval expedition's presence when U-boats had failed to immediately target the *Llanstephen Castle*.

"The most dangerous animal I have ever had to face was a rhinoceros," Spicer-Simson was telling his listeners. "The charges of the lion and the elephant have caused the deaths of many hunters, but the rhinoceros is the only animal who rushes upon a man without the least warning or provocation. This one, I remember—it was in wild, broken country near the banks of the Gambia River. The first I heard of him was a violent snort. I recognized its significance at once. It meant that he was suspicious but hadn't yet got wind of me. Quickly and silently—one learns to move silently when one's life depends on it—I moved to one side of the track where the undergrowth was thickest. Gently, I parted the bushes with my hands, and there he was in full view, not fifty yards away, his head raised, listening. Soon, reassured, he began feeding again, tearing up the ground with his huge horn, unearthing roots which he seized with thick, prehensile lips and crushed with his teeth. Then I noticed the birds. They were fluttering round him, alighting on his back and picking out insects from the thick folds of his skin. It was those birds which nearly cost me my life."

Hanschell had heard this particular story before, and he was only half listening as Spicer-Simson worked up to the climax, but his hazy attention was brought back to the here and now by a sound from the man at the next stool.

"Harrumph."

The grunt wasn't loud enough for Spicer-Simson or his listeners to hear, but it attracted Hanschell's attention. The man sitting there was Mr. Hyde. Hanschell knew Hyde didn't care for the commander. He didn't quite know why, though he understood it had something to do with a complaint that the ship was carrying war materiel and naval personnel as well as civilian passengers.

"Beg your pardon?" Hanschell prompted.

"I said, 'harrumph,'" Hyde replied without amplification.

"Are you referring to the commander's story?"

"To the commander, himself, if you like," Hyde said, taking a sip of his drink.

"You don't find his story credible?"

"I find it complete balderdash," Hyde said. "I've spent some time up along the Gambia, myself, and know the area well. I can assure you that there are absolutely no rhino for many hundreds of miles."

"Suddenly, one of the birds let out a screech, and they all flew away," Spicer-Simson continued. "The rhino threw up his head, breathing heavily with rage, and then he scented me! In an instant, with a sound like thunder, he was pounding straight for me! I have my ammunition specially made for me, you know, and my finely tempered bullet went right through the animal from stem to stern, piercing lungs, heart, spleen, and liver on the way. I was really sorry that I had to kill him, for these animals are becoming increasingly rare, but of course I had to in self-defense. His horn was exceptionally long—about twenty-seven inches, as I recollect. Anyway, it was so remarkable that I presented it to the Natural History Museum."

"Harrumph," Hyde commented.

5

The next night, Hanschell again was in the lounge, at the end of the bar, and Hyde again was next to him. In the middle of the room, Spicer-Simson was telling another story to the same crowd. It was as if none of them had moved in all the previous twenty-four hours. Perhaps they hadn't, since there was very little else to do aboard the ship.

"I came on the spoor of a waterbuck just after noon," Spicer-Simson said. "I knew he was out there in the bush, not fifteen minutes ahead of me. I tracked him as silently as any native and came upon him feeding, completely oblivious to my approach."

"Harrumph," Hyde said.

"What?" Hanschell wanted to know. "No waterbuck along the Gambia?"

"Oh, they're there, all right," Hyde said. "But not in the bush. Along the river." He took a swallow of his drink. "That's why they call them *water*buck."

"Unfortunately, he caught wind of me before I could get a clear shot, and he charged me through the bush," Spicer-Simson went on, ignorant of the side conversation between Hyde and the doctor. "Not as frightening as a rhino, I assure you, but just as deadly with those sharp antlers. My trusty rifle spoke, and he went down in a heap. In a moment I had him on my shoulders, and after an hour or two, I reached camp. We feasted that night, I can tell you."

"Harrumph," said Hyde, glancing sideways at Hanschell. "Know anything about waterbuck, Doctor?"

"Only that they live by the river, not in the bush," Hanschell replied.

"They're about the size of a pony," Hyde said, then nodded in Spicer-Simson's direction. "You think he could carry a pony on his shoulders and hike a couple of hours through the bush?"

"He's a pretty strong fellow," Hanschell said, realizing that his defense of his commander was too weak to carry even a hint of conviction.

"Harrumph," was Hyde's only comment.

"I'd have had the natives carry it for me," put in one of the men listening, who didn't seem to have any of Hyde's skepticism. "That's all the bloody blighters are good for, anyway."

"Is that so?" Spicer-Simson turned to the speaker—a Mr. Richardson, who was traveling alone to Cape Town as a representative of a trading firm. "I was under the impression that this is your first visit to Africa."

"So it is," Richardson said. "But I've done all the reading."

"Burton, Livingstone, Stanley, and all that?"

"Among others."

"So you've become quite the expert on the indigenes."

The snideness in Spicer-Simson's tone was an unmistakable challenge.

"You named Burton," Richardson said, rising to the occasion. "I believe he said it all, and he certainly knew them."

"Ah, yes," Spicer-Simson said dryly. "And perhaps you will inform us of Burton's opinion."

"Certainly. He classified the native stares, for example, into a number of categories, including furtive, stupid, flattering, and greedy. Not to mention cannibal."

"I'm familiar with his classification," Spicer-Simson drawled. "I believe you omitted several: open, curious, intelligent, and perhaps the most significant—contemptuous. Do you suppose they turn that on arrogant foreigners?"

Richardson bristled. "I resent the implication of that remark."

"I'm sure you do," Spicer-Simson replied nonchalantly, and an uncomfortable chuckle passed around the room. "Besides, Burton's description could certainly be a catalog of English stares."

"Are you saying you put the primitive races on the same footing as Europeans?" This from one of the ladies present.

"Madam," the commander replied. "I simply have lived in too many places in the world to consider customs other than our own to be particularly primitive."

"But the black man is lazy," Richardson resumed the argument. "Nor has he a disposition toward inventiveness and high culture."

"We Europeans make much of our own great technological advancements, especially those in the realm of modern warfare," Spicer-Simson responded. "But the African knows that a simple, tiny dart can be just as deadly as a machine gun—in some cases even more so. And as for high culture, look at the state Europe is in as we speak. Untold numbers are being slaughtered in muddy ditches all across France, and for what? Pride? Sovereignty?" He stared hard at Richardson. "Trade?"

"Your words smack of treason, sir!"

"You will notice that I wear a uniform of His Majesty's Royal Navy," Spicer-Simson pointed out. "I am going to lead my men into combat. You, sir, wear the uniform of a business that neither creates nor destroys but merely passes whatever is most profitable from one hand to another. When you put off your suit and put on a military uniform, then you can speak to me of treason."

Richardson, flushing, pushed his chair away from the table and stalked from the room. Watching him go in the sudden silence, Spicer-Simson called out, "Waiter! Another vermouth, here." Then he turned to his audience and said, "Met another fellow just like him up on the Gambia. Had a canoe seriously overloaded with manioc—cassava, you know. Well, none of the paddlers...."

Hanschell stopped listening and turned back to his drink. As he picked up the glass, Hyde caught his eye and, with a smile, raised his own glass.

"I believe your commander just bagged himself another waterbuck, and we're not even in Africa."

Hanschell couldn't help but return the smile as he tipped his glass toward Hyde then tossed off the contents.

"Excuse me, Mr. Hyde. I think I need some air."

6

Out on the deck, Hanschell wondered if there was anything he could do to halt his commander's barrage of half-baked tales or his peculiar and often insensitive behavior. Spicer-Simson's own men already were questioning his judgment after he'd given the gunboats those ridiculous names and then come up with that outlandish new uniform. Hanschell worried about Spicer-Simson—worried that he would not have the strength or weight of authority to carry through this hazardous and farfetched mission, especially since he seemed determined to undermine his own authority with his men by carelessly appearing ridiculous in their eyes. Unfortunately, the commander had made it clear to Hanschell that the friendship they'd once enjoyed was, if not over, then on hiatus until the war's end. Spicer-Simson would brook no criticism from the doctor—or anyone else, for that matter—and that was that.

The next night, Hanschell arrived early at the bar. He hoped to get his drinking done before either his commander or Mr. Hyde appeared. Just outside, a small group of passengers was gathered around someone who was pointing at the skies. The man, Hanschell knew, was the Astronomer Royal from Cape Town, who was returning from England with new equipment for his observatory. On several occasions, the passengers had delighted in his pointing out and naming various celestial features and telling stories relating to their naming or discovery.

"You can see it there, just a few degrees above the horizon," the astronomer was saying. "The Southern Cross is one of the most prominent features of the skies of the southern hemisphere and has long been used as a celestial beacon by mariners. It nestles right in the loins of Centaurus, that group of stars right...."

"Harrumph," came a voice from the darkness. Everyone turned as Spicer-Simson stepped into the light from the lounge door.

"I beg to differ with you, sir," the commander said. "That is not Centaurus but Sagittarius. Quite simple to mistake the two, really, if you aren't familiar with the constellations."

"You must forgive me if I don't agree," the astronomer said. "Stars are my line of work."

"Is that so?" was Spicer-Simson's haughty reply. "I certainly wouldn't know it from what you've been telling us."

"And you would know, sir?"

"Why certainly, man. I am a navigating officer of His Majesty's Royal Navy. It is my duty to know." Spicer-Simson snorted and turned toward the lounge door. "Celestial beacon, indeed."

As he passed Hanschell, he nudged the doctor and said in a loud, sneering tone, "He'd make a damned poor navigating officer, eh, Doctor?"

Hanschell followed Spicer-Simson into the lounge, not wanting to be the object of the attention of the startled passengers attending the Astronomer Royal, and when they'd finished drinking, he followed the commander out of the bar. The astronomer and his entourage were gone by this time, and the two men stood by the rail and stared into the starry sky. The deep silence was suddenly broken by a murmur from the direction of *Mimi* and *Toutou*, and Spicer-Simson headed off in that direction to see what it was about. Having nothing better to do, Hanschell trailed behind.

As they neared, the murmured voices grew distinct, and the first clear words Hanschell heard were, "Lee's expedition."

"Attention!" Spicer-Simson barked, and the half-dozen men huddled in the lee of the cradled boats leapt to their feet.

"Did I hear you men referring to our adventure as 'Lee's expedition?'" Spicer-Simson asked, his voice tense.

None of the men had any reply except murmured denial.

"I thought not," Spicer-Simson said. "And I had better not. This is *not* Lee's expedition. Lee neither is here nor is he the commander. Furthermore, this is no man's expedition, it is the Royal *Naval* African Expedition, and I am in command, personally chosen by Sir Henry Jackson, and I'll thank you not to forget it. Do I make myself perfectly clear?"

"Yes, sir," the men said in unison.

"I thought as much," Spicer-Simson replied, spun on his heel, and strode off into the darkness.

This time, Hanschell did not follow the commander. Instead, he went to the forward rail, pondering Spicer-Simson's words. The more he thought on them, the more right they seemed. Even the tone—harsh and demanding but not strident—was proper.

All his doubts of the previous evening vanished. No matter what the men thought of him, Spicer-Simson was their commander, and he would never let them forget that fact. It also seemed to Hanschell that his friend's entire demeanor had acquired an edge it previously lacked, and he didn't have to think long on that to understand why. Spicer-Simson had been given two chances at command in the past and had botched both as nearly fully as a man could manage. This expedition was Spicer-Simson's final hope, the last opportunity for him to vindicate himself and prove his mettle in the eyes of the Admiralty, his colleagues, his friends, and his wife. Perhaps even himself. If he failed, there was nothing left but to ponder a life spoiled by fickle winds and washed up on the shoals of personal oblivion. Spicer-Simson had to succeed with the Royal Naval African Expedition. There was no alternative.

Hanschell's only worry was Lieutenant Commander Lee. Or rather, Spicer-Simson's attitude toward his second in command. Lee had left for Africa before Spicer-Simson invited Hanschell to join, but the men Lee had chosen—Wainwright and Eastwood among them—seemed to be forthright, and they had nothing but good words to say about the big-game hunter. And their trust of Lee had automatically spilled over to the other expedition members. Hanschell was anxious to meet the man who had originated the plan for the expedition and even now was paving the way for it across a bleak and dangerous landscape.

But there was one dissenting opinion on Lee's character, and that came from the commander himself. The afternoon of the day following the scolding he'd given the seamen for referring to the enterprise as Lee's expedition, Spicer-Simson seemed anxious to make his position more clear to the doctor, though Hanschell wasn't sure that it didn't sound more like a personal justification than an explanation.

"One meets Lee's type all over Africa," Spicer-Simson said, removing the cigarette holder from his teeth and tossing the butt into the sea. "They do a bit of shooting, prospecting, contracting for native labor, and so on, but really they're little more than tramps."

"Even so," Hanschell said, "he's come up with a marvelous plan."

"Tosh, Doctor. Lee is a johnny-come-lately. The Admiralty had me studying the feasibility of a similar expedition long before Lee showed up. That's why they chose me to lead."

Hanschell, remembering that Spicer-Simson originally told him that he'd overheard the plan when Admiral Gamble offered leadership of the expedition to the major of the Royal Marines, started to say something, then thought better of it. But he did not think better of Spicer-Simson for usurping Lee's role in creating the expedition.

That afternoon, just before the evening meal, he was leaning against the starboard rail, staring at the horizon, when Tubby Eastwood came up.

"I say, old man," Eastwood said. "I heard that the commander has ordered the men not to refer to our endeavor as Lee's expedition."

"That's true," Hanschell said. "I was there."

Eastwood shook his head. "What a curious state of affairs. I can only think that the hand of God is over this expedition."

"I have no information on that," Hanschell replied, "but of one thing I am sure—Spicer-Simson's hand will be over it."

7

Spicer-Simson was the last person on Lee's mind. There simply was too much to do to think of anything but the tasks at hand and those on the immediate horizon. After a few days in Elizabethville organizing supplies and hiring bearers, he, the other officers and road crew supervisors, and two dozen bearers boarded a train headed for the railhead at Fungurume.

Elizabethville was a meager settlement by Cape Town's standards, but Fungurume was little more than a dilapidated station that had been rude even when new. Next to it was a small house for the stationmaster and single large shed, both bearing rusted tin roofs. Off to one side stood a gaggle of huts, rapidly being swallowed by the bush. Most of the huts had been erected to house the workers who had built this last leg of rail line. Once that was completed, the huts had been abandoned and remained so except when temporarily occupied by tribesmen from other regions who were just passing through. That wasn't often, though. There was nothing to attract anyone to Fungurume and even less reason for anyone to remain. After checking in with the stationmaster—an educated African whose father was French—and sending a telegram to Cape Town stating that they were about to depart, Lee and his group trekked off into the bush.

Lee had been through this area several times before the war, and he'd given it a cursory scouting as he was devising his plan, but he hadn't had an opportunity actually to survey a feasible route until now. About one hundred and fifty miles lay between Fungurume and Sankisia, and even discounting the 6,000-foot Mitumba Mountains, the terrain was rough. Gullies, dry washes, and the occasional small gorge shouldered by low, rocky hills crosshatched the countryside, and the whole was littered with boulders that didn't seem respectable unless they weighed half a ton.

Dense, low bush covered everything, and whole stretches were overarched with trees so tangled they blotted out the sunlight, while others were nearly devoid of trees. But even where there was thick growth, everything was dry as tinder since the rainy season was still a couple of months away. To make matters worse, the last few years had yielded less rainfall than normal, so everything was more parched than it had been for half a decade. But the rare wet stream or lowland marsh was a mixed blessing because of the malaria-carrying mosquitoes and sleeping-sickness-laden tsetse flies that bred at the slightest provocation of moisture.

It was Lee's job to find a way across this blasted landscape. The route had to be as free as possible of obstacles but near to water sources to feed the boilers of the traction engines that were to draw the boats. And supplies of wood had to be laid by to fire the boilers along sections where the trees were few. Steep hills would have to be rounded, boulders blasted to gravel, and giant trees felled. And even if he couldn't begin to build any bridges since the bridge engineers were with the main party, he'd still have to prepare approaches to any depression deep or wide enough to need a span.

Lee, the ten Rhodesian supervisors, and the bearers he'd brought with him from Elizabethville couldn't even begin to do the work necessary. Hundreds of men would have to be hired from the villages along the route, and so, even as he began the task of surveying, Lee opened negotiations with the several local tribal leaders for labor and supplies, aided by an introduction from the vice governor general.

They would be building a road where none had existed, Lee realized. Maybe where no one ever thought one could exist. The thought brought him a touch of sadness. He loved Africa as only an Englishman bit with the African bug could, and he knew that what he was about to do would as irrevocably alter the landscape as the Mombassa Railway had in Kenya, or, for that matter, the rail line they'd just departed. But such was the way of nature, he supposed—human nature, at least, which often seemed so intent on leaving its mark on Earth that it rarely considered whether the mark was a good thing or bad.

Lee mentally shrugged. There was little he could do about it. Africa had been discovered by Europe, and there was no turning back. At least he could do what little he might to save the continent from the Hun, who would be as bad for the natives as the Belgians had been. Or worse. And with the Germans quelled, maybe there would be a shot at driving out the Belgians, too. Anyway, he could take some solace in the fact that Africa's slow pace virtually ensured that, for the rest of his life, he would be able to disappear into the continent's hinterlands whenever he liked. Once the war was over, maybe he'd journey northward. He'd never been across the Sahara. Now that would be an adventure.

A few days later, a runner came up from Fungurume bearing a telegram. It proved to be from Spicer-Simson, and it ordered Hope to return immediately to Cape Town to meet the expedition when the *Llanstephen Castle* arrived. Lee passed the information on to Hope, and soon after, the sublieutenant was heading back down the trail to the railhead.

Frankly, Lee wasn't sorry to see the man go. Hope had done practically nothing, and Lee couldn't understand why Spicer-Simson had insisted he come along. The idea that Hope was to examine the suitability of the rail line seemed a little farfetched, but maybe the commander was simply being cautious. At any rate, Lee had plenty to do, and he soon forgot about Hope as he turned to the tasks at hand.

8

"There," Hanschell said. "All done."

"Thanks, Doctor. You've got a nice touch with a needle. Didn't hurt at all." Lieutenant Wainwright rolled down his sleeve. "I'd rather have that little prick now than catch whatever it is you inoculated me against later."

"Smallpox," the doctor said.

"Damn," Wainwright said. "I was hoping you medical men would have a vaccination for malaria by now."

Hanschell smiled. He rather liked Wainwright and appreciated the lieutenant's quick wit and the fact that he knew a lot about a great number of topics since most of the other men, while expert in their tasks, were not informed conversationalists. Wainwright told him that he'd always liked to read, and he'd gotten a lot of opportunity while he'd been a locomotive engineer after he'd first arrived in Africa.

"I'm afraid you'll have to continue to take quinine to prevent that."

"Too bad. I don't quite like the taste."

"One of the bitter pills we must swallow on this journey," Hanschell said, almost without thinking. He looked quickly at Wainwright to see how he would react, but the lieutenant merely chuckled then nodded thoughtfully.

"I think it must be hard for you, Doctor. The commander makes a lot of the fact that you and he were friends before the war, but I can see that there is some distance between you, now."

"He is the commander, after all," Hanschell said. "He has to lead, and it's not my place to do more than assist him."

"He's an odd duck, isn't he? Has he always been this way?"

"As long as I've known him," Hanschell said. "Most likely his whole life. I've met his mother, you know, and she's the same way, so I guess he got it from her. She's like a force of nature, and we both know that nature can be a tempestuous and demanding mistress." He chuckled. "The first time I was introduced to her and the commander told her I was a doctor, she told me, 'Of course none of *us* have been doctors,' as if we were the disease instead of the cure."

"I'm surprised you remained friends with him all these years."

"Oh, he's not a bad sort really, once you get past the surface. And he truly is dedicated to the service."

"We are all dedicated," Wainwright observed, "but that doesn't make some of us any less foolish in the eyes of others."

Hanschell laughed but quickly grew serious.

"I have to tell you, Wainwright, that I really think he is the man to lead this expedition."

"You think he can make a show of it?"

"I do."

"He's a martinet and a braggart and a stuffy snot, and most of the other officers and men find him insufferable, but I suppose I'll have to give him the benefit of the doubt for now. I don't think I could give my all if I thought at the outset that our leader would fail. There is something about him, though." Wainwright shook his head. "I don't know that I can define it."

"I try not to," Hanschell laughed. "It's better for my peace of mind."

"Probably the best course of action. I think I'll follow your example."

"I'm all done here. Care to visit the lounge for a drink?"

"I'd like nothing better, but duty calls, and I must be getting along."

"I thought duty was suspended until we reach Cape Town."

"Except for the daily parade, you mean?"

"Yes, I don't suppose I could forget that," Hanschell admitted, thinking of the full-dress parade that Spicer-Simson insisted on each morning after mess, much to the mocking delight of the *Llanstephen Castle*'s officers and crew. The commander had ordered the men to stop wearing their old uniforms and appear only in the blue ones he had designed, which caused some grumbling but not much discussion. Considering the lambasting the men had received for calling the mission Lee's expedition, that was probably wise. Oddly, the two who should have complained most loudly were Tait and Mollison of the London Scots, who certainly had to give up the most distinctive uniforms, but the two had stoically abandoned their kilts and now wore only the new garb.

"Actually," Wainwright said, "the commander is having an engine testing today, and all the boat crews are to be present. Cross may know internal combustion engines, but his engine room artificers don't. They're steam engine men. The commander wants Cross and Cobb to give them some lessons."

"What time?"

"About an hour."

"I might come up and watch," Hanschell said.

"I don't know why you'd want to do that. It'll most likely be a bore."

"Well, I suppose I could pass the time by giving you another shot."

"I know what you mean. Not much else happening, is there? Perhaps we can meet later for that drink."

After Wainwright left, Hanschell cleaned his instruments and tidied up before he went onto the promenade deck where it overlooked the boats resting in their cradles.

9

Both boats had their tarpaulin covers removed, and Cobb was in *Mimi*, lifting the access panel that covered the engine. Just behind him stood Wainwright and Mollison. Of Cross and his engine room artificers, there was as yet no sign.

Spicer-Simson, flanked by Eastwood, Rickson, and Sublieutenant Tyrer, stood at the rail of the promenade deck, looking down into the boats. Nearby but slightly separated from the officers, a handful of civilians watched the proceedings with interest born of boredom.

Hanschell circled around a small group of ratings and joined the commander and the other officers just as Cross and his two ERAs, John Scotland Lamont and Edward Berry, emerged from a hatch and walked over to *Mimi*. In short order, they'd clambered up the short ladder leaning against the boat and into the cockpit with the others.

By now, Cobb had removed the access hatch and was pointing out features of the engine to the others.

"You sure that's the distributor, Mr. Cobb?" Lamont asked. "I don't think that's right."

"Of course it's the distributor," Cross cut in.

"I thought you said you'd never been to sea, Mr. Cross," Lamont said.

"That's correct, Mr. Lamont."

"Well, then, how is it you know so much about boats' engines. I've been an engine room artificer for twelve years, and that don't look like no distributor to me."

Cobb, who was an expert petrol engine mechanic and knew he'd correctly identified the distributor, started to say so, but Cross didn't give him the opportunity.

"I dare say you've been working on steam engines all those twelve years, Lamont. What would you know about petrol engines?" Cross gave a half-face, completely derisive smirk he was in the habit of making whenever he found something equally annoying and ridiculous.

"Me? What about you. You're not even a real ERA."

"I've won the Grand Prix twice, Lamont," Cross snapped. "I think I ought to have learned something about petrol engines."

"Just 'cause you drove one don't mean you know how they work," Lamont retorted.

"Silence!" barked Spicer-Simson, leaning over the rail. "Don't ever again let me hear you use that tone of voice with a superior officer, Mr. Lamont. And Cross, erase that sarcastic expression from your face. It's unseemly."

"Yes, sir," Lamont said, paling. Cross said nothing, but simply turned back to the engine.

"The next sound I hear better be the sound of that engine, Mr. Cobb," the commander said.

"Coming right up, sir," Cobb called out, thankful that the outburst had been squashed. *Mimi*'s engine roared to life, sending a cloud of blue smoke wafting over the *Llanstephen Castle*'s wake.

The ship's captain joined the passengers on the promenade deck.

"No smoking here!" he ordered loudly. A couple of the passengers who had lit cigarettes went to the side and tossed them into the ocean.

"Why ever not?" Spicer-Simson asked with exaggerated surprise. His cigarette holder was clenched in his teeth, but at the moment it was empty, and none of the other expedition officers or ratings were smoking.

"Because of the danger of igniting petrol fumes," the captain answered.

"Well, I'm damned. What nonsense!" Spicer-Simson snorted. "We're far out of reach of any vapors up here."

"No smoking here!" the captain yelled. "Those are my orders!"

Spicer-Simson looked taken aback at the captain's vehemence, and he suddenly flushed crimson. But he said nothing in reply, giving only a tight-lipped smile that drew his upper lip away from his teeth and a shrug that seemed to say it was all a small matter and beneath him.

The rest of the engine testing went without incident, and the passengers, having witnessed the best part, drifted off to other areas of the ship. Hanschell returned to his cabin and read up on several more truly horrific African diseases before the dinner bell chimed.

During the meal, Spicer-Simson recalled the incident with the captain, and he commented that he had tolerated the fellow's insolence only out of good manners.

"You know what these merchant service fellows are like," he said loudly enough for the ships' officers and passengers at the other tables to hear. "Actually, I could have ordered the captain off the deck there and then. As a commander of the Royal Navy on the active list in time of war, I can order any merchant skipper to turn his ship over to me."

The gasp from the passengers and the angry murmur that rose from the ships' officers masked the disbelieving sounds that came from Spicer-Simson's own officers. Spicer-Simson ignored them; however, he did see that Cross was giving his derisive sneer.

"For God's sake, Cross, please rearrange your face," he snapped, his dislike for the lieutenant plain. "I suppose you won your two Grand Prix victories by frightening your opponents with that frown." He laughed scornfully.

Cross's face fell, and an angry light flashed in his eyes. Spicer-Simson seemed not to notice, however, just as he shrugged off the silence that his previous comments had left in their wake.

Soon after the meal was over, Spicer-Simson retired, leaving the rest of the expedition officers to bear the deadening glare from the ship's officers.

"Bloody liar!" Cross spat.

"I'm not so sure," Hanschell said, trying not to make his words sound like a reproach. "The commander had considerable training when he was a cadet, and he's made the navy his career. Perhaps he knows what he's talking about. After all, the Admiralty...."

"With all due respect, Doctor," Cross interrupted, "The Admiralty must be completely daft to place that buffoon in charge of this expedition."

"Give him time, Mr. Cross," was all Hanschell could say, hoping it didn't sound like begging. "Give him time."

10

Miraculously, the *Llanstephen Castle* managed to avoid German warships, though it couldn't dodge a second squall as it rounded the Cape of Good Hope. Hanschell again was violently ill—he realized he'd probably never get his sea legs—but it soon was over, and not long after, the ship made port at Cape Town. It was the afternoon of July 2.

The passengers quickly debarked. Despite the fact that Spicer-Simson had the men parading on the dock just a dozen yards from the gangplank, the civilians hurried by without a word, friendly or otherwise for the naval expedition. All were glad to have made it to their destination in one piece and thankful to be out of Spicer-Simson's range. The *Llanstephen Castle*'s officers may have been equally grateful, but they weren't about to ignore the expedition, and most of the parading was accompanied by their jeers and catcalls that took in just about everything from the new blue uniforms with their pith helmets to the wagging tail of Spicer-Simson's saber.

Spicer-Simson seemed completely oblivious to the remarks tossed by the *Llanstephen Castle*'s crew. Instead, he was intent on his own men, who were standing rigidly at attention.

"Men," he said in his best nasal tone. "We are about to embark on the greatest adventure of our lives. I wish to remind you of several points.

First, we are, one and all, sworn to secrecy. Now that we are in Africa, that oath is more important than ever. Understood?"

He paused imperiously as the men shouted somewhat discordantly, "Sir!"

"Excellent. My second point has to do with this." He reached up with the tip of his swagger stick to touch the commander's insignia on his shoulder. "You are not once to forget who is in command of this expedition, and I expect each and every one of you to act accordingly. From this moment hence, any time I approach, you are to stand at attention and salute. And you will unfailingly address me as 'sir' or by my proper rank of commander. Is *that* understood?"

"Sir!" came the chorused reply.

"Again, excellent. And now, my third and most important point. I expect—no, demand!—your unwavering loyalty. My wish is your command; my command is the very dictate of your soul."

"Sir!" the men shouted.

Spicer-Simson nodded with a self-satisfied air as he looked over the ranks. "As long as you obey my orders implicitly as well as explicitly," he went on, "our victory is inevitable."

"Sir!"

"Mr. Waterhouse," the commander ordered. "Parade them a turn or two."

Sublieutenant Hope, accompanied by an African, met them at the dock shortly after Spicer-Simson finally dismissed the men to gather their belongings and prepare to go ashore, and the commander had a private conference with him that lasted nearly an hour.

Hanschell already had packed his medical supplies and personal gear, and he stood on the promenade deck, relieved that the difficult and tense voyage was finished. He hadn't been there long when he spotted Hope departing alone down the gangplank, and he left the deck and went toward the commander's cabin. The African who had boarded the ship with Hope was standing in the corridor outside the open door of the commander's cabin. Inside, Spicer-Simson was glancing between a piece of paper and a map, but when he saw Hanschell, he folded the paper and tucked it into a pocket in his tunic.

"There you are, Doctor. I was just about to go looking for you. We're off now, so go grab your kit and meet me on the dock in ten minutes."

"We're leaving?"

"Of course, old man. Don't want to spend another night on this insufferable vessel, do you?"

"No, but what about our supplies."

"They'll be unloaded by morning."

"But my medical supplies... I can't have them just lying around on the dock all night."

"I don't suppose you can carry them yourself," the commander said, raising his eyebrows.

"No, of course not."

"Well, I certainly am not going to carry them for you, so be a good fellow and get your kit and meet me on the dock in ten minutes."

Hanschell, seeing there was no point in arguing, did as he was told. The commander showed up a couple of minutes later, followed by the African, who was carrying his two bags. As they came up to Hanschell, Spicer-Simson pointed with his cane at the doctor's valise"

"That one, too, Tom," he said to the African.

Hanschell winced at the commander's supercilious tone. It was one thing to hear him use it on shopkeepers and waiters back in England, but it somehow seemed more offensive when he directed it against a native.

"I can get it," Hanschell said, a little embarrassed to be carrying nothing while Tom carried three heavy suitcases.

"Nonsense," Spicer-Simson said. "You must remember that we are now in Africa, and things are done differently here. There is a certain decorum to be maintained. White men simply don't carry, unless it's a weapon." He turned to the African and gestured vaguely with his cane. "Lead on, Tom."

Tom, it turned out, had been hired by Hope to serve as Spicer-Simson's valet. He had a small oxcart, and once Spicer-Simson, Hanschell, and their suitcases were inside, he urged the ox forward.

"Where are we going?" Hanschell asked.

"Hope has taken rooms for us at the Mount Nelson Hotel. It is quite elegant, he tells me. The best hotel in the city."

"Will the other officers be joining us?"

"Certainly not, old man." Spicer-Simson gave Hanschell a smile that was at once amused and patronizing. "It's much too good for them. But never fear—Hope has arranged accommodations for them at a hotel on Adderly Street. He assures me it is quite satisfactory."

Hanschell wanted to ask about the ratings, but he was afraid he'd hear that Hope had assigned them to native huts. He later learned that Eastwood spent much of the afternoon and early evening finding various places in town for them to stay.

That evening over dinner, Spicer-Simson asked Hanschell how he liked his room.

"It's quite nice," the doctor said. "Better than anything I've stayed in back home."

"Enjoy it while you can," Spicer-Simson chuckled. "All too soon we'll be off on our trek, and things aren't so pleasant in the bush." He gripped his silver-headed cane and smacked it against one of the table legs to catch the attention of a passing waiter. "Boy, bring me another vermouth. And you, Doctor? Another whiskey?"

The next morning, just as Spicer-Simson and Hanschell were finishing their breakfast, Sublieutenant Tyrer, wearing his ribbons and pins and monocle, came into the restaurant, spotted them, and hurried over to their table.

"Reporting for duty, Commander," Tyrer said. He saluted with an exaggerated flair, turning a few heads in their direction, which seemed to please Spicer-Simson.

"Very good, Tyrer," Spicer-Simson said, standing. "We're off, Doctor. Official calls and all that."

Followed by Tyrer, he left the hotel. Hanschell watched through a window, shaking his head, as the two stepped into the bright sunlight and entered a waiting cab. What a pair—the commander with his jaunty cigarette holder and cane, Tyrer with his medals and monocle and yellow hair, and both in those odd blue uniforms. He hoped that they didn't cause too great a stir as they made their rounds.

Amusing as the thought was, he shook it off. He had his own rounds to make, and he'd best be at them. He had the clerk at the front desk find him a cab of his own, and he told the driver to take him to the hotel where the other officers were billeted.

Eastwood was waiting for him in the lobby.

"I've spoken to the chief medical officer at the naval base here," Eastwood said. "He recommended we talk to a medical supplier named Lennard. He's well regarded in town and reportedly has equipped numerous safaris."

Hanschell was pleased, when they arrived at Messrs. Lennard, Pharmacists, to see a placard on the door proclaiming that the company was the local agent for Burroughs Wellcome & Company, which increased the likelihood that the shop would be well-stocked with first-rate medicines and equipment.

Inside, they asked the clerk to fetch Lennard, and a smiling man in early middle age emerged from the rear of the establishment.

"I'm Mr. Lennard."

Hanschell introduced himself and Eastwood.

"The Royal Navy, eh?" Lennard squinted at their uniforms. "There was some talk about a military bunch who just came in on the *Llanstephen Castle*, but no one quite knew which branch they were with. Don't think I've ever seen uniforms quite like those."

"No," Hanschell said, feeling his face flush slightly. "They are rather unique."

"Well," Lennard said, ignoring Hanschell's momentary discomfort. "What can I do for you, Doctor?"

"I'm in need of some medical supplies," Hanschell said. "Quite a lot of them, in fact."

Lennard laughed. "I suppose they gave you *Medical Stores as Supplied to Gunboats on the West African Station*."

"They did," Hanschell admitted with a dry chuckle. "How did you guess?"

"Oh, I know the miserable thing by heart," Lennard said. "But only because nearly every ship's doctor who arrives here comes to me to remedy its deficiencies."

"I'm afraid they only gave me about a quarter of what I requisitioned."

Lennard shook his head. "They simply don't know Africa," he said. "Well, I take it you'll be going on up the west coast."

"Not exactly, Mr. Lennard. We're heading into the interior, and I'll need everything you can supply for thirty men."

"The interior?" Lennard looked puzzled. "Don't tell me you're going to Tanganyika."

Hanschell and Eastwood exchanged a glance.

"I'm afraid we can't divulge...," Eastwood began, but Lennard held up a hand.

"Gentlemen, I've lived here for more than fifteen years, and I know Africa quite well. If a Royal Naval expedition is going into the interior, then it must be going to one of the great lakes. As Britain presently commands all of those except Tanganyika, it only stands to reason that must be your destination. Especially from this direction."

"Tanganyika, it is," Hanschell admitted.

"Don't worry, Doctor. Your secret is safe with me. Now, let's go in the back and see what we can do for you. Do you have a list?"

Thank God he'd found Lennard, Hanschell thought as the pharmacist showed them around his warehouse. The man not only had everything Hanschell lacked, he understood the contingencies Hanschell would face.

"How long will you be gone?" Lennard asked.

"I'm not certain," Hanschell said. "The Admiralty has planned on six months."

"What the Admiralty knows about Africa could be stored in one of those." He indicated the several dozen zinc-lined, waterproof, and ant-proof portage boxes stacked neatly against one wall. "Everything takes longer in Africa than anywhere else. Except," he chuckled, "enlightening the minds at the Admiralty. That may never happen. I suggest you plan on a full year." He glanced at Hanschell's list. "I see you brought a surgical kit. Good. But you'll need two."

"Why two?"

"You're going to need at least two of everything," Lennard said. "And your stores are going to have to be parceled out among the boxes.

Native bearers are like people anywhere. Some are brave and honest; others are not. You don't want one of the latter running off into the bush carrying your only surgical kit or your only supply of quinine."

"I see what you mean."

"And I notice that you haven't included a dental kit."

"I was too worried about malaria, sleeping sickness, and yellow fever to even think about it," Hanschell admitted.

"Fortunately I have one, but not a second, so you'll have to be careful that it doesn't disappear. And," he said, moving over to an array of medium-sized wooden barrels and slapping one on the top, "you'll need two or three of these."

Hanschell peered at the label and laughed.

"Three barrels of purging pills? For thirty men?"

"Not your men," the pharmacist explained. "The natives. They tend to rely on their witchdoctors for most conditions—not that it does them much good—but they have learned the positively irresistible power of white medicine when it takes the form of purging pills. There'll be no end of them coming to you for them, and you'll want to have plenty on hand."

"All right. Three barrels, then."

"Excuse me, Doctor," Eastwood broke in. "A little of this, a few barrels of that, and all these boxes to carry them in is going to cost a pretty penny."

"Don't worry about the expense," Lennard said. "I may disparage the Admiralty's propensity for mental stagnation, but I never doubt its ability to pay its bills. I'll put together everything you need, and you can sign a draft. By the time I present it for payment, you'll be gone with the supplies you need, and there won't be any argument."

"Not with us, at least," Hanschell laughed.

"There is one more item you'll need," Lennard said, "and this maybe even more than the rest."

"And that is?"

"Rupia!" Lennard called out, and an African in his mid twenties hurried over. "This is Rupia," Lennard explained. "He can serve as your assistant."

"Assistant?" Hanschell asked. "I hadn't thought I'd need one."

"Trust me, Doctor. You're going to need him."

"Doesn't he have a say in the matter?"

"Oh, he's willing. Rupia may be young, but he's a seasoned traveler, aren't you, my boy?"

"Rupia travel everywhere, Bwana Doctor," the young man said. "Rupia like to travel."

"We're going a long way for a long time," Hanschell said, studying Rupia and taking in his sharp eyes and quick smile.

"The longer, the more pay," Lennard said. "There's nothing better than steady employment, especially while one is on a journey."

"What will he want in return?" asked Eastwood, thinking as a paymaster should.

"His rates are quite reasonable," Lennard said. "And worth every penny. Between trips, he works here, so he is familiar with medical equipment and supplies. I believe you'll find him invaluable."

"Very well," Hanschell said, recalling that Spicer-Simson had just employed a personal servant. "Welcome to my medical team."

As Rupia beamed, Lennard stuck out his hand, which Hanschell heartily and gratefully shook.

11

Meanwhile, Spicer-Simson, accompanied by Tyrer, met with Vice Admiral H. G. King-Hall, the naval commander-in-chief at the Cape.

The initial outlook was good. Lee's advance road-building party was already well into the bush beyond the railhead at Fungurume. They had built road twenty miles out from the railhead, pushed on through the brush for another thirty miles, and were in the process of building the next section.

"May I ask, sir, why have they left a thirty-mile gap?" Spicer-Simson asked.

"Apparently, they were uncertain of the terrain ahead and did not at first know which direction the road would take," King-Hall replied. "When they finish the section they're building now, they will return and complete the gap. It should be done by the time your main party arrives at Fungurume."

Early the next morning, the docks were bustling with activity as the expedition prepared for the journey. The first task was to hoist the boats from the deck of the *Llanstephen Castle* and deposit them on a flatcar that would be hitched to a specially commissioned South African Railways train for the journey north. The process took much of the day and was overseen by Wainwright, who wondered why Hope, who ostensibly had more experience at this sort of thing, wasn't on hand.

"He's gone north," the commander informed him. "I've sent him to check the rail line one more time and to assist Lieutenant Commander Lee."

Wainwright could do little more than shrug, though he wondered why the commander would send a railroad man to assist a scouting and survey mission.

Inside of a handful of days, the boats and all the supplies were loaded and the expedition was ready to move, but on the eve of departure, Vice Admiral King-Hall called in Spicer-Simson to deliver some bad news.

"We have just had word from the Admiralty," he began. "They've been in contact with the Belgians on the lake, and it seems they have an engineer, a man named Le Plae, who says that carrying the launches over the Mitumba Mountains will be impossible."

"What does that mean, sir?"

"You are to wait."

"My God, sir! We've come all this way to be bogged down now on the word of an unknown Belgian engineer? I wouldn't be surprised if he was singularly ill trained."

"I understand your impatience, Commander. But Monsieur Le Plae helped build the Katanga Railway, and his words carry some weight. For the time being, the matter is out of our hands. Lee has gone ahead to ascertain if Le Plae's claims are valid. Until he's finished his survey, you'll just have to wait."

12

When Hope arrived in Elizabethville, he learned from the vice governor general that Lee and his party had pressed on and, at last report, were at Mobile Kabantu, just south of the Mitumba Mountains. As that news was several days old, undoubtedly they were already at the mountains.

Hope went straight to the train depot and collared the telegrapher stationed there.

"I want to use your telegraph," Hope told the man.

"Certainly, sir. If you'll just write out your message on this piece of paper...."

"I prefer to send the messages myself," Hope said.

"That's not possible. "Company regulations prohibit...."

"You know John Lee?" Hope asked.

"Sure. He and that other fellow, Magee, they've been through here several times. Fine fellows, from what I could see. Lee knows his business."

"You are aware that they are on a highly secret mission?"

"I've heard talk."

"I am part of the same advance party," Hope said. "The war in Africa may hang in the balance. Now, I have a number of confidential messages to send to the expedition commander, and I want to send them myself. I assure you, I am quite proficient at the key. Now if you will just give me a few minutes alone...."

Hope urged the telegrapher out of the tiny office, shut the door, then sat down at the key and began sending his first message.

13

On July 6, Spicer-Simson was in King-Hall's office, sadly shaking his head as the vice admiral surveyed the three dozen telegrams on his desk. And these weren't the whole lot, just the worst of the ones that had come in over the past couple of days from Elizabethville, Fungurume, and various other telegraph stations in the Congo's Katanga Province.

"It's a bad case, Admiral," Spicer-Simson said. When King-Hall did not immediately reply, he tried a more expectant tack. "Perhaps we should do something?"

It was a bad case indeed, thought King-Hall. The messages, which apparently had spread like a brushfire throughout the territory, had even had passed through the hands of General Edwards, the commandant general of Rhodesia, as well as the office of the British administrator in Livingstone, both of whom were highly incensed. And now, even the high commissioner in Cape Town was involved. No, bad case did not adequately define the situation—it only could be called a grim picture.

Instead of preparing the route as they should have been doing, both Lee and Magee had spent most of the last ten days drinking and carousing at every known tavern and bar in the region. At last, their behavior had gotten so completely out of hand that they'd been arrested by the police. King-Hall wasn't entirely sure exactly where that had occurred or what had occurred, but the arrest was probably the best thing that could have happened since it seemed that the liquor they'd consumed had loosened their tongues, and every person this side of Tanganyika probably now knew the purpose of the expedition.

Worst of all, they'd not only used their besotted tongues to liberally insult every Belgian officer and colonial administrator they'd encountered, they'd even impugned King Leopold's right of accession. As a result, one series of messages stated, the Belgian vice governor general was demanding that the British remove Lee at once from the province or he would consider derailing the expedition at Elizabethville.

"What are we going to do, Admiral?" Spicer-Simson prompted again.

"Apparently, the man is not only unreliable and unstable, but he's a socialist and may have gone native. No wonder he's left huge gaps in the road. He's even dragged Magee into his debauchery. I know Magee, sir, and he wasn't like that before he met Lee. It's positively shameful, sir."

"Yes, shameful," King-Hall sighed. "The question is, what do we do now?"

"Why, carry on, sir," Spicer-Simson said, thankful that the vice admiral was so shocked by the extent and number of allegations that he failed to consider checking their veracity.

"Carry on?"

"Certainly, sir," Spicer-Simson said in a confident tone. "The expedition is ready to proceed, and I will be happy to take on the task of surveying. After all, I do have extensive experience in these matters, and it is *my* expedition."

"Carry on, then, Commander. I have too much to think about here on the coast to pay much attention to what happens inland. I leave it in your capable hands to sort these matters through. If there have been improprieties, you are authorized to deal in whatever way you see fit with the person or persons responsible."

"Thank you, sir." Spicer-Simson saluted and left.

The next morning, Spicer-Simson approached Wainwright, who was supervising the packing of the supplies and gear onto the train cars.

"I'm off to Salisbury to meet with General Edwards. I'm leaving you in charge of the expedition. As soon as our train is ready, take everything up the line. I will meet you in Bulawayo."

Spicer-Simson arrived in Salisbury, the capital of Southern Rhodesia, two days later, and he went immediately to General Edwards's headquarters.

"I have heard that someone is talking about your expedition," Edwards said after Spicer-Simson gave him his rendition of Lee's perfidy, "but I haven't been given to understand that Lee is the culprit. And these charges of drunkenness don't sound right to me. I've known Lee for some time, and I've never seen him in his cups."

"Nevertheless, sir," Spicer-Simson said forthrightly, "there are the reports. The vice governor of Katanga himself has lodged a complaint."

"Yes," Edwards said. "So it appears. I admit that I recently received a message or two regarding Lee's bad character, though I find them hard to believe. But I haven't seen him in some time, and men do change."

"I have to say, sir, that I wasn't impressed with Lee when I first met him, and I'm even less so now that we have these sorts of telegrams from every place that he's been. Vice Admiral King-Hall has authorized me to arrest him and confiscate any documents he may be carrying."

"A sad state of affairs," Edwards said. "I hate to see a man's character besmirched, but we *are* at war."

"I'll be leaving here tomorrow, sir. My plan is to go up to Elizabethville to investigate."

"Surely you could stay here for a day or two," the general said. "Mrs. Edwards is in dire need of news from England."

"It would be my pleasure, sir," Spicer-Simson replied. "After all, I don't suppose Lee can do any more damage than he's done already."

14

When he left Salisbury, Spicer-Simson did not go to Elizabethville as he'd told General Edwards he would. Instead, he simply returned to Bulawayo and waited for the expedition to come north on the rail line. He'd been there three days when, on July 16, a local boy approached him at the bar where he'd been spending much of his time.

"Telegram, sir," the youth said, handing over a folded piece of paper to the commander.

"That will be all," Spicer-Simson said, shooing the boy away and opening the paper. It was a message from Lee.

"Have found good route over Mitumba," Spicer-Simson read. "Survey of remainder of route complete and will send a packet on tomorrow's train. Good traveling. Will meet you in Elizabethville."

Spicer-Simson reread the message, then he carefully folded the paper and tore it into confetti, which he dropped into the ashtray. After lighting a cigarette, he touched the match to the shreds in the ashtray and watched them flare briefly before turning to ash. A smile on his face, he signaled the bartender for another vermouth.

The Saturday train did have a packet for him, whose contents he quickly absorbed. That evening, he carefully composed a lengthy message to the Admiralty in London, which he took to the telegraph office the following morning.

"Send this message," he told the telegrapher. "Urgent."

The telegrapher bent to his key and set to work. By the next day, the message was in the hands of Admiral David Gamble, who took it to Sir Henry Jackson.

"It says here," Sir Henry said, scanning the message, "That Mr. Lee has fallen down on the job."

"Fallen down drunk, more like it, according to what I've heard from King-Hall," Gamble amplified. "Spicer-Simson says Lee and Magee have disappeared entirely and can't be found. I've had confirmation from King-Hall, who says Lee's drunken and erratic behavior is the talk of the Congo and Rhodesia."

"So," Sir Henry said, reading further. "Spicer-Simson himself has personally surveyed ahead and found a feasible route over the Mitumba Mountains." He looked up at Gamble. "It appears you've made an excellent find in this man, Sir David. He's got the skills, and he's resourceful."

"Thank you, Sir Henry. And, I might point out, he is loyal to his own men. You'll see a little further down that he was accompanied on his survey by Sublieutenant Hope. He's asking that Hope be promoted to full lieutenant to take Lee's place as advance man. From what Spicer-Simson says, Hope knows the lie of the land and the locals."

"Done," Sir Henry said. "Wire Spicer-Simson that Hope is to be promoted to lieutenant, and tell him we are giving the expedition the go-ahead to proceed north."

Chapter IV: Cape Town to Fungurume

ON JULY 20, THE TRAIN carrying the Royal Naval African Expedition pulled out of Cape Town on the first leg of its 3,000-mile journey across Africa. It was a narrow gauge, with rails a mere meter apart and a small locomotive barely large enough to pull the train at any decent speed.

Mimi and *Toutou*, racked securely in their cradles and covered with tarpaulins, were chained to flatbed cars immediately behind the passenger car carrying the expedition members. That, in turn, was behind the locomotive. Following the boats, were two flatbed cars, the first holding the wheeled undercarriages, stacked one atop the other, and the other loaded with the motor lorry and more gear. Bringing up the rear were five flatcars piled high with supplies, mostly petrol, also covered with tarpaulins.

As the outskirts of Cape Town disappeared behind the train, the expedition members were in a fine fettle. They were off, at last. In no time at all, they'd have the boats in the water and would be giving the Germans on Tanganyika hell.

Wainwright, who had been making sure that the boats were secured, entered and sat next to Hanschell. He pulled a flask of whiskey from his kit and poured a round for the two of them.

"To the journey," he said, and they downed their drinks. Afterwards, Wainwright settled into his seat and watched the scenery roll past, happy that the work was done until they reached the end of the line.

But not ten miles out of Cape Town, cry rose from the back of the coach. "The boats! They're on fire!"

"What the hell!" Wainwright shouted, turning. "What's going on, Waterhouse?"

"I don't know, sir," the chief petty officer called above the rising cries of the other men.

Wainwright shoved his way to the rear of the coach, where one of the men had opened the door. As soon as he emerged, he could see smoke rising from the tarpaulin-covered boats.

"Get the men out here!" he yelled to Waterhouse. "Get buckets, barrels, anything. Piss on it if you have to, but get that fire out! Lamont! Tell the engineer to stop the train!"

Lamont hurried toward the locomotive, and a few moments later, it squealed to a halt, nearly throwing everyone to the floor. Waterhouse had a fire brigade organized in short order, and the first of the men arrived, sloshing water in their tracks from the buckets they carried. As they doused the flames, Wainwright felt a huge sense of relief wash over him —the tarpaulins covering the boats had been on fire, but the boats were barely touched, sustaining little more than a few blisters on their varnish. He shuddered to think what might have happened had the flames reached the petrol tanks.

The fire was soon out, without anyone having to resort to pissing. It had been started, it seemed, by embers flying from the stack of the wood-burning locomotive.

Wainwright surveyed the damage, then turned to Waterhouse.

"Put a guard on," he said. "I should have thought of it before."

"Yes, sir," Waterhouse replied. "I'll arm the men with buckets in case of more sparks."

As the chief petty officer organized a watch of two guards to stand on each of the flatbeds carrying the boats, Wainwright ordered the rest of the men back to the coach then waved to the engineer to resume the journey.

As the train moved steadily north, Hanschell enjoyed several long conversations with Wainwright. At one point, the lieutenant leaned close and asked, "What about those tattoos? How much of him do they cover?"

Several of the men had noticed that Commander Spicer-Simson had a ring of tattoos at the base of his neck and around each wrist.

"I really don't know," Hanschell admitted with a smile. "I've seen him in short sleeves, so I know they cover his arms, and there seems to be some beneath his collar. I haven't had the privilege of seeing our commander in the buff, though, so I haven't the answer."

"Some of the men have a little wager going. About the tattoos."

"What kind of wager?"

"A pool."

"What's the pot?"

"I couldn't say. Eastwood is holding it. His religious convictions prevent him from gambling but, apparently, not from aiding others in that vice. But it must be considerable by now. All the ratings and most of the junior officers are in."

"And who wins?"

"Whoever gets closest to the amount the commander is covered. Some say it's just his arms and neck, some say its all over his chest, too. You want in?"

For a moment, Hanschell was nonplussed as he realized, now that he was asked, that he'd never made a wager in his life.

"How much do I have to put in?"

"The ante's twenty quid."

"I suppose it can't hurt," Hanschell said, pulling out his wallet. He took out a twenty-pound note and passed it over to Wainwright.

"What do you say?" Wainwright asked, pocketing the bill. "Arms and neck or chest, too."

"I told you that you should have faith in our commander," Hanschell said. "I said that because Spicer-Simson strikes me as the sort of man who does not live by half measures if he can have the whole. I'm going to say he's completely covered."

"Completely?" Wainwright looked shocked. "Even his...you know, his...."

"His balls, you mean?" Hanschell chuckled and shook his head. "No, perhaps not there. He may not go for half measures, but he's certainly no fool. He won't fight a battle he can't win. So no, not his balls or cock, but I'll vote for everywhere else. What about you?"

"I haven't made my guess, yet. I wanted to hedge my bet by seeing what you know."

"And now that you know I know nothing?"

"You just may be right, anyway." Wainwright smiled. "I think I'll join you. Your opinion, I'm sure, will cause quite a stir among the ratings."

"They're good men," Hanschell said with a smile, "but precious few have any imagination."

2

The slowing of the train drew Hanschell out of a shallow slumber. He opened his eyes, squinted against the glare, then tried to stretch, his butt numb from the thin leather upholstery covering the wooden seat. His mouth was gummy, and he felt a definite need to bathe. He sat up just as the train rolled to a halt.

"Let the men off to relieve themselves," Wainwright ordered Waterhouse, then he strode to the end of the car and went out the door. Hanschell grabbed his cap and hastily followed before the crush of petty officers and ratings could press toward the exits. He trailed the lieutenant down the steps and up the tracks toward the locomotive. Long before they reached it, the problem that had halted the engineer was manifestly apparent.

It was a bridge over a river. And not just any bridge. It was a covered bridge, and the tunnel looked barely tall enough to accommodate the locomotive. Even a cursory glance told all. The boats on their flatcars were simply too high to make it through.

The engineer and his fireman were standing beside their locomotive as Wainwright and Hanschell walked up.

"Where the hell did this come from?" Wainwright demanded.

"It's always been here," the engineer replied.

Wainwright stared at the bridge, wondering about Hope. Wasn't he a railroad man, and hadn't Spicer-Simson sent him on ahead to scout the suitability of the line? Why hadn't he reported this obvious obstruction? And if he wasn't checking the line, what exactly *had* he been up to?

"What are we going to do?" Cross asked. He'd come up with Rickson and Waterhouse.

"We'll think of something," the lieutenant replied, and he stalked off toward the mouth of the tunnel.

The other men went with him, but Hanschell remained by the locomotive. This was an engineering task, and he thought he'd better stay out of the way, though he was curious to see what the outcome might be. Would they have to back the train all the way to Cape Town? He watched as the men entered the tunnel, becoming walking shadows as they made their way down the tracks toward the backlit far end. There they emerged into the sunlight, turning into men once again.

"What river is this," Hanschell asked the engineer.

"The Modder," came the reply, and the man gestured vaguely toward the north. "Kimberley's about forty-five mile that way. If we can get the boats past this bridge."

"Why is it covered?"

The driver shrugged, and seemed to think that answer enough.

Several ratings had clustered nearby, and Hanschell heard one of them remark, "Well, lads, we're finished already, and we've hardly begun. We'll never get through that thing."

The others voiced agreement, and the group went off to lounge in the shade cast by the flatcars and their cargo.

By now, Wainwright and his companions again were shadow men, and in another minute, they were standing beside the locomotive, except for Waterhouse, who went back toward the cars holding the boats.

"What are we going to do?" This time it was Hanschell asking.

"Wainwright's got a plan," Cross said.

"Sleepers," Wainwright said.

"What?"

"Sleepers."

"You mean you're going to bivouac here?" The doctor didn't see what good *that* would do. The bridge still would be here in the morning.

"Waterhouse is measuring the boats from deck to keel. We think that we can set them on sleepers—timbers laid across the rails—and they'll be short enough to fit through the bridge. We can use the loco to drag them along."

"But how do we get the boats off the trailers?" Hanschell asked.

"Getting them off's not the problem, Doctor," Cross said. "Getting them on again might be."

They went back to the boats just as Waterhouse hopped off the first flatbed, which held *Mimi*.

"I think we have the necessary clearance, sir," the chief said.

"You think?" Wainwright asked with a bemused frown.

"It's a pretty close thing, sir."

"Well, we've no choice but to try," Wainwright said, and he demonstrated how, using the locomotive and all the manpower at their disposal, they could lower the boats in their cradles onto timbers placed across the tracks.

They began the operation immediately. First, they unhitched *Mimi*'s trailer from the rest of the train behind it and locked all the cars' brakes. Then, they attached cables they'd brought along between *Mimi*'s cradle and *Toutou*'s flatbed. While this was being done, a crew of ratings felled four large trees near the tracks. Two of the stripped trunks were propped against the back end of *Mimi*'s flatbed like a ramp, and the top ends were made fast with more cable and telephone line that came from the gear. One of the other two timbers was laid across the tracks at the foot of the ramp and attached to the lower ends of the two ramp logs.

Then the locomotive, inching ahead, rolled *Mimi*'s flatbed forward. The boat and cradle, cabled to the rest of the train, began to slip off the rear of the flatbed as the car was pulled from beneath them. There was a collective intake of breath as the boat tipped toward the log ramp and its bow swung high in the air, but it didn't topple. In a few seconds, the rear end of the cradle bumped down onto the log ramp and then rolled across the first sleeper, which had been dragged along with the car and ramp.

Cross signaled the engineer to halt, and the sleeper was detached from the ramp and tied to the cradle. Then the locomotive again inched forward until the bow-end of the cradle was almost at the bottom of the ramp. The engineer stopped the locomotive, the second sleeper was slipped between the rails and the cradle, and the locomotive moved forward just enough to bring the cradle down onto the second sleeper.

It, too, was tied to the cradle, then the cables that connected the rear of the cradle to the rest of the train were brought forward and attached between the now-empty flatbed and *Mimi*'s cradle and the sleepers.

While the sleepers and the rails were being well greased, Wainwright went to the locomotive.

"Go slow, now," Wainwright ordered the engineer. "And stop at my first signal."

The engineer nodded and mounted his engine. Wainwright motioned for him to advance, and the engineer eased down on the throttle. With a lurch, *Mimi* broke free of inertia and began to slide grindingly forward on the greased sleepers and rails. The locomotive slowly chugged forward while Wainwright, Cross, Rickson, and Waterhouse paced alongside the boat, keeping keen eyes on the sleepers, the fastenings, and the progress to make sure *Mimi*'s cradle did not slide sideways off the track. Before long, the locomotive disappeared into the dim of the tunnel, and Wainwright ran to the cab.

"Slow!" He glanced back to see *Mimi*'s bow cut into shade. "Stop!"

The engineer obliged, and Wainwright trotted back to the boat. Its hull seemed to completely plug the hole, and Wainwright gestured to a huge red-headed rating who had been a fisherman in Donegal before the war.

"Come over here and give me a boost."

The big man's thick hands lifted Wainwright, who peered over the top of *Mimi*'s deck.

"How's it look?" Cross asked after the rating had lowered Wainwright to the ground.

"There's about six inches between the deck and the roof joists. I think we're all right. Go up and tell the engineer to proceed as slowly as he can."

This time, as the boat ground off, it was at slightly slower pace. Wainwright ignored the sleepers and concentrated on the narrow margin between the roof joists and *Mimi*'s deck, expecting at any instant to hear the crashing sound of splintering timbers and planking. Then, suddenly, sunlight burst upon them. The boat was through, and it was in one piece. A cheer rose from the other side of the river, and Wainwright nearly fainted with relief.

Dragging *Mimi* back onto its flat-bed was almost a reverse of the process they'd used to lower it to the tracks, and when it was done, dusk lay over the landscape. In a sense, Hanschell's question about bivouacking for the night had come true.

The next morning, *Toutou* was brought through the covered bridge, and even before it was reloaded on its flatbed, Mullen trundled the lorry through under its own power.

The entire operation was accomplished by midday, and the train began to roll again as if nothing had impeded its progress. But something *had* changed. It was the way the men looked at Wainwright. Before, he had simply been another officer apt to give unnecessary and bothersome orders. Now, however, he was a man who had the ingenuity to transform a superhuman feat into a simple problem of engineering. Even better, he'd been right there with the men, sweating and getting his hands dirty with the rest of them. After that, the "sirs" he received were not grudgingly given.

Just an hour and a half later, the engineer stopped the train in Kimberley to take on water and fuel. The delay wasn't lengthy, and soon, the train was rolling across the high veldt.

3

The expedition was two and a half days out of Cape Town, and, except for the fire and the covered bridge, everything had gone without a hitch. South Africa lay behind, and Bechuanaland lay ahead, just beyond the border town of Gaberones. After a short stop there to take on water and fuel, the train chugged on, and then there was nothing but arid countryside out of the passenger coach windows as the train skirted the edge of the Kalahari Desert, which spread out to the west. Fourteen hours of wasteland later, the train rolled into Bulawayo.

Stretching their cramped limbs, the men were glad to learn that there would be a lay-over while the train was fueled and a relief crew found. The stationmaster came over to Wainwright and informed him that Spicer-Simson was at the hotel, and he was to bring the men for lunch.

Wainwright told Waterhouse to organize the men and take them to the hotel, then, accompanied by Eastwood and Hanschell, he went on ahead. The hotel where they found Spicer-Simson was the largest and best in town, though that wasn't saying much. The commander was in the bar, and the officers couldn't help but notice that he had a new medal on his chest. Wainwright forbore to mention it, but Hanschell was curious.

"It's the Africa General Service Medal," Spicer-Simson said in reply to the doctor's query. "Presented by General Edwards."

Wainwright couldn't remember hearing of such a medal before, but he was no expert in military honors, so he shrugged it off and brought Spicer-Simson up to date on the journey so far—prudently leaving out the fire and earnestly hoping the commander would not immediately notice the holes charred in the tarpaulins.

By the time Wainwright was done, most of the other men had finished eating and were outside, lounging on the veranda. As he trailed the commander and doctor through the door, he heard a voice call out, "Lieutenant Cross of the Royal Seahorse Engineers!" Laughter followed, and he soon saw why.

Cross was astride a mangy looking pony that was staggering around in the street beneath its rider's weight. More hooting laughter followed while Cross, a stupid grin on his face, put his overburdened beast through its paces.

Anger clouded Spicer-Simson's brow, and he stomped down the steps bellowing, "Get off that horse, Cross!"

The catcalls from the other men ceased instantly, and Cross, the silly smile wiped from his face, dismounted, led the pony back to a post, and tied the reins.

"Is that your horse, Cross?" the commander demanded as the lieutenant came over to him. "No, sir."

"Where the hell did you get it?"

"It was tied to that post, there, sir."

"Tied to the post. Was it you who tied it there?"

"No, sir."

"I thought not. Don't you know that horse thieves are hanged in this country?"

"I didn't steal it, sir. I just borrowed it."

"I don't think they'd make such a fine distinction, Cross. Nor would I where you are concerned, considering your continued irresponsible behavior." Spicer-Simson turned to Wainwright. "Have the men return to the station. I doubt that Mr. Cross will find any further mischief there, and perhaps he might even comport himself as a proper officer should."

They'd nearly reached the station when they saw a teenager hurrying toward them.

"Captain Cross!" the youth called out, looking from one of the men to another. "Telegram for Captain Cross!"

Color rose in Spicer-Simson's face like mercury in a heated thermometer. "Give me that!"

He snatched the telegram out of the boy's hand, and as he read it, Eastwood gave Hanschell a nudge.

"You don't think Cross has been talking about the expedition, do you?"

"I don't see how anyone would know he was here if he hadn't," the doctor replied.

Spicer-Simson finished reading the telegram and thrust it at the now completely abashed Cross.

"I can readily understand that an engineer lieutenant would want to be thought an army captain," Spicer-Simson said with a sneer, "but as he is now serving under a Royal Naval commander and in a Royal Naval expedition, I, Commander Spicer-Simson of the Royal Navy must order that in future, Engineer *Lieutenant* Cross will bear that in mind and keep his army preferences to himself until he has left the Royal Navy."

With that, Spicer-Simson jammed his cigarette holder between his teeth and strode off toward the station, jauntily swinging his cane.

"That bastard's got no right to talk like that," Cross said after the commander was out of earshot. "I never said I was an army captain, even if that's what I would be if I was in the army."

"Who's the telegram from, Cross?" Tyrer asked.

Cross looked at the paper that had gotten him roasted.

"It's from the family I dossed with in Cape Town," he said. "They wish me good luck." He spat in the dirt. "Better they'd wished me another commander."

4

The stretch between Bulawayo and Livingstone was a ten-hour ride, but it was less than half over, and Hanschell was regretting taking the seat next to the commander. The dry air only seemed to enhance Spicer-Simson's already ebullient mood, and he'd kept up an almost steady stream of pretentious bombast all day. It didn't appear that he would wind down anytime before midnight.

For the past hour, he had been reminiscing about killer Buddhist monks along the Upper Yangtze. Spicer-Simson claimed that these monks had been trained from childhood in the arts of war and could easily defeat any normal man in hand-to-hand combat. He even insisted that some of them had developed fantastic physical, mental, and spiritual powers from their exercises, and that he'd actually seen one of them levitate. Frankly, Hanschell thought the entire concept sounded a little too much like Daniel Dunglas Home, and he said so. Spicer-Simson simply snorted that Home was a publicity-seeking charlatan while the monks were spiritual beings. He went on to relate how he had been befriended by these monks and taught some of their most secret war arts, and that training had enabled him to muster the courage he'd needed in facing many of the dangers he'd had in His Majesty's service.

Hanschell couldn't argue that point, though privately he doubted the complete veracity of his commander's many extraordinary exploits. But for some reason—he wasn't certain why—he couldn't doubt them completely. Perhaps it was because Hanschell saw something beneath Spicer-Simson's patently absurd appearance, his belligerently offhand style of command, and his irritating and grating patrician tone. The more he was around the man, the more Hanschell saw that the commander's appearance was not a costume but an external portrayal of his essence—that the officious dictates masked a sort of desperate need to succeed, and that the tone was.... Well, Hanschell couldn't excuse the tone, though he'd heard something like it in the voices of Americans who'd transplanted to Britain. Maybe it was simply that, despite Spicer-Simson's flaws, Hanschell actually admired his optimism, his drive, and his verve.

Even so, Spicer-Simson's tales of martial monks began to wear thin, and Hanschell found himself staring absently out of the window, his mind occupied by the baobab trees that dotted the dry plains.

"You seem peculiarly interested in the terrain, Doctor," Spicer-Simson said, noticing at last that Hanschell's attention was out the window instead of on his reminisces.

"The baobab, Commander," Hanschell pointed. "A fascinating plant."

"Indeed it is," Spicer-Simson replied. "Did you know that the Arabs have a legend concerning its peculiar shape?"

Hanschell had just been wondering about that shape—like a brandy bottle whose neck had grown branches—and he said, "What do they say?"

"They say that the devil plucked up the baobab, turned it end-on-end, and thrust its back into the ground. What you see in the air are really its roots."

"Interesting," Hanschell murmured, but Spicer-Simson didn't seem to hear as he'd already returned to the Yangtze and the warrior monks. It seems that they could perform prodigious feats of strength, and some could even break bricks and stones and such with their bare hands and suffer no harm.

Lulled by the rhythmic rocking and clatter and the drone of his commander's voice, Hanschell began to drift off to sleep. But his slumber was rudely shattered by a whistle and shout from the front of the car.

"Look at that, boys!" the voice called out excitedly.

Hanschell opened his eyes to see everyone in the coach, the commander included, craning toward the windows on the right side. The car took on a dangerous cant in that direction as the men on the opposite side left their seats and leaned over their mates for a better look out of the windows. Hanschell peered around Spicer-Simson's broad, round shoulders to see what the excitement was about.

It was an African girl on a motorcycle, though the whistles and shouts aboard the coach nearly drowned out the sputter of her machine. She was young and pretty enough, and she had a stunning figure. It was that figure the men approved of so vocally, and its contours were quite obvious since the girl was almost completely naked. As the motorcycle jounced down the rutted road that ran roughly parallel to the railroad tracks, her breasts bounced most splendidly.

The whistles and shouts grew in intensity, and some of the men even stretched out of the windows, waving at the girl. She grinned, lifted a hand from the handlebars, and waved back. The whistles and shouts peaked.

"Quite a nice thing, eh, Hanschell?" Spicer-Simson said, elbowing Hanschell without taking his eyes off the girl.

"I'm not quite ready to go native," Hanschell responded, though he had to admit he was intrigued by the girl's bobbing presence.

Spicer-Simson turned abruptly and fixed Hanschell in a gaze that, though brief, withered the doctor in an instant. Then the commander turned back to the window to watch the girl drift back behind the train as the locomotive drew it relentlessly onward. In minutes, she was merely a dark speck along the receding rails, but she'd left everyone in the coach in better spirits.

Everyone, that is, except Hanschell, though Spicer-Simson did not mention what Hanschell had said. He simply resumed his talk about the martial monks, and Hanschell had the strong impression that he'd already forgotten the doctor's comment. But Hanschell could not dismiss so easily the look the commander had shot him. Shot was the right word—the doctor actually had felt pierced. What had that look—astonishment with a veiled hint of contempt—meant? Hanschell wasn't sure, but he thought it had given him a momentary insight into the commander's nature, and what he'd seen there was an odd sort of egalitarianism. And perhaps it had given him an insight into his own nature as well, and what he'd seen there wasn't so noble.

"The old fellow couldn't resist teaching me," Spicer-Simson was saying. "He showed me the rudiments of what he called 'iron vest,' which bestows unusual toughness of flesh to a man who practices it. He marked me here as a sign that I'd passed his tests when he'd finished teaching me. See?"

Spicer-Simson unbuttoned his uniform tunic, bared his left breast, and indicated a simple tattoo of a circle divided into two halves by a wavy line. On one side of the line, the circle was colored black, and on the other, there was no pigment. The tattoo was nestled among the rest of the incredible menagerie on the commander's chest, and Hanschell couldn't help but stare with a sort of morbid curiosity, thinking of the pool concerning the actual extent of the commander's illustrations.

Oddly enough, the question about the extent of Spicer-Simson's tattoos had a salutary effect on his command. The men grumbled at the commander's arrogant style, at his outlandish stories, and at his increasingly unusual behavior, but they seemed to feel that any man brave enough to so thoroughly mark himself in such a painful manner was deserving of some consideration.

5

"What? What's that?" Hanschell was awakened by a change in the sound the car made over the rails—the wheels had taken on a sudden hollow tone. But after a few moments, the sound ended as abruptly as it had begun, and he realized that the train had gone over a long bridge. It was still dark outside, though the eastern sky was lightening, and he could see nothing through the windows, so he fell back asleep.

Not long after, though, the train slowed to a stop, and the cries of the railroad men as they loaded wood for the locomotive and filled its boiler with water woke him completely. A few of the naval expedition further disturbed him when they climbed down from the coach to relive themselves in the right-of-way along the track and then lingered outside, smoking and still joking about the naked young Nubian woman on the motorcycle. Hanschell was trying to pretend their chatter was the soothing clatter of the train on the tracks when a rude hand gripped his shoulder and shook him.

"Come on, man!" It was Spicer-Simson, looking unusually animated despite the fact that he'd been up all day and half the night. "Rise and shine!"

"What?" Hanschell asked sleepily.

"Up, man!" Spicer-Simson shook him again until he was fully awake. "We're in Livingstone!"

"Livingstone?"

"Don't you know what that means?" Spicer-Simson stepped back and surveyed Hanschell, wonder mixed with exasperation in his eyes.

"No," Hanschell replied uncertainly.

Spicer-Simson smiled tolerantly, then spread his arms wide and said, "Victoria Falls." He lowered one arm, used the other to gesture for Hanschell to follow, then headed toward the end of the coach.

When they were outside, he said, "I've sent Wainwright for transport."

"You mean we're going to see Victoria Falls?" Hanschell was still groggy and a bit bewildered.

"Good God, man! Could you get so close and not take the few extra steps? We may never pass this way again in our lives. We can't miss the opportunity." Spicer-Simson paused, and his brow darkened as he said in a quieter, more serious voice, "I know missed opportunity all too well to willingly face her again." Then he brightened as he saw a handcar wheeling toward them from the gloom down the tracks at the rear of the train. It was pumped by Wainwright and an African dressed, except for sandals, in old European clothes. A dirty Panama hat slouched on his head.

"Is this fellow our guide?" Spicer-Simson asked as the handcar rolled to a stop.

"This is Masheke," Wainwright said, and Masheke nodded and smiled, showing broken teeth. "Masheke, this is our commander."

"Bwana Commander," Masheke said. "You wish to see Mosi-oa-tunya?"

"What's he say, Wainwright?"

"Mosi-oa-tunya, Commander. It's their name for Victoria Falls. Means 'the smoke that thunders.' They call it that because of the mists that always hover over the falls."

"Mosi-oa-tunya," Spicer-Simson repeated, rolling it on his tongue like a fine wine. "Yes, Masheke. I want to see Mosi-oa-tunya."

Masheke showed some more of his ragged teeth. "We go now, yes, Bwana Commander? We hurry, we get there when sun come up."

"By all means," Spicer-Simson said, gesturing toward the horses. "Let us hurry."

He and Hanschell mounted the handcar, and Wainwright and Masheke took hold of the pump handles. In a few moments, the car was clattering merrily down the tracks. The two maintained a steady pace for some minutes as the sky reddened, sweat beginning to stain their shirts.

"Masheke," Wainwright called out after more time had passed, suddenly breaking rhythm. "Can we get this thing off the tracks?"

"Why?" the commander asked before Masheke could respond.

"I believe I hear a train approaching, sir."

"A train?" Spicer-Simson cocked his head, listening, and so did Hanschell, who heard a distant rumble.

"Masheke," Spicer-Simson said, eyes roving down the track. "We hear a train coming. Is there a siding nearby?"

Masheke just laughed.

"No train, bwana. Mosi-oa-tunya speak with voice heard even here."

"You mean that is the sound of the falls?"

"Yes, bwana."

"How far away are they?"

Masheke glanced around at the landscape, then turned back to Spicer-Simson. "Five mile, bwana."

"You mean we can hear the falls for five miles?" Hanschell asked.

"More in summer," Masheke replied. "Very dry right now. Water very low."

"That's right," Wainwright nodded. "Summer is around the first of the year, not that it matters much in this climate."

"That's power," Spicer-Simson said, obviously little interested in the scientific niceties. He impatiently waved at Masheke. "Onward, Masheke. Take me to meet Mosi-oa-tunya."

Obediently, the lieutenant and Masheke bent over the pump handles, and in another few minutes, the handcar arrived at a short siding. About half a mile away, down a shallow incline, Hanschell could see a long trestle. It must have been the bridge that had awakened him just before the train pulled into Livingstone.

After the car coasted to a halt, Masheke hopped to the ground, heaved the lever on the siding switch, and waited while Wainwright pumped the handcar onto the spur, then swung the lever back into its original position. He beckoned, and the three expedition members joined him on the ground and followed him along a path to their right that wound over a low hill.

They proceeded for another half hour, during which the roar of the falls grew in volume. Before long, the men could not hear themselves speak except at very close range. Masheke led them up an incline beyond the top of which rose swirling, rainbow-laced mists, golden in the light of the new sun. To their left lay a dense forest, but ahead and to the right, the land was more sparsely wooded. Abruptly, as they reached the top of the rise, the ground dropped away before them and the falls were completely visible.

Hanschell was stunned. He'd never seen anything so magnificent, so awe inspiring. He simply stood there, staring for several long minutes. The falls were as interesting as they were shockingly beautiful. As the Zambezi River approached the falls, its clear blue waters widened, spreading out across the landscape for what must have been nearly two thousand yards. And there it met its nemesis—a tremendous chasm. The rent, from its narrowest of about eighty feet, widened to about two hundred and fifty, and it was well over three hundred feet deep. The small party was standing on the edge of the chasm directly opposite the center of the falls, just a stone's throw away. Into this huge chasm, the Zambezi poured its heart and soul.

Steam from the boiling fury hung hundreds of feet above the race. There were actually four distinct falls spaced by islands that hung on the chasm's lip. From the west bank, extending for about a hundred feet, was the first cataract. A small island separated this first fall from the main fall, which looked to be three thousand feet in width. A larger island separated this from the next part of the falls, which were nearly two thousand feet wide, and a smaller island gave way to the last fall on the eastern end, which was, in size, much like its counterpart of the western end.

The water that poured over these four cataracts fell into the narrow gorge with a fury that was practically indescribable before sluicing off through a narrow ravine cut in the barrier wall to their right, about three-fifths of the way down the chasm. The ravine couldn't have been more than one hundred feet wide and one hundred and fifty feet long, and the entire river boiled madly through it.

Hanschell realized suddenly that Spicer-Simson had taken a dozen steps toward the precipice. He went to stand by the commander, Wainwright following. Masheke hung back. By now, everyone was dripping from the spray flung into the air by the raging turbulence below.

"I've seen Niagara," Spicer-Simson shouted, leaning toward Hanschell. "In America."

Hanschell nodded vigorously, as if the gesture needed to be amplified as much as a voice to be comprehended over the overwhelming thunder.

"Niagara is magnificent," the commander shouted. "But this.... This is majestic. This is the kind of thing that shows man for the pale creature he is. This," he continued to shout, but he straightened and stepped closer to the precipice and his voice grew fainter. "This is like a personal message from the gods that inspires a man to deeds worthy of...."

Though Hanschell followed, Spicer-Simson's voice was completely lost in the din as he took several more steps toward the edge. Wainwright clutched at his arm to restrain him, but Spicer-Simson pulled away and stepped closer yet. Hanschell followed as near as he dared, but the commander drew beyond his reach. Wainwright stopped, as well, and he and Hanschell watched their commander halt at last, on the very lip of the slippery precipice above the writhing tumult hundreds of feet straight beneath him. His lips still moving, Spicer-Simson raised his arms, and his chest heaved as he shouted out in...wonder? Praise? Sheer abandon?

Soon it did not matter. The thunder of the falls so filled Hanschell's body, his mind, his very soul, that he lost all sense of himself.

6

Spicer-Simson and his companions arrived back at the train station in Livingstone in mid afternoon. The commander and Hanschell dismounted from the handcar and watched Wainwright pay Masheke. The guide departed, pumping the handcar tiredly down the tracks.

"See how things are with the train, Lieutenant," Spicer-Simson told Wainwright.

"Yes, sir." Wainwright turned and headed toward the locomotive.

"God, that was magnificent," Spicer-Simson said, stretching. He slapped Hanschell on the shoulder. "Not apt to see the likes of that again, are we, Doctor?"

Hanschell didn't have time to answer before Spicer-Simson spun and quickly mounted the steps to the passenger coach. At the top he turned again, just as quickly, and stared down at Hanschell. "How about a drink before mess?"

"I'd like that," Hanschell agreed. He needed one.

Spicer-Simson nodded and entered the coach, Hanschell following wearily after him. He didn't know what could possess the commander to keep him so eternally charged with energy. For his own part, Hanschell was dead tired. Half the day had been spent riding the handcar, half of it amid the tumult and roar of Victoria Falls, and almost all of it under the sun. He hadn't had a bite to eat and precious little water to drink, and it seemed as if every nerve in his body had become screamingly sensitive. Thank God he'd had the foresight to bring along a hat or he'd be sunburned as well.

Two hours later, with a meal and a couple of numbing whiskeys in his belly, he had mellowed considerably. Spicer-Simson hadn't said much since they entered the coach, and that was unusual enough that it caught Hanschell's curiosity. As he sat and surreptitiously studied the commander, who was staring out the window, watching the final preparations for departure, Hanschell wondered what was going on in his complicated head. He soon gave up. Spicer-Simson wasn't the sort who readily lent himself to analysis, and Hanschell was a doctor of tropical disease, not one of those adherents of Sigmund Freud who seemed to have sprung up all over in the last decade. Yes, Hanschell thought, looking out the window. It would take one of those psychoanalysts to get at the true Spicer-Simson, and he pitied the doctor who would make the attempt.

At last, at sundown, the train pulled out of the station. There was nothing to see outside of the windows but darkening and dreary prairie, and the rocking, clacking ride soon lulled Hanschell into a doze that was broken by a mutter from the commander.

"Creation must have thunder in its heart."

At least that's what the doctor thought he heard. He glanced at Spicer-Simson, but the commander's eyes were closed, his face was almost totally devoid of its usual animation. What was he doing in there, behind those closed lids?

Hanschell closed his own eyes, and it seemed is if the darkness there was filled a sound beyond the subdued talk of the men around them, deeper than the clatter of the wheels on the rails and the distant chuff of the locomotive, more pervasive than the night growing around their conveyance.

It was the sound of the falls, though they were too far away to actually hear it. And Hanschell again felt his nerves sing with the overpowering symphony that was as much a sensation as sound, and he actually could visualize the vast panorama of falling water emptying into the womb of the earth only to snake wildly off into dizzying distance. He was surrounded by steam, buoyed by it and the incessant thunder and power that were palpably heavy in the sun-drenched air. He felt as if he, the train, the boats, and the men were ethereal, as if they would continue to drift irrevocably up the continent on the power of the commander's will alone.

Abruptly, the train lurched up a short, steep incline, and the vision behind Hanschell's lids vanished as, startled, he opened his eyes.

He looked again at Spicer-Simson, and the commander's eyes were still closed. All around them, the men were settling into their seats for the long night ahead.

"Did you say something, Commander?" Hanschell tried tentatively, but Spicer-Simson's eyes remained shut. Perhaps he's sleeping, Hanschell thought, though the hint of a smile at the corners of the commander's lips suggested that something cognizant still passed through his mind. Perhaps, even yet, he was listening to the powerful symphony of the falls.

Hanschell closed his eyes and listened for the roar, but all sounds beyond the train seemed to have vanished in the night. There was nothing but the murmur of the men, the chug of the locomotive, and the gentle clack of the wheels on the rails. And the swaying motion of the train was like a gentle rocker, rocking....

Hanschell slept, and by dawn of July 26, they had rolled across the border into the Congo and reached Elizabethville.

7

There was no fanfare at the station as there had been for Lee because the town was just now waking. The vice governor general of the province, looking a little bleary, did meet them, though, and he told Spicer-Simson that a large party was scheduled for that evening, and all the men were invited.

Shortly after noon, John Lee showed up, accompanied by Magee, an Arab carrying a Tower musket, and two tribesmen with spears. Hope was not with them. By Spicer-Simson's orders, he'd gone on ahead to Bukama, the tiny port on the Lualaba River, to make arrangements for a river steamer to meet them when they arrived.

Lee's party went first to the rail station, where they found Wainwright, who told them that Spicer-Simson was, at that very moment, at the home of Colonel Sawyers. Wainwright offered to show them the way, and though Lee knew well how to get there, the lieutenant seemed eager to go with them, so Lee and his party followed. Spicer-Simson, attended by Tyrer, was lounging on the veranda, chatting with Sawyers and his wife when the officers arrived, trailed by the Arab and the two tribesmen.

"Ah, Lee," Spicer-Simson said without ceremony and without standing. "It's about time you showed up. Who are these blighters?" He indicated the Arab and the two Africans.

"My scouts," Lee answered.

"I'm not a blighter," the Arab growled. "I am Ibn-el-Mani. You are the leader, Spicer-Simson?" His English was heavily accented but clear.

"I am *Commander* Spicer-Simson."

Ibn-el-Mani stepped up to the edge of the veranda in front of Spicer-Simson. He brandished the old musket with a movement that was so quick and belligerent that Mrs. Sawyers gasped and pulled back in her chair.

"I am not at your service," Ibn-el-Mani shouted, an angry frown contorting his face, "without a repeater. The bush is dangerous, and I am in danger there. For you! You must thereby supply protection. For me and my men." He waved at the pair of tribesmen without looking at them.

Lee stepped forward, put a hand on the Arab's arm, and said, "I'll take care of it."

But Ibn-el-Mani was not so easily placated. He shrugged Lee off and shook his fist at Spicer-Simson.

"Give me a repeater now, this instant, or I go and take my men, and you will be lost and never see the waters of Tanganyika alive."

"What the hell is this man talking about, Lee?" Spicer-Simson demanded, though he seemed more amused by the Arab's behavior than upset.

"He wants a Lee-Enfield rifle."

"I will step no farther without a Lee-Enfield in my grasp," Ibn-el-Mani stormed. "Now! You must give it to me now! Immediately or I will go. I must have protection in the bush if I am to kill wild animals and Germans."

"It is not likely that you'll be shooting at Germans," Spicer-Simson tried to say, but neither the Arab's spirit nor his words could be quenched.

"Without a Lee-Enfield, Ibn-el-Mani will not guide you. He will not die in the bush like a dog. He will be safe in Elizabethville, laughing at you because you were foolish enough to sell all your lives for one repeater!"

"For God's sake, Lee, can't you shut him up?" Spicer-Simson called over the Arab's shouting.

"I'm afraid I haven't a spare rifle," Lee said, chuckling. "That would seem the only thing that will keep him quiet."

"Is he worth a rifle?"

"I am worth twenty rifles," Ibn-el-Mani shouted. "A hundred rifles. I am worth...."

"Best give him what he wants, Commander," Lee said, "before he inflates his worth beyond our ability to pay."

"Lieutenant," Spicer-Simson called down to Wainwright as Ibn-el-Mani continued to rage. "You have a rifle. Give it to this man."

"But what shall *I* do, sir?"

"Have Eastwood issue you another from stores. For God's sake, man, be quick about it before this fellow broadcasts our position to the Germans in Kigoma!"

Wainwright unslung his rifle and passed it to the Arab, who snatched it from his hands then made gestures to Wainwright's bandoleer of ammunition. The lieutenant gave him that, as well, and almost without glancing, Ibn-el-Mani tossed his Tower musket to one of the tribesmen, who in turn handed his spear to the other African.

"A singularly intense fellow," Spicer-Simson commented as the Arab, the two Africans in tow, hurried away to examine his prize.

"He's useful," Lee said.

"His English is reasonable."

"He also speaks Swahili and various Bantu dialects. The two natives speak Swahili and Bantu."

"But can they be trusted?"

Lee frowned. "Trusted? You can trust them to do what they're paid for. If they don't want to do something, they won't take the money. They'll just go."

"Then we must be certain to pay them in advance."

"Would you like a drink, John?" Sawyers asked as Lee mounted the steps into the shade.

"If you will pardon me, Colonel," Spicer-Simson said. "Perhaps that should wait. Time is precious, and Lee and I have business to attend to. Is there some place where we can speak in private?"

"You may use my office, if you like," Sawyers offered, and in short order, Spicer-Simson and Lee were alone.

"Well, man," the commander said, sitting in the chair behind the desk, leaving Lee standing awkwardly. "Let's have it."

Lee was taken aback at Spicer-Simson's abruptness. He'd spent the better part of the past six weeks surveying the entire length of the overland route from the railhead at Fungurume to Sankisia on the far side of the Mitumba Mountains. He'd also hired local villagers all along the way who, even now, were making a road where there had been none across a landscape strewn with boulders, crisscrossed with ravines and gullies, and partially covered in a dense growth of trees and brush. Stockpiles of firewood had been amassed, waterholes mapped, and village chieftains had signed contracts to supply food. And to top it all off, he'd even taken the extra effort to explore and map alternate routes.

Damn, if he didn't need a drink, and he told Spicer-Simson so.

The commander remained unperturbed.

"I suppose you bear the maps and contracts as fruits of your efforts?" Spicer-Simson asked.

"Here," Lee answered shortly, starting to feel angry. He slipped a canvas musette bag off his shoulder, and tossed it onto the desk in front of Spicer-Simson. The commander set his drink aside, opened the bag, and glanced through its contents. When he was satisfied, he replaced the documents and set the bag back on the desk before looking up at Lee with narrowing eyes.

"I have been ordered to inform you," Spicer-Simson said, "that you are hereby charged with failing to carry out your duty as an officer in His Majesty's service, with divulging military secrets to the enemy, and of behavior unbecoming an officer, including being drunk and disorderly."

"What?" Lee exploded, his face flushing beneath his deep tan. "That is preposterous! Why would I sabotage my own expedition?"

"I remind you, sir, that this is *my* expedition, not yours," Spicer-Simson replied coolly. "As for your reasons, that is for a military tribunal to decide. In any case, if the charges aren't true, how is it that half of Africa seems to know we are going to Tanganyika?"

"Where the hell else would we be going?" Lee ask cuttingly.

"Us," Spicer-Simson said. "Not you. You are hereby ordered to return at once to Cape Town, where you will report at once to Vice Admiral King-Hall."

Lee's mouth opened once, twice, then snapped shut. Almost immediately he said, "Very well. But I'll take that with me." He stepped forward to reach for the musette bag but suddenly found the metal tip of Spicer-Simson's cane planted squarely in the center of his chest.

"I'll keep the bag."

"But you can't," Lee protested. "If I'm to clear myself, I need that for proof that I've done my duty."

"This bag contains important military secrets pertaining to the Royal Naval African Expedition," Spicer-Simson said, "and which I require to do my duty to His Majesty's Royal Navy and England. And even if I were not the commander of this expedition, what makes you think I would simply hand over this information to a man suspected of treason? That would be treason itself."

"Why, you pompous shit," Lee spit, stepping forward. "I ought to...."

Spicer-Simson's facial expression did not change, but his eyes glittered like steel marbles. He slammed his cane down on the desk top and with a sudden, catlike surge stood and came around the desk. In a flash, he was facing Lee, not inches away. Spicer-Simson's usual rather mincing and, at the same time, sharp behavior gave an impression of a smaller man, but now that Lee was facing him toe-to-toe, he realized for the first time that Spicer-Simson was just as tall as he, and that the commander's round shoulders packed considerable bulk. In fact, the commander's entire persona seemed to slough away, revealing a tiger poised to pounce and rend.

Lee suddenly felt tired.

"Have it your way, Commander." He backed off a step. "Keep the documents. You'll need all the help you can get, especially when you meet *von Götzen*."

"And, pray tell, who is that?"

"You'll find out soon enough, and I wouldn't want to spoil the surprise."

"If you're keeping something from me, Lee...."

"You haven't anything left to threaten me with," Lee said with a bitter snort. "You'll do well to remember that in Africa you must always keep something in reserve."

"I know Africa...."

"You know the Gambia River. I know Africa. And I'll still be here long after you're gone."

"As long as you're gone from this part of Africa, posthaste."

"Don't worry. I'll catch the next train to Cape Town."

"As you leave, tell Magee to report to me," Spicer-Simson said, an expression of smug triumph crossing his features. "After that, do not speak to any of the men. You are to consider yourself under house arrest until you report to the vice admiral. And leave the Arab. He's in my service, now, not yours."

Lee turned and left the office. As he emerged from the house, he told Magee that the commander waited for him inside. Then, without another word and leaving Magee and the others gaping after him, he trudged off down the dusty street toward the station.

"I wouldn't keep the commander waiting," Wainwright warned Magee, and the journalist nodded and hurried inside. He found Spicer-Simson sitting behind Sawyers's desk, face set with stern authority.

"Mr. Magee," the commander said, "I must inform you that the Admiralty is gravely disappointed with Lee. He is suspected of treason."

"Sir?"

"It is my sincere hope that I need not send you packing in the same direction as your erstwhile companion."

"No, sir." Magee seemed less than emphatic, but that was probably due to his precarious situation rather than to anything he suspected he might have done to displease the king of England. He couldn't mistake the dangerous glint in the commander's eyes.

"I told you back in England that things might come to an impasse and that you would be wise to make the proper choice."

"You can rely on me, sir," Magee said quietly. Damn if he hadn't felt safer in the bush surrounded by deadly wildlife and thousands of Africans than he did at this moment.

"Good," said Spicer-Simson. "To show that I am willing to give you the benefit of the doubt, I am hereby promoting you to warrant officer."

"You are, sir? Thank you, sir." Magee couldn't believe his luck at escaping Lee's fate, whatever that might be, and now, to come out with a promotion.

"I'm glad that we understand one another," the commander said. "Let's return to the veranda."

Magee meekly followed him outside.

"Wainwright," Spicer-Simson said, when they emerged from the house. "The Admiralty has recalled Mr. Lee. It seems they suspect him of speaking with too much liberty about our purpose. His actions may even have given aid to the enemy."

This last brought a gasp from Mrs. Sawyers, and the colonel raised his eyebrows.

"I don't understand, Commander," Mrs. Sawyers began. "We've known Mr. Lee for years. I can't believe he has a traitorous bone in his body."

"I am not the one to judge that, Mrs. Sawyers," Spicer-Simson said demurely. "You know how it is, though, I'm sure. The mission must come first. In desperate times we sometimes must put aside our finer feelings."

"But we're friends with Mrs. Lee's family...."

"I'm sure John will clear things up," Sawyers told his wife. "In the meantime, my dear, let the men fight the wars."

Spicer-Simson turned back to Wainwright and Magee.

"With Lee gone, I need a new second in command. That is you, Mr. Wainwright."

"Me, sir?" Wainwright asked.

"You, sir. And I've given Mr. Magee a promotion to warrant officer. Now, I want the both of you to return to the train and make certain those fresh supplies are shipped properly."

"Yes, sir," the two men said in unison. They saluted and hurried off across the dust that had settled in Lee's wake.

"Now, Mrs. Sawyers," Spicer-Simson said, putting on a pleasant smile and turning to the colonel's wife. "What was it you were saying about this little soiree that you and the other fine inhabitants of Elizabethville have in mind for my men?"

8

While Spicer-Simson was sacking Lee, Hanschell was conferring with the Belgian medical officer. He discovered that the Belgian doctor was an alumnus of the Liverpool School of Tropical Medicine, and the two fell to an earnest discussion of various tropical diseases.

"Have you heard of *Trypanasome rhodesiense*?" the Belgian asked.

"The new variety of sleeping sickness recently identified by your school?" Hanschell nodded. "I read about it on the way over."

"Invariably fatal. It's been wreaking havoc all along the Loanda Valley. It comes from a variety of the tsetse fly common to grasslands rather than marshlands. Look."

He showed Hanschell a wall map bristling with colored pins.

"The blue pins mark spots where we're seeing large numbers of cases of *rhodesiense*, and the red pins mark the areas stricken with *Trypanasome gambiense*, which is much more common here than the *rhodesiense*."

"We're going to have to pass right through some of the worst of it," Hanschell noted in a worried voice. "From your map, it looks like the *gambiense* is particularly bad along the Lualaba, and we're going to be on that for some time."

"Whole tribes in the area have been wiped out," the Belgian said. "But at least you can treat the *gambiense* if you catch it in time."

"I'd rather not catch it at all."

"Several cases of *rhodesiense* are here now. Would you like to see?"

"Please."

There were three victims, all local people, though the Belgian doctor told Hanschell that, just two weeks earlier, he'd sent two stricken Europeans home. Two of the Africans were in the final stages of the disease, feverish and emaciated, with lymph glands bulging. Both had sunk into comas. The third was shambling around the room, moaning and clutching his head with trembling hands.

"How long have they been here?" Hanschell asked, horrified despite his attempt at clinical objectivity.

"These two, a week. They won't last another day. That fellow over there has been showing signs of infection for just a few days, though he likely contracted the disease as long as eighteen months ago."

"My God. It's hard to believe it can do this kind of damage. And I thought *gambiense* was bad."

"I wish I could offer you some advice," the Belgian said. "Unfortunately, all I can do is wish you luck."

"Could I bother you for a copy of your map?"

"Certainly. I'll have one prepared. Have you read these?" The Belgian doctor waved toward a four-volume set of books.

"Sir Arthur Bagshawe's monumental work on *trypanosomiasis*? Of course. I have a set with me."

"We are greatly in debt to Sir Arthur. No one has done more for our understanding of the disease."

"It seems that there remains much work to be done, judging by your patients with the new strain."

"Yes," the Belgian agreed. "But there is one good thing about the region through which you'll be traveling."

"I can't imagine what."

"The earth there is relatively clear of tetanus and gangrene, so your wounded will need very little tetanus antitoxin."

The thought of wounded on top of the diseased shattered any calm Hanschell had before visiting the Belgian hospital. Shaken, he went to find Spicer-Simson. He learned from Eastwood that the commander was with Colonel Sawyers, and he hurried toward the colonel's house, arriving not long after the incident with Lee was over. But he was too agitated to notice the tension in the air.

"Commander," he said, "I must speak with you."

"Can't it wait, Doctor? I'd like you to meet Colonel and Mrs. Sawyers...."

"It's really very important, Commander. It's about sleeping sickness."

"What? Please excuse me Mrs. Sawyers, Colonel Sawyers. I must confer with my doctor."

When the two men were in Sawyers's office, Spicer-Simson said, "Now what's this about sleeping sickness?"

If Hanschell had thought that Spicer-Simson might dismiss his worries about the disease, he was mistaken.

"I've seen what evil the bloody stuff can do up on the Gambia," the commander said, a frown on his face.

"The Belgian medical officer here has shown me a map," Hanschell said, and he went on to describe what he'd seen.

"What can we do?" Spicer-Simson asked.

"At present, there are no drugs to fight this new strain," Hanschell said. "At best, we can merely help prevent the flies from biting. I think we should take a page from the Africans' book and issue fly whisks to all the men."

"If that's the best we can do, then we must do that," the commander said. "Have Eastwood purchase some at the market, and tell Wainwright to order the men to use them."

Thankful that Spicer-Simson had at last seriously taken note of the medical dangers they faced, Hanschell followed him outside to meet the Sawyers.

9

"Here's the telegraph office, sir."

"What?" Spicer-Simson stared uncomprehendingly at Hanschell, then at Wainwright. It was dusk, and the three men were walking toward the vice governor general's home for the party Mrs. Sawyers had organized for the expedition's men. The rest of the expedition were there already, eating and drinking and vainly trying to impress the four eligible young women—two English and two Belgian—among the European inhabitants. Spicer-Simson, as was customary, was wearing his sword at one side and swinging his cane at the other.

"I thought you were going to send a wire to the Admiralty."

"What about?" Then understanding lit the commander's eyes. "Oh, you mean Magee's promotion." He glanced at the telegraph office and nodded. "Excellent idea, Hanschell. Wait here. I won't be long."

Leaving Hanschell and Wainwright outside, Spicer-Simson entered the tiny building and told the clerk he wanted to send a wire to Vice Admiral King-Hall in Cape Town.

"What's the message?" asked the clerk, a youngish man with a clipped mustache and balding forelocks who did not look at all happy about having to run the office when all of his fellow Europeans were enjoying themselves just a few blocks away.

"Have arrived in Elizabethville. Stop. Have personally completed survey of route and put local tribesmen to work clearing the overland road and preparing fuel dumps for the traction engines. Stop. Have also personally surveyed alternate routes and obtained contracts with local native chiefs. Stop. Lee ordered to return to Cape Town. Stop. Situation regarding Lee more grave than realized. Stop. Suggest immediate investigation into his activities. Stop."

He paused, stroked his mustache, then gestured toward the paper on which the clerk had been writing the message. "Let me see it," he said. The clerk slid the paper across the counter, and Spicer-Simson read it through. "That's good," he said. "Send it on."

"Yes, sir," said the clerk, and Spicer-Simson left the office.

"Done," Spicer-Simson said to Hanschell and Wainwright as he stepped into the dust of the street. "Now, let's go and have a drink."

The three men arrived at the vice governor general's large, Mediterranean-style villa a few minutes later, but despite the fact that the British residents were excited and happy enough to see new faces and hear some news of the home country, the officers looked despondent. Spicer-Simson didn't seem to notice, and immediately headed to the bar for a vermouth.

"What's going on?" Hanschell asked Wainwright.

"It's Lee," the lieutenant replied. "The men can't understand what happened."

"I'm not sure any of us do," Hanschell said.

"The commander told Eastwood that Lee came down with fever."

"If he did," Hanschell said, "I think I'd know about it."

"Some of the men are saying that he's been accused of being drunk and giving away the expedition."

"Drink can sometimes get the better of a man."

"I've known the fellow for several years, and I always thought he was all right. Never saw him drunk. I hate to see a man like the commander ruin his reputation."

"I'd be careful of talk like that, Wainwright," Hanschell warned.

"I didn't mean anything by it, Doctor," Wainwright said. "Believe me, I want this expedition to succeed as much as any man. I've got more to lose than just a war. I'll lose my home and possibly my family. So I'd do my best even if it wasn't a duty—though that's enough excuse for me. We're part of the empire, too, you know."

"I never doubted you for a second. Not after that first covered bridge back down in South Africa. But I wouldn't let the men hear you say things like that."

"It wouldn't matter if they did. They're saying a lot worse."

"Such as?"

"Some of them," Wainwright replied with obvious reluctance, "say the commander's a touch loose. You know...." He pointed to his temple.

"And what do you think?"

"I can't say I'm comfortable with the situation. After all, we're here in the middle of Africa and in the middle of a war, and we've got a commander who.... Well, a commander like ours."

"I admit he's odd," Hanschell agreed. "And I also admit we haven't come far, really, though we've crossed an ocean and half a continent. But as I've said before, for some strange reason I have faith in our commander. I don't know what lies ahead, but I'm sure he'll pull us through. And you should believe in him, too. He has enough faith in you to promote you to second in command."

"I'm not sure that's a good thing," Wainwright said. "Look at what happened to his last second in command." He and Hanschell both chuckled.

"Too true," Hanschell nodded.

At that moment, a couple of the Belgians came over and introduced themselves as Philippe Roche and Oliver Soren.

"We had a similar party when your Monsieur Lee was here last month," Soren said.

"Yes?" Wainwright responded cautiously.

"Very fine gentlemen," Roche said. "I personally aided Monsieur Lee in obtaining an excellent Arab guide."

"I've met him," Wainwright said.

"Why are Lee and Magee not here?" Soren asked. "I heard they arrived."

"I'm afraid that Mr. Lee had to return to Cape Town suddenly," Hanschell said. "And Lieutenant Magee was feeling a touch feverish and said he wasn't up to festivities."

"Lieutenant?" Roche asked. "I believed him a sublieutenant."

"He's been promoted," Wainwright said.

"How fortunate for him," Roche said. Then he gestured with his glass toward Spicer-Simson, who sat on the other side of the room drinking vermouth and regaling anyone who would listen long enough with his tales of derring-do. "And the commander? I am curious. Why would he accept such a difficult mission?"

"He didn't accept," Hanschell said. "He volunteered."

"I am surprised." Roche's eyebrows went up.

"We are all volunteers," Wainwright added.

"Ah," Soren put in. "Then you must be very certain of success."

"Can anything be certain in war?" Wainwright asked.

"I think not," Soren pressed, "but perhaps you are more sure of failure."

"I don't believe I'm getting what you mean," Wainwright said, bristling slightly. "I hope you're not implying...."

"Please, gentlemen," Roche said, holding his hands up in a placating gesture. "My friend does not mean to impugn either your intent or courage but merely points out the possibility that you may not complete such a difficult journey."

"Oui," Soren said. "When we feted Monsieurs Lee and Magee, we also had another visitor—a Monsieur La Plae. An engineer for the Katanga Railway. He knows this region intimately, and it is his opinion that, with luck, you may reach the Mitumba Mountains, but there you will be halted."

"Have you seen these mountains?" Hanschell asked.

"I have, Doctor. Very steep and rocky, with many large trees blocking the slopes. There is an escarpment at the top that is not tall but nearly vertical."

"Lee has assured Commander Spicer-Simson that there is a way," Wainwright said.

"Your Monsieur Lee is an estimable guide," Soren said. "He is not an engineer, as is Monsieur La Plae."

"Maybe not," Wainwright said a trifle defensively, "but he has surveyed the route and knows what we have to take over it. I have faith in him."

"Ah, well, monsieur," Soren said with a shrug. "But you must have faith, even if it is an empty hope. You are soldiers, are you not?" He took a sip of his drink.

"Royal Navy, monsieur, not soldiers."

"Sailors in the middle of a barren continent," Roche replied.

"We'll do what we came to do," Wainwright said stiffly.

"I think," Roche said, turning to Soren, "that I must believe our Monsieur La Plae. They must fail, oui?"

"Oui," Soren agreed. "If La Plae says their mission is impossible, then...." He shrugged again.

"I don't give them one chance in one hundred," Roche said.

Suddenly, Hanschell, who had let Wainwright carry the conversation with the Belgians, smelled a rat. He wasn't sure what was going on, but something definitely....

"Those are the odds in town," Soren said.

"One to a hundred against us?" Wainwright frowned.

"It is common knowledge," Soren said, looking wolfish.

Ah, Hanschell thought. It's not a rat, but cheese. Bait. Before he could caution Wainwright, the lieutenant spoke.

"Well, it's not common knowledge to us. I don't suppose you gentlemen would care to show some cash on those odds?"

Giving his companion a sideway glance, Soren nodded slowly and said, "That might be interesting. But who would hold the stakes?"

"Wainwright...," Hanschell began, but the lieutenant silenced him with a casual wave.

"Don't worry, Doctor, it's just a friendly little wager." To Soren and Roche he said, "What about Colonel Sawyers or your vice governor? I'm sure one of them would be glad to do us the honor."

"An excellent suggestion," Roche said. "Your Colonel Sawyers will do, and he has a safe in his office. Shall we speak with him?"

"See you later," Wainwright tossed over his shoulder at Hanschell, winking, as he followed the two Belgians across the room. "Now, sirs, what sum did you have in mind?"

Hanschell watched them go, then smiling, he shook his head and headed toward the bar.

The officers' morale may have been low at the beginning of the party, but it picked up considerably as the food filled their bellies and the liquor their heads. By ten o'clock, the four young ladies had been danced sore and were demurely declining further invitations, most of the news from home had been told from one end of the vice governor general's villa to the other, and the wager with the Belgians had spread like wildfire. From what Hanschell could tell, the commander was the only expedition member who had not put in money. Even he had relented and practically emptied his wallet, though he was not normally a betting man. He laughed at the thought. Two weeks in Africa, and already he had two wagers under his belt.

The entire time, Spicer-Simson, buoyed to the brim with high spirits—including vermouth—sat at a table near the bar, surrounded by many of the older and younger residents, telling a story that Hanschell had not yet heard, of how, armed only with a Webley & Scott pistol, he had shot a man-eating tiger that had attacked him in China.

That story and two others lasted through another hour and three additional glasses of vermouth, at the end of which, the commander was still sitting and still talking but looking pretty glassy-eyed. By that time, news of the wager finally reached his ears, but apparently he was too drunk to realize it was a wager and not simply criticism.

"It doesn't really matter about the mountains," he said with a slurred sneer. "I've served at sea for so many years that I'm not likely to find any of these landsmen's problems too difficult."

Just then, he noticed that the music had stopped, and he got the musicians to strike up again despite the fact that they were too tired, or too amateur, to remain in key. Spicer-Simson insisted on singing, but the only song he and the musicians had in common was "Swanee River."

"Play it!" Spicer-Simson shouted, and the musicians played, and Spicer-Simson sang. Several people accompanied him through the first verse, but then the words escaped them. It seemed, however, that Spicer-Simson had the entire song at his ready disposal, and he insisted in singing all the verses with the chorus in full at the end of each verse. Nor, embarrassingly, did he neglect to use the dialectic inflection written into Foster's lyrics.

By eleven-thirty, after Spicer-Simson had sung the song through three complete times, every one of the expedition's officers was feeling as mortally humiliated as he had ever been in his life—or probably would ever be again.

"That man's a damn fool," Cross muttered to Rickson, giving his half-face sneer.

"It's a blasted American song," the bridge engineer replied. "How'd he know all the words?"

"I don't know," Cross said, "Just thank God there aren't more."

Spicer-Simson was ready for another round of drinks, but Hanschell and Wainwright managed to shepherd him out of the villa. As the doctor waved goodbye, he saw relief on Mrs. Sawyers's face and thanks in her husband's eyes. Then he and Wainwright had the commander outside, and after a few difficult moments, they managed to guide him down the dusty street toward the train station. He hummed "Swanee River" all the way.

10

"Damn it!" Spicer-Simson crumpled the telegram and threw it to the floor. "What is it, sir?" Now that Wainwright was second in command, he felt it his duty, and right, to know details about the expedition. Spicer-

Simson obviously didn't think *that* highly of Wainwright, but this time he told him.

"It's those bloody traction engines. They're still down in Rhodesia."

"You mean they haven't even shipped them out?" Wainwright was shocked.

He thought the success of the expedition was of paramount importance to the African campaign. After all, that's why they were here, wasn't it? Why they'd gone to all this expense and trouble to get this far? Now to hear that time would be lost just because some shipping clerk had got his orders all bollixed up, well, it was unthinkable.

"The man responsible ought to be whipped," Spicer-Simson spat.

Spicer-Simson's vexation was understandable. The expedition had lingered in Elizabethville for a week longer than necessary waiting for the traction engines to arrive. The expedition could have gone on to the railhead at Fungurume and waited there, but Fungurume was little more than a rail depot equipped with a couple of rail sidings and accompanied by a handful of mud and thatch huts. No one wanted to wait there, least of all Spicer-Simson.

Elizabethville wasn't much better in the view of several of the men, but at least there was the contingent of Europeans, and because of them, the town had a few amenities, such as a supply of liquor. But with the twenty-odd members of the Royal Naval African Expedition as well as the Europeans tapping into the liquor stores, even those were drying up.

Wainwright knew things would go from just a little bad to a lot worse when the local provisioner informed the commander that he was out of vermouth. That had happened just this morning, and that was why Spicer-Simson was in the telegraph office. He wanted to send an order to Cape Town for a couple of cases of his favorite drink. He'd handed the requisition to the telegraph clerk and the clerk had handed back the message about the delayed traction engines.

"Send another message," Spicer-Simson ordered, and the clerk picked up his pencil. "Tell them to get those goddamn engines up here on the next freight. Otherwise, we'll just sit around this hellhole using up all our goddamn supplies while they sit around on their asses."

Having finished dictating his message, the commander wheeled and stalked out of the building. The clerk, who had stopped writing at the sixth word, looked askance at Wainwright. Wainwright, realizing that now was as good a time as any to assume the duties and obligations of his position as second in command, wrinkled his nose to keep from smiling.

"I think you have the gist of the commander's message," he said. "Feel free to word it however you think necessary."

"Very good, sir," the clerk replied and set his pencil to paper. Wainwright followed Spicer-Simson into the dusty street.

"Perhaps the engines will arrive this week," he said as he caught up with the commander, but Spicer-Simson's only reply was a grumpy snort.

"Look!" Wainwright exclaimed as the two men reached the station. Beyond the tracks, from somewhere in the middle of the huts, shanties, and poor buildings in the African part of Elizabethville, a not-too-thick but vile column of smoke was rising. "What do you suppose it is?"

"A fire, Wainwright," Spicer-Simson replied snappishly. "I would have thought that completely obvious, even to a dolt."

Wainwright knew perfectly well it was a fire, and he really didn't appreciate the commander's implication that he was a dolt. Or worse.

"What I meant to say, sir, was, I wonder what is burning."

"Don't be pedantic, Wainwright. Or argumentative. Neither becomes an officer in His Majesty's Navy." But the commander said this with less vitriol, and the light of curiosity now shone in his eyes. "Let's go see, shall we?"

Spicer-Simson hurried off across the tracks, Wainwright close at his heels. The source of the smoke was not difficult to find. About half a mile from the depot, a large hut was blazing. Perhaps two hundred people stood around, yelling, laughing, and watching it burn. None of them seemed particularly interested in making an attempt to fight the flames. Not that it mattered. By the time Spicer-Simson and Wainwright arrived, the roof was gone and the walls were caving in. The sole fire fighting effort was on the part of close neighbors, who were flinging bucketfuls of water onto their own roofs to douse flying sparks and keep their huts from burning.

Wainwright stared at the flames then became aware that the commander was moving off toward half a dozen expedition members who clustered in a group off to the side. He followed, and as he got closer to the group, Wainwright began to notice several peculiar details. Leighton's trousers were unbuckled and his shirt sloppily tucked in. And his feet were bare—his boots dangled from one of his hands. A couple of the other men were in similar disarray. The only one who seemed to be properly dressed was Doctor Hanschell. In his left hand was a kerosene can. And all the men seemed to be arguing with Hanschell, and the doctor, in turn, was vehemently arguing back.

Suddenly Leighton spotted Spicer-Simson, and a shadow of fear crossed his face.

"Ten-hut," he shouted, and all the men, except for the doctor, froze in attitudes of guilt.

"Tell me, Wainwright," Spicer-Simson said in a tone oozing scorn as they stopped in front of the men. "Did I call for a full dress parade this morning or a fully undressed parade?"

"Full dress, sir." Wainwright kept control of his face, though underneath he wanted to laugh aloud at the way the features of the other men withered beneath Spicer-Simson's gaze. For the first time, he thought he might actually come to enjoy the commander's raking temper. As long as it wasn't raking over him.

"So I thought." Spicer-Simson walked around the cluster of men, swinging his cane with an exaggerated swagger and sharply looking the men up and down. "In either case, Wainwright, I'd have to give them poor marks. They're neither dressed nor undressed." He stopped again, having made a full circuit. "Would someone mind," he said, his voice going shrill as it rose rapidly in register, "telling me what the hell is going on here?"

Everyone started talking at once, most of them waving at Hanschell, who yelled as loudly as the rest and waved back.

"Quiet!" Spicer-Simson bellowed, and even the villagers in the streets around them fell silent. They began to gather around the group of British, curiosity replacing their various other expressions.

"You," Spicer-Simson pointed at Leighton. "Speak."

"It was the doctor, sir," the seaman blurted out. "*He* burned it."

"You?" Spicer-Simson turned to Hanschell, a touch of amazement in his eyes. "Did you burn this building, Doctor?"

"I did, Commander. I went in, poured kerosene," he raised the can, now obviously empty, "onto the filthy piles of rags that passed for beds, and lit it. As you can see, I managed to chase out the rats." He waved contemptuously toward the half-dressed men.

The men started up again but quieted quickly when Wainwright barked, "Shut up."

"Thank you, Lieutenant," Spicer-Simson said, then he turned back to the doctor. "Would you mind explaining why you burned this house?"

"Filth. This expedition is here to wage war on the Germans, but they are not our only enemy. Until we meet them, the enemy we must defeat is disease. It is so small we cannot see it, but make no mistake, it is there."

"Are you saying you burned this building to prevent the men from fornicating inside."

"This building was a whorehouse," Hanschell amplified. "And, yes, I burned it to save the men. I've tried all week to dissuade them from coming here. I've pointed out the dangers. Some of them listened, but, as you can see, some did not. They left me no choice but to find a way to permanently rid them of the temptation."

"Surely you can cure a small dose of the clap," Spicer-Simson said.

"In one man," Hanschell agreed. "Maybe two or three. But I simply haven't the supplies to cure twenty. Anyway, gonorrhea isn't the worst of it. There's African sleeping sickness, dysentery, and God knows what else, and it's up to me to make sure these men are in shape to fight when you need them."

"Sir," spoke up Leighton. "I respectfully note that diddling the local beauties is a sailor's prerogative whenever he's in port, even if it's in the middle of Africa."

"Look!" Hanschell barked. He reached out swiftly and jerked at the loose waistband of Leighton's trousers. The breeches came down amidst a roar of laughter from the surrounding villagers who could see the spectacle. The ass revealed in the stark sunlight shone as white as the moon, and it was just as cratered with splotches and pimples of bites old and new.

"No telling what this man and his mates have become infected with. And they could pass it on to the others." Hanschell let go of the sailor's pants, and while Leighton hastily re-covered his sorry behind, the doctor took a determined step toward Spicer-Simson and stared him in the eyes. "If we are not wary, Commander," he said, "if we do not take precautions, your expedition may never reach Tanganyika much less be in shape to fight."

"They'll be in shape to fight, Doctor," Spicer-Simson said, a steely edge of determination in his tone. "They'll be in shape because the next man caught with his pants down outside of the privy will go up on charges of treason and be ridden back to Cape Town in irons." He glared at the chagrined group of offenders and shouted in a harsh voice, "Is that perfectly clear?"

"Sir!" they responded as a bunch.

"Now get the hell back to the train. You are confined to quarters until we leave Elizabethville."

"Dismissed!" yelled Wainwright, and the men scurried off.

As they disappeared around a mud hut, Spicer-Simson looked at the remains of the burning hut. With the walls caved in, the flames had died down, but the ruins probably would smolder on into the night. Nearby he saw four women, in as disarray of dress as had been the departed sailors. And they weren't looking any happier. A couple of them were staring with open hostility at Hanschell.

"You may have done the right thing for the expedition, Doctor," Spicer-Simson commented, "but to them, you've been an unchivalrous cur."

"Commander?" Hanschell was taken aback.

"Oh, I suppose you acted within reason, though perhaps if you had come to me I could have handled the situation without taking such destructive means."

"They're whores," Hanschell said. "Living in squalor…."

"Right now, they're poor women without a roof or belongings."

"Well," Hanschell said, shamefaced. "What would you have me do?"

"Why, pay them, of course."

"Pay them?" Hanschell said. "I'll do no such thing."

"Lieutenant Wainwright."

"Yes, sir."

"Place Doctor Hanschell under arrest."

"Arrest!" Hanschell gasped. "What for?"

"I believe looting, pillaging, and rapine will do for starters."

"Rapine! I did no such thing!"

"Look at them." Spicer-Simson pointed to the four women. "They look angry enough to accuse you of just about anything. And certainly half the people gathered around here will testify that it was you who deliberately and wantonly burned the house."

"But I did it to save the expedition."

"And I thank you for it," Spicer-Simson replied, and the words sounded as sincere as anything Wainwright had ever heard him say. "Just as I'm sure those women will thank you for reimbursing them for the loss of their possessions."

"Oh, all right," Hanschell said poutingly. "How much?"

"Call them over, Wainwright. Ask them how much."

Wainwright beckoned to the women, and they approached reluctantly. They still shot angry glances at Hanschell, but they looked on the commander with thanks touched with awe.

"They say they want forty pounds each," Wainwright said after a brief conversation.

"Forty pounds each!" Hanschell blurted, then his shoulders sagged as Spicer-Simson's glare fell over him. "I haven't that much cash," he said, though he wasn't about to admit that he was low on funds because he'd spent almost all he had on the two wagers. "I've only got about ninety pounds total."

"Give it to Wainwright. Wainwright, divvy it up evenly between them and make up the difference from your own wallet. The doctor can reimburse you when he gets paid again."

Wainwright took Hanschell's money, added some of his own, and passed it out to the women. They fell on their knees before Spicer-Simson, chattering excitedly.

"What do they say?" the commander wanted to know.

"They say they would gladly, ah...um...ah, accommodate you anytime you wish, Commander. Either singly or, ah, all together."

"You don't say." Spicer-Simson grinned.

"I'll say they will," Hanschell said grumpily. "You've just given them a couple of years' wages."

"Not I, Hanschell. You." The commander laughed and slapped the doctor on the back. "I never took you for a missionary, Doctor."

"Nor I," Hanschell replied with a rueful grin.

"Give the ladies my regrets, Wainwright," Spicer-Simson said. "I haven't had a good dalliance since we left London, but I think I'll take a pass on red spots on *my* arse. Come, gentlemen." He turned and started back toward the depot.

Wainwright leaned close to Hanschell and whispered, "If he's afraid of red spots on his arse, does than mean the tattoos don't go that far?"

"I don't know," Hanschell shrugged.

"Care to rescind your wager?"

"No," Hanschell said, starting after the commander. "I'll let my money stand."

When the two men caught up with Spicer-Simson, the commander said, "I don't like languishing. It'll make the men soft, and we'll only see more trouble like today. Wainwright, make preparations for them to ship out."

"What about the traction engines, Commander?"

"They'll arrive eventually, but in the meantime, we can unload the boats and supplies. It will give the men something to do and give us a chance to survey the progress the advance party has made on preparing the route. Gentlemen," he said, still looking straight ahead, "we leave for Fungurume at dawn."

Chapter V: Fungurume

As Hanschell stepped off the train and looked around, he thought Fungurume the most miserable town he'd ever seen, though he'd seen a few nearly as bad along the Gold Coast. There was no real station, just a couple of crude shacks with rusting tin roofs near the tracks. One belonged to the stationmaster, a light-skinned and obviously educated African. East of the railway buildings were a half-dozen filthy huts—little more than hovels, really. Piles of rails and ties lay between the shacks and around the railhead. Only the winding Dukuluwe River, which nestled in a green valley slightly below the railhead, prevented a picture of complete desolation.

The village's sole reason for existence, they learned from the stationmaster, was that this was as far as the railroad construction had reached. Here the project had been balked by the Dukuluwe, which, due to the dryness, was now little more than a wide stream that snaked between somewhat steep banks that were about thirty yards apart and eight deep. But in the wet season, the stationmaster assured them proudly, the river was much more impressive and often overflowed onto the surrounding countryside. Fungurume was simply a temporary haven for the railway construction materials and equipment until beams to construct a proper bridge across the river could be brought up from Cape Town, though that might have to wait until after the war.

"Well," Spicer-Simson told the stationmaster. "You needn't worry about the bridge. We'll be building one ourselves and taking two road locomotives over it. After we're gone, you can just lay your track across it and carry on."

The remainder of the afternoon was filled with establishing camp alongside the crude depot. As soon as he saw the wretched huts next to the station, Hanschell approached Spicer-Simson.

"We are now in the danger zone," the doctor told the commander. "The huts are likely to be infested with pests—ticks and scorpions and even snakes. We don't want our men stung or bitten by any of them."

"I'll issue a standing order for the men to avoid entering any of them," the commander assured him.

"We probably won't encounter any tsetse fly until we reach the Lualaba," Hanschell went on, "but with this river nearby, mosquitoes will be rife. The *Anopheles* carries malaria and the *Stegomyia* yellow fever...."

"You may spare me the medical lecture, Doctor," Spicer-Simson said. "The men all have mosquito netting, and I will instruct them to use it."

"I have some oil of citronella," Hanschell said. "It protects from mosquitoes for four or five hours and might be good for the night watches, who can't use their nets."

"Excellent. Tell Waterhouse, since he is in charge of the watches."

The rail crews had cleared most of the underbrush from around the railhead, but several tall trees still stood, and the ratings, under Waterhouse's direction, strung tarpaulins between them to make simple shelters for the men. This was necessary because Spicer-Simson had ordered that only two tents be erected—one for himself and one for Hanschell.

"The doctor may have an emergency and will want a place to work," the commander explained to Wainwright. "But we shall only be here a couple of days—not long enough to warrant erecting a proper camp. The men's sleeping bags should be sufficient, and anyway, they'll be sleeping out of doors from now on until we reach the lake. They might as well get used to it. Make sure they use their mosquito nets."

"Very good, sir," Wainwright said, but before he could go, Spicer-Simson had one further command.

"And no lights except campfires," he said. "I want the men to bed early and up early. We have much to do tomorrow."

This last caused Wainwright to frown inwardly since he well knew that lighting lamps would be impossible since they didn't have any lamps to light. The commander had eliminated all but a handful of lamps and oil from the stores before leaving Cape Town. "No sense in dragging along any unnecessary baggage," he'd said at the time. But Wainwright didn't let the frown emerge. He had no intention of mimicking Cross's half sneer, even if he felt that Spicer-Simson often deserved one.

Before sundown, the expedition was joined by about two hundred Africans dressed in everything from animal-skin loincloths to blankets draped over their shoulders to skimpy kilts and g-strings made of reeds and straw. These men were members of the party of nearly six hundred that John Lee had hired to aid the expedition. Most of the rest of the African crew were still in the bush north of Fungurume, hard at work hacking out and dynamiting the narrow track that would serve as a road.

Wainwright thankfully surveyed the newcomers. They would be sorely needed in the morning to help unload the supplies and equipment from the train. That would be a monumental task not simply because of the size and weight of the boats, since the rail yard had a crane, but because the weapons, ammunition, petrol, food, and other stores and gear totaled about one hundred and thirty tons. Then there would be the work of helping carry all that overland from camp to camp.

Evening mess provided the expedition's first obstacle on the overland journey—the cook's "bread." Until now, the cook had merely had to keep track of his supplies but hadn't been called on to perform his culinary duties to any great extent. The men quickly learned that he could open tins of bully beef and stir up a reasonably decent meal, but his attempts at making bread left much to be desired.

"What's this?" PO Flynn asked, holding up a hard little lump that attended the bully beef on his plate. "I thought hardtack went out of style forty years ago."

"That isn't hardtack, Flynn," answered PO Cobb. "That's ballast for the boats for when we get 'em to the lake."

"You're both wrong," Wainwright said, examining his own lump of ossified dough. "Cook has been assigned to manufacture cobblestones to pave our way."

Hanschell, normally serious, couldn't help but put in his own comment. "As your medical officer," he said, "I warn you against consuming this stuff. But if you must, I have brought along several barrels of purgative pills to help you get rid of it again."

Spicer-Simson had the final word. "Every proper seaman knows how to make decent bread," he remarked, setting his own biscuit aside. "Our cook obviously is a land crab. Cook, see me after mess, and I'll instruct you on the proper recipe."

"Sir," the cook said. He had turned a deep shade of red and gone through several phases of embarrassed anger before realizing he'd do better to keep his mouth shut. He simply returned to his pots and stirred the stew with petulant vigor.

And since at least half the expedition members were land crabs, themselves, conversation was effectively ended—that night, at least—by the commander's statement.

After mess, Hanschell surveyed the medical supplies that his assistant, Rupia, had installed in his tent and was surprised to find everything in perfect order. Even the single hospital cot was made up with tight precision.

"Very good, Rupia," he told the African. "I think we're going to get along famously."

"Yes, Bwana Doctor. Rupia is good assistant."

"Yes, you are. I'm glad Mr. Lennard recommended you." Hanschell looked around. "I don't see my sleeping bag."

"You not sleep here in hospital?"

"Not tonight. I think I'll go find Lieutenant Wainwright. See you in the morning."

The sky was darkening as Hanschell came over to the spot Wainwright had chosen to sleep and spread out his own sleeping bag.

"Not sleeping in your tent, Doctor?"

"I don't think so. The night's too nice to be shut up in stuffy canvas walls."

It *was* a beautiful evening. A cool breeze quickly carried away the day's simmering heat, bringing with it the pungent odor of native tobacco and wood smoke from the expedition's African camp a hundred yards to the west. Fireflies flickered about the trees and brush nearer at hand, and the insects were, in turn, mirrored above as stars erupted in the growing darkness. Prominent was the Southern Cross, poised like a sign pointing toward their destination. Hanschell was reminded of Spicer-Simson's disagreement with the Astronomer Royal from Cape Town.

He did not relate the incident to Wainwright. Instead, the two men talked about the day's events. Finally they fell silent, listening as the air filled with a symphony that only could have been composed by Africa. It began with the chitter of crickets and other insects, followed swiftly by the cries of birds that took to the cooling air in search of those same insects. But all too soon, other instruments voiced their wild refrains: the whirr and croaking of small frogs, the basso boom of bullfrogs, the cries of monkeys, the screams and snarls of lion and leopard, and the uncanny laughter of hyena. In all that great wash of sound, the only human element was the beat of elephant-ear drums that, despite their rhythmic reverberations, seemed just as untamed as the wildlife and only heightened the raw harmony of life and death that rose around it.

Finally, Wainwright broke the silence between them.

"Still believe in the commander?" he asked.

"I do," Hanschell replied after a pause. "I admit that he's not the same man I thought he was back in England, but then I'm finding that neither am I."

Wainwright nodded. "The heart of the bush changes the heart of a man."

2

The long-drawn notes of the African dove woke them just before dawn. Immediately after morning parade, the task of unloading the train began. Thankfully, the stationmaster had a five-ton crane that was used for moving rail supplies around the yard, and by early afternoon, a crew supervised by Wainwright and Cross had used it to swing the boats' trailers off their rail car and position them beside the flatbeds that held the boats. He also detrained the lorry. Meanwhile, a crowd of tribesmen under the co-direction of Waterhouse and Eastwood unloaded supplies off the other cars, the rhythm of their work chant keeping time.

Spicer-Simson spent most of the morning swaggering up and down the tracks, swinging his cane, chafing at the slightest delay, and shouting orders that often contradicted those Wainwright or Waterhouse had given just moments earlier. By midday mess, though, he seemed bored, and he retreated to his tent, where he sat in the shade of the front flap, drinking vermouth, and scowling at anyone who happened to pass by.

Everything was unloaded by late afternoon except for the boats, and huge piles of supplies stood beside the tracks. Wainwright took a last stroll beside the supplies and found Eastwood still there beside stacks of munitions boxes and petrol tins, all marked "Keep Cool." Eastwood was carefully ticking off box and tin numbers on an inventory sheet.

"What's this?" Hanschell asked, indicating several unmarked crates and a large trunk.

"The commander brought them," Eastwood said.

"What's in them?"

"You know those monogrammed cigarettes he smokes?" Eastwood was referring to Spicer-Simson's special cigarettes. "Cmdr. G. B. Spicer-Simson, R.N." was imprinted in blue on each one. "Well, this crate's full of those, and the other three are full of vermouth. They came up from Cape Town just before we left Elizabethville."

"He plans on doing some drinking, I guess," Wainwright said. "What about the trunk?"

"Don't know about that. He's got the key, and I haven't seen inside."

"I hope he's paying for the extra porters out of his own pocket," Wainwright said in a disgruntled tone. He was thinking of Spicer-Simson's professed desire to take along as little additional weight as possible. Apparently, lanterns weren't as necessary as his vermouth and cigarettes.

"Speaking of porters, are you certain we have enough for all of this?" Eastwood waved at the huge mounds of supplies and equipment.

"I should think so," the lieutenant replied. "We'll be hauling some of it in the boats and the lorry, and anyway, I don't suppose we'll be traveling all that far any one day. The bearers can leap-frog things forward if they have to. An African trek isn't like what you might imagine, Tubby. When you read about Livingstone and Burton and all the others, you might visualize a group of men all walking together, but it isn't like that at all. A real expedition of this size is likely to be strung out for miles along the way."

"I've been meaning to ask why we're using men instead of draft animals to carry everything. I'd have thought a few oxen or mules would be more efficient."

"You won't find many draft animals where we're going," Wainwright said. "You know the sleeping sickness that Doctor Hanschell keeps going on about? Most domestic animals are incredibly susceptible and die quickly. About all you'll ever find in central Africa are waterbuck and buffalo and the like because they're immune. Horses and mules, though...." He shook his head. "One that's infected is a really pitiful sight —wasted, swollen-bellied, just staring dumbly at you with bloodless eyes. They simply don't last. That's why we use men. Only a couple of strains of sleeping sickness strike down men, so they've become the most reliable beasts of burden here."

"Well," Eastwood said, "thank God for them, then." He waved his tally sheet at the stack of petrol tins beside him. "There are seven thousand gallons of this stuff, and I'd hate to have to carry all of it myself." He patted the tin on top, and immediately jerked his hand back.

"My God!"

"What is it?" Wainwright asked, and he touched the petrol tin himself. "Christ, they're hot as hell."

An entire day in the African sun had made the petrol tins hot enough to blister skin, and both men immediately reflected once more on the fact that there were, as Eastwood had just noted, seven thousand gallons of it. And the munitions cases, also metal, were no better off. The "Keep Cool" stencils on the sides seemed a mockery.

"First thing in the morning, we'd better get these out of the sun, or we'll see the whole expedition go up in flames," Eastwood said.

"The whole province, more likely," Wainwright replied.

Leaving Eastwood to finish his tally, the lieutenant went back to the camp, which was unusually quiet. Since beginning their journey in England, the men had done practically nothing but ride, gawk at the scenery, eat, sleep, and participate in regular morning parades, but lately, even these had grown lax. A full day of hard labor had quieted their energy and would put most of them into an early bed.

He found Cross and explained the situation with the petrol tins and ammunition.

"But where can we store everything where it won't be in the sun?" Cross asked. "There's no warehouse here. Even if we use all those nasty huts, there won't be enough room."

"Best stay away from the huts," Wainwright warned. "We'll just have to put everything here." He gestured around at the space beneath the tarps where the men slept.

"The men won't like it."

"Like it or not, they'll have to vacate. The petrol and ammunition need the shade more than they do. After all, an exploding case of rifle shells can kill just as readily as an enemy's gun. I'll leave the details up to you and Eastwood. I have to detrain the boats tomorrow."

The next day came all too quickly, and the expedition members were noticeably stiff during morning parade. They also were grumbling at having to move their belongings from beneath the tarp. Nonetheless, they did so, and soon after, Cross had a crew of bearers hauling the supplies into the shade. In the meantime, Wainwright took most of the ratings and twenty or so workers to the train to begin the task of moving the boats to their trailers.

He hadn't been at it for long before Spicer-Simson came up.

"What the hell's going on here, Wainwright? What are these men doing?" He jabbed his cane in the direction of the crew moving the supplies into the shade.

Wainwright explained, and the commander nodded. "I was going to have them do that myself," he said. "Glad you saved me the trouble."

"Yes, sir," Wainwright said, fervently hoping the commander would go back to his tent and not interject confusion as his crew unloaded the boats. Sadly, he was disappointed as Spicer-Simson took up where he'd left off the day before, swaggering and shouting orders, unhindered by Wainwright's better grasp of logistics.

They had *Mimi* off the flatbed and fastened to her trailer by noon mess, and Wainwright went over to Spicer-Simson, who by now had tired of shouting and was simply standing there, idly swishing his cane back and forth.

"*Mimi* is ready, sir," the lieutenant said, and he was suddenly startled when Spicer-Simson laid a hand on his shoulder and stared him straight in the eyes.

Wainwright fought hard not to blanch. The commander was difficult enough to take at a distance, but at this range, Wainwright almost felt himself wilt under the intensity of the commander's gaze. It seemed to the lieutenant that behind that enthusiastic gleam lay more than a touch of madness, but there also was something intensely interesting, something that attracted more than it repelled. Yes, Spicer-Simson may have had something of the martinet in him, and a shrewishness, too, but he also had a quality that was as powerful as it was convincing—the commander truly believed in himself. Perhaps he hadn't always done so, but right here, right now, he was the supreme commander, as if he was destined to lead this expedition to greatness.

For the first time, Wainwright really understood Hanschell's confidence in Spicer-Simson. It wasn't rational, and it wasn't even emotional; it was as if some sort of destiny was at work that Wainwright was suddenly aware of in the core of his being. It didn't matter how difficult the journey might prove or how the battles yet to come might play out or even how odd the commander's behavior was. All that mattered was that every man in the expedition continue to place one foot in front of the next, pull mightily along, and obey this strange man who was their leader. In the end, if they were lucky, there might be some sort of illumination.

"Excellent work, Wainwright. I trust you can handle the rest by yourself?"

"Yes, sir. We'll have *Toutou* off by sundown."

"I know you will, Lieutenant."

Spicer-Simson let go of Wainwright's shoulder and turned to leave but was halted when the lieutenant said, "Commander?"

"What is it, Wainwright."

"I just wanted you to know, sir, that I'm with you all the way."

For an instant, Spicer-Simson looked completely surprised—not shocked or startled or even pleased, but more than a little puzzled over a conundrum he'd never before faced.

"Yes, Lieutenant," he said, equanimity replacing the puzzlement. "Of course you are. Get on with it, now."

Suddenly the commander's eyes strayed, staring down the tracks. Wainwright turned, shaded his eyes, and looked, too. There, in the distance of the dirt road that paralleled the rail line, he could see a peculiar figure, wavering with heat and dust. At first he thought it was a tree, narrow at the bottom, fatter at the top, but it was dwarfed by the trees that stood at a similar distance. And it was approaching rather rapidly, trailing a tiny puff of dust.

"What the hell's that, Wainwright?"

"I'm not sure, sir." Wainwright stared harder. "Is it.... No, it couldn't be."

"What, Lieutenant?"

"It looks like a man on a bicycle, sir."

It was. And he was in some kind of uniform. In another couple of minutes Wainwright could make out what sort.

"He's wearing the uniform of the Rhodesian Rifles," he told the commander, which meant khaki shorts and shirt and a floppy bush hat.

In short order, the man pulled his bicycle up next to Spicer-Simson and Wainwright, dismounted, lowered his vehicle to the ground, and snapped to attention.

Spicer-Simson, thoughtfully stroking his goatee, inspected the new arrival. He was in his early thirties, trim, and of medium height, and he seemed bursting with nervous energy. His coloring was hard to determine beneath its heavy coating of road grime, but despite the fact that the dust said that he'd ridden a long way on his bicycle, he didn't look in the least bit tired. Spicer-Simson reached out and brushed fastidiously at the insignia on the man's shoulder.

"At ease, Sublieutenant. What is your name?"

"Dudley, sir," the man said, relaxing. "Arthur Darville Dudley. Are you Commander Spicer-Simson?"

"In the flesh. Have you come to see me?"

"I've come to join you, sir."

"Who sent you?"

"General Edwards, sir. John Lee recommended that he transfer me to your expedition."

"Is that so?" Spicer-Simson's brow clouded. "And when was this?"

"Nearly a month ago, now, sir."

"A month!" Spicer-Simson's barked laugh neatly masked his relief. "Took you long enough, didn't it?"

"Sorry, sir. I was on temporary leave at my farm in the Western Province when I got the word. I tried to catch a train at Livingstone but you had the only one going this far north for the next month."

"Livingstone? You came from Livingstone?"

"From Salisbury, actually, sir. General Edwards's headquarters."

"Wainwright. How far is Salisbury?"

"By road and rail? About a thousand miles, commander."

"Quite interesting. You mean to tell me, Dudley, that you rode a bicycle one thousand miles to volunteer for my expedition?"

"Not really, sir. I picked up the bicycle in Livingstone."

"Six hundred miles," Wainwright amended dryly.

"I almost caught up with you in Elizabethville, but your train had pulled out the day before I arrived."

"You must want to join my little expedition pretty badly," Spicer-Simson asked with narrowed eyes. "Why?"

Dudley looked perplexed, then he shrugged.

"I suppose it seemed the thing to do, sir." His expression suddenly took on a hurt look. "You will have me, sir, won't you? I saw action in the Boer War, and I've fought German askaris along the border of Northern Rhodesia."

"Ever been aboard a naval vessel, Dudley?"

"Yes, sir. I hold papers as a second mate in the Merchant Service."

"Indeed?" Spicer-Simson seemed pleased. "An enterprising fellow, don't you think, Wainwright? Spunk, verve, and energy. And a sailor, as well."

"Then you'll have me?"

"I will. What do you think you would be best suited at?"

"I don't know, sir. Most anything you'd like."

"You live in Rhodesia. Do you speak any of the local languages?"

"Swahili," Dudley said. "And I can muddle through several kinds of Bantu."

"Excellent. I think I have just the position for a man with your talents and energy."

"What, sir?" Dudley looked expectant.

"You'll be my executive officer, Dudley." Spicer-Simson jerked his head in Wainwright's direction. "This is Lieutenant Wainwright, my second in command. He'll introduce you around camp and get you fixed up with quarters. You will begin your duties tomorrow morning. Now, I must attend to some urgent business. Wainwright, see to Sublieutenant Dudley."

Then Spicer-Simson stalked off toward his tent.

Dudley, who hadn't even had a chance to snap a parting salute, stared after him, then turned to Wainwright.

"Did Commander Spicer-Simson just make me his executive officer?"

"He did," Wainwright affirmed.

"But I only just arrived."

"No matter, old man," Wainwright said with a wry smile. "So have we."

"It's an honor, but I'm not sure I understand."

"I'm not sure any of us understand," Wainwright told him. "And as for the honor, well, let that sit awhile before you savor it."

Dudley gave him a curious look, and Wainwright laughed, slapped him on the arm, and gestured toward the bicycle.

"Did you really ride that thing all the way from Livingstone?"

"Yes. And I'll tell you my arse was pretty sore for the first couple of hundred miles."

The two men laughed.

"You must be hungry, then. Let's get Cook to fix you up with something. We weren't expecting another officer, but as you can see," Wainwright waved around the open-air camp, "we're not cramped for space. You can doss with me and Doctor Hanschell if you like."

"Thank you, Lieutenant."

"Call me A. E. That's what my friends do."

"Art," Dudley replied, and the two men shook hands, then Dudley picked up his bicycle and followed Wainwright toward the camp.

"Don't worry," Wainwright said. "The good doctor and I will fill you in tonight after mess."

Wainwright's crew finished moving *Toutou* from the train just before evening mess, and Wainwright sought out Spicer-Simson to apprise him of the situation. He did not find the commander at his usual location at this time of day—his table in the mess area—but in his tent, hunched over a bunch of maps and papers spread across the table he was using as a desk. At Wainwright's entry, the commander hastily folded the maps and papers and stuffed them into a well-worn musette bag. He didn't look guilty, but his haste betrayed him, and anyway, Wainwright immediately recognized the bag as the former property of John Lee.

"Well, Lieutenant," Spicer-Simson snapped. "What is it, now?"

"Both boats are on their trailers, sir," Wainwright announced, what pride he felt stifled under his embarrassment at catching the commander rifling Lee's plans.

"Very good," Spicer-Simson said, casually sliding the bag under his cot. "I suppose there's still no word from down the line?" The normally sharp edge in his voice was blunted, and even his nasal tone was absent.

Wainwright knew what the question meant. The expedition was now facing the toughest leg of its journey, and while the boats were ready to roll, there was nothing with which to pull them. The traction engines and oxen were late, and there as no word on where they were or when they might arrive.

"Not yet, sir," Wainwright said. "But we still have to build a bridge over the Dukuluwe. Rickson has been surveying the site and assures me that the construction will take most of the week. Surely the traction engines and oxen will arrive by then."

"Surely, Lieutenant," the commander replied colorlessly. He rose. "Shall we eat?"

In silence, Wainwright followed Spicer-Simson to mess.

3

After morning parade, Spicer-Simson called Wainwright, Cross, Dudley, and Hanschell over to his tent. The commander's table was set up in front of the tent, and Spicer-Simson spread out a map from the musette bag he'd taken from Lee.

"Now that the boats are off the train, Wainwright, it's your job to work with Rickson to get the bridge built. While you're at it, Doctor Hanschell, Dudley, and I will go off on a little jaunt."

"A jaunt?" Hanschell asked apprehensively.

"Yes. We've nothing to do until the bridge is built and the locos arrive. We might as well go up country and check on the condition of the road and the progress of the advance party."

"But they're halfway to the Mitumba Mountains by now, Commander. I don't see how we can walk...."

"Walk!" Spicer-Simson snorted. "Who said anything about walking? We'll ride in the lorry. Wainwright, make sure that Mullen has it ready immediately. We'll take along a couple of the ratings. They won't be any use to you in building the bridge, and we might need their assistance along the way."

"But Commander," Wainwright said. "What about the Dukuluwe?"

"What about it?" Spicer-Simson asked.

"What I mean, sir, is the lorry's with us on this side of the Dukuluwe, and you'll want it on the other side, and we haven't built the bridge yet."

"You'll just have to think of something, then, Lieutenant. Hop to it."

After Wainwright and Cross left, Spicer-Simson leaned over the map.

"Look here, Dudley," he said, pointing. "We are here, here is where we want to be, and this in between is probably godforsaken country. As you can see, a route is marked through this godforsaken country. You'll ride with us as far as the advance party, then I want you to take your bicycle and proceed along the route, gaining intelligence. Note any impediments to our progress that are not marked on the map. The road goes only to Sankisia, but don't stop there. Take the rail line to Bukama, again looking for impediments. You weren't with us on the rail journey from Cape Town, but we had a narrow scrape at one bridge. Sublieutenant Hope should be in Bukama arranging for a shallow-draft steamer to take us down the Lualaba, and I want you to find out about her if you can. But spend no more than a full day in Bukama, even if you learn nothing of the steamer. I will need you back here with the information you've gathered on the road. Tell Hope to come back as soon as he's contracted with the steamer, if he hasn't already. Are these orders clear and understood?"

"Yes, sir."

"I believe that yesterday I noted that your bicycle has a cyclometer."

"It does, sir."

"Good. Please make a note of the total mileage on both your outward and return journeys."

"Yes, sir," Dudley said expectantly.

"Well?" Spicer-Simson said with a dismissive wave of his hand. "Snap to it, man."

Dudley wasn't sure what he was going to snap to since Wainwright had yet to transport the lorry over the Dukuluwe, so he went off to find the lieutenant to see if there was anything he could do.

Wainwright was rounding up a crew of laborers, and under his supervision, they quickly constructed a makeshift raft with timbers that had been cut for bridge supports. When the raft was finished, they shoved it into the sluggish current and dragged it to a relatively gently sloping section of bank. Mullen got behind the wheel of the lorry, and after a silent prayer, tipped it over the edge and gasped as it slid rather than drove toward the raft.

Miraculously, he managed to bring it to a halt just before it would have plunged into the shallow water. After some wrestling, the African crew hauled it aboard the raft, which swayed so dangerously that Mullen leapt out of the cab in panic. But the raft remained upright, and in a couple of minutes, the crew pulled it across the narrow channel to the other bank.

There was no way that the lorry could drive up the far bank, so the crew hauled it up with block and tackle, and by eight-thirty, it was ready for Spicer-Simson's journey, with food for several days, spare petrol, weapons, and ammunition.

"Very good, Wainwright," was all that Spicer-Simson had to say about the ordeal as he, Hanschell, and Dudley boarded the raft to cross the river. With them were two of the engine room men: Berry and Lamont.

As the lorry started out, Hanschell was glad he was riding instead of walking, but after an hour of jouncing and jarring along the rugged track that masqueraded as the road the expedition would take, he felt pretty shaken up. Not only that, but now that the sun was well up in the sky, the crowded, windowless metal box of the lorry's body was beginning to feel like an oven. The commander, of course, rode up front with Mullen, where he could catch some breeze. Eventually, the ratings propped open the lorry's rear doors, allowing a breeze to flow through the truck. It was a hot breeze, but at least it stirred the stifling atmosphere inside.

When Spicer-Simson called a halt at noon, the doctor was thankful, and so were Dudley and the ratings judging from the groans they made as they exited through the rear doors. As he emerged, stretched, and stared around at the fierce terrain, he wished that Mullen had found a nice tree to park under. But the heat was as fierce as the terrain, and it was the dry season, and the blasting sun and lack of rain had half-denuded the foliage from the trees and bushes, leaving only tentative and frustrating hints of shade. The arid atmosphere seemed to suck the moisture right out of the men's bodies. They sat around in the shadow cast by the lorry, sipping their ration of water and flapping the sweat-soaked tails of their shirts in the almost nonexistent breeze to cool themselves off.

So far, there was no sign of the road crew, though the work they'd left in their wake was impressive. The thirty-mile gap had been filled in, although the road itself was rough and often twisted in and out of deep gullies whose steep walls had not been sufficiently graded. But the three-ton Dennis could climb steep gradients, and except for half a dozen rather scary moments where the wheels spun dirt and gravel as the lorry lurched upward, they had no difficulties. Strewn all along the way were great stacks of brush and trees that had been cleared from the road to be used as firewood for the traction engines.

The afternoon ride was even hotter, and at three, Hanschell was dazedly thinking that he was half-cooked. But by then, they'd run across the tail end of the road crew, and before another hour passed, they found the main group, which was under the direction of a man named Locke.

Locke was puzzled that Lee wasn't with them, but as no one volunteered any information, he seemed to understand that the subject of his former boss was taboo. As he showed Spicer-Simson and his party around, it was obvious that he ran his camp and crews with military efficiency, and the commander said so.

"It's the only way to keep things going out here," Locke said confidently. "I have an arrangement with the local king. Fellow named Kasavubu. He comes from a family of witchdoctors, and he's using his clansmen as supervisors. I call them my capitos. All the rest of the tribesmen hereabouts are in mortal fear of them and do whatever they tell them."

"I say, Hanschell," Spicer-Simson said jocularly. "Looks like your colleagues run things around these parts."

Hanschell could only smile sourly.

"Would you like to meet Kasavubu?" Locke asked.

"I would," the commander answered, and an hour later, Locke led them into a village that looked a lot more prosperous than Fungurume. It was surrounded by a thorn boma to keep out lions, leopards, and hyenas, and they were met just outside the entrance by several warriors who plucked sheaves of grass and waved them at Spicer-Simson's party.

"It's a sign of welcome," Locke explained. "Kasavubu has known we were coming nearly since we left camp."

"I heard no drums," Spicer-Simson said.

"They don't use drums for signals in these parts," Locke said. "They have this system of whistles, and they can send messages from one end of the district to the other nearly as fast as a telegraph."

After they crouched through the boma's low entrance, they found themselves surrounded by a crowd of curious warriors, many painted and smeared with combinations of a kind of whitewash, red clay, and grease. They each bore a hide shield and an assegai, a short spear with a broad iron tip. The women, who kept to the periphery, had some of the most amazing hair fashions that Hanschell had ever seen—like fantastic sculptural works—and each woman's hair was as individual as the men's markings.

Had Hanschell not spent time on the Gold Coast, he might have expected Kasavubu to be attired in some sort of royal garb such as a lion or leopard skin robe and a headdress of tropical feathers. But Hanschell knew that image sprang more from the Victorian imagination than from reality. In central Africa, the chiefs usually dressed much the same as their subjects, which is to say in practically nothing at all.

He was surprised, then, when they were presented to the king and found the old man decked out in an obsolete British militia uniform, complete with spats on his bare feet. On his head sat an incongruous and battered silk top hat, and in one hand, he clutched the shaft of an even more incongruous pink parasol.

Hanschell had to force back a laugh that threatened to undermine the dignity of the situation. Civilization, he reminded himself, works in strange ways. Was this barbaric chief any less entitled to shade himself with a pink parasol than were the ladies of gloomy London? Or any more ridiculous?

Locke spoke briefly to Kasavubu, answered one of the old man's queries, and turned to Spicer-Simson.

"I told Kasavubu that you are a chief of the English tribe. He invites you to sit with him and enjoy a bowl of pombe."

"Pombe?" the commander asked.

"A sort of native beer," Locke answered. "Usually its brewed from bananas. It's abominable stuff but potent enough."

"Tell him we'll be pleased to accept his hospitality."

The pombe was truly nasty, Hanschell thought, after the first tentative sip from the ladle made from a calabash. Even so, he found himself getting tipsy after just a few swallows. The commander didn't seem to mind the taste, though, and he enjoyed a full gourd with Kasavubu, answering questions and laughing at the old chief's jokes.

Suddenly, the chief stopped and stared as Spicer-Simson set his empty calabash onto the mat next to him. Touching the commander's wrist where the jacket sleeve had hiked up to expose a darkened ring of tattoos, he asked Locke something.

"What is that on your skin?" Locke translated. "Kasavubu wonders if you have a white face yet are black like him beneath your clothes."

"Would he care to see?" the commander asked and, to Hanschell's horror, began stripping off his jacket and shirt.

"Blimey," said Lamont under his breath, and he poked Berry in the ribs. "Get a look at the commander."

Spicer-Simson's muscular, round-shouldered torso was completely covered, every inch of it, with tattoos.

"I guess that lets you out, eh, Berry?" said Mullen, good-naturedly patting the engine man's shoulder.

"And Lamont and about half the rest of the blokes, too," Berry agreed. "I never thought they'd go past his arms."

"Whadya think, Lamont?" asked Mullen. "D'ya think they go all the way to his ankles?"

"Don't know," Lamont replied. "But I'm likely to change my bet."

"Can't do it," Berry said. "Bettin's closed. We agreed that, soon as we got a look at even a part of him, bettin' would be closed."

"That's right," affirmed Mullen. "Fair's fair. We've all had a chance to make our guess. Those who guessed wrong will just have to accept it."

"That's all well and good for you to say," said Lamont. "You bet they'd go on down to his arse."

"Hush," said Berry. "Look at the king."

Kasavubu appeared mortified and fascinated at the same time as he stared at the jungle of vines that covered Spicer-Simson's torso and the menagerie of birds, snakes, and other animals that inhabited it.

"The chief seems interested in my art," Spicer-Simson said without the faintest trace of self-consciousness as he turned to display his back, equally dense with illustrations. "Something I picked up here and there in the East. This fellow was the first one." He pointed to a portrait of a tiger that graced the outside of his bicep, "Got it in Shanghai. Chap who gave it to me was a real artist. Couldn't speak a word of English but knew well enough what I wanted. Probably took the guinea I paid him with and went off and bought some opium. Never saw him again, but there are plenty of tattooists in China and elsewhere who are good. I got most of this thicket," pointing at a clump of foliage with snakes and lizards cunningly worked in among the branches and leaves, "in Borneo. Another Chinaman. He had a bit of English. Told me he'd been in Borneo for twenty years and wouldn't go back to China for a keg of gold. Took him a damn long time to finish, I can tell you."

The commander stopped abruptly. He'd heard a steadily rising murmur from behind him, and he turned to see what the commotion was about.

It was about him.

The entire village was crowding forward, pointing at Spicer-Simson, and jostling one another for a closer look.

"What do they say, Locke?"

"They are remarking on your tattoos," Locke said after listening to the villagers. "They are arguing about their meaning. Some are of the opinion that you must be a great chief to be so marked by the gods, while a few others go further by insisting that you must have great juju to be able to endure marks that were surely placed there by the gods."

"What do you think, Lamont," Spicer-Simson asked the rating. "Am I a great chief or do I have great juju?"

"What's a juju, sir?"

Spicer-Simson's laugh boomed out over the dry earth, and Lamont grinned foolishly.

"Tell Kasavubu," the commander said to Locke, "that in our country, all our great chiefs have powerful juju. Our highest chief can, with a mere wave of his hand, cause whole villages to be destroyed with thunder and lightning, or he can cause the harvest to spring from barren earth."

Locke translated, listened to Kasavubu speak, than said, "He wonders if you can do these things."

"Tell him that I am our chief's spear in this land, and he has given me a part of his great power—the power to lay villages to waste."

Locke spoke, listened, then spoke again. "He asks if you will show him this power."

"I will show him," Spicer-Simson promised in a tone that made Hanschell believe him. "But tell him that I have no desire to prove my power against him or his villages. I desire his friendship, not his blood. Instead, I will work my power against the Germans, who have attacked his villages and killed his people."

"And the Belgians?" Kasavubu wanted to know, speaking through Locke. "They have been more evil to us than the Germans."

"I know that is true," Spicer-Simson said. "Perhaps the Belgians, too. After the Germans."

"When will you do these things," Kasavubu asked.

"When we reach Lake Tanganyika," Spicer-Simson said.

Kasavubu thought about this for a few long moments, still staring at the commander's torso.

"How do we know your promise does not carry lies like the words of the Belgians?" he asked at last. "How do we know you have this power?"

"You come from a family of juju men."

Kasavubu nodded, eyes narrowing.

"I will show you." Spicer-Simson turned to the crowd gathered around and jabbed his cane toward a man in the front row. "Step forward."

The man held back, obviously reluctant to come closer, but a snapped order from Kasavubu made him approach the commander.

"Look," Spicer-Simson commanded, holding his arm close to the man's face. "Stare into the eyes of the tiger."

The capito blinked quickly two or three times as Locke translated the order, then he forced himself to stare at the tiger.

"You know him?" Spicer-Simson asked.

"Chui," the man muttered before Locke could translate.

"The leopard," Locke said.

"Yes, chui," Spicer-Simson said, and he flexed his bicep.

The commander was right, Hanschell thought. The man who had etched the great cat into his skin had been a true artist. He was standing close enough to see that, not only was the tiger immaculately wrought, it was cleverly placed on the underlying musculature. When Spicer-Simson flexed, the tiger's eyes seemed to glower, its nostrils to flare, and the corners of its mouth curl up in a hungry smile. At the same time, a deep growl rumbled in Spicer-Simson's thick chest.

The man staring at the tattoo tiger reacted as suddenly and violently as if he'd been confronted by the beast in the flesh. In an instant, he leapt backwards and stumbled into his companions, who fell away with a frightened gasp.

"Do you believe now that I have the power?" Spicer-Simson asked Kasavubu.

"Yes, bwana," Kasavubu replied, eyes still narrowed but now showing a hint of fear.

"Know this, then, Kasavubu—my promise and my power are as one."

4

Back at Locke's camp, they learned that the forward party under a man named Davidson was doing the preliminary clearing of the track by dynamiting boulders and large trees, but that it was another ten miles farther on.

"You might as well stay here for the night," Locke said.

"We have another two hours of clear daylight," the commander replied. "Surely we can get there by dark."

"Not in that." Locke pointed at the lorry. "There's no real road, yet, and the terrain is much rougher, and as you've probably noticed, we haven't built any bridges. If you want to get to Davidson's camp, you'll have to walk. And maybe swim a little."

Although he wasn't particularly anxious to spend another stifling, bone-jarring day in the lorry, Hanschell didn't want to make a two-day forced hike— one going to Davidson's camp and a second coming back —so he was relieved when the commander said they'd return to Fungurume instead. Dudley, however, mounted his bicycle and began pedaling up the track. He soon was lost to sight.

Spicer-Simson and the others arrived back in Fungurume by mid-afternoon the next day to find that the crew under Wainwright and Rickson was well along in constructing the bridge across the Dukuluwe. The thing looked like a squat railroad trestle supported by tree trunks. About half the superstructure was in place, and a second crew was felling smaller trees to pave the roadbed. It was amazing how efficient the local workers were, though they were armed only with primitive axes.

The work went well, and the roadbed was completed on August 11, two days ahead of schedule. Wainwright had Mullen test it out by driving the lorry back to the Sankisia side of the river, then he hurried to tell the commander that the bridge was ready.

"Well, I suppose I should look at the thing, even if we can't use it yet," Spicer-Simson said, his usual grating tone turned grumpy.

He followed Wainwright to the river, stared at the span for a minute, then without saying anything, turned and stalked back to his tent, where he sank heavily into his canvas camp chair.

"Drink," he ordered Tom. "A double."

"Double, stationmaster," the manservant replied and hurried to obey.

By the time Spicer-Simson finished half of his drink, he'd managed to regain something of his normal mood, but he couldn't feel fully well with the world. Here it was already the middle of August, and he remained stalled in this hellhole of a railhead. Where the hell were his traction engines? Where the hell were the oxen?

At least his fresh cases of vermouth and cigarettes had arrived in Elizabethville before the expedition came up to Fungurume. Thank God for small favors.

He took another sip, held up the glass, and stared at the clear liquid then suddenly became aware of a distorted figure that shone through the glass. It looked like a dun-colored stork perching atop a single leg. He lowered the glass.

"Goddamn it, Dudley!" he shouted as the stork resolved itself into the lieutenant mounted on his bicycle. "What the hell kept you?"

"Sorry, sir," Dudley said as he pulled up in front of the commander and dismounted. "It was a long way, sir, and the riding wasn't easy."

"Well, give me your report. How far was it?"

"Near as I could judge by the cyclometer, sir, it's about one hundred and forty miles to the railhead at Sankisia and another fifteen miles by rail from Sankisia to Bukama."

"And the road?"

"It isn't much of a road, commander. I'd call it more of a wide path. But it's been cleared almost all the way to the mountains."

"What of the mountains? Do they look crossable?"

"They're pretty steep, sir, but we'll have to get to them first, and that won't be easy."

"Speak up, man! What's the problem? Cliffs? Swamp? Hostiles?"

"Gullies."

"Gullies?" Spicer-Simson was contemptuous. "Did you see the bridge over that stream?" He gestured toward the bridge that Rickson had engineered and Wainwright's crew had spent the better part of a week constructing.

"Yes, sir. I came across it. A very nice bridge it is, too."

"Don't pander to me, boy! Of course it's a nice bridge. And a damn sight longer, I'm thinking, than any we'll need to cross your gullies."

"For the most part, sir."

"You mean we'll need longer ones?"

"Only one or two, sir."

"No matter." Spicer-Simson dismissed the gullies with a wave of his hand. "Do you think I've come this far to be stopped by a few gullies?"

"There's more than a few, Commander."

"How many more?"

"A hundred and fifty, at least, if you count those of only six feet or wider. Maybe eighty or ninety over ten feet. I lost count after that, but there couldn't have been more than a couple of dozen small ones. Maybe a hundred and eighty in all."

"A hundred and eighty?" This actually gave Spicer-Simson pause, and he even looked worried. "With water?"

"Most are dry, sir. Bone dry. I fear we will have trouble finding enough water for ourselves much less the traction engines." He blinked. "By the way, sir, have the locos arrived?"

"Blast it, Dudley, do you see any damn locos?"

"No, sir."

"Furthermore, do you think I'd be sitting around on my arse if the engines had, by some miracle, arrived?"

"I suppose not, sir."

"Suppose nothing, Sublieutenant. Knowledge is power, supposition is doom.

What about Hope?"

"I saw him in Bukama, sir, and relayed your orders."

"Was the steamship there?"

"Not yet, sir. Hope said he might have to go down river and get it."

"Goddamn him!" the commander snarled. "I explicitly told him to return here."

"Yes, sir."

"Dudley, go get Tyrer. Then get something to eat. And bathe yourself, for God's sake."

"Yes, sir."

Dudley started to push his bicycle toward the mess tent, when the commander halted him.

"Dudley!"

"Yes, sir?"

"Call a parade after morning mess. I want to speak to the men."

"Yes, sir." Dudley moved off, leaving the commander to brood over his glass of vermouth.

"Where are those damn engines?" Spicer-Simson growled. "I'm losing time."

5

Tyrer met with Spicer-Simson, and first thing the next morning, he was off with a pair of bearers to find Hope and bring him back to the commander. He didn't even stay for morning parade, and it was just as well for him.

Apparently Dudley took his new role as Spicer-Simson's executive officer very seriously. More seriously, at least, than the ratings liked. As soon as morning mess was done, he had them assemble in full dress uniform in the open ground between the tracks and the area where they slept. The officers stood to one side, while Spicer-Simson paced impatiently in front of the assembly. Everyone was impatient, actually, but none of the men or officers showed it. They were waiting for two of the ratings, under Dudley's supervision, to finish hauling the expedition's Union Jack up a line thrown over the branch of a tree.

"For God's sake, Dudley!" Spicer-Simson yelled. "Will you hurry up!"

"Yes, sir!" Dudley called, then to the two ratings he snapped, "Hurry it up, men. You're keeping the commander waiting."

At last they were done, and they joined ranks with the rest of the ratings and POs. Then Dudley put the men through their paces, marching back and forth across the uneven ground and presenting arms. All the while, the officers, Hanschell included, held a stiff salute. When it was all over and everyone was standing at attention, Hanschell's arm was trembling from having held the pose so long. Spicer-Simson strode back and forth in front of them, glowering and brusquely swinging his cane.

"I want to remind you men that we are here on serious business. His Majesty's business. I've seen entirely too much lax behavior since I went up country. If you think we are here on holiday, let me remind you to the contrary. We are here to take our boats to Lake Tanganyika and, once there, to sink the *Hedwig von Wissmann* and the *Kingani*. My job is to see that this is done as expediently as possible. Once we are victorious, then we may consider the time we have left here a holiday, but not a moment sooner. Until then, I expect every man here to do his duty to God and King George, and I will consider anything less to be treason. Is that clear?" He slapped his cane against his boot in emphasis.

"Yes, sir!" the men chorused.

"I have heard some of you men talking about the natives in distinctly uncomplimentary terms, even to their faces," the commander said. Hanschell found himself feeling warm with embarrassment as he recalled his unthinking comment about the girl on the motorcycle.

"I am quite aware that the peoples of Africa do not adhere to European standards of dress and behavior," Spicer-Simson continued. "While many of you may find that amusing, I do not. Nor will I tolerate open rudeness. You men may consider yourselves more civilized than these people, but I do not see them treating you with the same rudeness despite the fact that you undoubtedly seem as ridiculous to them as they do to you. Or more so. What man in his right mind would go about fully dressed in this climate as we are doing? If anyone here is the fool, it is we. And I want you to consider this: If we are to complete our mission and retain our good names, we will do so only with the help of the people along our route. Remember that we will depend upon them to build our road, to carry our supplies, to bring us water and fresh meat, and a dozen other things. They may seem childlike enough and deferential to the might of our weapons and machines, but they are fierce and proud warriors nonetheless. Remember that, as well, the next time you think you can speak down to one of them. I will consider any breach of politeness to be a contravention of my direct orders. Am I clear?"

"Yes, sir!" the men said, some, like Hanschell, reddening.

Spicer-Simson paused and surveyed the ranks before going on.

"And last," he said, "I want to remind you that we have another enemy more deadly than the Germans. By that, I mean tropical diseases. Very special precautions are necessary which will be explained to you by Surgeon Hanschell. Every man in this expedition, in matters of health and hygiene, will unquestioningly follow his instructions. Each of you will report daily to the surgeon for a dose of quinine. All drinking water will be boiled. Fly whisks will be issued today, and it will be the duty of every man to whisk flies off his neighbor, irrespective of rank. You, being untrained, can't tell the difference between household flies and the dreaded tsetse flies which carry sleeping sickness, so all flies are to be treated as dangerous. All ailments, all cuts, scratches, and abrasions will be reported to the surgeon. We cannot burden ourselves with useless invalids. If any man falls sick, he will be given a supply of food and water and left with one of the natives to get back to Elizabethville under his own power. Am I clear?"

"Yes, sir," everyone barked.

"Dismissed!"

Hanschell was relieved to hear the commander give him carte blanche in medical matters. Unfortunately, the doctor himself was the first to fall ill. It began the very next day, and he stayed in his tent all afternoon trying to hide the fact, but Dudley discovered him there and told Spicer-Simson, who came immediately.

"I should have warned that becoming ill was a direct contravention of my orders, as well," the commander said in a disgusted tone that couldn't mask his concern. Hanschell looked awful, pallid, sweating, weak, and clutching at his gut. "What's wrong with you, Doctor? You haven't got something serious, I hope."

"Not really," Hanschell said from his cot. "It's a touch of amoebic dysentery. I had it once up on the Gold Coast. Nothing to worry about, really. I recognized the symptoms well enough and took immediate action. I'll be all right in a week."

"You'd better be, or you'll make a fool of me," Spicer-Simson groused. "I said I would leave behind any man who became sick, but I can't very well leave our only doctor. Once we get started, you can ride in the lorry until you're well enough to walk."

Hanschell groaned inwardly. He felt bad enough as it was, and the prospect of jolting across the countryside in the back of the Dennis brought a fresh flood of gorge to his throat. As he leaned over the bucket next to his cot, the commander beat a hasty retreat. At last, drained and trembling, Hanschell sank back on the cot and fell into a fitful slumber, thankful that he had Rupia to take care of him.

6

"Are you sure this is wise, sir?" Wainwright asked.

"Are you questioning my orders, Lieutenant?"

Wainwright knew his commander well enough to recognize real anger. "No, sir," he said.

"If Dudley is right about the terrain ahead, we have our work cut out for us to make it to Sankisia before the rains begin. At any rate, I have no intention of rotting in this sinkhole any longer than is absolutely necessary. We leave this morning, traction engines or no traction engines."

"I'll get the lorry ready, sir." Wainwright hurried off to find Mullen and Waterhouse. Privately, he had doubts that the Dennis could pull the boats very well, but he, too, was sick of Fungurume, and he was as anxious as the commander to be out on the road.

It was August 13, and Spicer-Simson was in a foul mood. He stalked toward *Toutou*, snarling at everyone within ten feet of his path.

Damn that tosspot Admiralty. They'd got him out here in the middle of things, close enough to scent his goal but far enough to deny him a taste of victory. What he couldn't understand was why they'd gone to the expense and trouble of getting him this far when they wouldn't take measures to ensure that he could go the whole way.

All the bloody way to victory, damn them!

But no. Them and their officious delays. The expedition was languishing here in oblivion, and all for the want of two measly traction engines. All he needed were those two engines, and he'd be set.

"Give me my traction engines," he growled, unaware that he'd spoken aloud, and he thwacked his cane against *Toutou*'s hull.

Almost immediately, he regretted the blow, not because the men might have seen and wondered but because it left a visible dent in the lacquered sheen. His boat, he thought, touching the wound. Scars should be left to the enemy to inflict, not to loved ones.

"The lorry is ready, sir," Wainwright told him half an hour later. "We should do all right—it's a slight downward slope to the bridge."

He gestured toward the newly built structure, which was nearly a quarter of a mile from where the boats stood beside the tracks.

Mullen had backed up the Dennis to the neck of *Toutou*'s trailer, and several ratings under Cross's direction were attaching the neck to the lorry's hitch.

"Very good," Spicer-Simson said, and he went to stand in front of the group of officers observing. Everyone in the camp was there, including the Africans.

Several more minutes passed while the hitch was secured, then Waterhouse and his ratings joined the rest of the expedition while Mullen looked expectantly at the commander. Everyone, in fact, was looking expectantly at the commander, who seemed oblivious to the attention.

But Spicer-Simson, far from oblivious, was only too aware of the British members of the expedition peering at him from behind *Mimi*'s rear and the great crowd of black faces staring at him with open wonder, and he couldn't help but feel a thrill of pride. Damn, but it was good. They were waiting for him. Him. His word. His word to propel two boats across an arid waste in search of deep waters. His word to press on into the distance where glory beckoned with a faint and plaintive sigh.

Sighing himself, feeling a sense of personal victory over personal defeat inflate his soul, Spicer-Simson signaled to Mullen.

"All right, Mullen!" he yelled. "Forward!"

Mullen revved the engine and eased up on the clutch. The Dennis heaved against the hitch, which grew taut, and Mullen pressed harder on the accelerator. The lorry shuddered, and suddenly the rear wheels began shrieking as they spun on the hard earth, sending up clouds of dust and the acrid smell of burned rubber.

Toutou sat as solidly still as if it had grown there.

After a minute of this spectacle of nothing happening but unnecessary tire loss, Spicer-Simson told Wainwright to have Mullen desist.

"Put them to work," the commander said, jabbing his cane at the crowd of tribesmen. "We're paying them, might as well get something out of them."

Another half hour passed while Waterhouse's ratings attached long tow ropes to both *Toutou*'s trailer and the lorry, and Wainwright assembled the African workers into several crews. That wasn't too long considering there were two hundred of them to organize. The men resisted attempts to marshal them as a crowd and randomly assign them to the dozens of tow ropes that Waterhouse's ratings had hooked up. In the end,

Wainwright had to explain to the local headmen what had to be accomplished and let them make the arrangements. After that, things went fairly smoothly as the men, who insisted on organizing themselves according to clan, grasped the ropes.

As soon as the Africans were ready and had taken up the slack in the ropes, the entire British party positioned themselves behind *Toutou* and leaned against the hull and trailer. Mullen revved the lorry's engine.

Spicer-Simson surveyed the scene, taut in its promise of power and momentum. He looked ahead. The near end of the bridge was a thousand feet away. He looked back at his party and saw that everyone was staring at him, awaiting his signal to move and conquer.

This time, Spicer-Simson drew his sword from its scabbard and raised its gleaming arc.

"Party!" he cried out. "Forward!" He slashed the air in a dramatic gesture that was lost on most of the men, European and African alike, as they bent their heads and leaned into their tasks.

Ropes and harnesses creaked and strained. The lorry engine roared, and exhaust blued the air. The African headmen shouted orders in Bantu, Wainwright shouted in English, and a great collective groan went up from the straining crowd. Suddenly, with a loud backfire, the lorry went dead, and the rising groan of effort abruptly died in a moan of defeat.

Toutou's trailer had moved less than three inches.

The Africans began to drop the tow ropes and wander from their clan structure, trying to get a look at what had been accomplished, and the expedition crew could be seen peering around the stern of the boat.

"Goddamn it!" Spicer-Simson screamed. "Get back in line!" He slashed his saber in the direction of the milling tribesmen, and the nearer ones retreated ten or fifteen feet. "Wainwright! Get these men back on those ropes!"

While they were urged back into line, Spicer-Simson stormed over to the lorry.

"What the hell's wrong with you, sailor?" he demanded. "I thought you knew how to drive."

"I do, sir," Mullen whined.

"Shut up. If you stall that truck again, you'll find yourself hauling on a rope with the rest of them. Understood?"

Spicer-Simson stalked away without waiting for Mullen's subdued affirmation.

"All right!" he shouted when everyone was ready again. "Forward!" He slashed the signal and watched the men and machine strain to move *Toutou*'s trailer ahead.

Harnesses and ropes creaked and strained. Tribal headmen shouted orders. Wainwright shouted orders. Spicer-Simson shouted orders. The crowd of Africans obeyed the orders shouted nearest to them, or at least the closest shouted in a language they could understand. The lorry snarled exhaust and spit dust and gravel. The sun climbed. The air heated, glare washed the scenery with starkness, and dust and smoke rose in a breathless simmer. And through it all, *Toutou* crept forward, one lonely inch at a time.

By noon, the party had managed to move *Toutou* two hundred feet toward the bridge. Only eight hundred feet remained. As soon as the noon mess was completed, the men reassembled. Spicer-Simson again slashed the air, and the whole straining, creaking, shouting procession inched onward in the heat, glare, and dust like some vision from a middle circle of Dante's Hell.

By two o'clock, *Toutou* was closer still to the bridge, but the men's strength and will had noticeably flagged. Curse and scream as he might, Spicer-Simson couldn't seem to encourage them onward at a pace he thought reasonable. At three, strain as the men might, *Toutou* had ceased to move. Spicer-Simson called a water break, and went back to his tent for a shot of vermouth. He stripped off his sweat-soaked shirt and sat despondently in his camp chair while Tom brought his drink. Wainwright arrived at the same time as the vermouth.

"I don't know what to do, sir. The men are played out."

"Dog shit they are," Spicer-Simson snapped. "I'm saddled with a bunch of sluggards, and that's God's truth."

"The men are trying, Commander," Wainwright insisted. "It's those damn trailers. They aren't designed properly."

"I designed those trailers myself, Lieutenant, and they are perfectly designed." Spicer-Simson didn't shout, but there was a dangerous edge to his voice.

"Sorry, sir, I just meant...."

"Shut up, Lieutenant. I don't care what you meant. I only care about one thing, and that is dragging those two boats to Tanganyika. If I have to do it with my own teeth, I'm going to get them there."

"Yes, sir. But...."

"But what, Wainwright? But the men won't work properly?"

"They might. I just don't think they've got the proper incentive."

"What the hell do they need? I've paid their king."

"True enough, sir. And I'm sure they appreciate it. But think of the men actually on the ropes doing the work. They're out there sweating in the sun, while their chiefs are in their huts drinking."

Spicer-Simson looked at Wainwright as if the lieutenant was going to go on, but he suddenly realized that Wainwright was done, and just as suddenly he noticed the glass of vermouth in his hand. As he looked back at the lieutenant, an expression of rage passed across his face and just as quickly vanished.

"If you ever make such a statement in front of the men, I'll have you horsewhipped. Is that understood?"

"Yes, sir."

"I expect so. Now listen to me, Wainwright. I've got an idea to make the men work harder. Tell them, the Africans as well as ours, that I'll offer every one of them a glassful of liquor if we make the bridge by sundown."

"We'll never make it that far by dark, sir. What if we manage halfway?"

"Very well," Spicer-Simson snarled. "Get moving!"

"Sir!" Wainwright hurried off.

Half an hour later, Spicer-Simson returned. He was still bare chested, and the sight of his tattoos caused a murmur to rise from the gathered Africans. Even the expedition members stared and nudged and whispered among themselves. The commander paraded himself in front of the work crews for two full minutes, and Hanschell noticed that the tribesmen tended to shrink back if the commander came too close.

"Wainwright!"

"Yes, sir?"

"Ask these men if they have heard of my juju."

Wainwright asked, and a hesitant mutter of assent came from the assembled Africans. Their knowledge had a modest genesis in the simple gestures and whispered words of one man. The bwana with the waving sword, this man said, was a great and resourceful leader. He had the spirit of chui beating within his chest. He had the power of storms. Look! Did he not have the marks of the gods and the spirits covering him. And some of these marks, this man told his companions, were alive. He knew this, for he'd seen one of them come alive before his own eyes. He whispered to his companions how he'd seen chui's eyes spark, seen its mouth curl in a cruel grin, and heard it growl.

Since none of them had ever heard of a tiger, they all took the cat to be a leopard, and it reminded them only too well of the feared cult of the leopard men—renegades who dressed in skins of the dreaded predator and stole through the night killing wherever they went, dragging the victims off into the bush, and eating them in a frenzy of ritualistic cannibalism.

Now the whispers about Spicer-Simson's menagerie were everywhere. And while many of the tribesmen held back in fear, all peered intently at the bwana with the sword so that they, too, might see one of his beasts come alive.

"Tell them if they work hard, they will be rewarded, but if they don't, chui will visit them in the night."

"Sir, are you sure...?"

"Lieutenant," Spicer-Simson growled, and seeing chui raise his ferocious head himself, Wainwright obediently translated, adding a few embellishments of his own.

Soon, every man was positioned at the ready, waiting for the sword to slash.

"Forward!" Spicer-Simson gave them the signal, and the assembled crowd leaned into their work anew. Spicer-Simson strode to the head of the straining crowd and stood there, shouting and waving his sword like a baton master, urging his parade onward.

The crowd pulled and ropes creaked, but it was not like before. Though the laborers were exhausted, there now was something besides mortal strength that ran like a strong undercurrent and unified their efforts, something that motivated the Africans beyond the promise of liquor. It was fear, to be sure, but it was pride, too, that the bwana they toiled for was so powerful and would make a great mark on the world, and that they were his chosen people.

By sundown, *Toutou* was halfway to the end of the bridge.

7

The next morning, when Spicer-Simson emerged from his tent, he was startled to see a group of about fifty tribesmen circled around, sitting on logs, stones, or their haunches. They stood hurriedly and, as he walked through them, opened a lane for him to pass. Several leaned forward for a closer look at the man who'd been touched by the gods with pictures on his skin that lived, but it was early yet, and the commander was fully clothed.

"Get back, you black devils," he said good-naturedly as he moved among them, perhaps sensing their curiosity and wonder. But he gave them little satisfaction as he headed straight for the mess area. Wainwright was there already, and as they ate, Spicer-Simson told him he wanted to try to drag *Mimi* all the way to the bridge.

"We have experience, now," he said. "We know how the trailers react."

"I expect you'll have to remove your shirt, Commander," Wainwright said. Hanschell, who'd just joined them, looked at the lieutenant as if he'd lost his mind, but Spicer-Simson simply nodded, his eyes narrowing.

"Perhaps you're right. My body art did seem to perk up the natives' vigor, didn't it?"

An hour later, the lorry was hitched to *Mimi*'s trailer, crew members were positioned behind, and the two hundred tribesmen were manning the pull-ropes. Then Spicer-Simson walked out in front of the procession.

The workers were instantly attentive. The commander was shirtless, and as he moved and turned, the shadows of his muscles made his tattoos seem to ripple as if a breeze had wafted across a living jungle.

The expedition crew took notice, too, only they seemed a trifle more shocked than the tribesmen.

"Christ!" Lamont swore sotto voce. "Is that what I think it is?"

"Looks to be made out of leather, don't it?" said Berry.

"Quite stylish, ain't it," Leighton commented with a twist of his lip. Seaman Hollis, who was younger and a bit slower on the uptake—and louder—than the rest, blurted out, "Lookit that. The commander's wearing a dress."

"It's a kilt," said Tait, shaking his head in disbelief, or maybe denial. "It a bloody kilt."

"Bloody hell it's a kilt," said Lamont. "I may not be a bloody Scotsman, but I know a proper kilt when I see one, and that there ain't no kilt."

"It's a dress," insisted Hollis. "My mum used to wear 'em all the time."

"That's 'cause they's easy to squirm out of," said Berry.

The rest of the men laughed, though Hollis just stared blankly. "Well," he said sullenly as the laughter died. "She did."

"Maybe," said Lamont, "but what the commander's wearing ain't no dress. It's a skirt."

"Dress, skirt," said Berry. "What's the difference?"

"A bodice, you fool," said Lamont.

"What's a bodice?" asked Hollis.

"The part that holds women's jugs, Billy."

The men laughed again at Hollis's expense.

By now Spicer-Simson had reached the front of the column. The men's eyes hadn't deceived them: He really was wearing a leather skirt that stopped a few inches above his knees, his boots, his sword, his hat, and nothing else. His legs—at least the portions that could be seen between the hem of the leather skirt and the top of his socks, were as tattooed as his torso.

Wainwright hustled over to him, obviously flustered at his commander's unorthodox dress but unwilling to acknowledge the skirt's presence by so much as a glance.

"The men are ready, sir," he said, keeping his gaze rock steady on the commander's face. Spicer-Simson's mouth stretched into a grin, and his eyes twinkled.

"What do you think of my uniform, Wainwright?"

The lieutenant hesitated, trying to decide if the commander was serious or making a joke, all the while wondering what the hell the man was doing wearing a leather skirt in front of everybody. Not only were his own men watching, so were the two hundred laborers and an even greater number of women and children. Of course, a lot of the African men wore a short, loose, skirt-like wrap, so he didn't suppose the Africans really noticed anything unusual. He quickly decided on a safe course of action.

"It looks quite comfortable, sir."

"Quite. Quite. It's my own design. Had my wife sew it up for me before I left. I've been to Africa. Knew what a beastly hot place it is."

His deadpan tone sounded serious, but that fiendish grin was still wiped across his face. Wainwright, in desperation to get away from the commander before he burst into laughter, said, "Do you think we should get on with it, sir?"

"Proceed." As Wainwright hurried over to the African headmen to let them know the procession was about to get under way, Spicer-Simson drew his sword, raised it over his head, and slashed it downward.

"Forward!"

The headmen, caught off guard, turned to shout orders to their men, but it wasn't necessary. The tribesmen had seen the signal, and they knew what the chief-touched-by-the-gods wanted. And by now they also knew he had promised to free them from the Germans who took so much delight in shelling their villages and setting the Ba-HoloHolo on them. And maybe even from the hated Belgians. So they took hold of the ropes and pulled.

The confused shouting of yesterday had subsided into a steady, low-murmured song that seemed to unify the entire crowd into a single, amorphous draft animal that created slack and drew it in as if the work was simply breathing. *Mimi* crept forward slowly, but faster than *Toutou* had the day before. The effort of the tribesmen was astounding.

By noon, *Mimi* was brought up nearly alongside of *Toutou*, when a loud cracking sound suddenly split the air, and the trailer buckled and sagged on its wheels. A loud moan rose from the workers on the ropes, and the Englishmen behind the boat cursed.

"Goddamn it!" Spicer-Simson spat, staring at the collapsed trailer. "I told them six inches wasn't enough beam! We need twelve inches for this terrain."

The center beam on *Mimi*'s trailer had broken, though the "terrain" to the bridge was scarcely rougher than an English country lane.

"What'll we do, sir?" Wainwright asked as he hurried over to the commander.

"Well, we'll have to fix the damn thing, Wainwright. Won't we?"

"That'll take days."

"Break for mess," Spicer-Simson said. "After that, we'll bring up *Toutou*."

When lunch was over, that is exactly what they did. They took *Toutou* in tow, and with their tattooed commander in his leather skirt smoking furiously from his foot-long cigarette holder, brandishing his sword, and bellowing unnecessary directions, they pulled toward the bridge.

They were still fifty feet away when *Toutou*'s trailer collapsed in the dust. As with *Mimi*'s, the center beam had snapped and the wheels buckled.

"Shit!" Wainwright yelled, but his voice was lost in the keening wail set up by the tribesmen. Even the women on the sidelines joined in the dirge.

Up in front, Spicer-Simson said nothing. He simply sheathed his sword and strode back through the wailing throng to where Wainwright stood next to the buckled trailer, shaking his head.

"We'll have to fix both of them, sir."

Spicer-Simson stared at the trailer, stroking his mustache and beard, then he looked at the sky all around. He shrugged.

"Give us something to do while we're waiting for those damn traction engines," he said.

"I'll get on it first thing in the morning."

"Very good, Wainwright. Thank the native headmen and dismiss our men." Spicer-Simson strode off toward his tent, and Wainwright walked over to where the crew were clustered behind the boat. As he approached, he heard arguing voices that suddenly quieted as he came up.

"Is there some trouble?" he asked, expecting to hear something about the shoddy trailers or having to spend two days shoving at the rear of the boats or, at the very least, the commander's new attire.

"Mullen says he's still in," said Berry. "But I remember it different."

"That's a lie, Berry, and you know it," Mullen said.

"I think you're the one ain't tellin' the truth," said Leighton.

"Whadya say, Lieutenant?" asked Mullen.

Wainwright realized that they were talking about the pool on the commander's tattoos.

"Let's not trust memory, men," he said, and he called Eastwood over. "Do you have the tally sheet with you, Tubby?"

"Ah, the tally sheet," Eastwood said when he learned the nature of the dispute. "Yes. It's right here." He pulled the paper from his pocket, unfolded it, and looked it over. "Yes. It says right here that Mullen bet the commander's tattoos stopped at his buttocks." He looked up at the lorry driver and shook his head. "Sorry, Mullen. You're out. Unless I'm completely blind, those were tattoos on his legs."

"Told you," Berry sneered.

"Blimey," said Mullen. "How many is left in? We know them tattoos cover near everything."

"Yeah," said Lamont. "Everything that weren't covered by that skirt! Wonder what's up there, anyway?"

Everyone laughed, even Wainwright and Eastwood.

"Well," Eastwood said, looking again at the paper. "It seems that there are three who say arms, chest, and legs—Hollis, here, Tyrer, and Cross."

"Any others?" Lamont asked, jostling forward to get a glimpse of the tally sheet. "Mr. Wainwright and the doctor?"

"Mr. Wainwright and the doctor," Eastwood affirmed. "And the two of them voted for the whole hog, all the way down, including is buttocks."

"How come its mostly you officers who are still in?" Leighton groused to Wainwright.

"Well, boys," Wainwright said with a grin, "That's why we're officers, don't you think?"

"What about Hollis?" Berry said. "He ain't no officer."

"Anything's possible on this jaunt," Wainwright said only half jokingly.

8

"I'm sorry, Commander," Wainwright said, bracing himself for another withering tirade from Spicer-Simson. "It's impossible."

"You'd better give me a damn good reason why it's not possible," Spicer-Simson said in a dangerous tone. "I'm not going to sit here on two broken-down trailers for want of proper beams to support them."

"It's not just the beams, sir," Wainwright said. "The wheels have collapsed as well. I think we should just wait for the traction engines. They'll be pulling heavy fuel trailers, and we can use them."

"And what if they're no better than what we've got?" the commander demanded. "Then we lose even more time. Forget it, Wainwright. I want the trailers we have repaired."

"All right, sir, but the hardwood trees in this area just aren't large enough to cut twelve inch beams," Wainwright said. "We used all the really big ones building the bridge. We can get six inch ones, but not twelve."

"We've seen what a bloody lot of good six inches will do," Spicer-Simson snarled.

"I suppose we could use two of them," Wainwright said. "Fasten them together. Might be as good, sir."

"Excellent idea," Spicer-Simson said, cracking the side of his boot with his cane. Although it was the day after the trailers had broken down and there was no immediate need to impress the Africans with his body art, he was again shirtless and dressed in his leather skirt. "Now," he continued. "We've got to do something about those wheels. I wish those fools at Thornycroft were here to see what miserable results have transpired from their work."

"The railway company has two sets of wheels," Wainwright said, ignoring the fact that the commander had bragged about engineering the trailers himself. "Came off a broken-down lorry. They use them to transport sleepers and rails around the rail yard."

"Commandeer them," Spicer-Simson ordered. "One set on each trailer. Have Chief Waterhouse reinforce the fore-carriages and put the larger wheels there. He can shift all the smaller wheels aft. That should help take some of the load. And make certain he reinforces both trailers all around."

Wainwright sought out the stationmaster, but when he explained what he was after, the stationmaster refused to part with the wheels.

"I have loaned you my crane," he told the lieutenant, "but I will not give you the wheels, or I will lose my job."

"What if we pay you for them?" Wainwright asked.

"Yes," the stationmaster said, an avaricious gleam in his eyes. "But it must be enough to satisfy the company."

"How much do you want for them?"

"The company would be satisfied with £200."

"That's ridiculous," Wainwright said. "Hell, I could practically buy a whole working lorry for that."

"By all means," the stationmaster replied, spreading his hands to take in the shabby station, the miserable huts, and the grim surroundings. "I know an excellent source—down in Cape Town."

"Let me remind you that we are at war. I can commandeer the wheels, and there's not a jot you can do about it."

"You would leave me destitute?" The stationmaster rolled his eyes, the sarcastic gesture of the moment before becoming one of supplication.

"Oh, all right, you conniving old cheat. I'll give you £100 for them, and not a penny more."

"I see I must agree. But you will pay me in cash?"

"I'll be back shortly with the money," Wainwright said. "Just have those wheels ready."

Leaving the stationmaster, Wainwright found Eastwood, explained the situation, and had the paymaster return with him bearing the requisite amount of cash. After the stationmaster counted the bills, he had several of his workers roll the wheels down to the boats, where Waterhouse's crew was preparing to use the crane to lift the craft and their cradles free of their trailers.

The work progressed smoothly until Spicer-Simson showed up and began barking impatient and useless orders, spouting caustic remarks, and generally making the men nervous and fumble-fingered with his angry and unwarranted attentions.

"I can get these boats off the trailers easy enough," Cross finally said to Wainwright, "if you can distract the commander."

"I'll do what I can," Wainwright promised, and he went over to Spicer-Simson.

"While the men are unloading the boats, sir, would you care to inspect the bridge?"

"I've already inspected the bridge, Lieutenant."

"But you haven't seen the final touches, sir. Or the road we made on the other side."

"Oh, very well," Spicer-Simson said. "It can't be any duller than watching these fools make a simple job impossible."

Although Rickson had done the principal engineering on the bridge, Wainwright had supervised the actual construction, and he was inordinately proud of the hundred-foot-long finished structure. Spicer-Simson strode out onto the planking until he was in the center, and he pounded the roadbed with his cane. The metal-shod tip thumped with a reassuring solidness.

"Admirable, Wainwright," the commander commented with a decided lack of interest. "Now, let's get back to the repairs."

"If you please, sir, I'd like to show you the road we've constructed on the far side."

"Very well," Spicer-Simson sighed dramatically. "Show me your road if you must."

Wainwright led the commander to the far end of the bridge. The land on this side of the river rose sharply, thirty or so feet higher than on the other side, where the men labored to repair the trailers. Spicer-Simson obviously was uninterested in the somewhat steep track Wainwright's crew of tribesmen had dug up the bank. There were too many other things to worry about, such as trailers that couldn't hold up for a thousand feet and might not be repairable. And even if they were, there were only another one hundred and forty miles to go. Only one hundred and eighty more bridges to build. Only a mountain range to cross. Only a river journey and a harbor to establish and battles to fight and win.

It all seemed like too much to do with too little time and too few resources. He couldn't let it show to his officers or men, but he was fast despairing that the whole enterprise was the worst sort of folly imaginable. What in God's name had he been thinking when he volunteered? But he knew all too well. His dead-end career, the monstrous unfairness of events that had continually conspired to raise his hopes and then dash them to pieces in the most ironic and publicly humiliating way possible. And now his trailers collapsing before he'd properly begun.

If only those blasted traction engines would arrive! Even if they could repair the trailers, without the engines, he was doomed to make, at best, five hundred feet a day. At that rate it would take them, let's see....

He performed a few rapid calculations in his head.

It would take them four and a half years to get to the railhead in Sankisia. Not counting the Mitumba Mountains.

By then the bloody damn war would be over!

He nearly jabbed his cane tip into the log decking of the bridge, but he caught himself. He couldn't let Wainwright see the depth of his frustration, his anger. He had to maintain his equipoise as well as his equilibrium. To hell with everything that was going wrong. To hell with the Admiralty's slipshod planning that had left him stranded here in the midst of Africa with a duty to perform but without the means to perform it. Goddamn them, and the hell if he wouldn't succeed despite them. The breakdown of the trailers was a test of his faith, nothing more, and while it was shaken, he'd survived on less. If he had to, he'd manufacture his own faith just as his men were manufacturing the new double center beams for the trailers. He'd create his own backbone, and it would be double-strength to withstand whatever fate threw in his way. Goddamn if he wouldn't!

Spicer-Simson lifted his darkened brow to the African sun and stared past Wainwright to where the men were lifting *Mimi* free from the trailer. As he did, Wainwright saw a light flare in his eyes, and the lieutenant wasn't sure if what he saw was anger or inspiration or madness. Whatever it was, it seemed as if the commander had been struck by a heavenly bolt.

And then, Wainwright heard the thunder.

For an instant, he was equally awestruck and dumbfounded as his spirit soared and his mind strove to understand how he could feel with his own senses what existed only in his commander's febrile spirit.

But the thunder did not vanish into rolling distances. It grew as if it rolled across the African plains toward them. Wainwright's mental faculties took over, and his ears unconsciously sought direction. And even as he turned to see what storm approached, the men up by *Mimi* were turning and a great cry rising from the Africans watching the men work.

It was not a storm that approached, but two storms, each with its own thunder, each with its own dense, dark cloud rising into the air. And as these storms rolled toward the crowd gathered at the boat, Wainwright felt a surge of elation tempered with thankfulness.

"The locos!" he heard Chief Waterhouse cry. "The locos are here!"

The locos—the road locomotives, the traction engines—had finally arrived. Miniature wood-burning steam locomotive of about ten tons in weight, with large wheels banded with ribbed steel instead of the driving wheels of a railroad locomotive, they'd been in service in South Africa for several years before the war broke out. On each, a canopy covered the engineer's station at the rear of the boiler and ran halfway to the front of the boiler, where a tall, pipe stem smokestack belched great clouds of dense, choking smoke. Each pulled a ten-ton trailer loaded with wood, and each rolled forward with a great pandemonium of snorting, chuffing, clanking, and clattering that grew to completely drown out the happy shouts of the British as they rushed toward the machines.

The Africans, in contrast, were stricken dumb. They'd all seen railroad locomotives, but these two behemoths, freed of their entrapping rails, almost seemed beyond them. Many ran into the fringe of bush surrounding the railhead, but after seeing the sailors' delight at the machines' arrival, the ones who had run away came back and, with their fellows, moved forward, quiet and cautious despite their great number. As the locos hissed and clattered to a stop beside *Toutou*, they were completely surrounded by the entire mob. Wainwright, who'd hurried across the bridge at the sight of the locos, was swept forward by the mass of Africans and soon found himself confronting the lead driver.

"I'm Harrison," the driver said. "Are you Commander Spicer-Simson?" Harrison was a stocky, weather-beaten man of about Wainwright's age, wearing khakis and a bush hat.

"I'm Lieutenant Wainwright. Commander Spicer-Simson is...." He turned and looked. Where was the commander? Then he saw him.

Spicer-Simson, a lone and distant figure, was still on the end of the bridge. He was too far away for Wainwright to be able to read his expression, but the lieutenant imagined that it would be one of relief.

"That's him," he told Harrison. "Out there on the bridge."

"What's that he's wearin'?" Harrison asked, squinting. "Is that a kilt? Must be a Scots, though his name don't sound like it."

"I can assure you he's pure English," Wainwright said, smothering a smile. "And that's no kilt."

By now Spicer-Simson had crossed the bridge and was striding toward the crowd. The Africans melted a path as he entered their midst.

"Sure it's a kilt," Harrison said. "I can see plain it ain't no britches."

"It's a leather skirt, Mr. Harrison," Wainwright said bluntly. "And if you know what's good for you, you'll like it."

"A skirt?" Harrison seemed exasperated. "And is that some kind of leggings he's got on?"

"It's too late," said Waterhouse. "He can't join the pool."

"Quite right, Chief," Wainwright agreed.

"Pool?" asked Harrison. "What pool?"

"I'm sure you'll hear all about it later, Mr. Harrison."

"That's a mighty fancy shirt he's wearin'." Harrison commented.

"You'll find out just how fancy in a second," Wainwright said, and he turned as the last of the Africans between Spicer-Simson and the lead traction engine opened to let the commander pass. The lieutenant may have been right about Spicer-Simson's expression, but if he had been, it had drastically altered in the couple of minutes since.

"This is Mr. Harrison," Wainwright began, but the commander cut him off with a curt slice of his cane.

"Where the hell have you been?" Spicer-Simson snarled, his brow darkened with anger. "The fate of Africa lies in the balance, and you've been off, meandering around."

"I.... I...." Harrison, eyes wide and darting between Spicer-Simson's exposed menagerie, his skirt, and the silver-headed cane being shaken in his face, clearly was at a loss for words. "I.... I mean, we was delayed in Livingstone, Commander. Then we had to drive up from Elizabethville 'cause we was too big and heavy for the trains."

"How long did that take?" The commander's brow remained dark, but there was a note of genuine curiosity in his voice.

"Four days, sir. We can do better on a good road, but...."

"How much better?"

"A hundred mile a day on really good road if conditions is perfect."

Spicer-Simson stepped back and surveyed Harrison's machine. It looked capable enough. Nonetheless, he snorted.

"We'll be lucky to make ten a day in this terrain," he said.

Harrison looked injured.

"I swear by my machines, Commander," he said. "They can go anywhere."

"We'll see about that," Spicer-Simson said darkly. He then turned to Wainwright.

"As soon as Waterhouse finishes the repairs to the trailer, we'll put Mr. Harrison's machines to the test. Get one of the locomotives in front of *Mimi*."

With that, he turned on his heel and stalked off toward the camp.

"That's an odd man," Harrison commented after the commander was a hundred feet away.

"May I remind you that we are a Royal Naval expedition in the middle of the African bush?" Wainwright asked. "What would you expect?"

"No offense meant," Harrison shrugged. "I admit to a bit 'o surprise, but I been in this godforsaken continent long enough not to hold surprise for long."

"No offense taken," Wainwright said. "As long as you follow his orders to the letter, you'll be all right. If you don't, you'll be hearing more from the commander."

"I haven't brought my machines all this distance to do naught," Harrison said.

"Good. Then you see that boat up ahead?"

"The one with the lorry parked in front?"

"The very one. We'll move the lorry, and your machine will take its place. Both boat trailers are broken down right now, but we should have them fixed in a couple of days."

While a grateful Mullen moved the lorry, Wainwright had Harrison show him the locos.

"I drove locos myself when I first came to Africa," Wainwright explained to Harrison. "I'm a farmer now, but I don't think I've ever washed the engine oil from my blood."

"I know how it is," Harrison agreed. "Even the steam smells good after a time. This machine here," he patted the side of the lead engine, "was built by John Fowler & Company of Leeds. D'ya know that big bronze statue of King Alfred in Winchester?"

"I've seen it."

"Well that 40-ton granite plinth it sits on and the 20-ton granite base that supports the plinth was hauled all the way from Cornwall by two Fowler engines workin' in tandem. This other was built by Charles Burrell & Sons of Thetford in Norfolk. The driver, there, is Bisho. From Cape Town. I trained him myself."

"What about the engines? They look to be eights."

"I can see you know your locos," Harrison said approvingly. "Both of 'em's eight horsepower with three forward gears. Low gears is for heavy haulin'. And we put on bigger wheels to set 'em up higher off the ground. Look at this." Harrison bent and pointed beneath the engine. "We built ash pans under the firebox so's we won't set the countryside ablazin' with sparks."

"Smart work," Wainwright nodded as he straightened, and he ran his eyes over the locos. Beneath the veneer of road grime, their brass work shone bright with polish. "They look to be in very good condition."

"That they are, Mr. Wainwright. I take as good a care of 'em as I do of my wife." He chuckled. "Maybe better. She's back at home with the kids while I'm off here gallivantin' in the bush."

"Gallivanting," Wainwright repeated with a bemused smile. "Yes, that's the word for it."

9

The trailers were repaired by the afternoon of the 17th. Each beam had been replaced by two fresh six-inch beams sistered together. The original wheels had been moved back a few feet from center and the new wheels installed forward to help shoulder the burden.

When Wainwright told the commander that everything was ready, Spicer-Simson nodded sagely and said, "We'll move out after morning parade, then. Make sure all the gear is packed up tonight. You might as well have the bearers load as much of the heavy equipment as they can in the boats. I want us to move as efficiently as possible."

Wainwright complied, and by evening mess, the boats were packed with the guns and ammunition, and there was even room left over for some of the heavier supplies like sacks of flour and meal. Considering the way the sun had heated the petrol tins, Wainwright thought it prudent to let the bearers carry those—a lone petrol tin that caught fire was replaceable, but the boats weren't, and he wasn't about to take chances. The ammunition wasn't as volatile, and if the heat got really bad, he would have the men douse the canvas sheets covering the boats.

The next morning, Cross directed the loco drivers and a handful of ratings in hitching the trailers to the traction engines. Actually, the boats were hitched behind the wood-carrying trailers, making two short overland trains with the boats as cabooses. As soon as the task was accomplished, Wainwright fetched Spicer-Simson.

"So, Harrison," the commander said as he swaggered up to the traction engine. He again was wearing his skirt. "You still think your machines up to the job?"

"Yes, sir."

"But your title is not."

"Sir?"

"It is unseemly that a man under my command fail to have a commission in the Royal Navy. I hereby commission you a staff sergeant. See Mr. Eastwood about a uniform."

"Yes, sir. But, sir? What about Bisho, sir?"

Spicer-Simson frowned.

"It would not be seemly for a native man to hold rank of any sort. But considering his position aboard your engine, he will be allowed to wear a uniform tunic. I trust that will do, Staff Sergeant?"

"Yes, sir."

"All right, then." Spicer-Simson unsheathed his sword. "On my signal, pull toward that bridge there."

"Shall I have the natives pull on the ropes?" Wainwright asked.

"No," the commander said. "Let's see what Harrison's machine can do alone."

Wainwright had the workers untie the ropes, and in a few minutes, the Burrell and its fuel car stood by themselves at the head of *Mimi*'s trailer. As the tribesmen moved back, chattering with expectancy, Spicer-Simson strode toward the end of the bridge, spun on his heel, and raised his blade, glittering, into the air. Harrison throttled up his machine, and a great roar was borne into the air on an asphyxiating spew of smoke.

Spicer-Simson slashed downward with his sword.

Harrison clutched, and with nary a slip, the traction engine rolled forward, drawing *Mimi* effortlessly behind. In seconds, more ground was covered than in hours of pulling with the lorry and hundreds of men.

And in those seconds, a roar greater than that raised by the traction engine drowned the machine's clatter and clank as the two hundred African men and their assembled wives and families shouted with wonder and excitement. Ten or twelve tribesmen, armed with drums, leapt out in front of the loco and lead the way toward the bridge, dancing, shouting, and singing.

His emotions riding the Africans' wave of awe, Wainwright tore his eyes from the traction engine and stared at his commander.

Spicer-Simson stood to one side at the end of the bridge, tip of his sword planted in the log roadbed, leaning on the pommel in a jaunty posture. He seemed to be smiling, but the distance was too great for the lieutenant to be certain.

In less than two minutes, the traction engine had reached the bridge and halted. Wainwright rushed forward just as Spicer-Simson stepped onto the ground at the side of the bridge and motioned for Harrison to continue across the span.

We're crossing, Wainwright thought with pride. We're crossing at last.

The traction engine, proceeding at a cautious pace, rolled across the log roadbed so painstakingly engineered above the muddy stream two dozen feet below. In a minute it had reached the center of the span.

Suddenly a splintering sound ripped the air. The traction engine's weight was too much for Rickson's creation, and a pair of support beams split and collapsed. Instantly, the roadbed beneath the traction engine's front wheels gave way, and with a lurch, the wheels crashed through, leaving the engine canted dangerously forward, lifting the trailer hitch high into the air. That tipped the fuel trailer backward, slamming the bow end of *Mimi*'s cradle against the roadbed. The boat's stern levered upward on the fulcrum of the wheels and swung sideways over the lip of the bridge. With the sudden shift in weight, the roadbed skewed sideways, shearing another support pillar with a tremendous cracking sound and pulling away some of the side beams. The idyllic crossing of the minute before was a mere memory as the bridge and its burden swayed precariously above the muddy stream below.

The Africans had stopped talking and laughing and singing when the first beam splintered, and as Harrison hurriedly shut down his engine, silence reigned.

10

"I'm sorry, Commander." Wainwright hung his head, unable to face Spicer-Simson. "I don't know what to say."

"Then don't say anything, man," Spicer-Simson said with surprisingly even composure. "Just get that boat back to safety. Call that other engine over here."

Wainwright hurried to obey. The second traction engine was brought up, and Bisho backed it to the very end of the bridge. From it, cables and ropes were dragged to *Mimi*'s trailer and fastened to every conceivable strut. Meanwhile the Burrell, the fuel trailer, and *Mimi*'s trailer were unhitched from each other, and Harrison, who had bragged that his engines could go anywhere, joined Bisho in the cab of the Fowler, thankful that his own loco hadn't gone into the river.

With steady deliberateness, the traction engine crew dragged, first, *Mimi* back to safety, then the Burrell's wood trailer, and then the Burrell itself. The whole process, accompanied by a loud uproar in Anglo-Saxon and Bantu from the assembled crowd, was finished within an hour, and when it was, Spicer-Simson stood on the bank with Wainwright and Rickson, surveying the damaged span.

"Well," he said in a surprisingly jovial tone. "We've spent a good portion of our morning trying to cross our first bridge, and not only have we failed to cross it, we now no longer have a bridge."

"I'm sorry, sir." This was from Rickson. "I guess I didn't expect the locos to be so heavy."

"I suppose it couldn't have been helped," Spicer-Simson said, waving it off. "We are in unusual circumstances, and our resources are meager. Even so, I hope you've learned your lesson from today's fiasco."

"I'm not sure what we can do, sir," Rickson said. "We can put in more supports and try to shore it up."

"Bravo," said the commander. "That's the spirit."

"It'll take some time...," Rickson went on, but Wainwright interrupted.

"Excuse me, sir, but it strikes me that we may be getting too complex here. No offense, Rickson, but your design might be well and good for heavy timber and steel, but we haven't got those, as the commander points out. There might be a simpler way."

"Go on," Spicer-Simson said.

"When I was driving engines down in Rhodesia, the crews used to make quick and simple bridges that would do until the real bridge crews could come along and make something sturdier. But of course, we won't need anything sturdier behind us, just something easy to build that won't collapse."

"I don't see how...," Rickson began, but Spicer-Simson stopped him.

"Perhaps the lieutenant will demonstrate. Just how long will this new bridge of yours take to construct?"

"We have a good crowd of men to work with," Wainwright said, gesturing to the multitude of Africans up on the bank behind them. "A couple of hours."

"Well, get to it, man!"

The bridge that Wainwright constructed, if that is the term, was quite simple. First, he had the workers hack down hundreds of small- to medium-sized trees and strip the trunks of their branches. Then he instructed the workers to lay the trunks lengthwise in the streambed itself rather than across it, and as more and more trunks were piled on top, the pile gradually rose, forming a crude but effective causeway. Because the logs were parallel to the flow rather than across it, the water just ran between the logs and went on its way. Amazingly, the bridge was completed by noon.

"Bravo, Wainwright," Spicer-Simson said. "Quick and easy, just as you said. Are you certain that pile will hold the locos, trailers, and boats?"

"We took whole trains across them down in Rhodesia," the lieutenant said. "It should do."

"Really brilliant," Rickson said. "We can set up a crew that goes ahead to make the bridges we need, and if we come to a stretch without enough trees, all we have to do is go back and collect our old bridges."

"And if we don't need them," Wainwright pointed out, "we can use them to fuel the locos."

"Let's give your contraption a try, shall we?" Spicer-Simson said. "Better have Harrison leave his wood trailer behind, though, until we're sure it's safe."

Harrison had been eyeing the bridge with skepticism while the workers dumped the logs into the gap, but Wainwright assured him it would hold. So the loco driver took his place in the cab of the engine, inhaled deeply, and gripped the clutch. With *Mimi* in tow, he inched his machine toward the pile of logs. Because the second bridge was so much lower than the first one and had been hastily constructed without a road graded down to it, he had to take the engine down the steep, slippery bank, and as *Mimi*'s trailer came over the edge and followed him down, it slewed sideways, accompanied by a roar of dismay from the thronged Africans and unheard groans from the Englishmen. But Harrison was a good driver, and his machine staggered onto the end of the bridge just in time to straighten out the trailer and set everything on an even keel as his little parade proceeded gingerly across the pile.

Wainwright, standing on the lower bank just downstream, kept a careful eye on the bridge, and he noted that the whole thing sagged and creaked beneath the combined weight of the traction engine and *Mimi*'s trailer, but the pile remained intact. Although Harrison took it slow, the bridge wasn't long, and in just a couple of minutes, his wheels were rolling onto the dirt of the far slope.

Only then did Wainwright realize with sudden chagrin how steep that bank was. He'd been so preoccupied in building the bridge that he hadn't given a thought to cutting a road up the bank as he'd done for the first bridge. But Harrison, supremely confident that his machines could go anywhere, simply poured on the steam and, with a clanking roar, started up the bank.

He got about halfway to the top before the engine began sliding backwards in the mud, right toward the river. Wainwright caught his breath as if that could catch the loco and halt its slide. *Mimi*'s trailer rolled back, too, but not with as much momentum. In an instant, the traction engine and the boat trailer jackknifed, and Harrison, in an effort to keep his machine from sliding farther and crushing the boat's hull, opened his clutch all the way. The driving wheels spun in the soft, damp bank, spitting out great chunks of muddy earth as the loco's tremendous weight bore them deeper and deeper. Finally, completely mired, the engine clashed to a steaming, sputtering halt.

"Goddamn it!" Wainwright ground out the breath that he'd caught. "Goddamn it!" Just as he saw Harrison dismount from the traction engine, he became aware that someone was standing at his elbow. He turned and felt his heart sink. It was Spicer-Simson.

"Commander. I'm sorry…."

To his surprise, Spicer-Simson laughed and slapped him on the shoulder. "It appears that your bridge is quite admirable. It held the engine and the boat to the other side."

"Perhaps, sir, but it seems I also have to make certain that the slope on the far side isn't too steep to get up."

"No matter, Lieutenant. I'm sure your next bridge will take that into account. Best get to it." "Yes, sir."

Building a third, higher bridge upstream from the first two took less time than the second had because they made use of the logs they'd already cut. Like the second bridge, the third was spongy under the weight of the Fowler, its wood trailer, and *Toutou*, but unlike the second, it had an adequate ramp up to higher ground excavated by hand and a few well-placed dynamite charges. Wainwright made certain of that. Two humiliations in one day were enough for him.

As soon as *Toutou* was safely on the other side, cables and ropes were strung between the Fowler and the Burrell, and the mired loco was dragged up the bank. *Mimi* soon followed, and by then, Mullen had driven the lorry over the bridge, trailed by a chain of bearers.

"We're across," Wainwright said, too exhausted, anxious, and embarrassed to feel any pride of accomplishment.

He was standing next to Dudley, watching Harrison and some of the ratings re-hitching *Toutou* to the Fowler.

"And only one hundred and seventy-nine left to go," Dudley said with a bright, humorously mocking grin. "I counted them."

"Thanks for the encouragement," Wainwright said, his tired face wrinkling with disgust.

"Call the men in," Spicer-Simson ordered, coming up to them. "We leave first thing in the morning."

Chapter VI: On the Road

NOW THAT IT WAS OFF, the expedition made six miles the first day, and Spicer-Simson seemed pleased with the progress. He even passed out an extra ration of whiskey that evening, and the expedition members began to relax a little after the snafus of the last couple of weeks.

They shouldn't have. Not an hour into the trek the next morning, while the Fowler was pulling *Mimi* up an embankment, the earth collapsed beneath the loco's wheels, and with ponderous majesty, the Fowler slid ten feet sideways down the bank. Harrison and his fireman leapt clear just as the loco tilted over and fell onto it side, its hitch luckily breaking free of *Mimi*'s trailer.

"Goddamn it!" Wainwright snarled when the loco started to slide. Then he saw Harrison and the fireman get up from where they'd fallen, rubbing at bumps and bruises but otherwise unhurt, and the epithet changed to "Thank God."

Nearly two days were wasted in rescuing the Fowler and regrading and reinforcing the road up the bank before the expedition could proceed.

But proceed it did. The days were long, hot slogs over rough road thickly carpeted with fine dust that rose chokingly with every step and billowed cumulus-like behind the locos and boat trailers. The nights were a din of chirps, hoots, cries, growls, snarls, and screams of pain from prey in the jaws of predator cats. The ratings and junior officers took nervous turns at night sentry duty to keep the big cats and hyenas at bay, then often were too tired the next day to complain about stumbling over rocks hidden in the thick dust carpeting the track.

Every mile found them facing one or more gullies or small ravines that caused them to pause while one of Wainwright's makeshift bridges was built. Almost all of the gullies were dry, though their bottoms sported growths of thorny acacia, bark mottled brown and yellow, giving testimony that water was not a complete stranger to this land.

After the second day, Wainwright assigned a bridge-building crew under Rickson to move out ahead and prepare as many bridges as they could so that the engines would not have to stop to wait for a crew to finish a span.

Riding in the lorry was as boring to Hanschell as it was hot and uncomfortable, but for several days, he was simply too weak to walk. The time did give him the opportunity to reflect on the differences in the two men who had become his closest companions: Wainwright and Dudley. Both men were intelligent and thoroughly loyal, and both loved Rhodesia and spent many an evening discussing the country, its resources, and its future, but there the similarities ended.

Wainwright, whom many of the men referred to as the Old Loco Driver because of his experiences before settling down in Rhodesia, was inventive and naturally optimistic, and he never seemed to forget his roots as a construction worker and engineer. And because he had been instrumental in overcoming the several major problems the expedition had thus far encountered, everyone—even Spicer-Simson—respected and deferred to him. Those traits inspired the men who labored for him to believe they could accomplish anything, even the impossible, but his wit gave him an ease with everyone that obviously was genuine and was much admired among the POs and ratings.

Dudley was a very different sort, nervous and serious where Wainwright was relaxed and humorous. More important to the men was the way the two regarded authority. In this, Dudley was more akin to the commander than to his fellow lieutenant. While Wainwright barely noticed—or cared about—the deference paid to him, Dudley was painfully aware that many of the expedition members considered him a newcomer. Consequently, he was somewhat jealous of Wainwright's rapport with the men, though he admired Wainwright as much as any of them.

Equally damaging to Dudley's standing with the men was that, being a bundle of energy, he couldn't stand to see anyone working at less than his own breakneck pace. He constantly hounded anyone he suspected of indolence, and too often, he was shrill with the ratings, which did little to endear him to them or give them reason to respect him as they did Wainwright.

Hanschell wondered if there was any way he could lend the young sublieutenant this insight, but he decided that it would probably be a futile effort. As futile as it would be to approach Spicer-Simson about the same matter. But Dudley, at least, had the chance to grow out of it, Hanschell reflected, particularly with a man like Wainwright as a model—unlike the commander, who was entrenched in his own eccentricities.

And those eccentricities were surfacing quite readily, these days, especially since he'd taken to wearing his skirt and going bare-chested to expose his illustrated torso. Although both Tait and Mollison, who were used to wearing kilts, seemed to accept the skirt without reserve, most of the other men thought the commander a touch daft at best.

Nor did Spicer-Simson's imperious, snobbish, and caustic attitude toward nearly everyone except Hanschell and Wainwright, help matters. It not only gave the men ample excuse for grousing, it often created unnecessary tension during the few hours each day when the expedition wasn't on the move—the very hours during which the men should be relaxing. Especially disruptive was Spicer-Simson's demand that, at his approach, every man snap instantly to attention until he had passed or given the order to stand at ease. Hanschell was half-tempted to count the cumulative hours lost to this ridiculous practice, but before he could start making notes, he saw that the men often ignored the order when they were engaged in laborious tasks, and the commander apparently wasn't noticing the breaches.

And so, mile by slow mile, the expedition rumbled in fits and starts across the rugged landscape.

2

"Christ!" Wainwright slapped a sand fly sucking at his arm, then wiped the bloody nubbin on his trouser leg, where it disappeared among the dozens of other smears. Then he slapped another.

"Malaria," commented Hanschell, who trudged along beside the lieutenant. "Cholera. Yellow fever. Sleeping sickness. No telling what other ungodly disease and sickness is out there in that mess." He waved at the dense brush around them.

"Are you trying to justify your presence, Doctor?" Wainwright asked.

"To myself, if no one else," Hanschell replied, slapping at one of the offending bugs himself and wiping the little smashed carcass on his own trouser leg. "What else can a doctor think when he's the only one who's gotten sick?"

"You seem to be doing much better now."

"Some, but to tell the truth, I just couldn't stand riding in that lorry one more mile."

"The walking'll be good for you."

"I hope something will. There has to be a reason I'm stuck in this place. Perhaps God is trying to punish me for forsaking Him for science."

"Perhaps the Admiralty needed a squadron of fools," Wainwright said. "Who else would help a tattooed dandy braggart in a leather skirt drag two boats across Africa?"

"Moan and complain," said Dudley, who followed closely enough to overhear the conversation.

"Well, aren't you sick of it?" the doctor asked.

"We've only been on the road a few days. Where's your spirit?"

"I might have some left if I didn't have the shits."

Dudley laughed. "Buck up. I rode six hundred miles on a bicycle to join this outfit, and that'll give you a sore arse for sure. And look at me now—I'm having a grand time. Besides, wasn't it you who gave me a pep-talk about our commander when I first arrived? He'd odd, to be sure. A real British eccentric. But damn me if we aren't on the move."

"I suppose you're right," the doctor conceded.

"I know I'm right."

"Easy enough for you to be rosy," Wainwright said. "The doctor has gotten sick, and I've built a couple of stupid bridges. We look like the fools, not you."

"Give me time," Dudley laughed. "I'm sure the commander will find fault with me yet. Anyway, it was Rickson who designed that first bridge, and your second worked fine. It was just the getting away from it that failed, and you've learned from your mistakes. We've gotten over all the rest of the bridges just fine."

Wainwright took heart at that, especially since he knew it was true. His design was just right for the terrain and materials at hand, and most of the bridges that Rickson's advance party had made had taken only an hour or two to construct. And since it was still the dry season, the majority of the gullies and washes they'd crossed were devoid of water, so the bridges would remain intact until the entire column had passed.

That often wasn't for some time. As the caravan lumbered forward through the dry brush and nearly leafless mahogany and rosewood forest and occasional open savannah, the traction engines and their loads could barely manage three miles a day. The locos' wheels constantly slipped and spun in the slick dust that blanketed the ground in thick layers, and each loco was attended by a troop of workers whose sole task was to cut bundles of crackling brush with their pangas and throw the branches beneath the spinning wheels until they caught and the engine lurched forward to another spinning stop.

Even worse, the road that the advance crew had spent months preparing was pitifully inadequate. That led to more jokes about the cook's hardtack, which still refused to be proper bread despite the fact that the cook was using Spicer-Simson's own recipe. But the joke soon grew old and was heard no longer amid the labor to maintain the expedition's forward momentum. The lead engine often had to stop while a section of the track was widened, or in some cases, entirely rebuilt. And even at its best, the road's surface was far from level or smooth. The traction engines were heavy enough that they did little more than lurch in their forward progress, but the boats on their trailers bucked and swayed as if in anticipation of Tanganyika's ferocious storms. Great care had to be taken to protect the rudders and propellers, which projected several feet beyond the sterns of the crafts and could easily be bent or snapped against an inconvenient hillock or rock.

The normal roughness of the terrain was exacerbated by the proclivities of one of nature's oddities—the antbear. This curious nocturnal mammal of uncertain relationship to the rest of the animal kingdom is about four feet long and two high and has a two-foot tail. Its piglike body is covered with a short coat of sandy-brown to black fur that is scanty enough to let the skin show through. It has the ears of a donkey, a long head, a longer snout, and a foot-long tongue it uses to lap up its major source of nourishment: termites. These it digs up with short, thick, and powerful legs armed with strong, dull claws, which it also uses ferociously to defend itself against lions and leopards, its principal predators. Crossword puzzle lovers and aficionados of the ABCs know it as the aardvark, which is Dutch for earth pig.

The antbear lives in the forest and plains from the Cape of Good Hope to the sub-Sahara region, but wherever it lives, it is known principally for four things. The first is its eating of termites. A termite Godzilla wreaking havoc on insect metropolises, it waddles into a termite mound, tears it up with its powerful legs and claws, and scoops up the scurrying bugs with its sticky tongue. It is prized, as well, for its tasty flesh, and third, its teeth are considered to be amulets that will ward off evil.

Last—and most noticeable to the expedition—it is known for the large burrows in which it lives during the day. These burrows literally pocket the countryside where food is plentiful, and food was plentiful between Fungurume and Sankisia. Almost anywhere the men looked, they could see termite mounds: concrete-like mud towers, really, some as tall as fifteen feet.

An antbear burrow is as much as three and a half feet in diameter, and usually they are buried quite deep, so a man—even a man bearing fifty pounds—could safely walk across one and be none the wiser. Sometimes, though, they are near the surface, acting like a pitfall trap. The traction engines might lurch into them and easily forge on, but the boat trailers were another matter entirely. All too frequently, one of the trailer wheels rolled over one, and the ground gave way, dropping the wheel into the hole. Pulling the wheels free wasn't a problem; the danger was that if a wheel dropped too deeply into one of these pits, the boat's propellers and rudder might be damaged.

Dudley came up with the solution when he learned that the bearers were used to hunting antbears for their flesh and could tap the ground with a stick and hear the hollowness that signaled a burrow. He drafted half a dozen tribesmen to walk in front of each of the locos, tapping the ground and marking the locations of burrows so that the locos could slalom around them. Though the trailers still occasionally dropped into a den, the boats suffered no harm.

Pests of a smaller sort were a constant bother, too, causing the men more discomfort than almost anything else. Just as every tree and bush had thorns, it seemed that every insect had a sting or bite. For each beautiful butterfly or iridescent dragonfly that were heaven to the eye, there were thousands of insects that were hell on the flesh. The expedition had yet to encounter tsetse flies, and mosquitoes were rare here, except near streams or waterholes, but there were plenty of bugs that bit—and sometimes were more than aggravating.

Take the bott fly. Similar in looks to the bluebottle, it has an iridescent blue body and a red head that should be a warning. This little devil bites a hole in its victim's skin, into which it lays its egg. As the egg develops, the bite grows into a boil, and in about ten days, the boil is full of a squirming larva. The boil can be opened and the larva removed with tweezers, but if it escapes back into the wound, it will die and fester, causing even more problems for the victim.

Added to the bott fly were grass ticks that crawled into socks and up trouser legs, sand flies that bit any exposed flesh they could find, scorpions that stung, centipedes that left swollen red trails across arms and legs, and that minor horror, the jigger flea. This nasty mite lurks in grass and dirt, hops aboard its victim, crawls into his socks, then burrows into the flesh around the toenails. Once there, it lays eggs by the millions in a little sack, causing excruciating itching. The sack could be removed by tweezers manipulated by a steady hand, but if it is broken open, abscesses and blood poisoning were frequently the result.

And even if some of the insects the men encountered didn't bite, they had some sort of unpleasant defense. Like the stink ant. When stepped on or disturbed—which occurred often as the column threaded its way across the countryside—stink ants produce an aroma redolent of rotting horse carcass.

Worst of all, though—worse than the insects, the slip-shod bridges, the bad road, and the near-impossible terrain—were the dryness and heat. The expedition had deliberately arrived in the middle of the dry season, otherwise they'd be attempting to drag the boats through thick mud instead of dust, and the dust was bad enough. But this year the drought had been particularly severe, and not only was the ground as dry as a bone, so were most of the waterholes. It seemed as if every drop of moisture had been parched out of the land and the bush. The sun scoured the stark and savage landscape with an unrelenting glare, and temperatures during midday were as high as 118 degrees in the shade. If the men could find any shade. Because they were cutting down trees for bridge-building and traction engine fuel—not to mention those already felled to make the road—what little shade that had existed before the arrival of the advance parties was gone by the time the main column passed.

And what a glorious column it was. Only a week out from Fungarume, the procession stretched for three miles, raising a long, thousand-foot-high pall of white dust that took hours to settle after the last footsteps had departed. This white dust not only covered everything it touched with a white film but made the dry heat and lack of water even more unbearable.

And that wasn't the half of it.

3

"Snow-blind!" Spicer-Simson bellowed. "Are you mad, man? We're in the middle of Africa, less than five hundred miles from the equator, it's the middle of the dry season, and it must be a hundred and twenty degrees!"

Nevertheless, sir, these men are snow-blind." Hanschell gestured to the six ratings and junior officers who lay in the shade of a canvas canopy, their eyes swathed in bandages.

"Malingerers is what they are," the commander snapped. "I need those men out there supervising the road crews and bearers."

"Give them two days in here, and I'll return them to you as fit as ever. But if you put them out on the road without adequate rest, we'll be sending them back to Cape Town within the week, truly blind," Hanschell warned.

"What the hell's causing the problem?" Spicer-Simson grated. "The nearest snow must be Kilimanjaro, and that's a thousand miles or more from here."

"Snow blindness is just a term," Hanschell said. "The condition can be caused by any excessively bright landscape. The area where these men have been working is practically barren, the soil is light colored, and the sunlight is fierce. And look at this." Hanschell reached down and scooped up a handful of dust from the ground and poured a little hill of it into the palm of his other hand. When he held it up to the light, the powder glittered with a thousand tiny sparkles. "See that? Flakes of mica. The dust in this region is full of it. The ground here is probably as reflective as any snowfield. Maybe more so."

"Damn it, Doctor, I don't need a lesson in natural history!" Spicer-Simson snarled. "I need these men working. The blacks are fine for labor, but I need these men for direction." His eyes suddenly narrowed. "Speaking of the blacks, why is it that none of them are snow-blind?"

"Possibly because they are used to it," Hanschell replied. "Or possibly because they *are* black. The dark pigment of the skin around their eyes would soak up excess light instead of reflecting it right into the eyes as the light skin of our men does."

"Wainwright!" Spicer-Simson yelled, though the lieutenant stood not ten paces distant. "Get out there and order our men to darken the areas around their eyes."

"Yes, Sir. But with what, sir?"

"Use the imagination God gave you as a human being!" Spicer-Simson said impatiently. "Tell them to use soot or charcoal. Anything. Bootblack, if they have to. Just tell them to do it. It is to become a regular part of the uniform. Any man caught without his eyes darkened will be reprimanded and his pay docked. Is that clear?"

"Sir!" Wainwright hurried off to spread the word that mascara was now an official part of the sailor's kit.

Spicer-Simson turned back to Hanschell.

"You said two days, Doctor. I expect these men back on the crew in that time."

"I'll see to it, Commander."

Hanschell started to turn back to his patients then hesitated.

"Something else, Doctor?"

"Yes, sir, there is. It's your skirt. Not exactly regulation."

"Your point?"

"I'm not sure it's quite safe," Hanschell said. "It gives your legs no protection from insects or the climate."

Spicer-Simson laughed.

"A little hot African air never hurt anything, Doctor." The commander looked down at his skirt with a quirky smile. "Besides, this climate was made for a skirt. You'll never see me getting crotch rot as will happen to a lot of the men. You'll have to watch out for that, yourself, Doctor, as I'm sure you know."

"Um, yes, I'm quite sure," Hanschell muttered.

"I do have an extra if you'd like to try it," Spicer-Simson offered with the aplomb of a true aristocrat. "Amy made them, you know."

"I didn't. But no thanks, sir. I think I'll stick to my trousers."

4

The African runner found Wainwright before he found Spicer-Simson, and the lieutenant led him to the commander, who was walking not far behind *Mimi*. Wainwright handed him the message the runner had brought, and they paused in a crescent of shade cast by a large boulder.

"That blasted idiot!" Spicer-Simson fumed when he finished reading.

"It's from Hope, sir?" Wainwright asked. "Trouble ahead?"

"If Hope is ahead of us, apparently so," Spicer-Simson growled. "He says here that the road is practically impossible. Listen to this drivel: 'Lee's deliberate purpose in recommending that route to the Admiralty had been to lead to expedition astray.' Baahh! No man sabotages his own work."

Wainwright pretended not to have heard the last.

"He doesn't mention Tyrer, sir?"

"Not a word. Where's Dudley? He's been over the route. I want to hear what he has to say."

Dudley was somewhere near the head of the column, nearly two miles away, but Wainwright promised to fetch him as quickly as possible. He didn't go himself but sent a rating, and he and the commander resumed walking behind *Mimi*, though the commander's mood was considerably darker than before he'd read the message, which he passed to Wainwright.

The report went on to say that the Lualaba, not very substantial this far up the Congo, was lower than it had been in a decade and was getting shallower by the day. Hope wrote he had to go some distance downstream to make contact with the captain of the steamer the Belgians were providing to carry the expedition to Kabalo. Kabalo was the town where the expedition was due to come out of the river and board another train for the remainder of the journey to Lukuga. His plan was to go all the way downriver to Kabalo, and from there, he would send a full report on the river's condition before taking the train to Lukuga.

Spicer-Simson was furious at this last. He'd already sent orders through Dudley for Hope to rejoin the expedition, which Hope had blatantly disregarded, and if Tyrer *had* caught up with him, apparently he was ignoring similar orders delivered by the sublieutenant. The commander ranted about Hope's insubordination until Dudley arrived. Hanschell, sensing something interesting, joined the little group in the shade of another boulder.

"We'll have our trouble, sir," Dudley responded to Spicer-Simson's query about the condition of the road. "No doubt about it. Much the same as we've had so far. And the mountains...." He shook his head. "Well, sir, it's a very steep track to the bottom of the escarpment. And the escarpment is almost vertical. But if we can make it over the mountains, the land beyond is not too different from this. Dryer, maybe."

"I didn't think that would be possible," commented Hanschell.

"That would agree with Hope's assessment of the Lualaba," Spicer-Simson mused, though his brow was still wrinkled with anger.

"But how can we trust anything the man says? He thinks we should return to Cape Town, ship the boats up the coast, and sail them up the Congo."

The commander snorted. "The fool's got no sense of geography, and he must be totally ignorant of Stanley's adventures. If that route was feasible, we would have gone that way in the first place."

"Well, sir," Wainwright began. "It doesn't really matter about Hope, does it. He's just the advance man...."

"Are you forgetting Lee?" the commander snapped. "That bastard nearly sank the expedition on dry land, and now Hope is following in his footsteps. Wouldn't surprise me at all if he's in his cups worse than Lee, and we're closer to German territory. Wainwright, get that runner over here and bring another. I want the first to take a message to Bukama to be sent on at once to Kabalo ordering Hope to report to me immediately. He's abandoned his post, and I don't want the bastard taking one step deeper into Africa. The second will take a telegram back to Fungurume. The Admiralty must know that Hope's a clear danger to the expedition as well as an affront to my authority as commanding officer. The gall, implying I have not carefully thought out the route and all possible alternates. At any rate, I don't need a bleeding land crab telling me how to navigate a river. I don't give a damn how low it is."

Having worked himself into a righteous rage, Spicer-Simson stormed off to pen his missives.

"I hope that there isn't any real trouble with Hope," Wainwright said as soon as the commander was out of earshot.

"Tyrer is with him" Hanschell said. "He's a straight sort. He'll keep him on track."

"Like Magee kept Lee on track?" Wainwright asked sardonically.

"I noticed that the other men don't like Magee very much," Dudley said. "What happened? He doesn't seem a bad sort."

"Before your time," Wainwright said. "Let's just say that he aided and abetted the commander in giving John Lee the boot, and now he's not trusted."

"You can't think that this is the same sort of thing," Hanschell protested, but his words sounded silly, even to himself.

"I can only say, Doctor," Wainwright shook his head, "that I hope it's not me who's the next to be sent out as forward liaison. There's danger enough in front, without having to worry about what's behind your back."

5

If Spicer-Simson really was worried that Hope would give away the expedition, he needn't have bothered. The Germans already knew about the it—had known, in fact, since May, a full month before the *Llanstephen Castle* had sailed from England. Luckily for the commander and his doughty crew, the Germans didn't know nearly enough—or perhaps they simply couldn't believe their ears—and they hadn't transmitted their knowledge to the African front. Captain Zimmer had to hear the news from one of his Ba-HoloHolo, who heard it from members of other tribes farther south, who'd heard it from yet other tribes.

"This many white men," Zimmer's informant said, flashing all ten fingers three times. "Many, many black men. Two trains without tracks. Two big canoes. All come this way."

"What kind of white men" the captain asked.

"White white men."

"What kind of uniforms?" Zimmer snapped impatiently.

"Uniforms, bwana?"

"Clothing." Zimmer plucked at his own sleeve. "What color?"

"Blue, stationmaster."

"Blue?" Zimmer couldn't think of any blue uniform on either side of the war. "You're certain they're blue."

"Like the sky."

"And two trains without tracks?"

"Yes, stationmaster."

It all sounded like nonsense to Zimmer, but he thought he'd better inform his commander.

"Big canoes?" von Lettow-Vorbeck asked when Zimmer brought him the news that an expedition of about thirty men in strange blue uniforms was on its way north from points unknown.

"That's what our scouts report."

"They can't be much," von Lettow-Vorbeck said contemptuously, studying a map. "There are no roads, and the Mitumba Mountains are in the way."

"He says they are accompanied by some sort of trains without tracks," Zimmer admitted.

"Could be land locomotives, but I never heard of any in that region." The colonel chuckled. "I don't see why they'd bother bringing canoes. You'd think they'd just have the natives make them some."

"These native informants are notoriously unreliable, sir. Do you think I should I investigate?"

"I don't think you need actively pursue it, Captain, but send me any reports on the matter that come in if it makes you feel better."

"Very good, sir."

"I've heard that the Belgians have installed several four-inch guns at Lukuga."

"The ones from Shinsakasa, sir," Zimmer affirmed. "We aren't sure what it portends, but we've had to give Lukuga wide berth since."

"It portends a pair of 105s for your *Graf von Götzen*," von Lettow-Vorbeck said.

The *Graf von Götzen* was the ship the British hadn't yet heard of because construction on it hadn't begun until after Lee had gone to England to approach the Admiralty. Steel-hulled and displacing twelve hundred tons, it was armed with the twin 88s from the *Moewe* and would be the deadliest warship ever seen on the lakes of Africa as soon as construction was finished. And now, with the promise of two 105s, Zimmer was elated.

"From the *Königsberg*, sir?" he asked, hiding his joy from his joyless superior.

The German cruiser *Königsberg*, commanded by Captain A. D. Looff, had guarded the mouth of the Rufiji River just south of Dar-es-Salaam, aiding blockade runners in reaching East Africa with ammunition and supplies. But about a month earlier, it finally had been cornered by the British and had been scuttled by its captain. The demise of the *Königsberg* might have ended the German threat to Allied shipping in the Indian Ocean, but that wasn't the end of the dangers posed by the cruiser's erstwhile ordnance. Von Lettow-Vorbeck planned to use most of it to bolster his land defenses, but he told Zimmer he would have a couple of the big guns mounted on Zimmer's new warship.

"Thank you, sir."

"Don't thank me, Captain. Just make sure you put them to good use."

"I will, sir," Zimmer promised, seeing visions of Lukuga being torn apart by the 105s. The new Belgian guns may have been larger, but they were stationary and thus easy targets, while his own would have the benefit of mobility. Besides, bigger didn't mean more accurate, and the Belgian gunners had proved themselves inept at every encounter.

"Dismissed."

Zimmer saluted and turned to leave, but von Lettow-Vorbeck's voice made him turn again.

"By the way, Zimmer," the colonel said. "If you do see these mysterious canoes, you have my permission to blast them out of the water as soon as possible."

"With pleasure, sir!"

6

Spicer-Simson had divided the overland journey into three fifty-mile sections, and at the end of each section, a depot was established. The depots were necessary because the majority of the petrol, ammunition, and food was labeled "Keep Cool"—not a simple task beneath the blazing sun. The depots were necessary also because the combined equipment and stores weighed one hundred and thirty tons and could not be transported en masse by a lorry marked, "Load not to exceed three tons." Fortunately, Dudley found an abandoned ox wagon in one of the villages they passed. Chief Waterhouse repaired the broken wheels and fitted it with a hitch that could be attached to the lorry's rear bumper. An additional ton could be loaded safely onto the cart.

The lorry and cart were severely overloaded at Fungurume—six tons of guns, ammunition, and stores in the lorry and another ton and a half in the wagon—and then run out to the first depot, where their loads were dumped. Then they went back to Fungurume for another load. Each trip—loading, transit time, unloading, and return—took a full day, and during the lorry's absence, the depot was guarded by three ratings and a handful of tribesmen under the authority of one of the expedition officers.

All the while, hundreds of tribesmen, each bearing a load of fifty or sixty pounds, wound along the road like a stream of ants from one depot to another, singing and chanting. One morning, about a week into the trek, Hanschell was walking along with Magee—the seemingly unneeded physician with the mistrusted journalist—when a small crowd of bearers trotted by, singing as usual.

"Listen, Doctor. Do you hear that?"

"The natives' song?"

"Don't you recognize it?"

The Africans always were chanting something or other while they worked, and these bearers were no different. Hanschell had never really paid much attention, but Magee's words made him listen more closely.

The general character of the song did sound familiar, but its jogging cadence seemed out of place and alien. Suddenly, Magee snorted out a quick laugh.

"And it's barely nine in the morning," he said

"That makes it funny?"

"Listen more carefully, Doctor." And Magee began to hum along with the bearers, who were now mostly past them, his harmony smoothing out the choppy rhythm.

Then Hanschell had it. The tribesmen were singing a hymn—"Now the Laborer's Task Is O'er," to be precise. And suddenly he saw the joke, and he laughed too.

"Thirty miles to go in the blazing sun with four stone on your head," he acknowledged. "More like the laborer's task is just begun."

"I doubt they know what the words mean," Magee said. "Probably learned them from some missionary near their village."

And, the laborers' task *had* just begun. Moving the entire load of expedition stores, weapons, and petrol to each new depot took more than a week, but it wasn't as if time mattered. The boats' daily progress was, at times, measured in yards instead of miles. And the slowness of the advance was not always due to problems with bridges or bad road. Every couple of days, the lorry had to be commandeered from shuttling supplies from the first depot to the second to send it ranging ahead in search of water for men and machines.

The men might have been thirsty, but the traction engines demanded that the parched countryside give up its last drop in the service of their boilers. And at last, the boats, too long out of their element, baked by the harsh African sun, and parched by the dry air, began demanding their fair share of water.

"Commander!" Wainwright shouted. It was the afternoon of the eighth day out of Fungurume, and he was trotting alongside the Fowler with *Mimi* in tow, nearly two miles ahead of *Toutou*, from which he'd just run.

Spicer-Simson turned and looked down from his perch on the seat next to Staff Sergeant Harrison. Spicer-Simson liked to ride on the traction engine. He deeply appreciated the huff and clatter and clank and sense of omnipotence that the shuddering machine sent up his legs and into his spine.

"What is it, Wainwright?"

Wainwright grabbed a stanchion, swung himself aboard the aft end of the loco, and climbed up to the cab.

"It's the boats," he gasped, wiping sweat from his face with a dirty bandana. "They need water."

"Hell, Wainwright, every damn thing in this country needs water."

"Seriously, sir. They've been out of water too long. Chief Waterhouse says *Toutou*'s hull is warping, and if it's not wet down soon, he says her seams'll open."

"Well, what are you waiting for? Get back there and fill her with water. Then bring some up here and fill *Mimi*, as well."

"What about the men?"

"What about them?"

"If we fill the boats with water, the men won't have enough to drink."

"Damn the men, Wainwright. If we reach Tanganyika, I won't need half the men I've got, but I'll need both the boats. Now get them filled."

"Sir!" Wainwright hopped off the rear of the loco and scuttled out of the way before *Mimi*'s trailer could mash him beneath its wheels, then he began trudging up the track. He was in no hurry. He would have to wait until the lorry passed by again on today's run out to the depot so that he could order Mullen to go all the way back to the Dukuluwe River to begin fetching enough water for the boats.

7

It was mid morning of the ninth day out of Fungurume, and everyone was dry as hell. What little water the traction engines' boilers did not greedily swallow went into the boats hulls, while expedition members and the tribesmen accompanying them were on a ration of a mere cup a day.

Wainwright wanted to send the lorry back to the Dukuluwe for another load of water, but Spicer-Simson refused. The lorry was needed to help move all the supplies and equipment that had been removed from the boats before they filled the hulls with water. Wainwright couldn't argue with the commander's logic, but even if he wanted to, his throat was too parched.

The lieutenant was, at the moment, riding on the Fowler, filling in Spicer-Simson on the conditions of the road ahead.

"We're coming up on a long, flat stretch," he said, "so we should make good time for a couple of miles."

As he said this, he glanced around the loco's boiler, and suddenly wondered if the thirst wasn't going to his head. He'd been seeing mirages ever since they'd started across this blasted wasteland, but they'd always vanished in the middle distance, evaporating like water off a hot skillet. But the one that lay across the road about a quarter of a mile ahead, glistening in the late morning sun, seemed to be getting larger.

"What the hell is *that*?" Wainwright pointed to the huge, wet-looking patch.

"Looks like another one of them bloody apparitions," Harrison commented. Spicer-Simson roused himself and peered ahead around the loco's boiler.

"Not apparition, Staff Sergeant," Spicer-Simson corrected. "Mirage."

Yes, Wainwright was certain of it. The mirage was growing, not evaporating. Even more astonishing, the tribesmen who preceded the loco, searching for antbear holes, had stopped at the mirage's rim, and it appeared that they were drinking and dousing water over their heads. Wainwright's tongue rasped over his dry lips.

"I'll see what its about, Commander," he said and hopped off the loco just as Harrison brought it to a hissing stop.

"What is this?" he asked in Swahili as he hurried up to the mirage.

"A wandering swamp, bwana," answered one of the natives.

Up close, Wainwright could see that it really was a swamp, roughly circular and about two hundred feet in diameter. "How did it get here?"

"It wandered here, stationmaster," the man said, and his companions laughed.

"You mean it just came here out of nowhere? How? There's been no rain for weeks."

"It did not come from nowhere," another of the men said, and pointed to the west. "It came from there."

And when Wainwright looked, he could see that the ground immediately on that side of the swamp was damp.

"I see, now," Wainwright said. "But I do not understand. How did it get here from there? I have never heard of a swamp that moved."

"Look," said a third man, and he gestured for Wainwright to approach the edge of the swamp. And edge is what it was. The swamp was actually a thick growth of water cabbage. Wainwright had seen the plant before, but only in lakes and ponds. It looks something like a water lily floating on the surface, and it sends out long, hair-like tendrils that grow up to six feet long. These tendrils do not root in the soil but draw their nutrients from the water itself. Here, the water cabbage had grown so dense and thick around the perimeter of the swamp that they formed a barrier that literally held the water in check like a huge rubber wading pool.

"It is heavy," the third man amplified. "It creeps across the ground."

Wainwright reached over the edge, dipped a dripping palmful, and swallowed it. The mouthful was more water than he'd had all day. Then he hurried back to Spicer-Simson.

"A wandering swamp?" the commander said when Wainwright had explained the nature of the phenomenon. "Why'd the bloody thing have to cross the road?"

"We can't go through it, Commander," said Harrison. "The loco's firebox is too low. It'll drag in the water and douse the fire."

"We can go around it easy enough," Wainwright said. "But think of the water, sir."

"Right you are, Lieutenant." Spicer-Simson waved toward the wandering swamp. "Have at it, and drain it to the last drop. In the meantime, have the men prepare a road around."

The wandering swamp delayed the expedition's progress several hours, but it was a delay that, for a change, was welcome. The water the swamp carried was more than enough to fill the traction engines, the boats hulls, and every spare container they could find. For the first time in nearly a week, everyone's thirst was thoroughly slaked, including that of the throngs of Africans who did the lion's share of the rough labor.

Amazingly, the African contingent had grown by another two hundred despite the heat, dryness, and exhausting drudgery. Even stranger, they seemed to be satisfied at the end of each back-breaking day of work to be paid with a pittance of water, a bowl or plate of food prepared by their own women, and a chance to see Spicer-Simson's tattoos and be near him. And the commander was only too glad to oblige them, doffing his shirt and dressing exclusively in his leather skirt. Hanschell tried again to warn him that such prolonged exposure to the unrelenting African sun could only do great harm to fair English skin, but the commander only laughed, held out his arms, and perused his own torso.

"Really, now, Doctor. With all my illustrations I'm practically as dark as the natives. And frankly, the sun feels marvelous. I love the heat."

Or perhaps it was the growing adulation he really adored, since it was obvious he did everything he could to foster it. One evening, before sundown, while the Africans observed him drawing on a map, Kasavubu approached and asked, through Dudley, what the commander was doing.

Spicer-Simson looked up from the chart. "Tell him I'm putting his village on the map."

The sublieutenant translated, waited for a reply, and said, "He wants to know what a map is."

"A map is a picture of the land," Spicer-Simson waved around, "and its features—mountains, rivers, and villages."

"Of what use is it?" asked Dudley after an exchange with the headman. "He says all men know where they are. They are right here."

"Tell him no, they are not all right here. Some are there," the commander pointed north, "and some are there." He pointed south. "A map tells us where they are."

"Of what use is this knowledge?" asked the headman. "Does it help us eat? Does it give us shelter? Does it appease the spirits? What does it matter that they are there and there?"

"Because they may not always be there," Spicer-Simson replied patiently. "We were there, and now we are here. We need your food, we seek your shelter, and we bring with us our own spirits."

Kasavubu pondered this then touched the edge of the map.

"What lies beyond your map?" The English word "map" sounded incongruous in the midst of the man's Bantu.

In answer, Spicer-Simson unfurled another map and spread it on the table.

"What lies beyond my map is this," he said. "This is the whole world." Taking a pen, he made a small circle in south-central Africa. "This is the extent of the map I am drawing. And this," pointing to England, "is our country."

"And your enemies? The Germans. Where is their land?"

The commander indicated Germany. The headman touched a finger to the Straits of Dover.

"With such a large river between your countries, still you fight?"

Spicer-Simson laid a hand on the Atlantic Ocean. "With the largest lake between us, we once fought this country." He pointed to America.

"Such powerful enmity," said Kasavubu, worry creasing his brow. "Who will win your fight?"

"We will."

"But the Germans are strong. My people say they control Tanganyika with large and powerful canoes with guns so big they can destroy whole villages though their walls are stone."

"But I am so powerful that I can destroy them with only these," Spicer-Simson waved toward *Mimi* and *Toutou*.

The headman glanced at the boats then back at the map Spicer-Simson had been drawing.

"Perhaps," he said, "you put our village on your map so that you may find us when you wish to fight us."

"No," Spicer-Simson said. "I put your village here to make you real for the rest of the world."

The headman turned to his people and spoke for a couple of minutes. "What's he saying?" Spicer-Simson asked.

Dudley stepped close. "He is telling them that you are helping create the world and that you have given his people a place in it."

"Tell him this, Dudley. Say that I am not simply creating the world, but I am helping create the future."

Chapter VII: The End of Bad Luck

"WHAT DID YOU SAY IS the name of this village up ahead?" Hanschell reflexively drew a sleeve across his forehead and was dismayed that it came away nearly dry despite the blasting heat. He simply hadn't had enough to drink to sweat. None of them had. It made him nervous. Tropical sickness was just one aspect of health that could go drastically wrong out here. Heat stroke and dehydration were just as likely. He steeled himself, promising not to be the first—and only—one to succumb.

"Mwenda Makosi," Wainwright told him. "Means 'the end of bad luck.'"

"If they have water," Hanschell said, "I'll believe it."

"Dudley says it's a good sized village, so they're bound to have water. But don't have too much faith in the name. The boat trailers have had it, and Cross doesn't think we'll make it that far."

But the expedition did manage to stagger into the village on August 28, ten days after the trek had begun. They'd traveled thirty miles from Fungurume.

Spicer-Simson was in Harrison's loco cab as the outskirts of the village came into view, and a small crowd of women, children, and old men rushed out to witness the approaching smoking metal monster that allowed men to ride its back. Suddenly from behind the engine came a grinding crash, and the loco gave a heaving shudder and fell to less than a quarter of its former speed. *Mimi*'s trailer had collapsed, and the loco was dragging it by its undercarriage. The thirty miles of harsh road, jostling boulders, rough bridges, and lurking antbear holes had proved too much for the trailer, even if it had been reinforced. The rear axel had snapped, placing undue strain on the front axel, which promptly gave out, too, followed almost instantly by the main support timbers.

Harrison brought the Fowler to a halt, and the men hurriedly got off and inspected the wreckage.

"The boat isn't damaged," Spicer-Simson said after a quick look. "Lucky for that."

"Luck!" Wainwright snorted in spite of himself. "Here we are in the middle of nowhere, and one of our trailers is gone. We haven't even reached the mountains. Worse, the rainy season is approaching. We've been lucky so far, but even a moderately heavy rain is going to sink us in mud."

"Don't worry so much, Lieutenant," the commander said in a tone that was as enigmatic as it was reassuring. "We'll get by."

"I'm not sure how, Commander."

"We will make do. Now run along to *Toutou*, and tell them we're having a spot of difficulty."

Wainwright ran along to *Toutou* and the Burrell, which were a mile and a half back. He found them just as the second boat's trailer fell apart.

"Goddamn," swore Bisho, sounding a lot like Harrison.

"Don't worry, old chap," Wainwright said. "Don't you know we've reached the end of bad luck?"

"I no believe in luck," Bisho groused.

"I didn't either, until I joined this outfit," Wainwright muttered. "Now I do, and it's all bad." He returned to Mwenda Makosi to inform Spicer-Simson of the latest events.

The expedition members were too hot and tired to raise a fuss. Most just sought shade beneath the walls of the village huts while Spicer-Simson, Wainwright, Cross, and Waterhouse surveyed the trailers. Hanschell, sensing that the expedition would be stuck here for several days at least, took a quick survey of the village, accompanied by Magee.

The place was a little larger than the occasional gaggle of huts they'd passed since leaving Fungurume. Most of the huts were rectangular in shape, instead of circular, and relatively well built, implying some sense of stability. In the center of the village stood a much larger round hut with open sides, which they understood served as a place for public meetings. At the moment, however, it was filled with children playing games in the shade and raising a cheerful racket.

There were many women, too, working singly and chanting or in various-sized groups, and a few old men lounged in the shade. But younger and middle-age men were conspicuously absent.

"Where are all the men, do you think?" Hanschell asked.

"Off building our road, I'd say," Magee answered.

"Quite right. What's that?" Hanschell pointed down a short street that led to the village perimeter. About thirty yards from the last hut was a walled, thatch-roofed hut nearly as large as the central meeting house.

As they approached it, the two men noted its dilapidated condition.

"Looks like a government rest house," Magee said. "They're built for travelers."

Wrinkling his nose, Hanschell inspected the structure's exterior, then he gingerly pushed at the flimsy thatch door with the toe of his boot. It swayed partially open then fell off its rotten hide hinges and clattered to the floor. The doctor stepped back as a heated, fetid dampness exhaled from the dark interior.

"I wouldn't care to stay in there," Magee said, waving a hand in front of his face.

Hanschell went in, anyway, while Magee waited outside. A rat scurried out of sight toward the rear, but the doctor ignored it as he scraped the earthen floor with his knife then inspected the rafters and thatching. When he emerged a few minutes later, he wore an expression of disgust.

"Bad news?" the journalist asked.

"The whole place is infested with just about every kind of pest imaginable," Hanschell answered. He showed Magee a cigarette tin and opened it to reveal half a dozen ticks and a centipede. "I saw rats and scorpions, too, but I hesitated to remain long enough to collect samples."

"I say, old man, you're not planning on burning it, are you?"

"Burning it?"

"Like you did that whores' hut in Elizabethville."

"I'd like to, but the commander wouldn't take it kindly. He warned me not to set fire to any more structures."

"Too bad. This one would make a dandy blaze."

"Is that a river over there?"

There *was* a river, about sixty yards away, through a fringe of trees. As they neared it, they could see a couple of young boys tending a herd of goats.

"Excellent," the doctor said.

"Thinking of goat stew?" Magee asked.

"Not at all, though you might mention it to the cook. Domestic animals are particularly susceptible to sleeping sickness. Observe how healthy those goats look. The area is probably relatively free of tsetse flies."

Though the "river" wasn't much more than a small stream trickling in a meandering spread over the rocky ground, it was steady and clear. Magee started to unship his canteen, but Hanschell stopped him.

"Not here," he warned. "Look at that."

A pile of feces trailed off a rock into the water, and a cursory inspection revealed more.

"Seems as if the villagers use this area as a latrine," Hanschell said. "Let's look farther upstream."

The feces abated after a hundred yards, but they went a little farther until they reached a spot where a curve had created a wide sand flat on the inner side of the bend. Trees neighboring the outer bend overhung a shallow pool.

"I'd say we've found a good spot for the camp," Hanschell said. "Let's go inform the commander."

They headed back downstream then cut over toward the government rest house. As they neared the structure, they saw Dudley walking around it.

"Hello," the sublieutenant said as the two approached. "I believe I've found our camping place."

"Here?"

Dudley couldn't help but catch the sharpness in Hanschell's voice.

"Why not?" he asked. "The government built it for travelers, and that's what we are."

"Have you been inside?"

"A little way. It's a bit rank, I know, but…."

"Diseased would be more accurate," Hanschell snorted.

"Come now, Doctor. It's nothing a little cleaning won't cure."

"Careful, Dudley," Magee warned. "You may not understand Doctor Hanschell's notion of cleaning."

"Anyway, it's infested with pests," Hanschell said. "You won't get rid of them with a simple airing."

"I still think we can make it habitable for the short time we're here," Dudley insisted. "Besides, the river is right over there."

"It's full of shit," Hanschell said.

"I suppose you've found a better spot?" Dudley asked, tone carrying a slight challenge.

"Right upstream. About half a mile from here."

"Half a mile? And we're that far or more from the trailers, which have broken down. And you want to drag them another mile? Besides, we'll have to make a road."

"Better than digging the graves you'll need if we stay here," Hanschell said.

"Gentlemen," Magee temporized. "We certainly can't make the decision among ourselves. I suggest we take the matter to the commander."

"I won't be overruled, Commander," Hanschell told Spicer-Simson when they'd found him and explained the disagreement.

"I'm surprised you even thought to contradict the doctor, Dudley," Spicer-Simson snapped. "I imagine that you would not want him to give you directives about how to wage combat."

"No, sir."

"Then why would you desire to have your way in a medical matter?"

"I guess I wouldn't, sir," Dudley said grudgingly.

"Of course not. Go over to Waterhouse and make yourself useful."

A hangdog expression lengthening his face, Dudley complied, and Spicer-Simson turned to Wainwright.

"We've dragged these contraptions," he indicated the boats, "for thirty miles. I suppose we can drag them one more. Accompany the doctor to the site he has selected, then form a crew to make the necessary track."

"Yes, sir."

As Hanschell led Wainwright to the camp site, he learned that the situation with the trailers was not good.

"They're simply shaken to pieces," the lieutenant said. "We'll be able to get them to your camp, but that's as far as they'll go."

"What are we going to do?" Hanschell felt a sudden sinking feeling. "We've come this far. We can't stop now."

"We've barely gone one-fifth of the way to Sankisia," Wainwright reminded him. "And we have yet to cross any really challenging terrain."

"But there must be something we can do."

"You once told me to trust the commander, and against my better judgment, I've begun to do just that. He's a bastard, to be sure, but something is driving him, and that something seems to be infecting us all, if you'll excuse the expression. Yes, there must be a way, and I'm beginning to believe that Spicer-Simson always will find it, whether it's because of planning, luck, or just sheer tenacity."

"So there *is* something?"

"The locos' wood tenders."

"Will they hold?"

"They're made to carry extremely heavy loads of wood. We'll have to replace some of the wheels and add the springs from the boat trailers, but there's a chance."

"How long?"

"Cross and Waterhouse think a week will do it."

The depression that had settled over Hanschell suddenly lifted, bringing a laugh to his lips.

"I'd like to say I can't believe it," the doctor said, "but I was down because we might be stalled."

"I know," Wainwright said. "I don't know why, but I feel as if we're in the grip of some kind of destiny, and it won't let go until it's done with us."

"Let's just hope it all ends well."

"We're led by Commander Spicer-Simson. How could it end otherwise?"

2

Even Dudley had to concede that Hanschell had chosen an excellent spot to camp. Here, upstream from the village, the water was clear and clean, and the trees, thick along the banks and scattered around the sandy flats, lent welcome shade that lasted until the evening's cool breeze. The locos and the boats were parked in a clearing on the bank, and the tents were arrayed out on the flats, where the sand provided a softer bed than the men had enjoyed since leaving Elizabethville.

By the time the camp was set up, Spicer-Simson had met with the local chief. The two exchanged gifts as well as compliments, the commander receiving a goat and the chief several bottles of whiskey. Once the treaty was in effect, villagers came into camp bringing fruit, vegetables, chickens, and goats to trade for whatever the Europeans might offer. The fresh food was a welcome break from the bully beef stew and hard tack that was their usual ration except when they managed to supplement it with a wild buck, pig, or guinea fowl.

Continuing to take his role seriously, Hanschell insisted on chaperoning the bearers assigned to bring water back to the camp from the diminished river. He took Rupia along to translate.

"Look," he told the twenty men who accompanied him, each carrying an empty water can. "Don't do this." He shook his head vigorously as he dunked his own can into the tepid stream and allowed it to fill. When it was full, he twisted the cap onto the mouth and made as if to carry it back to camp.

"No!" he said again, then he gestured for one of the bearers to come over. He had the fellow set his empty can on the bank and open the top. After opening the top to his own can, he dug a rag from his pocket and stretched it over the empty can's mouth. Then he motioned for the bearer to pour the full can into the empty one, straining the water in the process. When the transfer was done, he screwed the top onto the full can, straightened, and waved triumphantly at it.

"There," he said, nodding emphatically. "That's the way to do it."

Several of the men returned his nod, but most still looked puzzled, and they spoke among themselves.

"Do they understand?" he asked Rupia.

"I tell them again, Bwana Doctor," Rupia said, and while Hanschell pulled more rags from his pocket and handed them around, Rupia explained once more what the doctor wanted.

Waving for attention, Hanschell repeated the operation, pouring the water through the cloth into a second can.

At that, the men began filling the cans on their own, and when nineteen were full, Hanschell led the procession back to camp. There, he had the water poured into large cooking pots, where it was to be thoroughly boiled before he would allow it to be used for cooking, drinking, or making tea.

While the bearers were filling the cooking pots with the contents of their cans, Hanschell said to Rupia, "Take these men back to the river and repeat what I showed them. They're not to vary the process in any way."

"Yes, Bwana Doctor."

"It's very important," Hanschell insisted.

"I do it, Bwana Doctor," Rupia said.

He spoke to the men, who stared suspiciously at Hanschell then quickly averted their eyes. In a flash, they were on their way back to the river.

As Hanschell turned his attention to the boiling pot, he noticed Wainwright looking at him, a wry grin twisting his lips.

Recalling the Africans' suspicious glances, Hanschell asked Wainwright, "What did he say to them?"

"He said you were the boss."

"Come on, A. E. I know there was more to it than that."

"Well," the lieutenant conceded, "They thought it was a lot of work just to get some water, and they wanted to know why you wanted them to filter it like that. You know that all these natives are awed by our commander, what with the expedition and all, not to mention his tattoos and skirt. They think he's got a lot of juju."

"But what has that to do with what Rupia said?"

"He simply told them that you're the commander's personal witchdoctor, and your juju is nearly as powerful as his."

"For Christ's sake, he didn't!"

"That Rupia is a smart boy," Wainwright said. "He knew it was the easiest way to ensure their complete compliance with all your hygienic orders, no matter how outlandish they might seem." Wainwright laughed. "You can be assured that they'll do everything you want to the letter, even if they don't understand it, since they now fear *your* wrath."

3

The next morning, work began in earnest on the loco trailers, occupying most of the engineers and engine room artificers. But there was no lack of labor for the rest of the expedition. The relay of stores to the advance depots proceeded, and the men not engaged in this work were sent on ahead to help supervise the road-building crews, which were now on the far side of the Mitumba Mountains.

Initially, Hanschell was at a loss for something to do, so he decided to take a stroll around the village and its environs. He soon had explored the village and, before long, wandered into the bush, following a winding footpath. The footpath was met by other paths and trails, and sometimes it diverged, and before half an hour passed, he was thoroughly lost.

As soon as he realized his predicament, the hot forest around him grew oppressive, its daytime silence preternaturally menacing. Half in a panic, he hurried on, and about twenty minutes later, as he was crossing a dry streambed, he noticed a set of tracks traversing the sand in roughly the same direction. Heart sinking, he realized they were his own footprints and that he was going around in circles.

"I have to get control of myself," he thought, fighting an even greater panic beginning to balloon in his chest. "If I don't, I'll be lost out here. And when the sun goes down.... That's it! The sun."

Thinking back on which direction the sun was when he left the camp, he turned until his shadow was roughly pointing in the right way, allowing for the difference in angle that his absence of two hours would make. When he had what he thought was a good bearing, he set off, and within fifteen minutes, he spotted three village women carrying loads of firewood on their heads, ambling down a path and chatting among themselves. He followed them, and in another ten minutes, he was back in Mwenda Makosi, thanking his lucky stars that he'd kept his wits about him.

That evening, after mess, Spicer-Simson called Hanschell over to his tent for a drink.

"How did the work go today?" the doctor asked.

"Well enough," Spicer-Simson replied nonchalantly. "We may be here for a week, though.

"Will the new trailers hold up?"

"That's not what I'm worried about. We're losing precious time, and the rainy season isn't far off. Here in Central Africa, it can rain for weeks at a time, and I don't mean anything like English drizzle. The stuff'll come down practically in sheets. All this dust will become muck, every gully will be awash, and that little stream over there will be a raging flood. Any low-lying land will be a lake, our road will vanish as if never had been, and all the stacks of cordwood we've laid by for fuel along the route will be too soaked to burn."

"You make it sound pretty miserable, Commander."

"Oh, I don't know, old chap," Spicer-Simson laughed. "Perhaps if everything becomes as flooded as all that, we might just fire up the boats' engines and motor across this wretched countryside."

"That certainly might be quicker," Hanschell agreed. "And easier on the feet."

"What did you do all day?" Spicer-Simson asked. "Not much to occupy you around here, I suppose."

Somewhat reluctantly, Hanschell related his adventure of getting lost in the bush. "I finally found my way back," he finished, "but it was a frightening experience, I assure you."

"I'm not surprised you got lost," the commander said. "You're not a seaman, are you?"

Hanschell wasn't quite sure why a seaman wouldn't get lost in the African bush like any good English doctor, but his thoughts were interrupted.

"By the way, old chap, since you don't have much to do right now except misplace yourself, I thought I'd give you an additional duty."

"A new duty, Commander?"

"Yes. It shouldn't take much of your time. I'm appointing you the expedition's official naval censor."

"Censor? Whatever for?"

"To censor any reference to the expedition that might appear in the men's mail." The commander gave him a sharp look. "You haven't forgotten that this is a secret mission, have you?"

"I wasn't aware that it ever really was a secret," Hanschell said daringly. "My wife knew about it before I did."

"She did?" Spicer-Simson's eyes widened. "How could she?"

"From your wife."

"Why that brazen hussy! I'll have to discipline her when I return. But that's neither here nor there. We are at war, and the closer we approach enemy lines, the greater is our need for circumspection."

"We've got thirty-five white men with hundreds of natives in tow. Our progress can't be any kind of secret."

"Nonetheless," Spicer-Simson said in a sharper tone. "You will read all outgoing correspondence and censor appropriately."

"Fine, sir. As you wish."

"I do wish. Here." He handed over a small wooden box.

Inside was a rubber stamp and a pad of red ink. Hanschell glanced at the stamp; it read "Approved."

"Thank you, sir," Hanschell said somewhat grumpily. "I'll do my best."

He was considering rubber-stamping everything without looking at it, since he had absolutely no desire to read the men's correspondence. Maybe the ink on the pad would quickly dry up in all this arid heat.

"I know you will, Doctor. I have complete faith in you."

4

It seems that everyone had complete faith in Hanschell. Word quickly spread that the expedition carried a great white witchdoctor who wielded powerful medicine, and as soon as he emerged from his tent for morning mess, he was beset by dozens of sick and lame villagers seeking treatment for various ills. Some even came from villages as far as twenty miles. Luckily, Lennard, the medical supplier in Cape Town, had convinced him to lay in the barrels of purging pills, and Hanschell dispensed these individually to the majority of his new patients. And as the line of patients grew longer, he thanked Lennard even more for Rupia, without whom Hanschell would have been inundated.

By midmorning of the next day, he'd diagnosed three of his village patients with terminal diseases—none, thank God, with sleeping sickness. Mwenda Makosi seemed blissfully free of that problem, most likely because the open sandy flats across which the little river flowed prevented any substantial swampy buildup of water, keeping a low incidence of malignant tsetse flies.

No, his dying patients didn't have sleeping sickness, but cancer and heart disease kill just as readily, and it tore at his conscience to plaster a smile onto his face and hand over the purging pills to them, just as he did to everyone else. But what else could he do? Even in London's best hospital, he would have been helpless in these cases.

He explained his discomposure to Wainwright as the two took morning tea together, and the lieutenant patted him on the back and tsked sympathetically.

"Don't worry too much about it, Doctor. If you can't do anything, that's simply all there is to it."

"Yes, but I still feel guilty handing them those pills as if they'll do some good."

"Look at it this way: Those people are down by the river taking a healthy shit along with their friends and neighbors. In fact, since you started handing out those pills, there's been a steady stream to the stream. There's a regular crapping party going on down there, and everybody is feeling right as rain."

Hanschell had to chuckle.

"Makes me doubly glad I had us locate upstream."

"You and the rest of us. Just hope the wind keeps its southerly direction."

"Thanks, old man," Hanschell said, rising. "Better get back to my patients."

When he arrived at the medical tent, the line was longer than ever, and he recognized a lot of familiar faces.

"Damn," he thought. "Wish I'd brought some sugar pills, too. Don't want these folks to get too dependent on laxatives."

But he didn't have a chance to deal with the situation, for just after he established himself at the table in front of the tent, Locke came up, leading a panting teenager covered with sweat.

"What is it, Mr. Locke?"

"This boy says there is a sick white man in a nearby village. Says the man is acting crazy and yelling at nothing and frightening everyone. His chief heard that the commander had a powerful witchdoctor and sent this lad to fetch him. I suppose that must be you."

"The man is delirious?" Hanschell asked the boy.

"Him make loud voice," the youth said. "Him...him...." His English failed him, and he lapsed into quick words directed at Locke.

"Well," Locke said when the teen was done, "He doesn't use the word delirious, but it sounds like that must be the case."

"Tell these people," Hanschell said to Rupia, waving toward his queue of patients, "that I have an emergency. Tell them to come back tomorrow. Wait here with the boy. Give him some water if he wants it." He looked back at Locke. "Have you seen the commander?"

"He's with the boats."

Leaving the now-murmuring line of patients behind, Hanschell hurried off to the worksite, where the engineers were rebuilding the trailers. Spicer-Simson was sitting in his camp chair beneath a patch of shade, a glass of vermouth in his hand, watching the work progress. Hanschell explained the situation to him.

"You don't have to go," the commander said, taking a sip from his glass.

"I don't see how I can refuse. A sick white man...."

"It's up to you entirely," Spicer-Simson broke in. "If you want to go, you have my permission."

Hanschell hurried back to the medical tent. The line of patients had dispersed, and Locke was gone, though the teenager remained in front of the tent with Rupia.

"I'm going inside to get my bag," Hanschell said to his assistant. "Tell the boy that he's to lead me to the sick white man."

Hanschell went inside, packed his bag with a little of every sort of medicine he carried, and came out a few minutes later.

"Him understand," Rupia said. "It not far—just two, three miles."

Hanschell nodded to the teen. "Lead on."

"Yes, stationmaster."

They set off, the teen in front and Hanschell close behind. The well-trodden path they followed seemed to be a major thoroughfare, and though it was joined by numerous other trails, Hanschell had no difficulty distinguishing it from the others. And he paid close attention at critical junctures, even leaving blazes on some of the trees, not wishing to get lost again, especially this far from camp.

The path led in and out of patches of dense brush and across other areas that were nearly barren—terrain that had been monotonously the same since the expedition had left Fungurume. The only difference now was that it was just the two of them rather than a great, clanking, murmuring, three mile column that raised a quarter-mile-high cloud of dust.

Actually, except for his brief stint the other morning, this was the first time that Hanschell had been this alone in the deep bush, and as the quiet and solitude finally impinged on his senses, he felt a momentary pang of fear. But it quickly passed, leached out by the fierce heat and the beauty and peace of the very quiet and solitude that had engendered it.

Raising his eyes to the tree canopy, he saw a wealth of birds moving about in the branches. Some were colorful, some drab, but most flittered silently through the bright heat. Down on the ground, the forest surrounding the path was equally muted, but Hanschell knew it was just as secretly alive, its hidden energy fully waking only after the sun went down.

"How much farther?" he asked his guide after an hour's walk.
"Close, stationmaster."

They trudged on for another hour.
"Are we near, now?" Hanschell asked.
"Close, stationmaster."
"I thought you said it was only a couple of miles."
"Close, stationmaster."
"I certainly hope so."

A third hour of walking elicited the same response.
"How close?" Hanschell demanded irritably.
"Very close, stationmaster."

When they arrived nearly an hour later, Hanschell judged they'd come at least eight miles.

But they were finally there, wherever that might be. The path ended at a woven thorn boma set in the middle of a large bare space among the trees. The boma surrounded a government rest house much like the one in Mwenda Makosi, and from inside the house, a thin, hoarse voice rose in the simmering air.

"I can find my way back," Hanschell told his guide, hoping it was true. "You go home."

The teenager hurried off, obviously thankful to be done with his duty, and Hanschell pushed open the gate, walked across the narrow compound, and cautiously entered the rest house.

The sole occupant of the dim interior wasn't hard to find—he was standing near the back wall, shouting incoherently, his bearded face raised to the rafters. His hair hung in damp strings down his neck, and he was dressed in little more than greasy rags.

Hanschell approached the noisy apparition cautiously, though he doubted that the man's gaunt frame held much real strength. But one never knew—madmen could sometimes command prodigious energy. The man seemed not to notice until Hanschell was within two yards of him, then he suddenly turned wild eyes on the doctor. The eyes widened in surprise, and the man raised protective hands to his face. Hanschell was now close enough to see the yellow pallor of his skin and the blackness of his long fingernails, and he knew the man was definitely too weak to pose a threat.

"Come, good fellow," he said soothingly, grasping the man firmly but gently by the shoulders and steering him toward the meager, filthy pallet that was the shelter's only bed. "You've got tick fever. I'm a doctor. I'll help you."

A scourge caused by a spirillum that uses ground ticks as hosts, tick fever is more violent than malaria, though sooner over. But the fever attacks at roughly weekly intervals for up to seven weeks, leaving the debilitated victim barely enough time to recover before the next bout.

The man, obviously exhausted nearly to death, submitted to Hanschell's ministrations, which included a hypodermic filled with Ehrlich's 606—a strong arsenical preparation that kills the spirillum. Thank God the man was in the throes of the fever because that was the only time the vaccine worked. After Hanschell made the man more comfortable and cleaned him up as best he could with water from his canteen, the doctor started a small fire and warmed a tin of milk that he spooned between his patient's parched lips. The tin was barely empty before the man fell into a profound slumber.

Hanschell then took a small brand from the fire and began searching the clay walls of the rest house. In just a couple of minutes, he found and caught eight ticks, which he collected in an empty cigarette tin. After tucking the tin into his bag, completely dousing the fire, and making sure his patient was still resting comfortably, he left the rest house and started the long walk home.

He didn't arrive until well after mess, but the cook fixed him something to eat, and afterwards, he went straight to the commander's tent.

"I see you're back, Doctor," Spicer-Simson said jovially. "How about a drink?"

"This," he said, holding up the cigarette tin, "is why I insisted that we not stay in the government rest house." He poured the ticks into his palm, and the little fellows immediately began crawling onto his arm, looking for a nice place to affix themselves. "Ticks," Hanschell continued, wrinkling his nose and starting to pick them off his arm one by one and crush them between the nails of his thumb and forefinger. "Diseased ticks."

"So these little fellows are what's causing all the commotion," Spicer-Simson said with a smile that Hanschell, tired as he was, thought a bit snotty.

"They are. The man is in a government rest house about eight miles from here, and the whole structure is infested."

"I trust you left him better off than you found him."

"That remains to be seen. I'll go back tomorrow to check on him."

"Eight miles, you say? Each way?"

"At least. My feet are sore as hell."

"I'm not sure we can do without you for another entire day, Doctor. Your native patients were most impatient after you left. We require their assistance, and it seems that you are a lodestone for their cooperation."

"But what about...?"

"I wasn't going to suggest that you abandon your new patient, Doctor. I simply think that you can make the journey much more rapidly if you use Dudley's bicycle."

"An excellent idea, Commander. I can get there, treat him, and be back in a couple of hours at most."

"That's right. Now how about that drink? Let me tell you how we've progressed while you were gone."

5

The next morning, Hanschell tracked down Dudley to borrow his bicycle. He found the sublieutenant with the commander, red-faced and waving his arms in gestures characteristic of an excitable man foiled at every turn.

"It's Tait and Mollison, sir," Dudley was saying.

"Tait and Mollison?" Spicer-Simson asked. "Not giving you trouble, I hope. Be fairly difficult to take on those two at once, eh, Sublieutenant?"

"Not trouble, exactly, sir. It's just that I can't get them to *work*."

"Nonsense," Spicer-Simson said. "I've seen those two lads do more work at one sitting than any other ten men on this expedition—Lieutenant Wainwright, the doctor, and myself excluded, of course."

"Maybe, sir, but that one sitting takes them ten times as long as any other man."

"Each at his own pace, Dudley."

"I suppose so, sir, but it might help if they'd at least argue back. If Tait utters one more word, it'll be twice what Mollison's said since we left Fungurume."

"I suspect they communicate with one another, somehow," Spicer-Simson said. "Perhaps you just don't understand the language." Then he looked at Hanschell. "What do you think, Doctor?"

"Solid workers," Hanschell agreed reluctantly. Dudley still was irked that Hanschell had won the argument over where to camp, and he didn't want to further incense the high-strung young sublieutenant. "But I can see how their pace might be frustrating. They are rather ponderous."

"Indeed," Spicer-Simson chuckled. "And what can I do for you, Doctor?"

"Not you, sir. I came to ask Mr. Dudley for his bicycle. For my reconnaissance of the sick European in the next village."

"You needn't ask Dudley," the commander said. "The bicycle is now an official vehicle of the expedition, and you may requisition it any time you like."

Trying to ignore Dudley's glower, Hanschell thanked the commander, and in a few minutes, he was pedaling off on the path to the remote rest house. He quickly realized what a good idea the bicycle had been. The path wasn't very wide, but it was hardened from decades, if not centuries, of foot traffic—animal as well as human.

He had gone perhaps two miles when he caught a sudden movement with his peripheral vision. Turning his head to the right, he saw a large baboon loping along parallel to the path, pacing the bicycle.

Without pausing, the baboon turned and looked right at Hanschell. It snorted and snarled and bared its teeth.

Damn peculiar, the doctor thought, and a chill ran through him as he took in the animal's powerful body and fearsome features, especially the long canines that looked like they could easily bite right through an arm.

Hanschell had to look back at the path he was following to keep on course, and when he glanced over at the baboon again, there were two of them. Then three. Then four.

The snapping sound of a breaking branch came from his left, and he jerked his head in that direction, only to wish he hadn't. There were five, six...no, seven apes on that side!

He looked back to the right, and the original four had doubled in number. The total seemed to hold steady, then, at fifteen, but would it have mattered if there were more? Hanschell had heard terrible things about baboons—they'd attack in groups with almost human predatory violence and tactics, but there was nothing human about the way they'd tear you to shreds with those huge fangs.

And the way these fellows were snarling and snapping, he wouldn't have long before finding out firsthand. God, why hadn't he brought a gun?

He began pedaling faster, but it did no good. Within seconds, he was going at breakneck speed down the rut of the path, but the baboon pack stayed with him, sometimes a little closer, sometimes a little farther depending on the curve of the track and the adjacent terrain, but always with him.

Maybe it was his speed that enraged them, goading them to the chase —like some dogs back in England that just had to chase automobiles. Forcing himself to slow down was difficult, even though his breath was coming in thick pants and the muscles of his thighs burned with effort. But he did slow, though not too much—he couldn't bear the thought of the beasts surrounding him and grabbing at him with those all-too-human hands and sinking those all too bestial fangs....

It made no difference. The baboon pack simply slowed down, too.

God! How long were they going to keep this up? How long before they attacked?

Unable to help himself, he sped up again. If he was tiring, then so must they. And he was on a bicycle, and they were afoot. He could hear their heavy breathing as they snorted and grunted at him. He would go as fast as possible, and they'd soon give up the chase. He pumped the pedals even harder.

How much more, God? How much farther? Christ, they must have chased him for five miles already! The damn rest house had to be just up ahead, just around the next curve. Desperately, he plied the pedals with his weakening legs and cursed the sweat that burned his eyes and the fear that seared his brain.

And then his front wheel hit a rut, and he nearly pitched off the bike as it jounced and swerved toward the seven baboons to his left. The nearer apes snarled, and he thought it was the end. But the baboons veered away from him just as he managed to right the bicycle and dive it back onto the trail.

Panting and drenched with sweat and fear, he willed himself to go more slowly. It would do no good to take a tumble. Then they'd be at him for certain. He had to stay on the bicycle, remain on the path, and keep moving. If panic caused him to deviate one whit from any of these imperatives, he was doomed. Doggedly he focused on the trail and kept his feet moving to the rhythm of his pounding heart.

At last, at long last, when he'd begun to think he'd taken a wrong turn somewhere and that the infernal baboon-infested path would go on forever like some kind of hellish penance, the clearing with the rest house came into view. He plunged through the gate and skidded to a halt beside the door. Leaping from the bicycle, he spun, expecting the baboon pack to swarm in after him. But there was nothing—nothing but some dust raised by the bicycle's sudden halt and a few bird twitters laughing in the heating air.

The eight miles had breezed by in half an hour.

Trying to calm his gasping breath, he propped the bike against the mud wall next to the rest house's entrance. Panic wasn't the best thing for a bedside manner, and it wouldn't do for his patient to see him this way.

"Hello?" he called out a few minutes later when he felt better, though his legs were still rubbery from the exertion. "Hello in there. It's the doctor."

He went inside, and this time, he snapped on the hand torch he'd thought to bring along. Its yellow beam played across the rude pallet, and his patient's eyes glittered tiredly from beneath their brows. The man's hand stirred at his side, and he lifted it in limp greeting.

"You are the doctor?" he asked with an accent that didn't sound French.

"You look much better," Hanschell replied, stepping over to the cot. Thank God the light was too dim and the man too sick to notice how sick Hanschell himself felt. "How are you?"

"Weak," the man said. "But I am myself, at least."

"You were delirious," Hanschell told him, busying himself with the items in his bag and trying to steady his hands. He was still in no shape to give the man an injection.

"What is it? Am I going to die?"

"No, you'll be all right in a few days, though the weakness might last somewhat longer," Hanschell said. "It's tick fever. This place is infested with the little buggers. We must move you out of here as quickly as possible. Is there some place in the village you can stay?"

"It is a poor village," the man replied. "Anyway, I was just passing through. That's why I was staying here."

"I'll go see if someone can accommodate you, at least until you're back on your feet. What's your name?"

"Spyridon, Loxias Spyridon."

"Greek?"

"Yes. I'm a trader."

"Oh?" Hanschell said absently, propping up each of Spyridon's eyelids with a thumb and peering at the pallid skin beneath them. "What sorts of things do you trade?"

"My goods are in those baskets over there." Spyridon waved weakly toward three round baskets leaning against the mud wall. Hanschell hadn't noticed them in the dimness. "Perhaps you would care to look them over. I'd be happy to trade for my treatment."

"That won't be necessary."

"Please, I insist."

"All right." Hanschell went over to the baskets, thankful for a slightly longer respite before continuing the examination.

One by one, he picked through the three baskets, finding on top of each bundles of what at first looked like rags but that proved to be filthy clothes worn to tatters. Beneath the clothes in two of the baskets was an assortment of oddities such as tin mirrors, spoons, shot glasses, and razor blades. The third basket concealed about twenty cans of potted meat. Even in the dim light, Hanschell noticed that the cans each had a brownish encrustation of about half an inch in diameter on the top, and he picked up one of the cans to examine it more carefully. As he brought it close to his face, the odor of rot assailed his nostrils, and he wrinkled his nose and snorted. The brownish spot looked like a patch of dried clay.

"What's this?" he demanded, holding up the tin.

"Potted meat," Spyridon said.

"What's this clay?"

"Oh, that. I'm afraid the tins are old. They were swelling, so I had to puncture them to let the air out. I used mud to patch them up."

"And you sell this stuff to the natives?"

"Trade it," Spyridon said. "They don't have any money to buy things. I get a little of this and that for it."

This last sounded evasive, and Hanschell wondered what could be found here that would entice a trader to risk tick fever, malaria, and worse.

"Don't you worry that you'll poison some of them with your rotten meat?"

"It doesn't seem to have harmed them much," the Greek said. "Actually, I think they really want the tins and don't care about what's inside." He chuckled weakly. "Besides, I'm never in one spot long enough to worry about what happens after I leave."

By now, the doctor's nerves had settled beneath an air of disgusted anger. He filled the syringe with an inoculant and injected it into the man's arm. Then he went about preparing a quick meal from the supplies he'd brought, and after feeding Spyridon, he repacked his bag.

"I'm going to the village to see if someone there will be willing to put you up," he told the Greek. "I'll be back soon."

He left, so glad to be out of the rest house's dim, nasty interior that he'd completely forgotten about the baboons that had chased him.

His bicycle, propped against the wall next to the door, reminded him. Pushing the bicycle over to the gate in the boma around the rest house, he stared around, looking for flashes of yellow-brown fur in the surrounding foliage, but all he saw were a few fluttering birds. He got onto the bike and pedaled off down the path in the direction his guide had disappeared the day before.

The village lay less than half a mile away, and it was as poor as Spyridon had implied. After a short hunt, he found the teen who had guided him, and the boy took him to the headman. Hanschell promised to give the chief fifty purgative pills in exchange for room and board for Spyridon for three days. The guide would come to camp to carry back the pills, and the chief would send several men to the rest house to pick up the Greek.

With the arrangements made, Hanschell returned to the rest house to tell Spyridon the news.

"You should be fine in a couple of days," the doctor told him. "If you feel worse, send someone to fetch me."

"Thank you again, Doctor," Spyridon said, limply shaking Hanschell's hand.

"You're welcome. And in the future, try to pick your rest houses more carefully. Or sleep off the floor, at least. Ground ticks carry the fever, and they can't climb."

"I will, Doctor. Oh, by the way," he halted Hanschell as the doctor was turning to leave. "May I ask one more favor?"

"What is it?"

"Do you have any sticking plaster?"

"Sticking plaster? Do you have a cut?"

"No. It's for the meat tins. The mud plugs don't last, and I have to keep replacing them. Sticking plaster will probably last longer."

"I'm afraid I neglected to pack any," Hanschell lied. "Now I must be on my way."

Outside, he mounted his bicycle and, after scouting one more time for the baboons, began pedaling back toward camp. He kept up a rapid pace almost the whole way, completely fearful that he would again be beset. But nothing interfered with his journey until he was about two miles from camp, when a hyena dashed across the trail. For an instant, he thought it was a leopard and that he was finished. Panting from the exertion and sudden fright, he vowed never again to enter the bush unarmed. The first thing he was going to do after he returned to camp was to ask the commander to have Eastwood issue him a Webley.

"A revolver?" Spicer-Simson asked when Hanschell sought him out. "Whatever for?"

"Baboons, Commander. I was nearly attacked by a band of them. I counted fifteen. They followed me for miles. I thought they were going to tear me to pieces."

To the doctor's amazement, Spicer-Simson burst into laughter.

"Baboons? Attacking? Ridiculous," the commander snorted. "We saw hundreds of them up by the Gambia River. They were a regular pest, raiding the plantations and digging up the ground nuts. We had to beat the bush to get rid of them. Of course, from my river craft, I shot more of them than anybody else. I got tired of firing at them with a rifle—only one round at a time, you know—so I changed over to my double-barreled shotgun. I bagged so many with buckshot that I had a letter of commendation from the governor. I preserved the pelts of the finest specimens in Cooper's Sheep Dip—there's lots of it out in the Gambia, you know—and I had a fur coat with a little cap to match made of them for my wife. It was the envy of all the other women! None of them could get one like it."

"Nevertheless, Commander, I would like a revolver."

"Oh, very well. Tell Eastwood to give you one. But mind you don't accidentally shoot anybody with it."

"Thank you, Commander," Hanschell said. As soon as he was away from Spicer-Simson, he began muttering, "Shoot myself, you pompous ass. Perhaps I'll shoot you for that asinine story about giving Amy a baboon hat and coat. If you did, she must be completely embarrassed by it since I never saw such a thing in all the years I've known her."

Suddenly the absurdity of the commander's tale cut through Hanschell's disgruntlement, and he couldn't help but laugh. And the really funny thing was that the commander probably half-believed that business about a baboon-fur coat.

But even if Hanschell was now in a better mood, he stayed on course to Eastwood's tent. Absurd story or not, he wasn't going unarmed into the bush ever again.

6

"You're certain you want this?" Eastwood asked for the third time, looking quizzically at Hanschell. "It doesn't seem right, somehow, giving a gun to a doctor."

"I assure you I don't intend to participate in the fighting, if there is any," Hanschell told the quartermaster. "But I nearly had a run-in with some baboon, and I don't intend to feel quite that vulnerable again."

"Point taken," Eastwood said, handing over a heavy Webley, a holster and belt, and a box of ammunition. "Sign here."

As they were leaving the supply depot, which was located in a shady grove near the clearing where the work on the boats was taking place, they noticed a hubbub going on at the edge of camp. They headed toward the center of activity, and as they did, one of the ratings hurrying in the opposite direction nearly bumped into them.

"What's going on, Berry?" Hanschell asked, but the fellow barely slowed enough to glance at him.

"Leopard, Doctor. One of the hunting boys caught it in a trap and brought it back to camp. Mr. Magee's got that camera of his and says he'll take photos of us with it like we was big game hunters. Don't worry, sir. It's dead. I got to go tell Mullen and Tasker. 'Scuse me, sir."

"A leopard," Eastwood said excitedly as Berry hurried off. "We haven't seen one of those yet. I fancy I might just get my photo made, too."

Hanschell was feeling more phlegmatic—the chase with the baboons and the brush with the spotted hyena had nearly depleted him of the emotion of surprise, at least for the time being. Nonetheless, his curiosity, piqued by a chance to glimpse one of the jungle's most secretive predators, made him direct his footsteps in the wake of the quartermaster.

By the time they arrived, a sizable bunch of the expedition members was gathered around the hunter and the carcass, and all Hanschell could see over the bobbing heads was Magee. The journalist was shouting for quiet.

"All right, men, listen," he said when the noise had died enough for him to be heard. "I'll take your photo in any pose you like, but it's going to cost you."

"Cost us?" Vickars called out. "Ain't you supposed to be recording the expedition for posteriority?"

"Indeed, Mr. Vickars," Magee agreed with a smile. "But seeing as this isn't official business, I'm making it my business. Take it or leave it. You won't likely have another chance to have a picture of yourself and a genuine African leopard to show your families and friends back home."

"And just how much will you be chargin'?"

"Five pound for ratings, ten for officers."

"What?" This from Lieutenant Rickson. "Why are you charging officers more?"

"'Cause they can pay more," Lamont snorted. "Sir."

"Mr. Lamont has it exactly," Magee said. "But if you think about it, the price is well worth the product. Imagine the stories you'll be able to tell your grandchildren about the way you hunted the leopard to its very lair and brought the snarling beast down with a single shot."

"Careful, Mr. Magee, you'll be givin' the commander ideas," Flynn said, and everybody laughed.

"Come on," Magee coaxed. "Who'll be the first? How about you, Flynn?"

"Five pounds is a lot of money," the gunner said.

"Go on, mate," Lamont said. "What else are you spending your money on out here, anyway?"

"That's right," Magee said as Flynn gave in and stepped forward. "Here." He handed the rating a rifle. "Take a good pose. Now, everybody stand back and give me room."

The small crowd parted as Magee backed into it, camera in hand, giving Hanschell and Eastwood their first glimpse of the leopard.

"Glory be," Eastwood whispered, leaning toward the doctor. "Do you see what I see?"

"I see spots," Hanschell said, "But not on a leopard."

"Don't these men realize that's a hyena?"

"Apparently not," Hanschell replied, smothering a smile. "Still want your picture taken with it?"

"With a hyena? Filthy beast. Do you think we should say something?"

"What, and spoil the fun? There's precious little of that around here, and plenty of trials ahead. I think it's worth the price they're paying."

"Perhaps so, Doctor, but it doesn't make me think any more highly of Magee, flummoxing the men this way."

"Desperate men do desperate things. Magee's been shunned since Lee's ouster, and this is the most friendly attention he's had since. If he can exact a bit of innocent revenge for his ill treatment and make friends at the same time, I salute him. You have to admit, it's a nice joke."

"It is," Eastwood nodded, smiling himself.

"See you later, Tubby. I think I'll go back to my patients. Undoubtedly, they are in need of more purgative pills."

Hanschell hadn't been at this task for long before Spicer-Simson showed up, trailed by Wainwright.

"I see you got your popgun, Doctor." the commander said. "Feel safer?"

"Much," Hanschell said.

"Well, don't hurt anyone with it."

"I know how to handle a pistol, Commander."

"I should hope so," Spicer-Simson drawled. "After all, you'll have to do more than make noise with the damn thing. By the way, did you get your picture taken with Magee's leopard?"

"You mean his hyena?"

"Hyena? Good God! You mean I'm stuck with a bunch of European fools who can't tell a hyena from a leopard. I bet they can't shoot, either. What do you think, Wainwright? One sporting man to another?"

"You're probably right, Commander."

"Damn right I am. We're likely to see action before long, and I want us to do the shooting, not be shot at. I think it's high time we had a little target practice around here. Wainwright, arrange it for one hour from now. The sandy flats out by the river should be a capital spot."

The appointed hour came, and Spicer-Simson's prediction proved correct. Not one in ten could hit the target.

"It's easily seen that none of you were trained in sail," the commander sneered. "When I was a midshipman on the *Voltage*, I would stand on the quarterdeck with a rifle and shatter a bottle, six times out of six, that was swinging from the weather yardarm of a topsail yard. Wainwright, give them each twenty more rounds. And gentleman, please strive to hit something smaller than the broad side of Africa!"

7

"I suppose I'll just have to show the men how hunting is properly done." Spicer-Simson made this announcement at the end of morning mess. "I'll return by noon," he continued, "assuming I can carry all the game I shoot."

"Are you certain you don't want some of the men to go with you?" Wainwright asked.

"After yesterday's sorry exhibition of marksmanship, it is obvious that none of them have anything to teach *me*," the commander said haughtily. "Just you make sure that work continues on the trailers. I want to see appreciable progress when I return."

He left the mess tent, and a few minutes later, Wainwright saw him stalk off into the bush carrying a rifle and a shotgun. For the hunting trip, he'd donned his trousers and tunic, presumably to protect himself from the thorny underbrush. As soon as the commander was out of sight, the lieutenant turned to Waterhouse and the others toiling on the wagons.

"You heard what he said. Work lively, lads. We wouldn't want to disappoint him, would we."

None of them wanted that, and by late morning, they'd accomplished much of what had been planned for the entire day. The noon mess bugle called them from the clearing, and Wainwright, approaching the mess tent, saw Dudley striding in from the other direction.

"Is he back?" Wainwright asked.

"Not yet," the sublieutenant responded. "But he must have gone some distance—I haven't heard any shots."

"Just as well," Wainwright said. "We're getting more done with him gone than with him crowding over our shoulders."

After the meal, Wainwright and the mechanics returned to their job, and Dudley, somewhat at a loss without the commander shouting his name every ten minutes, paid a visit to the doctor's tent.

There, Hanschell was still giving out purging pills to the men and women lined up to see him, though occasionally he took one inside the tent for a more specific treatment. He almost never gave the pills to the children but tried to do the best he could for them short of depleting his precious supply of drugs. A basket beside the tent door was filled with fresh fruit and vegetables that the villagers had brought in payment for the doctor's services, and Dudley speculatively eyed the contents, thankful for the supplement to the diet of game, bully beef, and hardtack that they normally had to put up with.

Leaving the medical tent, he continued his rounds of the camp. All was well and quiet. He had to admit that Hanschell had picked a capital spot with plenty of shade and water and with the noisome village generally downwind. At last, he checked his watch and was amazed to see how time had rushed by. It was nearly three o'clock.

He returned to the center of the camp where the commander's tent was situated. There he found Spicer-Simson's servant, Tom, polishing the commander's sword.

"Is the commander back?"

"No, sir, Bwana Dudley," Tom said, pausing in his work and waving the rag he held toward the bush. "Him still hunting."

"Hang it all," Dudley said, and he hurried off to the clearing where Wainwright's crew was repairing the trailers.

"He's not back?" Wainwright glanced at his watch. "It's three. He said he'd be back by noon."

"Well, he's not here," Dudley affirmed. "You don't think a lion or leopard got him, do you?"

"Or baboons?" Wainwright quickly sketched Hanschell's experience of the previous day.

"Something's gone wrong."

"It'd be a hell of a note if we came this far only to have our commander eaten by wild beasts."

"Should I go find him?"

"I think you'd better. Take some of the Africans who know this area. Work your way around the camp in an outward spiral. Maybe you can pick up his trail."

Dudley organized the search party and headed out in search of Spicer-Simson, leaving the camp abuzz behind him. As it turned out, the search didn't take long.

"Say, Mr. Dudley," George Tasker, the signalman, whispered after they'd been circling the camp for about an hour. "Do you see what I see?"

He pointed through a gap in the bush. There, on top of a ten-foot termite mound with a shape reminiscent of the Matterhorn as El Greco might have painted it, sat the commander. He was facing away from the search party, his head nervously swinging back and forth as he scanned the bush in front of him.

"Commander!" Dudley called as the search party approached.

"Blast!" Spicer-Simson bellowed as he jumped at the sound of Dudley's voice, and he nearly fell off the mound. He did drop the shotgun he was holding and probably would have dropped the rifle if it hadn't been slung across his thick shoulders. Luckily the shotgun didn't go off when it fell, and Spicer-Simson scrambled down after it.

"I say, Commander. Are you all right?"

"Do I look all right, you idiot?" Spicer-Simson spat

Dudley had to admit he didn't, but he refrained from saying so. The blue uniform was dirty and torn, and scratches coagulated with dust hatched his face and arms.

"Why has nobody heard my shots?" Spicer-Simson demanded. "Doesn't anybody keep watch?"

"I was listening," Dudley tried to say, but the commander cut him off with a jerk of his hand.

"Shut up, Dudley, and lead me back to camp. I dare say, it must be miles away."

Again the sublieutenant refrained from speaking and simply led the way, which wasn't far—Dudley calculated that the termite mound was only half a mile from the edge of the camp. Spicer-Simson seemed nonplussed when they entered the camp only fifteen minutes after having been found, but he quickly recovered.

"Dismissed," he said. "And next time I put you to the test, I expect you to find me in much shorter order. Any fool could have accomplished the feat in no more than half an hour."

"Yes, sir," Dudley said. "I'll let Wainwright know you're back."

"He was lost?" Wainwright chuckled when Dudley told him the story.

"Not half a mile away," Dudley said. "He was up on a huge termite hill looking for us the wrong way."

"Makes you wonder if we're taking the boats in the right direction."

"Oh, I guess Lee steered us right. And when we get to the Lualaba, the river will be a path even the commander can't wander off of."

At dinner, Wainwright made the mistake of saying, "I'm glad that Dudley found you in good order, sir."

"Found me?" Spicer-Simson bristled. "Are you implying that I was lost?"

Oh, shit, Wainwright chided himself. Aloud, he said, "He did say you had climbed a termite hill. I only assumed it was to gain a better vantage of the terrain."

"Indeed," the commander said in a frosty tone. "For your information, I had run across a wonderful specimen of waterbuck, but the blighter managed to elude me in the bush. I simply ascended the termite hill to get a shot at him."

Hanschell, who'd overheard, recalled Mr. Hyde, and thought, no waterbuck in the bush.

8

The commander's disposition hadn't improved much by morning, and all during mess he complained about how long the repairs were taking.

"At this rate," he said, "we'll be washed out before we ever get to the Mitumba."

The concern was very real. Every day, the warmer weather brought the rains closer, and this arid landscape would instantly become a muddy hell washed through by torrential gullies and raging rivers or inundated by shallow lakes.

"I've never seen the dry season last so long," Wainwright said. "We should be all right for a few more weeks, but after that, we'd better not camp in places such as this."

"What of the trailers?" Spicer-Simson demanded. "I trust you made good progress."

"We did, sir," Wainwright answered hopefully. "We're almost ready to mount the wheels. Another good day, and we might even be able to transfer the boats."

"That's excellent news, Lieutenant," Spicer-Simson said brightening. "I'm ready to be out of this wretched place. Show me what you've accomplished."

An hour or so later, when the commander strolled up to Hanschell's tent, he seemed in good spirits.

"How are the trailers, Commander?" the doctor asked, taking a break from his purgative patients.

"Wainwright is correct, as usual," Spicer-Simson said. "It appears that we will test the mettle of our new trailers tomorrow."

"Do you think they'll hold up?" Hanschell asked somewhat hesitantly and was surprised to hear Spicer-Simson laugh.

"Of course they will, Doctor! They're splendid creations. After all, I engineered them, didn't I?"

"Yes, sir, I suppose you did," Hanschell replied, keeping his smile to himself. By now, he didn't suppose that the commander would ever change.

"What the hell is that?" the commander asked, suddenly bristling.

"What?"

"That dirty stiff walking about the camp over there." Spicer-Simson pointed with his cane.

Hanschell followed the gesture and saw a scraggly, gaunt scrap of a man entering the camp, followed by three Africans, each carrying a basket on his head.

"Why that's Spyridon, the Greek trader I've been treating for tick fever."

"You don't deserve much credit for keeping a thing like that alive," Spicer-Simson snorted contemptuously. "Good God! He's coming this way!"

Spyridon had spotted Hanschell and was waving and leading his little train of bearers toward the medical tent.

"You'll excuse me, Doctor, if I leave you to your patient. Please get rid of him as soon as possible—preferably before evening mess. I don't think I could bear eating with him anywhere in sight."

The commander hurried off in the opposite direction just as Spyridon reached the tent.

After an effusive greeting, the Greek trader again offered Hanschell some of his wares in recompense for his treatment, but the doctor refused. He took Spyridon into the tent for a thorough examination and saw that the man's condition had dramatically improved. Just to be certain, he gave Spyridon an inoculation against cholera.

"That should hold you. Tell me, how are your accommodations in the village?"

"The village? Oh, yes. Well, Doctor, I was going to stay there as you suggested, but I just couldn't see myself paying them for a place to sleep when the rest house was there for free."

"You mean you're still there? But the place is crawling with the same ticks that gave you this fever."

"But you've cured me. Yes?"

"For now," the doctor said. "But we'll be leaving in a couple of days, and eventually the inoculation will wear off. I suggest you go back there immediately and burn the rest house to the ground."

"Then where will I stay? And I'm afraid if I burn it, the government will eject me from the territory. Besides, I won't be there any longer than I must to sell off all my wares."

"Why don't you try selling them here?" Hanschell offered. "We've got some four hundred workers and their families. Surely you can sell it all at once."

"An excellent idea." Spyridon's eyes lit up. "I'll start immediately."

"Yes, but first let me take you to the mess tent. It wouldn't do to have you show up during regular mess, but I think I can persuade Cook to fix you a meal. After that, I'll show you where the native camp is located."

The cook suspiciously eyed the Greek but finally assented to make him a meal, which consisted of bully beef and hardtack. Spyridon chewed the meat appreciatively, but wrinkled his nose at the hardtack.

"What?" asked the cook. "Not good enough for you?"

"Good enough to build a wall with," the Greek replied. "But eat it?"

"I suppose you're an expert in bread?"

"As a matter of fact, I am."

The cook scratched his chin and gave a disbelieving look, but Hanschell perked up.

"Why don't you give him a try?" he asked the cook. "After all, even you won't eat this stuff the way you make it."

"I guess you're right, sir," the cook reluctantly agreed. He wrinkled his nose at Spyridon. "Show me what you know, then, but you'll have to wash up first."

"I'll leave you gentlemen to your baking," Hanschell said. "When you're done with him, point him toward the native camp." Then he leaned closer so that Spyridon couldn't hear. "But for God sake, don't let him see the government rest house."

Hanschell left the mess tent and hurried back to his own tent, strapped on his gun, then went to find the bicycle.

An hour later—with nary a sight of baboons to distract him—he'd reached the rest house. He quickly made up a fire brand, lit it, and rushed around the perimeter of the large hut, setting the eaves on fire. Almost instantaneously, the entire structure was ablaze.

"Take that, you little buggers!" Hanschell shouted with satisfaction as he backed away from the heat. The plume of smoke quickly brought couple of dozen villagers, who stood around laughing and pointing as the flames consumed the rest house. At last, all that was left was a charred ring of stubbled mud that had been the foundation of the wall, and Hanschell again mounted his bike and pedaled back to the camp.

Spyridon was nowhere in sight when he arrived, but something was happening. He propped the bicycle against a tree and joined the group gathering at the western edge of the camp. Magee was on the fringes when Hanschell walked up.

"What's going on?" the doctor asked.

"One of the boys brought in an ox for our dinner, and the commander's going to shoot it," Magee replied. "Waterhouse has gone for a rifle."

Craning his neck, Hanschell could just see the ox. It was fairly young, and a rope around its neck was attached to a nearby tree. It was plucking with its thick teeth at the skimpy grass around its feet, seemingly oblivious to the growing crowd gathering to watch its demise.

"Shoot it?" Hanschell asked. "Why doesn't he just let the slaughtering crew deal with it?"

"I suppose he's going to teach us how to hunt," the journalist said with a smile.

Just then, Waterhouse arrived with one of the Enfields, which he handed to Spicer-Simson.

"Back, now, boys," the commander warned. "These beasts can be dangerous if aroused."

The ring of men gained several feet in diameter and, fortunately, opened up entirely behind the ox. The commander chambered a round and, from a distance of about ten yards, took aim and fired.

The ox started at the sound, then went back to plucking at the grass.

No one dared comment aloud that Spicer-Simson had missed, but there was some whispering among the crowd of Africans.

"Blast!" the commander said, squinting along the barrel then jerking the rifle as if to shake it into alignment. "Wainwright! Why haven't these rifles been properly sighted?"

Hanschell saw the lieutenant across the ring of men, looking resigned.

"I'll see to it, sir," Wainwright said. "Would you like me to sight that one right now?"

"No need, I'll just close in on the beast."

And Spicer-Simson did, though the animal took no notice. At ten feet, the commander sank onto one knee, brought up the rifle, and pulled the trigger. The ox started at the close report, and its head jerked sideways because the bullet had glanced off its left horn before whining off into the dry, dusty air. The ox looked at the commander then dropped its head to pluck some more grass.

An angry scowl flushing his face, Spicer-Simson took three quick strides right up to the ox and shot it point blank in the head. The ox collapsed in an inert heap.

The scowl disappeared from the commander's face, and he smiled as he returned the rifle to Waterhouse.

"Just the same with buffalo," he announced. "You've got to face up to them. It's only when they lower their heads to charge that they expose the vital spot."

No one complained that evening at mess about the fresh meat, but a considerable amount of humor was passed around about the method of its procurement. At the officers' table, Dudley was in the midst of making a joke about the ox when Spicer-Simson arrived.

"What's this about the ox?" the commander demanded, brow darkening.

"Ox, sir?" Dudley stammered, but Wainwright saved him.

"Not ox, sir. Ox teams. Mr. Dudley was just saying how it's odd the ox teams haven't arrived yet."

"Quite right," Spicer-Simson said, taking his seat. "We should have seen them before now. But no matter. The trailers are ready, and I suspect we can get along without the oxen until we reach the Mitumba."

The cook and his ratings began bringing the plates, and he set the first one before the commander.

"Your game cooked up right nice, sir," he said.

"What's this?" Spicer-Simson asked, pointing.

"Why, that's bread, sir."

"Can it be?" Spicer-Simson asked, rolling his eyes to the heavens. He prodded the bread with a forefinger. It was soft and light. He broke off a morsel and nibbled at it with exaggerated timidity. "By God, it is bread. Cook has finally mastered the art of bread making!"

By now, the rest of the officers had their plates, and each had sampled the miraculous loaf.

"Well, now, sir. It waren't much. Dr. Hanschell brought me this...."

"Oh, nonsense, Cook," Hanschell interrupted. "What I did was nothing more than bring in the vegetables. You should take all the credit for the bread." He gave the cook a quick wink.

"Thank you, sir."

And for the next half hour, all else was forgotten as the expedition dug into the best meal they'd had in weeks.

9

Thursday, September 2, dawned bright and clear, with little hint of the massive cloud banks that would build up by late afternoon. The new trailers were ready, and the men began the task of transferring the boats from the skeletons of the broken-down Thornycroft trailers.

With Wainwright directing the operation, the new trailers were parked directly in front of the old and everything staked down until it was immobile. Then, the locos were brought to the head of the new trailers and cable and block-and-tackle run from the winding gears to each boat. The trailer beds were greased to smooth the transfer, and the locos were brought up to steam.

With a great huffing and chugging of the locos and much shouting among the men, the boats were dragged, inch by inch, off the old trailers and onto their new conveyances. The operation was accomplished by noon and capped off with a cheer not only from the expedition members but the tribesmen, whose ranks had swelled to more than six hundred—moths drawn to Spicer-Simson's flame.

"Excellent work, Wainwright," Spicer-Simson said when it was all finished. "Have the men start packing up immediately after noon mess. We leave at first light."

Early afternoon brought a temporary disruption to the preparations when Hanschell spotted smoke to the south. After observing the growing cloud, he muttered, "That's odd."

"What?" Wainwright looked at the doctor, who pointed at the advancing haze.

"That fire."

"Brush fires are common this time of year. Don't worry. It won't bother us. The wind's wrong. It's coming from the north."

"But if the wind's wrong," Hanschell persisted, "Then why is the smoke advancing?"

The answer was soon forthcoming as they saw Dudley hurrying toward them from the same direction as the smoke.

"The ox teams," Dudley explained, and in short order, the beasts arrived, trailed by a massive dust cloud.

"Sixty-four head," Wainwright pronounced after a quick count. About half the dove-colored animals were strolling along in a loosely-knit herd, and the rest were hitched in teams to three wagons. The head wrangler, a young Boer shaded by a hat with a wide brim, a band of giraffe hide, and flat crown, snapped his long whip over the heads of the lead team and shouted at them to "Gi-up!" Behind the teams, more drovers switched at the beasts' hocks to keep them grouped and moving forward.

Wainwright sent one of the ratings to find Spicer-Simson, and when the commander reached his subordinates, he immediately went over to the young Boer and told him to keep his bovine moving to camp. Then he joined his officers to watch the oxen and their drovers amble through the village.

"How do you like them?" he said grandly, hands on his hips, staring after the herd. "Magnificent beasts, aren't they?"

Apparently he didn't expect an answer, for he immediately strode off in the wake of the herd.

"Let's catch up," Wainwright said. "But mind your step." He pointed to the generous piles of dung that littered the oxen's wake.

Mess that night again was excellent, but instead of joking about oxen, the men talked with subdued excitement. The rest at Mwenda Makosi had been good, but everybody was ready to be on the move. The whole expedition, Hanschell noted, seemed to have been infected with Spicer-Simson's unbounded enthusiasm and optimism, even if it often was driven to sarcastic ruminations on the mental condition of the commander.

Hanschell was particularly glad to be leaving. He was exhausted with treating patients who did not understand what he was saying and who only wanted purging, even if their problem was something completely unrelated to the digestive tract. He also was thankful to be rid of Spyridon, who insisted on keeping company with the doctor as often as possible. Hanschell tried to alienate the Greek by telling him he'd burned the rest house, but Spyridon merely laughed and said he didn't plan on returning there, anyway, since Mwenda Makosi was a much wealthier place.

"I'm thinking of going to Fungurume. I understand your expedition has left a good road behind it. Maybe I'll even go to Elizabethville. I'm nearly out of trade goods and need to purchase more."

"I'm curious," Hanschell said despite himself. "Just what do these natives have that you find so valuable? They have no money."

"Money, no," Spyridon said, then he glanced around suspiciously. "You promise to keep what I tell you to yourself?"

"On my word."

"I trust you, Doctor. You cared for me and asked nothing in return, so I will show you."

The Greek hunched over and secretly drew a leather pouch suspended by a thong from beneath his shirt. Carefully undoing the lacing that held the bag closed, he dumped the contents into his hand. There were eight chunks of what looked like dirty green glass and an even larger number of metallic nuggets that gleamed with a dull golden sheen.

"Is that gold?" Hanschell asked, amazed.

"And emeralds."

"Where do they come from?"

"Here and there in the jungle and bush. The natives think they're pretty baubles but don't place much value on them. It is simple to trade what I offer for such as these."

"How long did it take you to collect these?"

"A few months." Spyridon glanced around again then leaned close to Hanschell and whispered, "I have a box in the bank in Elizabethville about so big." He indicated a cube ten or twelve inches square. "Nearly full. In six more months, I will return to Greece a wealthy man."

He smiled, and Hanschell couldn't help but smile with him, thinking of Spicer-Simson's description of the Greek as a "dirty stiff." Dirty now, perhaps, but no stiff and certainly no fool. He clapped the man on the back and wished him well and even relented enough to accept a packet of razor blades that Spyridon insisted he take.

Chapter VIII: Back on Track

THE EXPEDITION RUMBLED OUT OF Mwenda Makosi at first light. Signal whistles from the village broadcast news of their movement before they'd traveled a hundred yards. The one person who wasn't with the expedition when it pulled out was Hanschell, although he joined it soon after, pedaling up on the bicycle.

"Tending last-minute patients, Doctor?" Spicer-Simson asked.

"Yes, Commander. Something like that." Hanschell surreptitiously glanced behind them, hoping that Spicer-Simson wouldn't notice the plume of smoke rising from the clearing where Mwenda Makosi's government rest house stood. Used to stand, he amended, smiling to himself.

Going up and down and around hills, into dales, and over bridges and avoiding, when they could, areas that Hanschell said were infested with tsetse flies, the main body of the expedition moved steadily onward, though pulling the boat trailers through the thick dust was a chore, and the loco wheels still frequently lost traction. On a couple of occasions, the oxen had to be brought up to lend their backs in pulling the boat trailers and locos through especially thick patches.

Here and there, they encountered swaths where fire had burned across their path, and several times, they even saw smoke rising above the bush to the north. But none of the fires seemed close, and the only occurrence of note came at mid afternoon, when they were joined by an escort of a couple of dozen Belgian askaris under the command of a young Danish-born lieutenant named Freiesleben. With him was a big-boned, loutish Belgian sergeant named Marmion.

Wainwright was the first to see them as he was well ahead of the main convoy, supervising the construction of a bridge.

The warriors were barefoot, draped with bandoleers, and dressed in ragtag uniforms whose only real similarity was a red fez. They were armed with obsolete European rifles, but the warriors appeared to be quite formidable, and each also carried an assegai and a large knife in addition to the Belgian-supplied armaments. Those were the things that Wainwright noticed as they approached, but as they got closer, he also saw that the askaris' teeth were filed to points.

"Cannibals," he muttered.

"I am looking for Commander Spicer-Simson," said Freiesleben after introducing himself in English that was reasonably good.

"He's with the traction engines," Wainwright said. "I'll take you to him."

Leaving instructions with the bridge-building crew, Wainwright led Freiesleben and his askaris down-trail to where the lead traction engine was hauling *Mimi* across the burn from a recent fire. Spicer-Simson was riding on the Burrell with Harrison.

"That's him," Wainwright pointed when they were still a couple of hundred feet distant. "On the engine with the driver."

"The African?" Freiesleben was surprised.

"He's not African. He's English."

"I never heard of a dark-skinned Englishman. And he seems to be dressed in African attire."

"It's a little difficult to explain. You'll understand when you meet him."

Frankly, Wainwright wasn't sure that Freiesleben *would* understand, but there was little he could do to prepare him, so he left it at that. But he watched the lieutenant's expression as they neared the commander, and his watchfulness did not go unrewarded. The Dane's eyes began to widen, then his training, or perhaps his worldliness, gained the upper hand, and he controlled his reaction. The sergeant and the troopers with him had a more difficult time restraining themselves. Marmion gave a smile that was a stupid as it was derisive, and the askaris stared in wonder and pointed openly.

Wainwright felt odd. To tell the truth, he barely gave his commander's appearance, or behavior, a second thought anymore. The man was a queer bird, all right, and dictatorial to boot, but his enthusiasm and energy were infectious. Despite the hardships the expedition already had faced, the spirits of the men were high, and progress undeniably was being made. It was only when a stranger—someone unfamiliar with Spicer-Simson's oddities of dress and skin—happened on them that Wainwright felt the shock of the commander's bizarreness. Perhaps it was embarrassment when he had to admit to a stranger that this cocky, goateed, tattooed, be-skirted rooster of a man sporting his sword, his swagger stick, and his foot-long cigarette holder was his leader.

But no, he realized. It wasn't embarrassment. It was a defensive feeling, knowing that the newcomers would see only the surface and not the core of dogged confidence that lay beneath. So what if that confidence masked, as Wainwright often suspected, a terrible desperation? That only gave a dimension of utter reality to the commander's self-assurance.

Nor would they see—and Wainwright was more loath to admit this—the almost supernatural way in which the expedition was progressing as it inched its way painfully across the continent. Spicer-Simson had an almost uncanny certainty that everything would work out for the best, and so far, despite the difficulties, that had proved to be the case. Even more amazing, considering the horrendous conditions that long since should have depleted the men and their morale, the expedition was gaining, not only in ground but in confidence and physical resources. They'd left Fungurume with two hundred tribesmen, and that number had tripled. And every one of the men who joined did so not for pay or because he were ordered to but because he believed in the power of Spicer-Simson's mana. To the Africans, the commander was akin to a great chief, only more powerful. He was like a holy man, only more transcendent. In short, he was a man, but at the same time, he was something like a brother to the gods.

"I'm Lieutenant Freiesleben," the Dane said, snapping a salute as Spicer-Simson swung down off the loco. His men did likewise, though they seemed uncertain.

"Welcome to the Royal Naval African Expedition," Spicer-Simson replied. "At ease."

Freiesleben and his men relaxed.

"Where do you come from?"

"Lukuga, sir."

"And what brings you so far south?"

"Commandant Stinghlamber, the commandant at Lukuga, wanted me to check on your progress and provide any assistance you might need. Oh, and I've brought several communiqués." He signaled to one of his men, who passed him a packet of envelopes that he, in turn, handed to Spicer-Simson.

"Thank you, Lieutenant," Spicer-Simson said, taking the messages. "We will be making camp in a couple of hours about a mile ahead. You and your sergeant are welcome to join us, and your askaris can eat with our natives. Now, if you will excuse me, I must rejoin my driver."

Without a backward look, Spicer-Simson strolled toward the traction engine, which had advanced about two hundred feet during the conversation, and climbed aboard.

"I understand what you meant when you said it would be difficult to explain," Freiesleben said to Wainwright.

"Don't let his looks fool you. The commander means business."

"Yes." Freiesleben's eyes narrowed, and a muscle in his jaw jumped.

"I must report in," Wainwright said. "See you at mess?"

Freiesleben nodded and Wainwright hurried to join the commander and Harrison in the cab of the traction engine.

Spicer-Simson was leafing through the envelopes and papers Freiesleben had given him.

"What do you make of them, Commander?"

"Bloody Belgians," Spicer-Simson said in an exasperated tone. "I'd been warned about this."

"Warned about what, Commander?"

"The fools think we're here to annex Katanga," Spicer-Simson snorted. "As if anyone but a fool would fight over this godforsaken land."

"But isn't that what we're here for, Commander? To fight for this territory?"

Spicer-Simson gave him a condescending stare.

"We are a naval expedition, Lieutenant. We do not squabble over patches of dirt. We seek domination over the waves. Ahh!"

The commander tore into one of the envelopes, and Wainwright could see it was from Tyrer.

"That bastard!" Spicer-Simson snarled, reading.

"Tyrer, sir?"

"Not Tyrer. Hope. He's bloody well disobeyed me again. Tyrer caught up with him in Kabalo, but Hope insisted on going on to Lukuga. Shit!" The commander crumpled the paper and threw it overboard, where it lay in an incongruous little white ball in the thick dust. Moments later, an ox team trampled it out of existence.

The expedition made six miles that day, when, about an hour from sundown, it came to a large burned-out place in the bush.

"This looks like an excellent spot to bivouac," Spicer-Simson exclaimed. "Almost as if someone knew we were coming and cleared it for us."

That evening, Freiesleben joined the expedition officers' mess, and after the meal was over, he produced two bottles of wine. The commander ordered that camp chairs be set up, and the officers drank to one another's health while watching the bearers and laborers tiredly drift into camp and set about eating their own meals cooked by the women who accompanied them.

"It is very good of your commandant to supply you and your men," Spicer-Simson told the young officer, "though I'm not quite sure that you afford us any real protection. Most of the natives in these parts are in my employ."

Freiesleben's white teeth flashed as he laughed heartily.

"You think we're here to protect you from the Congo?" He shook his head of dark hair. "No, Commander, we are here to protect the Congo from you!"

"I don't think the Congo is in much danger from us," Hanschell put in quickly. "We're nearly all amateurs, you know, except for the commander."

"Exactly!" the lieutenant said, his eyes still twinkling. "That's precisely the point. You English have a genius for amateurism. That's what makes you so dangerous. It's almost always obvious what professionals are going to do, but who but amateurs could have dreamed up an expedition like this!"

"You appear to think better of our prospects than most of your colleagues," Spicer-Simson said blandly.

"I do," Freiesleben replied. "I think two dozen English amateurs with guns in their hands are capable of any folly—or of any heroism—and no government in its right senses would allow them to wander about unwatched. We'll be very relieved to get you out of the Katanga again, I can assure you. You British nearly took it over once, with fewer men than this!"

"You can't really believe that we have any designs against our own allies," Dudley said, bristling.

"Perhaps not," Freiesleben said. "I can only say that you English amateurs are unpredictable and have been quite good in the past at taking other people's colonies." He gave a bright grin and turned his eyes on Hanschell. "And when we heard that the good doctor was busy burning down government property, the vice governor general thought that it was time someone should investigate."

At this, Spicer-Simson guffawed good-naturedly and slapped Hanschell on the shoulder.

"I told you your pyrotechnics would get you into trouble." He looked at Freiesleben. "I assure you there will be no further conflagrations, either of government rest houses or of anti-colonial passions within the Katanga Province."

2

In the morning, Wainwright left at daybreak for the next major village, Mobile Kabantu, nearly 30 miles ahead, where he was to supervise the building of the largest bridge the expedition would have to cross. Even with him gone, however, travel went just as smoothly as the day before, though there were several more bridges to cross, which slowed them down slightly. By late afternoon, they'd gone about five miles, and the commander called a halt. The Fowler, with Harrison at the controls and *Toutou* on the trailer behind, halted in a large clearing, and the ox teams followed. The animals were as tired as the men, and the young Boer called to his wranglers to unyoke them and find water.

The Burrell and *Mimi* hadn't yet arrived when, half an hour later, Hanschell hurried over to Spicer-Simson, his face flushed.

"We can't stay here, Commander," Hanschell stated urgently.

"Why not, Doctor? There is good forage for the oxen and Ibn-el-Mani reports that we're close to a small stream that's still running."

"But look at the oxen."

Spicer-Simson complied, glancing at the animals before looking back at Hanschell.

"What of them?"

"Look again, sir. Please."

"Yes. They're switching their tails and moving about as if anxious for their forage."

"Did you see that?" The doctor pointed quickly.

One of the oxen had quickly doubled and nipped at its own flank. As Spicer-Simson watched, he saw more of the beasts doing the same.

"They're not restless for food, Commander. They're being bitten by tsetse. Look." He took a folded sheaf of paper from his pouch and began to unfolded it. It was a map.

"Where the devil did you get that?" Spicer-Simson demanded. "Let me see that." He practically snatched the map from Hanschell's hands and quickly perused it.

"The Belgian medical officer in Elizabethville gave it to me," the doctor said. "He marked areas where tsetse fly infestation is particularly bad. If you'll look right there," he pointed, "you can see that this area where we are...."

"I think I can do without your advice on how to read a map, thank you."

"This area is infested."

"Really, Doctor," Dudley put in. "Don't you think you're reacting a trifle strongly? Others obviously have camped here before us. We're safe enough."

At that instant, a tsetse fly landed on the center of Dudley's forehead. The commander spotted it just as Hanschell's fly whisk shot out and flicked it off.

"I say!" Dudley's eyes flashed. "That was uncalled for. You saw what he did, didn't you, Commander?"

"I saw the tsetse fly on your forehead," Spicer-Simson said. "You should thank the doctor for giving you preventive treatment."

A sudden yelp came from one of the ratings, who quickly slapped his arm.

The bit of a tsetse fly is not a pleasant thing, and the men's clothing provided little protection since the flies could poke their proboscises right through the cloth. Even worse, tsetse are tough little beasts notoriously difficult to smash and impossible to shake off. But they could be brushed away—hence the invaluable invention of the fly whisk.

"I suggest you order the men to deploy their fly whisks as vigorously as possible," Hanschell warned Spicer-Simson.

"Yes," the commander said, tugging his own elaborate, ivory-handled model from his belt and applying it absently about his head and shoulders as he scanned the map further, his eyes darting ahead along the proposed route. "It seems that we shall be safe enough as long as we avoid camping here for the night. Dudley, I know the men are tired, but we won't get much rest here, not with these flies biting us all night. We shall move on to here to set up camp." He touched a spot on the map about two miles farther along the trail.

"Are you sure, Commander?" Dudley asked.

"Doctor Hanschell is certain," Spicer-Simson said a trifle hotly, shoving the map back at Hanschell. "As I've told you before—and hope I don't have to repeat—Doctor Hanschell is the sole arbiter in medical matters. My job is to get this expedition to Lake Tanganyika, and his job is to get us there without illness. I won't let him tell me how to set up fortifications, but likewise, I won't diagnose his patients. Not that I'd want to," he chuckled, "what with all the beastly things than man can be infected with here. Now move along, Sublieutenant, and tell the men that the faster we proceed, the sooner we shall reach a safe place camp. And Dudley, mind the men use their whisks."

Dudley hurried off to inform the men of the changes in plan, and as he left, Hanschell folded the map in silence, knowing better than to thank the commander for defending him against Dudley's impatience with his caution. But he gave thanks silently.

The new camp was in the shallow valley carved by the Panda River. As with most of the waterways that still flowed through this sun-blasted landscape, it was reduced to little more than a muddy stream, although its bank was thickly lined with trees and dense, thorny underbrush.

"I hope there are no tsetse here, Doctor," Spicer-Simson said as the expedition ground to an exhausted halt. "If there are, I fear we must simply deal with them since we can't go on in darkness. I'll order the men to stay close to camp tonight."

"The flies are more prevalent during the daytime," Hanschell replied. "We should be all right as long as the men keep their whisks handy and don't sleep without their mosquito netting."

As it turned out, they seemed to have passed through the tsetse infestation, though Hanschell had to treat several of the men for bites. Untreated, the bites could fester into painful sores or abscesses.

"Hurts like hell, Doctor," Lamont told him, baring his arm for the inoculation. "Like being stuck with a red-hot needle."

"I'm afraid this won't feel much better," Hanschell warned, jabbing him with the hypodermic. "But at least you won't get sleeping sickness."

3

The next day, the expedition continued down the Panda River Valley, which ran roughly parallel to the Mitumba Mountains. The valley floor was relatively even, and the gullies and watercourses were shallow and broad, so they had to build few bridges since most of the steeper banks already had been dynamited into submission by the advance road crews. The mountains were still ten miles off, and from this distance, they didn't seem impressively high, but the escarpment looked ragged enough.

At last, they crossed the river and came up out of the valley onto a flat plain whose hard-packed reddish earth sported a dense hair of waist-high grass interspersed with skeletal bushes, everything crisped to death by the dryness and heat. Here and there, the plain was scattered with groves of thorn trees, but little else seemed alive beneath the scorching sun. The hard ground made for good travel, and the river was still close enough that a train of bearers could keep the Burrell and Fowler well supplied with water.

"At this rate, we'll be in Mobile Kabantu by evening," Spicer-Simson said proudly.

He and Hanschell were walking in the wake of the Burrell, which was pulling *Mimi*'s trailer, and about two hundred feet behind them was the Fowler and *Toutou*. The ox teams trailed by another two hundred feet. It was nearly time for noon mess, and the blazing sun had reached its peak, which meant that the heat's molten zenith was not far behind. But the temperatures were eased by a brisk breeze off the mountains to the west that, if it didn't exactly cool the air, at least kept it in motion.

Suddenly, Spicer-Simson tilted his face to the wind and sniffed.

"We must be near a village," he commented.

"How can you tell?" Hanschell asked, though he really wasn't much interested. "I can't imagine anyone living in this godforsaken place."

"I smell their cooking fires."

Almost immediately after, a figure leapt off the Burrell and ran toward them.

It was Dudley, who'd been keeping Bisho and the fireman company. "Commander!" he shouted as he neared. "Look!"

Spicer-Simson and Hanschell squinted in the direction he pointed, which happened to be upwind. A gray haze hovered over the brush there.

"The commander smells smoke from native cooking fires," Hanschell said.

"Cooking fires my arse!" Dudley yelled. "That's a brush fire, and its coming this way!"

The three men rushed to the traction engine and climbed aboard. From the slight elevation, they could see more clearly over the vegetation, and it took but a glance to realize that Dudley was right. It *was* a brush fire. The upper tongues of flame were clearly visible despite the distance, and the haze had thickened into a definite pall of smoke.

"It's movin' pretty fast, sir," Harrison said anxiously. "It was much farther when Mr. Dudley went to fetch you."

"Looks to be three or four miles," Dudley said, a touch of hysteria in his voice. "My God, but it must be moving twenty miles an hour!"

"That'll bring it on us in minutes," Hanschell said. "What can we do?"

"Takes water to fight a fire," Harrison said. "And we're fresh out."

"Find a clearing, Sergeant," Spicer-Simson ordered crisply. "And quickly." That was easier said than done. The grass and bushes around them grew thickly, brushing right up against *Mimi*'s hull, and if the dryness of the season had made the bushes relatively barren of foliage, it also made them, and the grass, as combustible as tinder. Harrison forged ahead as fast as the traction engine could pull *Mimi*, *Toutou*'s engine following. Meanwhile, Dudley clambered up onto the canopy that covered the engine and cab and cast about frantically for a clearing, or at least a wide place in the road where they could stop and pray.

But it seemed that luck was not with them. No clearing was to be had, and if there was, the growing screen of gray smoke driven before the fire by the wind was sweeping across their course, making visibility—and breathing—increasingly difficult.

"Faster!" Spicer-Simson bellowed before his voice choked off in the dense smoke.

The rest of the men were coughing, too, though they pressed handkerchiefs to their faces. In seconds, only Dudley from his vantage on the cab could see anything at all, and what he saw wasn't reassuring. The fire, burning everything in its path in great sheets of flame, was headed right toward them. He watched tongues of fire leap from brush to tree and treetop to treetop in an instant and knew that any wide place in the road they might happen on in these final few minutes would be no impediment to the conflagration.

"It's coming on, Commander!" he yelled. "Four hundred yards."

"Faster!" Spicer-Simson screamed at Harrison. "Faster!"

"I'm going as fast as I can, sir! The trailer's starting to shake apart!"

"Damn the trailers! Full speed ahead!"

"Three hundred yards," Dudley yelled. What he didn't yell was the prayer that passed silently across his lips. Even at this distance he could feel the heat from the blaze, and all he could see was flame as the smoke tore in great ribbons and billows across the landscape. The irony of it all, he thought. To roast to death in a boat riding a sea of flames.

"Halt!" the commander yelled, and if Harrison thought the order as insane as both Hanschell and Dudley did, he ignored his own judgment and brought the engine to an abrupt stop.

"Quick!" Spicer-Simson ordered. "Grab some faggots from the fire and set fire to the grass on the other side of the road!"

Following his own order, he grabbed a burning brand, leapt off the Burrell, and swept it through the tall grass to the east of the road. The grass gouted into flame, and only Dudley, who quickly followed the commander, made any difference with his own torch. In seconds, the wind carried the fresh blaze off to the east leaving burned ground behind shimmering with heat and dancing with still-live sparks.

"Harrison! Go!" Spicer-Simson waved to the freshly burned area. "Dudley, get *Toutou* and the oxen!"

By some miracle, the whole expedition was able to reach the fresh burn and crouch as the massive brush fire bore down on them. It was so huge that it created its own moaning wind that was laced through with sprites of flaming insects and screaming birds. Even at a hundred yards, the heat was practically unbearable.

"We're done for!" cried Dudley, who had climbed back on top of the Burrell's canopy. In the few seconds before the flames reached them, he wondered if he could make it to *Mimi* and submerge himself in the water in her hull and thus, by a miracle, survive. He began feeling his way along the canopy toward the rear of the loco, hoping that he wouldn't break a leg when he leapt blindly onto the boat's tarpaulin cover—or bounce off it right into the conflagration.

He found the end of the canopy and crouched there unsteadily, choking and gasping, his eyes burning and running with tears. In just a moment he would leap. In just a second....

In just that second, the breeze that had been blowing the smoke and heat in their direction shifted slightly, and he gratefully grabbed a lungful of clean air and managed to see the outlines of the boat below him through the watery haze of his eyes. He poised to leap.

And halted. The sudden change in the direction of the breeze blew more fresh air over him, and he rubbed the tears from his eyes and stared across the bush toward the brush fire. There it was, less than one hundred yards distant, its sheets of flame as vicious as before, but now, with the change in the breeze, it was moving across the path ahead of them rather than directly at them. Dudley stared in amazement as the fire whipped across the road and passed on, leaving a massive charred swath in its wake.

Thank God, he thought, and not simply that they'd been saved, but that he hadn't actually made it to *Mimi* and hidden in the water-filled hull. He'd have looked like a damned fool after the fire had passed and the others saw him emerge from the water looking like a drowned rat.

Laughing nervously, he swung down off the canopy and into the cab to tell the commander that they were safe, but when he landed in the cab, it was empty.

"Commander?" he called out. "Doctor Hanschell! Harrison!"

A splashing sound to the rear caught his attention, and he turned in surprise. A portion of the boat's tarpaulin cover had been peeled back, and through the gap emerged the three men he'd just called, accompanied by the fireman, dripping wet. Dudley stared, realizing that he could have been burned to death.

"The fire's passed," he said lamely.

"I can see that, Sublieutenant," Spicer-Simson said, then he laughed and turned to Hanschell. "Felt damn good, didn't it Doctor. First bath I've had in a week."

"Quite refreshing, Commander," Hanschell agreed. "Quite refreshing."

"Too bad you didn't join us, Dudley," Spicer-Simson said. "Ah, well, no time now. We must press onward. Give the order, will you?"

"Yes, sir."

Smudged nearly as black as the Africans, Dudley climbed down off the traction engine and headed off to the Fowler to tell Bisho it was time to move on.

4

They made a spectacular fourteen miles that day in spite of the fire—or perhaps because of it and the mad dash for safety that had carried them along faster than ever for the few minutes it lasted. They wound up in bush country once more, only five miles from Mobile Kabantu, where Wainwright, who had been overseeing the construction of the huge bridge just beyond the village, rejoined them.

At first, he went unnoticed in the argument that was taking place over the three government rest houses located where they'd stopped.

"They are some of the best in the region," bragged Lieutenant Freiesleben, and they *were* impressive—larger by far than any they'd seen, with new walls of mud thatch coated with whitewash. Even the thatch on the roofs looked fresh. They clustered in a largish clearing with a common ground between that held several benches and a cooking pit.

"Looks to be an ideal spot," Dudley declared.

"Ideal for tick fever," Hanschell said loudly. "Commander, we simply cannot stay in these places. They are breeding grounds for disease and infection."

"Lieutenant Freiesleben says they're new," Dudley pressed, pointedly ignoring Hanschell. "They should be safe enough."

"They are new," Freiesleben said. "Built just six months ago."

"Commander, I must insist…."

"That's enough, Doctor. I relented in the past to your better judgment because the evidence was clearly in your favor. But Lieutenant Freiesleben assures us that these rest houses are of recent construction and are safe. And considering how we've pressed on today and the little daylight that remains, I suggest we take him up on his offer to avail ourselves of ready accommodations."

"Very well, sir," Hanschell said, since it was obvious that he'd lost this particular battle. But damn it, didn't Spicer-Simson understand that this wasn't like a conflict between men where one side could lose a battle yet win the war. One night in the wrong place here, and the expedition might be nearly wiped out. Or at least incapacitated.

Leaving Spicer-Simson to hobnob with Freiesleben and ignoring the glance of triumph that Dudley tossed in his direction, Hanschell went over to the nearest of the rest houses, the stiff breeze off the mountains tugging at his hat and clothes. Inside, he found half a dozen ticks in less than two minutes, even in the dim light. Shaking his head, he returned to the doorway and saw that the commander was still engrossed in conversation with the Belgian lieutenant while Dudley stood attentively by. He also noticed that Wainwright and Cross had joined them. He slipped out of the door and went to the windward side of the first hut, dug in his pocket, and produced a packet of matches. Lighting one, he touched the flaring tip to the eaves of the rest house. The thatch may have looked fresh, but like everything else for hundreds of miles, it was extremely dry. In seconds, flames were licking toward the roof peak.

Hanschell hurried to the second hut and then the third, grinning fiendishly as he torched each in turn. Just as the third one caught fire, a cry came from the other side of the rest houses. It was Freiesleben.

"Merde! He's at it again! Amateur! Amateur! You want to burn up all the Congo?"

The alarm had come too late. The site had but one miserable well, and that would practically be drained by morning to fill the traction engine boilers, the oxen's bellies, and the men. The expedition members just stood helplessly, watching the structures burn.

"Goddamn it, Hanschell!" Spicer-Simson roared as the doctor hurried around the blaze and rejoined the officers. The commander's face was as red and hot as the fires.

"The ground is sterilized, now," Hanschell said, saluting. "You may camp here if you like, sir."

Spicer-Simson looked as if he were about to explode.

"Didn't I give you a direct order to leave these rest houses alone?"

"I was appointed medical officer in charge of keeping the expedition safe from disease and injury," Hanschell retorted. "I was simply doing my duty."

Spicer-Simson sputtered then abruptly turned his back and stalked off. Dudley hurried after him and, after a shooting a bemused glance at Hanschell, so did Wainwright, Cross following.

"Sorry about your rest houses," Hanschell said to Freiesleben.

"Don't concern yourself," the lieutenant replied with a sigh and a wave. "I confide in you that I agree with your diagnosis. The rest houses are dismal places and home to a great many pests. I, for one, would never stay in one, and I never let my men stay in them, either. But I do remind you that they are the property of the Belgian government."

"Perhaps the government should burn them, then, instead of leaving the job to an English amateur."

"Perhaps," Freiesleben laughed. "But why should we waste our efforts when we have you?"

As keyed up as he was, Hanschell had to laugh, too.

That evening, after mess, the officers were sitting around, chatting over drinks. Wainwright had finished telling them about the new bridge and then listened attentively as Dudley recounted the drama of the grass fire.

"What do you think is causing the fires?" Hanschell asked when the story was done.

"You're a fine one to ask that," Spicer-Simson snorted, then he said, "Set by the natives, I suppose. Most likely some of those damn BaHoloHolo who are working for the Germans."

"Kasavubu says they're being set by followers of a rival headman," Wainwright said. "At least that's what they're saying in Mobile Kabantu. It seems that a lot of the other headman's people have joined Kasavubu."

"Don't you mean they've joined Commander Spicer-Simson?" Dudley put in.

"Quite right," Wainwright laughed. "The commander is making quite a mark for himself among the natives."

"I should expect so," Spicer-Simson replied somewhat grandiosely.

They were joined, then, by Freiesleben, who refused a drink but invited them to view the prowess of his askaris.

"Very well," Spicer-Simson said. "It's the least we could do in repayment for Doctor Hanschell's rashness this afternoon."

In short order, the entire expedition was gathered around to watch Freiesleben put his men through their paces.

"An unimpressive looking lot," Waterhouse remarked quietly to Flynn as some of the ratings set up practice dummies.

"Bloody cannibals," Flynn agreed, and, apparently, others in the expedition did too, if the muttering and laughter that ran through the ranks was any indication. The rag-tag uniforms, the bare feet, and the ridiculous fezzes were decidedly unmilitary, as were the tribal scars. Then Freiesleben shouted orders, and the muttering and laughter stopped, leaving mouths agape.

Even the Beefeaters couldn't have been more sharply focused and perfectly in time, and they certainly couldn't have been as deadly with their pikes as these men were with their bayonets. Fearsome shrieks and bright blades slashed through the dusk air, leaving ragged tatters in their wake and newfound respect on the faces of the Englishmen.

"Thank God England's got a navy," Flynn breathed. "I'd hate to face men like that."

Waterhouse said nothing, but he noticed his mouth was still open, and he closed it.

"That was quite wonderful," Spicer-Simson commented when the askaris were again standing in ordered rank. It seemed as if they'd barely broken a sweat. The commander's mouth smiled, but his eyes didn't.

"All hand-picked men, recruited only from the cannibal tribes," Freiesleben said with his own tight smile and eyes that matched Spicer-Simson's. "They're fiercer. And better nourished."

Nervous laughter ran through the expedition, and despite their exhaustion, the men didn't sleep quite as deeply that night as they had in the past.

5

Luckily, they didn't have far to go the next day—Mobile Kabantu was only five miles distant. They'd barely made it there by mid afternoon, when the sky darkened as a towering black thunderhead, shot through with bolts of lightning, rolled over the Mitumba escarpment. In little more than a minute, the temperature dropped twenty degrees, and the clouds closer to the ground took on a dangerous greenish tinge as if bruised by the ragged gusts of wind that suddenly kicked up, whirling dust before their threatening breath. Thunder cracked and rattled the air.

"Oh, shit," Wainwright muttered, looking up and expecting at any moment to feel heavy drops spatter his face. He was thinking of his magnificent bridge and how all the work that had gone into it would be washed away in one overwhelming instant once the rains came.

"Better uncover the boats," Spicer-Simson said nonchalantly. "If it rains, we'll want to catch what water we can."

All Wainwright could do was stare at him. Didn't he realize the danger?

"Oh, don't bother, Lieutenant," the commander amended. "It's not going to rain. That would make a mess of your bridge and our plans." He grinned at Wainwright. "Let's see your bridge."

To Wainwright's amazement, the sky was completely clear by the time the commander had finished inspecting the bridge.

"Fine work," Spicer-Simson said as they walked back toward the waiting tractors. "But we best make camp here for the night before we attempt it. "Tom!" he called to his servant. "Set up my tent over there." He pointed with his swagger stick. "And make me a drink."

"Yes, bwana."

"Not much of a village," Spicer-Simson said as Tom hurried to obey.

"It means 'two people,'" Wainwright informed him.

"Seems that's about all that lives here," Spicer-Simson commented jauntily, looking around at the small cluster of huts. "And no wonder, there's not a tree in sight."

No wonder. The bridge that Wainwright had spent the previous week building stretched one hundred and eight feet long and thirty two feet high over the steep, nearly dry river bed, and five hundred tons of timber covering two square miles had been felled to build it.

Despite the bridge's length and height, Wainwright had few misgivings. In the three weeks and sixty-two miles since the expedition had left Fungurume, he and Rickson had supervised the construction of nearly one hundred bridges of various sizes in this same style, and he felt so much like an old hand at the game that he barely remembered the fiasco at the first bridge. And his latest and greatest creation proved equally worthy. The next morning, the Burrell, towing *Mimi*, crossed without mishap, though the whole bridge sagged alarmingly beneath the weight of the loco and boat. But Harrison kept a light hand on the throttle and a sharp eye on the tiller, and in two minutes, he was heading up the incline on the opposite bank made gentle by excavation and blasting.

What no one had taken into account was the worn condition of the towing cables. The rough terrain that had shaken the original trailers apart after only thirty miles had been no kinder to any of the rest of the towing gear, and so, near the crest of the far bank, the cable linking *Mimi* to the Burrell snapped. Suddenly relieved of its load, the traction engine surged to the crest while *Mimi* and its improvised trailer took a roller-coaster dive backwards toward Wainwright's engineering masterpiece.

"Good God! Stop her!" Spicer-Simson shouted, little thinking how anyone could stop twenty tons of boat and trailer rampaging backwards down a river bank. The men who had been following on foot behind the trailer ignored the commander's ill-conceived order and leapt out of the way.

Mimi's careening ride took it, miraculously, right back onto the bridge, though the trailer was by no means well-aligned when it jounced out onto the timbers. It bounded for thirty feet before its rear wheels slipped off the bridge and the modified trailer's undercarriage hit the timbers with a grinding, splintering crash. The whole mess slid sideways along the bridge for another fifteen feet, plowing up logs, before it came to a shuddering halt, its rear wheels dangling over the shallow gorge.

The air reeked with apprehensive silence that was broken by the commander's voice.

"Not as bad as at the first bridge, eh, Wainwright?"

A titter sounded from somewhere in the middle of the expedition's assembled men, and within seconds, they all were laughing aloud. Someone translated for the Africans, and they set the air roaring with mirth.

Wainwright felt a hot flush rush up his cheeks, then he, too, burst into laughter.

"No, sir. Not as bad." Then he gestured to several of the men. "All right!" he shouted. "Let's get a new cable on that trailer!"

While it was being done, Spicer-Simson took Wainwright aside.

"We're getting tired," the commander said. "And that makes men careless. We must be diligent."

"Yes, sir. Perhaps a short rest will restore us."

"We cannot afford to rest. Not now. I don't want inertia to set in or make the men lazy and complacent. That won't get us to Tanganyika. Only work will do that."

"Yes, sir," Wainwright replied. "We'll pay extra attention from now on."

"Excellent, Lieutenant. It looks as if the men are waiting for your command."

Wainwright returned to the task at hand, and it was accomplished quickly enough. Soon, *Mimi* was safely across, followed by *Toutou*. But if the men had enjoyed a good laugh and felt a sense of accomplishment, it didn't last long. Immediately ahead loomed their greatest challenge yet —the rugged mass of the Mitumba Mountains.

Chapter IX: The Mitumba Mountains

"YOU MUST BE JOKING, SIR," Wainwright said to Spicer-Simson as they looked at the Mitumba plateau spread before them like a gigantic crumbled birthday cake.

"I never joke," Spicer-Simson replied tartly.

"No, sir. But isn't there some way around them?"

"This *is* the way around, Wainwright. Just get us over, there's a good chap."

While Wainwright lifted his field glasses and scanned the slope, Harrison said, "They look mighty high to me, sir."

"Tosh," retorted Spicer-Simson. "They're only six thousand feet. Why, I could walk all the way to the upper escarpment in a brisk morning."

"Perhaps you could, sir," Harrison said dubiously, "but they's a fair pitch for the locos."

"A sheer wall, if you ask me, sir," Dudley muttered.

"Well, I didn't ask you, Sublieutenant, did I?"

"No, sir."

"Look here, Staff Sergeant," Spicer-Simson said as he turned back to Harrison.

"We've come across desert, gorges, rivers, God knows how many damned gullies, brush fires, and wandering swamps. Good God, man! Wandering swamps! I'm not going to be stopped dead by an over-grown pile of dirt. Now get up a head of steam and take my boats up that slope!"

"Yes, sir," Harrison responded glumly.

The commander gestured to Wainwright. "Follow me, Lieutenant. We'll probably need the natives and the oxen." He stalked off toward the Africans massed near the base of the mountains, Wainwright in his wake.

"Yes, sir," Harrison repeated when Spicer-Simson was out of earshot. "We'll get up a bloody head of steam and we'll start up that bloody slope, but I can tell you one bloody thing for bloody damn sure." He glared at the escarpment high above them at the top of the steep, rugged, tree and boulder-strewn mountainside. "We won't be draggin' your bloody boats over *this* bloody pile of dirt."

Bisho and both firemen followed his gaze with skeptical eyes and nodded in agreement, but Dudley just laughed.

"Perhaps not, Staff Sergeant," he said, "but I think you'd better give it your best try."

Harrison glowered at him then nodded.

"Right you are, Mr. Dudley. It's not for us to question the commander's orders, is it?" He turned to his crews. "All right, men. You heard the commander. We're going up that mountain." He pointed with a dramatic gesture, as if drama might reduce the grade or the escarpment's height. "Let's get up a head of steam."

"Make that a really big head of steam, Mr. Harrison," Dudley said with a chuckle.

"You heard him, men. Fire 'em 'til the rivets is ready to pop!"

The crews dispersed to their engines, and the firemen tossed their best wood into the fireboxes until the boilers nearly swelled with pressure. Then Harrison blew a single short blast on the Burrell's whistle and opened the throttle.

The engine, *Mimi* jouncing in its wake, charged toward the slope. It hit the incline at a good clip and managed to drag *Mimi* about twenty yards upward, spitting dirt and gravel from beneath its spinning wheels. But it could go no farther, and it grunted to a halt.

"Bring up the other engine!" yelled Spicer-Simson when it became evident that Harrison's machine alone was no match for the Mitumba Mountains. In short order, the second engine was coupled to the first, and on the commander's signal, both drivers poured on the steam. More dirt and gravel spewed, and the racket was thunderous, but the headway could be measured in just three yards before both engines wheezed and stalled.

"Bring up the oxen!" yelled Spicer-Simson, and the entire herd was urged forward and hitched in front of the two engines. At the commander's signal, the engineers poured on the steam and the drovers whipped their beasts, but despite the thunder in the air, the panicked lowing of the animals, and the excavation of more mountainous material from beneath the locos' wheels, the train budged not an inch. The fault lay, apparently, with the oxen. They had proved wonderful draft beasts across the rocks, sand, and dust of the deserts, but when it came to pulling uphill, they balked. The ox drovers whipped and yelled and kicked at the beasts' hocks, but they would not strain.

In the cab of the Burrell, Harrison could be seen grinding his teeth in red-faced fury at the recalcitrant animals. He bellowed profanities that were mercifully swallowed by the din of the two locos until he realized that the effect of his anger was being lost. Loathe to waste the momentum of his spirit, to say nothing of the momentum of the train of machine and beasts that was trying to drag an impossible weight up an impossible incline, he grabbed the pull-chain of the steam whistle, and jerked it hard. The shrill blast cut through the bedlam of thundering machines and shouting men, right into the minuscule brains of the oxen. The beasts may not have wanted to pull uphill, but the whistle startled them into reflexive action. In fear, trying to escape from the sound, the oxen lurched forward, and *Mimi* was dragged another ten feet upward before the animals relaxed in their harnesses.

"Excellent, Staff Sergeant!" Spicer-Simson yelled. Then he waved at Wainwright. "Bring up the natives!"

With much to-do and shouting, Wainwright got the six hundred Africans manning ropes and cables, ready to add their strength to the effort. Spicer-Simson signaled the traction engines, and the engineers poured on the steam. He signaled the Africans, who leaned into their ropes and cables, muscles straining. Then he signaled again to the engineers, who tugged their whistle chains, sending the oxen lurching forward in fear. Heaving and swaying, *Mimi* moved upward another fifteen feet before the beasts quieted and the Africans moaned with effort.

"Again!" Spicer-Simson yelled, and again the Africans strained, again the engineers opened their throttles and shrilled their fury at the beasts, again the oxen surged forward in panic, and again *Mimi* staggered upward a few feet.

"Again!" yelled Spicer-Simson.
"Again!"
"Again!"

But at last, there were no more agains. A hundred yards had been conquered, but the oxen had grown used to the whistles and now balked completely, as if deaf to anything but the call to rest. And they weren't the only ones. The gradient was now steeper than ever, and the Africans and even the engines gave out entirely.

As Wainwright looked over the Burrell, Spicer-Simson approached.

"Well, Lieutenant," the commander said, removing his cap and squinting at the loco and boat on its trailer. "That was a miserable effort, wouldn't you say?"

"Yes, sir."

Spicer-Simson scowled up at the steepening slope and, at the top, the almost sheer escarpment.

"And this is the lowest bloody spot on this end of the range?"

"I don't know, sir. You have all the maps."

"Don't get smart with me, Wainwright."

"No, sir."

"At this rate, we'll make it to the top by the end of the war." Spicer-Simson turned the scowl on Wainwright, who blinked back.

"If we make it at all," Harrison whispered to Bisho, but apparently Spicer-Simson failed to hear him because he continued speaking to Wainwright instead of chastising the loco driver.

"I expect better progress than this, Lieutenant. Tomorrow, we'd better get cracking."

With that, he stalked off toward his tent.

The locos and *Mimi* were belayed to trees, the wheels of the engines and trailers were chocked to prevent any loss of hard-gained ground, and the ascent was abandoned for the day. Tiredly, everyone returned to the base camp for evening mess.

On the way, Freiesleben and Marmion caught up with Wainwright.

"You made little progress today," the Dane said.

"No," was all Wainwright could answer—all he was capable of answering.

"It seems your oxen aren't adapted to pulling uphill," Freiesleben persisted, the sergeant grinning sadistically beside him. "Perhaps they will do a better job pulling downhill on the other side of the mountains."

"If they get to the other side," Marmion said with a derisive chuckle.

"Excuse me," Wainwright said. "I'm tired and need to get some sleep."

"Yes," Freiesleben said. "They may need your muscles tomorrow as well." With that, he and Marmion headed off toward their own camp.

Later, at dusk, Wainwright, carrying a flask of whiskey, went back up to make certain that the lines holding the engines and boat were secure. After he checked them, he spent many long minutes sipping from the flask and looking at the equipment, at the way it was belayed to the trees above, and at the slope above that.

"Damn him," he muttered.

"You should not go far from the firelight, tonight, sahib," came a voice from the growing darkness. Wainwright jumped and spun, seeing Ibn-el-Mani emerge from the deepening gloom. The Arab was smoking a pipe and trailing the pungent odor of bhang. "My scouts saw a hunting pair of lions."

"You startled me." Wainwright took a drink from his flask.

"You speak of your leader?"

"I do," Wainwright admitted. "You try to do what the bastard wants, and he twists you the other way. It's like his spirit works oppositely from everybody else. I can't do a damn thing that pleases him."

"But, sahib," the Arab said. "He has led you this far. Would you not have thought that impossible when you unloaded the boats from the train in Fungurume?"

"Before that, my friend," Wainwright said, taking a slug of whiskey. "Long before, I assure you. In Cape Town. That's where I thought we'd fail."

"Yet you are here."

"Yes. Here. Here I am." Wainwright waved around. "And here we all are. And where's that?"

He meant it rhetorically, but Ibn-el-Mani answered anyway.

"In the middle of Africa, facing mountains. And we all rely on you, not just your commander. We all know who it is who really has brought us this far."

"The bastard," Wainwright said, not seeming to have heard the Arab's compliment. He planted himself desultorily on a small boulder and stared at the stars popping out in the darkening sky, preternaturally sharp in the dry, rarified air. "It's bad enough that I have to put up with him without that damn Freiesleben and Marmion making backhanded remarks."

"What did they say, sahib?"

"Something about the oxen pulling better downhill on the other side of the mountains."

Suddenly, Wainwright's eyes lit up. "Yes," he muttered. "Yes."

"You have found a way?"

"Maybe so."

"Will it work?"

The lieutenant turned and gave the Arab a look at once thoughtful, anxious, and bold.

"I certainly hope so," he said. "I haven't a better solution."

"Does 'better' mean 'alternate' in this case?"

"I suppose it does, Ibn-el-Mani," Wainwright said, taking another drink. "We should go back to camp before simba comes." The Arab turned to leave. "You go ahead. I'll be along shortly."

Wainwright continued to stare at the engines, boats, tackle, and slope for a few more minutes, but soon the darkness forced him to follow after the Arab. On his way, he spied several shadowy figures ahead in the darkness.

"...give you odds," one was saying as Wainwright came into earshot. The voice was Freiesleben's.

"What kind of odds?" asked another voice Wainwright recognized as Lamont's. The third man was Marmion.

"Say, one hundred to one," Freiesleben replied. "I've heard those are the odds against you in Elizabethville. I'll bet your commandant does not even make it halfway up the mountains."

"Not likely," snorted the rating. "I take my bets on something that at least has a chance."

"Lamont!" Wainwright said, and the rating stiffened. "Did I hear you turn down that man's wager?"

"Yes, sir."

"That's not very sporting."

"Ah," said Freiesleben. "Lieutenant The commandant's trusty engineer."

"Lieutenant," Wainwright nodded. "I take it you have no real faith in our English amateurism."

"Let us just say I have more faith in the mountains' stubbornness. Besides, Monsieur La Plae says it is impossible, and he helped build our railroad."

"We've gotten this far," said Wainwright. "There's no reason to believe we won't go the whole way."

"Your amateurism may have gotten you this far," the Dane said, teeth gleaming in the dim light, "but your engines cannot pull uphill as well as they could across the desert and the oxen...." He shrugged as if to dismiss the beasts. "They, too, are worthless on the mountains."

He and Marmion laughed and, in their mirth, missed the sly look that crossed Wainwright's eyes.

"Yes, of course," Wainwright said. "You're perfectly right." He waited until the two had quieted enough to take him seriously before he said, "Nevertheless, gentlemen, I believe I will take the bet." He looked at Lamont. "And you'd best do the same."

"But, sir. He's right. We barely got a hundred yards, and that was the easiest go. The rest's like the Great Wall 'a China, itself."

"Do as you like," Wainwright said. "But don't say I didn't warn you." He turned to Freiesleben. "I believe you did say odds of one hundred to one. What amount would you like to lose?"

The Belgian laughed and produced a wad of money and peeled off a couple of bills.

"Let's see," Wainwright said, squinting at the money in the dim light. "That looks like it might amount to about forty quid." He laughed. "Be a sport, man!" he said to the Dane. "What do you stand to gain from such a paltry sum at one hundred to one? Make it a hundred quid and I'll take you on."

"Agreed," Freiesleben said. "I'll do it. One hundred pounds."

"Shake on it." Wainwright stuck out his hand, and the two men shook. "Now, let's find some honest soul to hold the funds, shall we?" He gestured to Freiesleben to follow. "I know our quartermaster will accommodate us." He disappeared into the dusk, Freiesleben and Marmion trailing behind.

"What about me, sir?" Lamont called after them.

"I have no problem if you take Belgian money," Wainwright's voice came back out of the darkness. "Long as they don't."

Freiesleben's laugh drifted back on the heels of Wainwright's words, and Lamont hurried after them.

2

"The English are overconfident," Freiesleben said in French to his men. They were standing just outside the perimeter of the English camp, watching the expedition fall in after morning mess. The Danish lieutenant laughed. "Because of that overconfidence, by sundown, we will line our pockets with English money."

"I just wish we didn't have to give all our money to the quartermaster to hold," said his sergeant. "I haven't a sou on me."

"Don't worry, Marmion, you shall get it back and more." The lieutenant patted the sergeant on the shoulder.

"Look, Lieutenant," said the chief askari. "The commandant."

Spicer-Simson emerged from his tent attired in his usual leather skirt, sword, and cane, all topped with his gold-braided cap. Wainwright went over to him and said something, Spicer-Simson nodded and said something in return, then Wainwright hurried off. Spicer-Simson stretched, then spotted the Belgian contingent.

"Le commandant à la jupe approaches," Marmion said.

At the epithet, the askaris laughed.

"I was going to wish you gentlemen a good morning," Spicer-Simson said as he came up to them, "but I see you already are having one."

"That is right, Commander. It is a beautiful morning, yes?"

"Glorious, Lieutenant. Just the kind of morning to make real headway."

"Le commandant à la jupe thinks he will make headway this morning," the chief askari said in French.

"Hush," Marmion replied, also in French.

"Say what you like," said Freiesleben in the same language. "These pig-headed English think their crude language is so important that they never bother to learn another."

"I understand you have been making wagers with my men," Spicer-Simson said without batting an eyelash.

"That is right. I hope you do not object."

"Not at all," Spicer-Simson said. "A wager now and again makes sport of tedious work."

"Good, good," said Freiesleben. "I'm glad you approve."

"I don't suppose you would care to make one last bet before we tackle the mountain once more."

"With you?" The lieutenant affected surprise.

"Oh, I have a modest amount of cash on hand," Spicer-Simson said. "And I understand you have wagered enough with my men to clean out the expedition until next payday."

Freiesleben laughed.

"What you have heard is exaggeration, Commander, but only just so."

"At odds of one hundred-to-one, you must have put up a heavy stake."

"We have been stationed here for three years," Freiesleben replied. "And there is not a place within a thousand miles to spend a tenth of the cash we accumulate."

"Ah, then you must have a bit more to cover my wager, eh?"

"How much would you care to put up, Commander?" the lieutenant asked, pointedly ignoring the sharp-toothed grins of his askaris.

Spicer-Simson dug a finger into the top of his boot and edged out a tight packet of bills. He thumbed the bills open and fanned them out. There was a hundred pounds.

Marmion whistled and said in French, "Le commandant à la jupe is rich." The askaris laughed, but Freiesleben silenced them, his eyes narrowing.

"That is a lot of money, Commander. Especially if I have to put up a thousand to cover my end."

"I thought you said you've managed to accumulate major funds," Spicer-Simson said. "But, if you can't cover it...." He shrugged and made to fold the bills and return them to his boot.

"One moment, Commander. I believe I have enough."

"You scoundrel," said Marmion. "You have a thousand tucked away, and you make us give up our last sou?"

"The privilege of rank," the lieutenant replied. "Hush now."

"Let him hear all he wants," said the sergeant. "You said he can't understand. I could call him something worse than le commandant à la jupe, and all he'd ever do would be smile stupidly and offer you even more money."

"You know, Lieutenant," Spicer-Simson said, showing his teeth. "I believe I might have a few more pounds to wager if you think you can cover it." He dug in his other boot and came up with another packet of bills which, when he fanned it out, proved to amount to another hundred pounds.

Freiesleben looked nervously at the money, but his men egged him on.

"What did I say?" Marmion said. "Take the wager, sir. If you've got the first thousand, you must have a thousand more."

Freiesleben glanced at the askaris, who were watching expectantly. Askaris though they were, he had to retain their respect. Licking his lips, he held up his hand.

"All right," Freiesleben said. "But only if you can take an IOU. I assure you it will be honored at Lukuga."

"Of course, Lieutenant."

While Freiesleben wrote out the IOU, Marmion said, "Don't worry, sir. By tonight, we will be lining our pockets with English wealth. Remember?"

"Eastwood!" Spicer-Simson yelled. "Come over here."

Eastwood hurried over, a musette bag bouncing heavily against his hip. "Another wager, Quartermaster. Here are my funds, and there," he pointed to Frciesleben, "are his."

Eastwood opened the musette bag, took out a small notebook and pen, and noted the wager on a page filled with other lines of names and figures. Then he stuffed the notebook, pen, money, and IOU into the already packed bag and closed the flap.

"Now, gentlemen," said Spicer-Simson expansively. "Shall we see to the operations?"

With the commander leading the way and Eastwood and the Belgian contingent following, they headed toward the scene of the lift. When they arrived, the traction engines were being unlashed from the trees. Harrison and Bisho worked up heads of steam in their machines, then urged their huge metal beasts forward, only to have them slip further down the slope than *Mimi* and slew completely around until they aimed downward, one on either side of *Mimi*. Men rushed forward and relashed the engines to new trees.

"Alas, Commander," said Freiesleben. "Things do not start out well for your engines. Even as they strive to gain, they lose."

"Such is the constant state of life, wouldn't you say, Lieutenant?" Spicer-Simson replied.

"Aptly phrased, Commander." Freiesleben pointed downhill. "Here come your faithful oxen. Perhaps they will not be so accustomed to the whistle today."

"Perhaps," Spicer-Simson said as they watched Wainwright direct the drovers to lead the beasts up toward the moored boat. The herd began the climb in a tight bunch, but by the time it reached *Mimi*, half of it was going to one side of the boat and half to the other.

"Ah, Commander, it seems as if even the cattle are uncooperative today."

"Perhaps they will regroup above the boat," said Spicer-Simson, a hopeful tone in his voice.

"I think not," said the lieutenant. "See? The herders cannot urge them past your traction engines."

The two halves of the bifurcated herd had stopped beside the traction engines, and the drovers quickly roped them together as if to keep them from straying.

"Lieutenant," Marmion said, pointing. "Look." On the slope twenty or thirty yards above the boat, several men were attaching two heavy blocks and tackle to a pair of thick tree trunks. In a few minutes the operation was complete, and heavy hawsers were passed though the blocks. One end of each cable was attached to *Mimi*'s trailer and the other to the rear of a traction engine.

While this was going on, the drovers had regrouped their two herds of oxen in front of their respective engines and were hitching them up to the fronts of the machines. Within half an hour, everything was ready: The traction engines were positioned one on each side of *Mimi*, with oxen in front of each engine, and the engines and the oxen were all pointed downhill. Wainwright made one last inspection, then came over to Spicer-Simson.

"Ready, Commander."

Spicer-Simson leaned on his cane, cocked his head, and said, "Goodness, Lieutenant. Whatever made you think of that curious arrangement?"

"Well, sir, I figured out some of it, but Lieutenant Freiesleben's suggestion that the oxen might more readily pull downhill was what gave me the idea."

Spicer-Simson shot Freiesleben a sardonic look and said, "We'll have to thank the lieutenant this evening if the plan works, won't we, Wainwright?"

"I'm certain we will."

"Splendid. Carry on at your command."

"Thank you, sir." Wainwright turned, looked at the expectant drivers and drovers, raised his hand high overhead, then chopped downward.

Drovers shouted their beasts into motion, and behind them, the traction engines' whistles startled them into greater effort as the engines belched smoke and roared into movement of their own. With the engines and oxen pulling downhill, the slack in the cables going through the blocks and tackles grew taut, and *Mimi* crept slowly but steadily up the slope.

In short order, the boat had been drawn as close to the blocks and tackle as was possible. The boat was belayed and its wheels chocked, the blocks and tackle moved to two more trees thirty yards higher up the slope, the oxen and traction engines repositioned, and the entire process was repeated. Neither beast nor engine had trouble, or reluctance, pulling downhill, and by late sundown *Mimi* was hauled over the top of the escarpment, 4,200 feet above the desert floor and 6,400 feet above sea level. It was September 8.

"I believe we have an interesting situation here, do we not, Lieutenant Freiesleben?"

The Belgian lieutenant's weary eyes peered at Spicer-Simson out of a wooden face.

"But Commander, I believe the wager was that your expedition would get both boats completely over the mountains, was it not?"

"Hey, now," Wainwright began, but Spicer-Simson laid a calming hand on his arm.

"Patience, Lieutenant. Patience. Do you foresee any difficulty in accomplishing the task?"

"No, sir. We should have *Toutou* up top by tomorrow this time."

"Nor do I. In fact, as Lieutenant Freiesleben pointed out to you yesterday, getting down the other side should prove to be far easier than getting up. I suggest we let the wager stand as our Belgian comrades have just now stipulated. That way there will be no quibbling or doubt. Am I right?"

"You are, sir."

"Very well, Lieutenant Freiesleben," Spicer-Simson said. "The wager stands as you propose. We must get both boats all the way over."

"The wager stands," said the Belgian lieutenant, face still wooden but a feral gleam growing in his eyes.

"Now, if you will excuse us," Spicer-Simson said, "we've had a tiring day, and we'll be off to mess."

"Until tomorrow, Commander," Freiesleben said, and he led his men toward their bivouac on the outskirts of the English camp.

Spicer-Simson watched them go, his face stiffening.

"Double the sentries on the boats and engines," he ordered Wainwright. "And post men outside Eastwood's tent."

"Yes, sir." Wainwright turned to go.

"Oh, Lieutenant."

"Sir?"

"For the sentries on Eastwood's tent, find the men who have wagered the most."

"I know just the men for the job, sir," Wainwright grinned, then he was off. If the Belgians had planned either larceny or sabotage, they evidently abandoned their plans, for in the morning, nothing was amiss. Just after morning mess and roll call, the expedition assembled at the foot of the slope and, using the same method that took *Mimi* to the top, tackled *Toutou*'s rise up to and over the escarpment. If anything, even less trouble was encountered, and by late afternoon, both boats and both traction engines were casting huge shadows across the plains below as they caught the lowering rays of the evening sun.

"Commendable work, Wainwright," Spicer-Simson said after evening mess as the officers and Hanschell sat around the fire, sipping drinks.

"Bloody marvelous, I'd say," Hanschell said. Wainwright flushed.

3

That very day, Lee was sending a message to the Admiralty. He was still in Cape Town, but instead of railing against Spicer-Simson and the way the commander had underhandedly wrested away his position, Lee was suggesting an augmentation to the original plan. Since Spicer-Simson was making headway—albeit not terribly rapidly—toward Tanganyika and should, if all went according to plan, engage the Germans inside of two months, Lee wanted to bring in a force of 1,000 men to attack stations on the German shore. The men could be drafted from existing regiments, such as Colonel Driscoll's Legion of Frontiersmen, or from men in South Africa and Rhodesia who were not yet engaged. Lee wrote that he was certain he could raise the force if England could arm and supply them.

Lee's logic was that attacks inland would pressure von Lettow-Vorbeck into abandoning his coastal positions on Tanganyika in favor of protecting the rail line from Kigoma to Dar-es-Salaam. This would open the countryside to British forces from the north and south and Belgian raids directly across the lake and around its northern end, all of which might squeeze the Germans right out of Africa.

The letter went through channels from Rear Admiral Gillatt, the senior naval officer in Simonstown, to General Edwards, who still gave credence to Spicer-Simson's disinformation campaign regarding the expedition's former co-commander. Edwards responded that he would not authorize such an expeditionary force, but that he would forward the request on to Sir Henry Jackson.

Sir Henry, also a believer in Spicer-Simson's lies, was not in the mood to answer at all.

"Have not the services of Mr. Lee been dispensed with, and has he not been so informed?" Sir Henry demanded of Admiral Gamble. Sir David had to admit that such was the case. "Well, then," Sir Henry said, "Do something about him."

Sir David quickly forwarded the letter and all previous correspondence regarding Lee's outrageous and possibly treasonous behavior to the Naval Law Branch, with specific instructions that, if Lee dared to show his face in England, a case should be brought against him for breach of duty as suggested by Spicer-Simson.

Spicer-Simson, having often been beached himself, knew all too well how to beach another.

4

The next day, the expedition set out for the far side of the Mitumba Plateau. Although the plateau spread in front of them wasn't flat but ranged in broad, grass-covered hills and vales dotted with trees, it was a far cry from the ragged corrugations of the land through which they'd come, and the twenty-mile journey to the western rim took only three days.

The only incident of note came on the afternoon of the second day. Hanschell was riding in the Burrell with the commander when the bearers ahead of them suddenly dropped their loads and scattered, yelling in panic and slapping wildly at the air and their bodies.

Hanschell started to laugh at their capering when he found himself performing the same antics as angry buzzing bees darted in and out of the loco's cab, leaving behind some nasty stings.

The swarm wasn't large, however, and with all the ready targets near at hand, nobody sustained more than two or three stings each.

"Shall I get my medical kit?" the doctor asked, seeing Spicer-Simson rubbing at a lump already rising on his forehead.

"Don't bother," the commander said as they descended from the cab and the Burrell kept moving. "I can stand it. But those were damn fine bees, eh, Doctor?"

He lit a cigarette, stuck it into his cigarette holder, and jammed the holder between his teeth. After a meditative puff or two, he eyed Hanschell and said, "Give our bees back home hell, what?"

Then he spotted a loose rope dragging in the dirt behind the boat and strode off toward it, pointing with his cane. "Here," he called out to Waterhouse, who was gingerly massaging a swollen cheek. "Chief! Man that dragging lanyard! Can we have a little efficiency, here, please!" Waterhouse hurried to order one of the seamen to obey.

The afternoon of September 12 saw them staring out over the northern slopes of the mountains, which descended into a drab land without the slightest hint of foliage and that looked more barren than the stretch between Fungurume and the mountains.

"It's the beginning of the Congo River watershed," Spicer-Simson said. "Look." He pointed, and in the distance, they could see a snaking line of green that wound off from the southwest toward the northeast. "The Lualaba River. When we get there, we'll be back in our proper element. No more of these blasted traction engines and seas of sand and rock."

"Only a third of the way to go," Dudley said with some satisfaction.

"We still have to get down," Wainwright reminded him.

"Don't worry, old chap," said Dudley. "Like the commander says, getting down will be a daisy walk compared to getting up."

In the morning, they learned just how easy it would be.

"All right, Harrison!" Spicer-Simson bellowed above the racket of the traction engine. "Proceed!"

The loco driver eased the drive lever forward, and the Burrell, with *Mimi* in tow, crept toward the edge of the escarpment and the long slope down to the plains below. Bisho, in the Fowler, followed suit, pulling *Toutou* toward the edge scant yards aft of *Mimi*.

"You know, sir," Harrison had told Spicer-Simson the night before, "I'm not sure that the engines brakes will be enough getting down."

"What?" The commander's face flushed and his eyes turned steely. "Why didn't you say something earlier?"

"I guess I just didn't think of it, sir. Not while we was havin' all that trouble getting' up here. Just didn't seem...what's the word, sir? Relevant. That's it. It just didn't seem relevant."

"Sergeant, if you've let me get my boats on top of these mountains with no way down, you'd better plan on remaining here with them."

"Don't worry, sir," Wainwright soothed. "Harrison has a plan."

"Right, sir," said the engineer, sounding enthusiastic enough. "I'll keep the brakes on, but the trick'll be to throw the engines into full reverse gear as soon as we make the head of the slope. That should slow us sufficiently until we make the bottom."

Spicer-Simson looked askance at Wainwright.

"You've driven locos. Will it work?"

"Should do the trick," Wainwright nodded. "We'll have both boats down by nightfall."

Now, as Harrison's engine tipped over the edge, gravity, insistent as only it can be when thirty tons of steel and boat and trailer are involved, began to pull the loco downward. Harrison quickly threw levers, and the huge drive wheels reversed and began chewing backwards against the slope, slowing the descent to a reasonable rate. In seconds *Mimi* was drawn over the edge and began to follow the Burrell along the downward road.

Bisho drew his machine up to the edge and looked after the Burrell. Seeing the lead engine and *Mimi* progressing easily down the incline, he geared up and started to follow Harrison's lead, but Wainwright waved him back.

"Give him a chance to see how it goes," the lieutenant yelled into the Fowler's cab, and Bisho nodded without taking his eyes from the Burrell.

Up close, the sound of the Burrell's descent was deafening. The engine was stoked to full capacity and was running full out in reverse. Added to the roar and racket the engine normally made was the clatter of dry earth and rock that jetted from beneath the backward-spinning drive wheels. In a few minutes, the whole scene was practically hidden from the watchers above in a cloud of dust. Even Harrison was having a hard time seeing. Maybe that's why he didn't notice the slight increase in the mountain's gradient. Or perhaps he simply felt his machine *could* conquer all. And maybe it could have save for the simple fact that the earth along this section of the path was as loose as a toupee on a bald man's head.

The first sign Harrison had that something was amiss was when the drive wheels, no longer catching anything worth gripping, let loose, and his engine revved almost instantly to its maximum RPM. The second sign was the almost simultaneous surge in the forward momentum of the engine. Unlike the engine's RPM, which quickly reached a peak, the loco's descent continued to increase sickeningly.

The third sign Harrison had that something had gone terribly wrong was when his fireman shrieked, "She's runaway!"

To his credit, Harrison tried the only thing he might have under the circumstances—he pulled as hard as he could on the reverse gear lever, as if extra pressure might make another, lower, gear magically mould itself into place in the gearbox. Being not only a burly man but a desperate one, he pulled hard enough to make the lever bend.

The fireman took one look at the crooked lever and opened his mouth in a scream that might as well have been silent, drowned as it was in the pandemonium of the loco's roaring downward slide. He turned and leapt overboard, trusting the flashing scenery more than Harrison's herculean strength. He was swallowed instantly in the billowing clouds of dust spewed up by the raging, shrieking wheels.

Harrison threw one glance back to see the fireman tumble over the rocky slope only to be brought up short by a tree, then his eyes were captured by the sight of *Mimi* jouncing like a storm-tossed cork in the loco's crazed wake.

In the seconds that elapsed since the fireman abandoned ship, the Burrell accelerated noticeably. Harrison gave one last, very quick glance forward—the only direction in which he could see anything at all—and saw nothing but greater speed and either devastating impact or a soaring but brief flight that could only end in something even worse. He'd quit the railroad to avoid head-on collisions, and damn if he was going to die on some godforsaken mountain in Africa in a head-on with a tree. But he wasn't about to lose his engine, either, or the boat, off the edge of some deranged cliff. If only he could capsize the engine, its mad descent would slew to a stop, and the boat, being attached, would halt, too.

With a single tearing jerk, Harrison wrenched the steering tiller then prepared to dive out of the cab as the engine twisted and fell.

Things didn't happen quite as he planned. The soft earth that kept him sliding also failed to provide sufficient bite to capsize the engine. The runaway behemoth only turned, and all the turn did was snap the trailer hitch.

The Burrell instantly picked up more speed and aimed straight for the edge of the escarpment. Fearing a head-on with a tree substantially less than a plunge over a cliff, Harrison did the only thing he could. With another titanic yank on the tiller, he aimed straight for the largest tree his panicked eyes could see close to his path. It was right on the lip of the escarpment. When he was reasonably certain the engine was going to hit the tree, he flung himself out of the cab.

Since the first thing Harrison encountered after he left the cab of his engine was a largish boulder that immediately rendered him unconscious, he saw nothing further. Just as well. Belatedly, the engine turned broadside and fell over onto its left side, roared on for another fifty feet, plowing up the ground and snapping off small trees, then came up short against the large tree with a horrific, splintering crash. The tree, large as it was, fractured, and the trunk and branches toppled over the edge of the escarpment, but the stump managed to hold the Burrell and keep it from following the tree into oblivion. The engine lay there in the dust and steam of the crash, still running, its left drive wheel slinging dirt and gravel as it endeavored to chew its way into the heart of the Mitumba Mountains.

If Harrison had been awake, he might have been pleased with the fate of the engine, but if he hadn't been unconscious, he surely would have fainted with shock and fright as one of the wheels of *Mimi*'s trailer, traveling at a good clip, passed less than one foot above his head before the boat met Scylla and Charybdis in the form of a pair of mid-size trees. Incredibly, the crash was loud enough to carry over the roar and ejecta of the fallen engine. Splinters and metal fittings flew, and the trees rocked and whipped, the few leaves that had been on them fluttering to the ground. The supplies that had been in the boat for the trip across the plateau leapt out and flung themselves in a wild tangle down the mountainside.

Then the fallen engine died in a great huff and sigh of expelled vapor, and its drive wheel stopped gouging the earth.

For an instant, excruciating silence ruled the north slope of the Mitumba Mountains before a single voice split the hushed air.

"Goddamn it!" Spicer-Simson raged. "What the hell have you bastards done to my boat!?"

5

Spicer-Simson's words echoed off the slopes and were probably heard, and their emotion comprehended, in the valley below. As if they were the pebble to start an avalanche, the excited babble of men shouting in English, Swahili, and a score of Bantu dialects, sprang up and chased them and the wreckage down the rocky slope.

"Wainwright!"

But Wainwright was already gone, loping down through the settling dust toward the fallen engine. African workers were gathering around the fireman and helping him sit groggily after his encounter with the tree, so the lieutenant bypassed them and headed for Harrison.

The engineer's head was covered in blood, and Wainwright thought he was dead. But then he detected a steady beat beneath the red glaze that covered Harrison's neck, and he breathed a sigh. A few seconds later, Hanschell arrived with Spicer-Simson and bent over the stricken man.

"To hell with him!" the commander yelled, a tinge of panic in his voice. "What about my boat?"

Wainwright left Harrison to the doctor and ran, skittering stones and loose earth, past the still-hissing engine, toward the wreck of *Mimi*.

The boat was not a pretty sight. Its bow no longer was a nice, rounded V, and the hull looked uneven on its keel as if the whole craft had been warped by insistent pressure. Even so, it wasn't as much of a wreck as he imagined it might be, considering what had happened. The trailer and cradle, both built to withstand heavy abuse, were gouged and battered and had taken the brunt of the damage. He breathed a sigh that had, perhaps, a trifle more relief than the one he'd exhaled over Harrison. The expedition had extra engineers, but only two boats.

He stood up, only to see the commander bearing down on him, face a mask of rage. But he sensed there, too, an edge of fear, and rather than bow beneath the blows of the rage that descended as Spicer-Simson reached him, he quickly assuaged the fear. The damage to the boat was not extensive and could be repaired. The men were relatively unhurt. And the engine, well, nothing could hurt those beasts. They had only to right the fallen, repair the stricken, and proceed downward using the tried and true reliable method of cabling and block-and-tackle—only in reverse—that had brought them to the plateau's table.

"See that you succeed, Lieutenant," Spicer-Simson growled, but Wainwright could tell he was mollified if not outright relieved.

The rest of the day was spent setting the ten tons of fallen engine onto its wheels, and never before on an expedition that had not been shy in its use of thick language was such blaspheming heard. But after storms of bluster and gallons of sweat, the traction engine was finally upright and belayed securely to Scylla and Charybdis. Then the day was done.

The next morning, the expedition turned to their block-and-tackle and began lowering *Toutou*. By midday, they passed *Mimi*, whose trailer was under hasty repair, but soon afterwards an unforeseen difficulty arose. Or, rather, didn't arise, for the next section of slope, perhaps a mile in length, was completely devoid of trees large enough from which to belay.

"What now, Lieutenant?" Spicer-Simson snarled, though beneath his harsh tone was a note of plaintive desperation.

"Dead man, sir."

"Are you describing me or yourself?"

"Neither, sir. A method."

The method was soon evident. Wainwright had a group of Africans cut dozens of twenty-foot lengths of timber from upslope. Meanwhile, another crew dug pits down the treeless section at thirty-yard intervals. These pits were twenty feet long, ten wide, and nine deep and situated with their long dimension perpendicular to the incline. As the pits were finished, log posts were planted along their downward edges and wire and cable laid across their bottoms.

"That is my telephone wire, Wainwright," Spicer-Simson said.

"Yes, sir, it is. But if we don't make it down the mountains, there will be little use for it."

"Very well," the commander conceded, stroking his beard. "Proceed."

Several logs were laid lengthwise in each pit, the wire and cable wrapped around them and secured, and the whole business buried under, leaving only the cable ends protruding and the bollards, or log posts, sticking up.

"That, sir, is a dead man," said Wainwright.

"An ingenious idea," Spicer-Simson said. "It had better hold."

"It will, sir," Wainwright said with more feeling than conviction. He knew what tremendous weight each dead man had to sustain.

Toutou and the engines were belayed to the cables and bollards of the first dead man and the lowering attempted. The dead man shifted and heaved in his grave but stayed buried, and the next did, likewise. At last the boat reached the trees farther down the slope, and at nightfall of the second day, *Toutou* was sitting securely at the Mitumba's feet.

By now, Mimi's trailer had been roughly repaired, and it followed Toutou. Harrison, claiming he felt fit enough to drive, insisted on taking the controls, though everyone knew he still sported a splitting headache. Using the path already traversed by *Toutou* and with the dead men already planted, *Mimi*'s descent took but a single day.

After mess that evening, Spicer-Simson, accompanied by Wainwright, Dudley, Eastwood, and Hanschell, paid a visit to the Belgian bivouac.

"I believe the terms of the wager have been fulfilled, wouldn't you say, Lieutenant Freiesleben?"

The Belgian lieutenant simply looked resigned.

"Yes, Commander," he said. "I must commend Lieutenant Wainwright's ingenuity. We did not think it possible."

"Everything is possible, Lieutenant. So, with your permission, I will have our quartermaster dole out the winnings to my men."

"Yes, Commander," Freiesleben nodded his assent.

Suddenly Spicer-Simson stepped forward and stared fiercely into the lieutenant's face. The flickering firelight seemed to make his tattoos crawl with independent life, and his eyes gleamed savagely. He let out a hissing, blistering stream of French, then he turned haughtily and stalked off, leaving Freiesleben ashen-faced. The other British officers followed at a discreet distance.

"What was that he said?" asked Dudley.

"I don't speak French," Wainwright said. "Do you, Doctor?"

"A little. I studied it in school, but that was some time ago. I think he said something about the English smiling when they know they're going to be rich and that the next time an English amateur offers them a wager, they'd better turn and run. Then he said that they'd better not ever again let him hear them call him 'le commandant à la jupe.'"

"What's 'lee commandant a jupe?'" asked Dudley.

"That's 'le commandant à la jupe,'" Hanschell corrected, then he laughed softly so that Spicer-Simson could not hear him. "That's the funny part. It means 'the commandant in skirts.' Seems he doesn't mind the skirt part, he just objected to being called a commandant. That's equivalent to a major, and I guess the commander didn't like being demoted. Said that if they have to use an army title, they are to call him 'mon colonel!'"

6

Freiesleben and his askaris left before daybreak, not bothering to say goodbye. No one missed them, since they'd left everything of importance behind—namely, Eastwood's satchelful of cash, which the quartermaster dutifully handed out to the winners.

It was now mid-September. The expedition had left the railhead at Fungurume a month earlier and traveled more than a hundred miles, or, a little over three miles a day, including the haul over the Mitumba Mountains.

"Marvelous!" exclaimed Spicer-Simson that evening at mess. "I never expected to make such good time."

The other officers looked at him, thinking he might be about to burst into a sarcastic tirade, but the commander appeared to be quite sincere.

"Only forty miles left until we reach Sankisia," Spicer-Simson went on, "and then it'll be smooth sailing down the Lualaba."

Actually, Sankisia was still fifteen miles from Bukama, which was on the river, but there was a narrow-gauge railway between the two villages that could cart the expedition. And once they reached Bukama, they wouldn't really have to sail the boats since Hope had commissioned a river steamer to ferry them all the way to Kabalo, where they could catch another train the rest of the distance to Lukuga.

Spicer-Simson might not have been quite so upbeat if he'd known why the slope they'd just descended had been far more devoid of trees than the upward climb on the other side of the mountains. Although,

269

technically, they were in the Lualaba River Valley, they were still a long way from the river, and the terrain was dryer here than anywhere they'd encountered since leaving Fungurume. Far dryer. It was absolutely dry.

And blazing hot. And treacherously rocky. And chokingly dusty. And suffocatingly still. And riddled with antbear dens. Every inimical characteristic of the desert on the far side of the mountains was magnified and protracted, making the forty miles seem twice the distance from Fungurume to the mountains. The sun-dried landscape was almost completely barren of life except for shrivels of sunburned grass and an occasional withered trunk of a tree whose branches contained not the barest hint or remotest promise of leafy shade. The earth was either baked and cracked or made of dusty sand so thick, slippery, and soft that the traction engines continually bogged down and had to be dragged onward by main force of beasts and men. And everywhere it was seamed with gullies and ravines so deep as to be nearly impassible. Worse, on the whole plain there were barely enough trees to make a single bridge, so they repeatedly dragged the logs from ravine to gully to dry wash and on again.

Explosives were used to soften the edges of the ravines and create ramps, but at the deeper ravines, they often had to resort to the block-and-tackle system Wainwright had worked out for the slopes of the Mitumba Mountains, using boulders and dead man weights instead of trees as anchor points. The only salvation was the fact that everything was so parched that the ravines contained no water and could be crossed without bridges.

The few waterholes they encountered were reduced to nasty, muddy sumps barely fit for beast, certainly not for men. Yet the oxen drank what they were allowed, which wasn't much, for the traction engines were greedier still. Wainwright often found himself dreaming of the wandering swamp they'd encountered between Fungurume and the Mitumba, wishing one would wander up now. But none did, and the heat and lack of water quickly worked disastrous results on the boats. Where the boats' hulls had been threatened with the arid conditions on the far side of the mountains, they now began to crack at the seams. Precious water for which men, oxen, and traction engines alike thirsted went into their hulls to keep the planks from warping entirely off their keels. What little water that remained went into the boilers of the locos. The twenty-odd white men and six hundred tribesmen were reduced to a water ration of a half-pint a day.

The oxen fared a little better than the men. They were given water enough to keep them moving, though for feed, the shriveled grasses of the barren plain had to suffice. But their brute muscle was still needed to pull traction engine and boat trailer alike from the antbear holes that proliferated inexplicably in the arid terrain where termites would have trouble finding wood to consume. Despite the men assigned to sound for the pesky dens, the holes sometimes trapped the engines or trailers for hours at a time, and even when the trailers weren't trapped, the constant jarring was shaking them apart.

The expedition crept forward as much from dull habit as out of duty or even desperation. Some days, they barely managed a mile or two, though they began at five in the morning and struggled until nine at night. The whites slogged along through the thick sand with footsteps barely active enough to raise dust on a landscape where even the slightest stir of air lifted a grey cloud, and the Africans began to look like scraps of living leather. No one spoke except when necessary, and the arid atmosphere swallowed even the clatter and chuff of the traction engines, so that the expedition seemed to move in a vast, ghostly, silent column through the white-hot air.

The nights brought little relief. The blasting sun's power invested the earth with a mighty heat that radiated constantly all night, and the air was almost totally still. And when there was a breeze that pretended to cool, it brought instead choking dust, day or night. Everyone stank, but no one noticed. They were simply too baked by the sun, dehydrated, and exhausted to do more than wolf their dole of canned rations, sip their meager ration of water, and fall into uneasy slumber on the inhospitable ground. If Hanschell gave any thanks at all, it was that his dreaded tsetse fly could never survive in conditions such as this.

But lions did, and they were as undoubtedly as starved as the countryside. Every night, they hung around the camp, roaring their desire for flesh, even if it be human. Since none of the expedition fancied becoming a meal, no one ventured beyond the firelight at night or went unarmed during the day. Fortunately, there were no casualties from predators, but lack of water was a more visceral threat.

After three days, their entire potable water supply outside of the few gallons sloshing inside the boats' hulls was reduced to a couple of hundred gallons. The daily ration was decreased to a single tablespoon, and the oxen had to go entirely without because even the mud holes of the past few days vanished entirely.

"What'll we do, sir," Harrison asked Wainwright as the two stared at the mud spot that was the last point of departure of a vanished transient lake. "There's no water for the locos."

"Dig up the mud, Mr. Harrison. Dig it up, and pack it into the boilers."

"Not good for the locos, Mr. Wainwright," the engine driver shook his head. "But I suppose we have no other choice."

"None at all," the lieutenant replied.

And so the expedition set about scooping sandy mud from the bottom of the deepest ravines and packing it into the boilers. Each load provided sufficient steam for as far as a mile, then the engines had to be shut down and cooled off and the dried sand scraped from the boilers. The former rate of three miles a day diminished to two. Then less. One day, they managed only 400 yards.

Then, on the seventh day, the expedition rested. It had no choice. Their boilers ulcerated with silt, the engines simply stopped.

"Ten gallons, Commander," Wainwright said through cracked lips, answering Spicer-Simson's query about how much water remained. "All in the bottoms of the boats." He resented the need to speak.

"Form search parties," Spicer-Simpson told him. "Take some of the natives with you. Find water."

"Water, sir? Where am I going to find water in this hellish land?"

"If I knew that, Lieutenant, I wouldn't have to send you searching, would I?"

"I'll do my best, sir, but it's damn dry out there. Even the villages are suffering."

"And so shall we," Spicer-Simson said, "but even more if we fail now."

"I'll go look, sir, but I can't promise I'll bring back good news."

"I don't need good news, Wainwright. I need water. I know you'll find some for me."

"You have a great deal of faith, sir."

"Bollocks on faith. Can't you smell destiny, man? We've not come this far to be stranded a mere five hundred miles short of our goal."

Wainwright formed three groups, with himself in charge of one and Hanschell and Dudley commanding the other two. Each party had two ratings and half a dozen Africans carrying empty petrol tins, and they set out just after midday, fanning out in the general direction of the river, with Hanschell following the road that the advance party had lackadaisically carved through this sear wilderness.

The doctor couldn't believe his eyes when, an hour and a half later, his party came on a small village nestled on the slopes of a ravine and surrounded by a thorn-bush boma. Through his interpreter, he dickered for nearly half an hour with the village headman, learning in the process that the village had a working well. The headman finally relented and sold him enough brackish, foul-smelling water to fill the empty petrol tins carried by the bearers, but it literally cost Hanschell his own blue shirt and those of the two ratings. But by now, even the whites were brown enough to withstand the sun for a few hours, and the doctor proudly led his foraging group back to the stalled expedition.

"Good work, Doctor," Spicer-Simson said. "Take four tins to the engines and boil the rest for the men. Do you think the chief there will sell us some more?"

"If he does, we'll be without uniforms, sir."

"I suppose so," Spicer-Simson chuckled, though his voice died in a cracked choke.

Hanschell sent the four tins to the engines, which had nearly been cleared of the silt packed into them, and took the others to the cook's tent, where he supervised straining and boiling it, though he doubted that anything would be able to rid it of its foul odor and presumably equally nasty taste.

Suddenly a ruckus broke out near the road. He looked up from his work to see Wainwright stride into camp, followed by a train of women wearing colorful cloth skirts but nothing above the waist, each balancing a large earthenware jar on her head. There were so many of the women that Hanschell soon lost count, though he later learned that there were nearly one hundred and fifty. And the jars were filled with water—not nasty bilge like he had brought home, but stuff that was almost potable right out of the jars.

As the camp rejoiced, Wainwright quickly told his story.

At two in the afternoon, the mirage that he had been observing for some time at last resolved itself into the huts of a moderately-sized village located eight or nine miles from the expedition. Wainwright and his band staggered into the village, expecting to see the inhabitants in as miserable shape as he and his men, but were surprised to be surrounded by the one hundred and fifty topless women. Even better, they all looked well watered.

"Water," Wainwright croaked in Swahili, unable, despite his thirst, to take his eyes from the three hundred breasts that surrounded him. A moment later, an undeveloped girl of about nine brought forward a calabash ladle filled with water. The water looked clean, it smelled fresh, and when he drank, tasted like nectar. Reluctantly he lowered the gourd and passed it to his companions, and while they drank, he asked where the water had come from.

The women led him to the center of the village, where a circle of stones surrounded a forty-foot deep rent in the earth. At the bottom, liquid gleamed with dark life. Wainwright drew up a bucketful and drank as he had never before drunk in his life. As the liquid coursed into him, he could literally feel his desiccated flesh swelling with gratitude.

The fact that the village was occupied completely and solely by women and children was soon explained. All the men had been hired by the advance road building party, which was closing in on Sankisia.

"Men," he said when all were satiated. "I think we've run into some good luck."

7

The first load of water brought to camp, carried in jars and gourds on the heads of the women, went to the traction engine boilers along with the evil-smelling brew Hanschell had been attempting to purify. When the boilers were about half full, Spicer-Simson ordered that most of the remainder go to the African contingent. Hanschell began to boil the ten gallons set aside for the expedition crew, and after he'd done that and the men had drank, Spicer-Simson sat down in the middle of the camp with the women's leader to bargain for more.

"What will we bargain with, sir?" asked Wainwright.

"As you gentlemen may know, my dear wife is from Canada," Spicer-Simson said. "The English domination of the North American continent is a fascinating story, and one that is yet not complete in some parts of Mrs. Spicer-Simson's native country. We shall endeavor to utilize a stratagem often applied by Europeans as they encountered aboriginals in their westward movement across that continent. I also found this method quite useful in my travels on the Gambia and employed it often."

"What strategy is that, Commander?" Wainwright wanted to know.

"We shall take unfair advantage."

"Is that ethical?" Hanschell asked rhetorically.

"War knows no ethics," the commander answered glibly. "But if we take unfair advantage, I assure you we will still leave these poor people happier than when we arrived." He turned to his manservant. "The trunk, Tom," he ordered, and Tom hurried into Spicer-Simson's tent only to return a moment later dragging a huge trunk over to the commander.

The whole expedition looked on expectantly. They'd seen the trunk often enough, but never its contents, which had been the cause of much speculation among the ratings. Many of the expedition were of the opinion that it contained items of apparel even more feminine than the leather skirt the commander wore, and there was a wager going around the camp between those in favor of that opinion and those opposed. The wager was certainly more modest than the one accompanying the speculation on the exact extent of Spicer-Simson's tattoos, but the winner was certain to collect a handy sum.

Hanschell was against the notion that the trunk contained more feminine clothing, deciding that if the commander was brazen enough to wear a skirt in public he would wear anything, and thus any other feminine clothing the trunk might contain would surely have made its appearance on Spicer-Simson's body before now. Wainwright, who was with the other camp, if only for the contrariness of it, argued that the commander was too proud of his tattoos to cover them in public with a corset or other feminine undergarment, but there was no telling what he might don in the privacy of his own tent.

"I take it you have the solution in your mysterious trunk, Commander," Wainwright said, gesturing to the locked box.

"Indeed. Look around you, Doctor, and tell me what you see."

Wainwright did as he was told, and so did all the men. What they saw was three hundred breasts. A few snickers rose from the ratings but were quickly silenced by a stormy look from the commander.

"There you have it, Lieutenant," Spicer-Simson said, returning his gaze to Wainwright.

"Have what, sir?" Dudley spoke up, eyebrows knit together in puzzlement, and Spicer-Simson turned his eyes on him.

"Really, now, Sublieutenant. Don't play coy. I asked a question, and the men answered plainly enough, though rather crudely. But you are young and perhaps too courteous or shy. Or inexperienced. Maybe the doctor, with his trained scientific detachment, will deign to be more loquacious."

"You wish us to observe that these women are not as fully clothed as is the wont among our own women," Hanschell said.

A Cheshire cat grin broke out on the commander's face.

"So. The doctor has articulated what we all can observe: In this place, a woman is regarded as over-dressed if she has two articles of clothing."

"But how will that help us, Commander?" Wainwright asked.

"Observe, then, Lieutenant, but don't forget to translate. Ask these women if they would be willing to carry more water to our expedition in exchange for payment."

"Is your chest filled with gold, sir?"

"Ask them."

Wainwright did and received a chorus of replies to which he listened intently before turning again to the commander.

"They said they would. Their men are out earning great rewards with the road-building crew, and they wish to have something to show them when they return."

"Tell the head-woman, then, that I shall reward each woman who, tomorrow, brings us enough water to fill four jars like that." He gestured to the earthenware jar next to a nearby woman. The jar, shaped like a large urn, would hold about two gallons.

Wainwright spoke, listened, and repeated what he'd heard.

"She says that will be a great effort and the reward must be sufficient."

"Tell that little girl to come closer." Spicer-Simson pointed at the prepubescent girl who had handed the gourd of water to Wainwright earlier in the day. Wainwright told her to come forward, and she did so, reluctantly, unable to take her eyes off the commander's tattoos. The commander impatiently beckoned her closer. "Come on, you little Amazon. I won't hurt you."

As if mesmerized, the little girl shuffled forward until Spicer-Simson took her gently by the arm and positioned her with him behind the chest. He then opened the chest. Hanschell and Wainwright quickly leaned forward to glimpse the contents, and if they were severely disappointed to see only a dozen bolts of cloth instead of lacy nothings, they concealed it. Each bolt was of a different pattern, but all were so gaudy that Hanschell thought a London tart would shy from them.

His movements hidden from the view of the surrounding women by the open lid of the trunk, Spicer-Simson unwound a length of cloth from the top bolt and twined it around the girl's shoulders, then, with a dramatic flourish, he slammed the trunk shut and hoisted the girl up onto its closed lid for all to see.

The women, who had been talking in curious tones among themselves fell instantly silent. Then a shrill voice stabbed the air. It was from the girl's mother, who rushed forward and plucked at the cloth, holding it up to herself. The voices of the other women rose in a combination of awe, jealousy, and excitement.

"I believe we have their attention, eh, gentlemen?" said the commander. "Wainwright. Tell them that the payment for four jars of water will be enough cloth to completely wrap the woman's body from head to foot, but the water must be delivered to the expedition by tomorrow's nightfall."

Wainwright passed on the message, and the excited chatter reached a crescendo of babble before the head-woman gave her assent. Then, in short order, the camp emptied of the women, though their voices could be heard for long minutes after, fading into the late afternoon heat.

"Where on earth did you get that abominable cloth?" Hanschell asked.

"We can thank Mrs. Spicer-Simson for that," the commander said with a sniff. "I believe she picked it up on the cheap somewhere." He chuckled. "And all the more a bargain here, eh?"

With water, the dark lethargy that the expedition had fallen into during the week since they'd left the Mitumba Mountains lifted. And the next day was even better. Not only were the traction engine boilers filled to capacity, the boat sloshed with as much as their cracked hulls would allow. Every empty water container was filled, and everyone's bellies were gurgling. There wasn't enough water left for the men to bathe and shave, but that didn't stop them from rejoicing. The Africans accompanying them were no less ecstatic as they celebrated the return of water to their parched souls.

Luckily, there was enough cloth, though even the commander was beginning to look worried toward the end of the reckoning with the women as the bolts quickly diminished. But there was just enough, and the women went away happy and left happiness behind them.

"You see, Wainwright?" Spicer-Simson said as the women disappeared. "What did I tell you? We managed to exploit them and make them happy at the same time."

"A remarkable feat, sir."

"Naturally, Lieutenant."

"Thank God the cloth held out," Hanschell said, looking down on the scattered remnants lying in the bottom of the trunk.

"Indeed, Doctor," Spicer-Simson said with a flashed grin. "One angry woman is bad enough. Can you imagine one hundred and fifty of them?"

"I think they would have done to us what the desert and mountains couldn't," Hanschell agreed.

"No doubt," Spicer-Simson said.

"What are you going to do with the remnants, Commander?" Hanschell asked.

"Oh, I'm sure I'll find a use for them," Spicer-Simson replied, then he strode off twirling his swagger stick.

"Lucky damn bugger, isn't he?" Wainwright said to Hanschell as soon as the commander was out of earshot. "Water when he needs it, oxen when he needs them, locos, you name it."

"Apparently he doesn't need luck when he has you," the doctor replied, but Wainwright merely shrugged.

"I'm not sure it would matter. It's all fallen into place, one damn thing after another. We shouldn't have made it beyond the railhead at Fungurume, but here we are, the desert and Mitumba behind us and the Lualaba just a week away. Remember what you once told me about the commander? That he was destined for something remarkable? I'm not sure I completely believed you then, but I do now." He gave a sudden, short snort of a laugh.

"What?" Hanschell asked.

"If only he wasn't such a smug, bull-headed prig."

"If only our boats had wings and could fly."

The celebration lasted until reveille and a little beyond, but Spicer-Simson tolerated the breach. All during evening mess, he congratulated himself on his own cleverness in so cheaply obtaining the desperately needed water, even admitting that, if the truth were told, he'd have paid a much higher price. But as it was, he'd gotten a bargain and wasn't shy about letting everyone know.

With everyone's spirits buoyed and sated, the next day, the expedition made excellent time and arrived in the village where Hanschell had gotten his brackish water. They were met there by Locke and his Rhodesians. The road was finished, and the tribesmen hired to do the labor were on their way back to their villages. Locke wished the expedition good luck before departing the next morning for home.

The next four days went well, too, each easier than the one before. The terrain leveled out as the road neared the river, and finally they encountered enough trees for modest shade.

On September 28, about midday, the expedition, still well watered but completely exhausted, filthy, and scruffily bearded, emerged from the wasteland at their destination of Sankisia. The village was a miserable little place, but it was a significant mark nonetheless. The overland journey of one hundred and forty miles had taken six harrowing weeks.

Book Two: On Water

Chapter X: The Lualaba River

WHILE THE CAMP IN SANKISIA was being set up, Eastwood paid the leader of the ox teams, and the Boer and his cattle quickly departed back the way the expedition had come. Harrison planned to follow in the morning as soon as he and Bisho had cleaned up the Burrell and Fowler, though he seemed less than anxious to reattempt the Mitumba slopes alone. Wainwright suggested he be allowed to take a set of block-and-tackle, and Spicer-Simson agreed. The lorry was simply abandoned.

The hundreds of Africans also left, reluctantly, it seemed, for they had become attached to the outlandish English commander. With them went Ibn-el-Mani and his two companions, the Arab stating that now that the boats had found water, there was no further need for a scout.

By two, Wainwright had organized a team to make preparations to transfer the boats and their cradles to the flatcars that were waiting in the rail yard. It was the last task the locos would help perform before they left.

"They must've thought we were bringing canoes," Waterhouse grunted dourly, staring at the small size of the cars.

"They'll do," Wainwright said, giving the flatcars a close inspection. "We only have to go fifteen miles, and look how far we came on those things." He gestured toward the now-dilapidated trailers they'd constructed in Mwenda Makosi.

"Quite right, Mr. Wainwright, though they don't look like they'd've held up much longer."

"They got us here, and that's what counts. Frankly, I think I'm going to miss the lumbering things. And the locos."

"I ain't gonna miss not having baths, sir. That's the first thing I'm going to do when we reach the Lualaba. And have a good shave."

Wainwright laughed. "Did we really survive for nearly half a week on a teaspoon a day?"

"Rightly so, as I recall, sir. God, but I was never so parched in all my life."

"Well, no more of that. We'll be on the river soon and then the lake. But first we have to get the boats onto those rail trucks."

The job was completed by sundown, and the expedition enjoyed an upbeat evening mess that even included some fresh antelope meat brought in by one of the villagers.

Later, Hanschell and Wainwright sat in front of Hanschell's tent, sipping whiskey and water, slapping mosquitoes, and watching the moon rise over the trees.

"To tell the truth," the doctor admitted, "I never believed we'd get this far."

"What happened to your faith in Spicer-Simson?"

"I believe it was leeched out of me somewhere back along the track. But we have come along nicely, haven't we?"

"We have Lee to thank for that," the lieutenant said. He raised his glass in a half salute and drank. "His route couldn't have been better chosen. I think we'd all have to agree on that point." Then he frowned and amended, "Well, maybe not the commander."

"I suppose so," Hanschell sighed. "I sometimes think he has been living in the borderlands all his life."

"Most likely," Wainwright agreed with a chuckle, then he sobered and shook his head. "I must admit, Doctor, that his enthusiasm is as contagious as the diseases you came to observe, but that doesn't make him any less puzzling."

"He's kept us guessing, hasn't he?"

At that very instant, Spicer-Simson was in his own tent, drafting a dispatch that he was going to send by courier to the Admiralty first thing in the morning. He paused in his writing, stared thoughtfully at the paper, then once again set his pen to it.

"Lee's route could not have been more thoughtlessly selected," he wrote. "The man's willful ignorance of both the territory and the terrain has caused us to blunder over half of Africa, losing precious time and wasting much of our supplies of explosives and telephone wire. These must be replaced as soon as possible. I hope to make up for the lost time once we reach the Lualaba River tomorrow."

Spicer-Simson paused again then scratched out the last sentence. Sankisia was still fifteen miles from Bukama, the tiny port village on the Lualaba toward which the expedition was bound, but he hadn't yet actually seen the river. No use in sounding too optimistic, he thought.

The fifteen-mile rail journey began at first light, just after Spicer-Simson held a briefing for his officers.

"When we arrive in Bukama," he said to them, "we will load the boats and supplies onto a shallow-draft steamer that Lieutenant Hope has commissioned. The steamer will ferry us downstream to Kabalo, a distance of two hundred and fifty miles as the crow flies, but more on the order of three hundred and fifty when the river's meanders are taken into account. At Kabalo, there is a rail line that will transport us the final one hundred and eighty miles, and then, gentlemen, we will be at Tanganyika." He paused for dramatic effect, but when the officers failed to respond, he snapped, "Well, get to it, then!"

The officers went and shouted useless orders to the men as the expedition boarded the train, and then they were off. The tiny engine could pull the heavily loaded train only about seven miles per hour, and they had to stop frequently to chop down trees that stood too close to the tracks to allow the boats' broad beams to pass unscathed. The trip took nearly until noon, but it was such a treat to ride rather than trek all day through sand and dust that nobody complained. The train brought them right up to Bukama's single wharf.

Spicer-Simson hopped out, followed by Wainwright, Dudley, and Hanschell, and he quickly scanned the Lualaba for the steamer.

The wharf was empty.

Face turning livid beneath his deep tan, Spicer-Simson stalked toward the only building that seemed to have substance, figuring it was the port master's headquarters. They found a weathered, barrel-chested man with ash-blond hair dressed in perfectly clean but simple white trousers and shirt sitting in a wicker chair on the front veranda, which overlooked the river. He was reading an English newspaper that was three months old while a young, bare-breasted girl wafted a banana-leaf over him. A drink sat on a small, round wicker table at his elbow, and a second wicker chair held a panama hat.

"See here, man," Spicer-Simson blustered. "Where the hell is my steamer?"

"You are?" the man asked laconically, though he could not have been unaware of who everybody was within a one-hundred mile radius of his veranda, especially the expedition. His accent marked him as northern European.

"Commander Geoffrey Spicer-Simson. And you'd best stand when an officer of the Royal Navy addresses you."

"To hell with the Royal Navy," said the man. He picked up the drink from the wicker table and raised it to his lips.

"You insolent foreign cur!" Spicer-Simson growled and raised his walking stick as if to strike the man where he sat, but a click that was both definite and deadly stopped him before he'd half-completed the upward swing. The man produced a revolver from beneath his newspaper, and it was cocked and pointed at Spicer-Simson's chest.

"Let me tell you something, Commander Spicer-Simson," he said in a level tone that was as deadly as the weapon in his hand. "You have dragged your boats over the deserts and the mountains, and that is an admirable, even remarkable feat. But remember that here you are in my land, and you are the insolent foreign cur. I rule this roost, and I've been here too long to have the patience for civilized twaddle or for men who want to bully and intimidate me, especially in my own place of business. Now put down that silly cane and have a drink."

Spicer-Simson lowered his cane, the man uncocked his revolver, and both men laughed. Gesturing to the girl, who ran off to get drinks, the man swept his panama hat off the second chair and motioned for Spicer-Simson to join him.

"You are?" the commander asked.

"Captain Jens G. Mauritzen. I work for the Congo Hydrographic Department. I have been sent to pilot the steamer you seek."

"You don't sound Belgian," Hanschell said.

"I am Danish, but the Belgians are the only ones who pay well in this jungle," Mauritzen replied. "You've been traveling with one of my countrymen, young Lieutenant Freiesleben. He's told me much about you." His eyes sparkled with humor, as he turned to Wainwright. "And you must be the remarkable Lieutenant Wainwright who cost Lieutenant Freiesleben so much of his hard-earned pay."

"It was nothing," Wainwright said, smiling.

"I think not," Mauritzen said, but Spicer-Simson impatiently cut off the niceties.

"My steamer. Lieutenant Hope was supposed to have it waiting for me."

"It was supposed to be here," Mauritzen agreed. "Unfortunately we are in the midst of a drought, and the river is lower than it has been in six years. They're even having difficulties farther down on the Congo proper. Your steamer is in Musanga, sixty miles downstream."

"The river looks wide enough from here," Wainwright said, looking toward the flowing water sparkling in the sun.

"Wide enough," Mauritzen nodded. "But it is only a few feet deep at the most. Lieutenants Hope and Tyrer went downstream to persuade the captain to come up here, but the captain has sent word that he cannot navigate the channel closer than Musanga. Or possibly he won't try."

"Won't?" the commander asked shortly.

"Captain Blaes is a most stubborn and spiteful man," Mauritzen said. "He regards the *Constantin de Burlay* as his private kingdom and me as a noisome pest. Perhaps it is because I am quite expert, and he couldn't find a clear channel if it was marked with men yelling and pointing." He shrugged. "I am afraid, gentlemen, that you will just have to wait here until the drought breaks."

"And how long will that be?" asked the commander.

"Who knows," the pilot shrugged again. "It's been six years in the making; it could be six more before it's over. In the meantime, let's have another drink, then you can go through that." He pointed to a bushel basket filled with envelopes and papers. "Communiqués," Mauritzen amplified. "They may take six years to read, as well."

2

Spicer-Simson read his communiqués first thing after morning mess. It didn't take him six years to go through them all since many were for other expedition members, and he had Eastwood pass those around. There was one from Amy, but he didn't open it right away. Instead, he turned to the two that were addressed to him from Hope.

One had been sent from Kabalo. Hope apologized for taking the initiative in leaving Musanga and going all the way to Kabalo, but claimed that Tyrer, bearing Spicer-Simson's orders not to leave Bukama, hadn't arrived before he left but had caught up with him in Kabalo. "And it's a good thing I did make the journey," Hope continued with what Spicer-Simson considered pitiful excuses. "The river has been well worth reporting as most of the previous reports are erroneous." He went on to give details of the river—where it was passable or too shallow or rocky and where there were villages large enough to barter for fresh food.

"At present," the message wound up, "there are no steamers prepared to make the journey upstream until the rains come, and I have a touch of fever, making my return to Bukama impossible at this time. Tyrer and I will proceed on to Lukuga to ascertain the condition of the rail line, to make initial contact with the Belgian authorities at Lukuga, and to scout out possible locations for our encampment on the lake."

"Why you presumptuous nit," Spicer-Simson spat as he read this last. "How dare you pretend to usurp my authority?"

He tore into Hope's second message with venom and was even less pleased. "No food stores have been sent for us," Hope wrote. "And the Belgians have made absolutely no preparations against our arrival. Major Stinghlamber, the commandant at the fort, tells me that he knew you'd been sent but that he has no great faith that you'll arrive."

"Bloody Belgian tosspot," Spicer-Simson growled. "We're sneaking up your backsides, and you don't even realize we're here."

There was a third communiqué from Hope, and it was the most revealing since it hadn't been addressed to Spicer-Simson but to Rear Admiral Gillatt, the senior naval officer in Simonstown. How it ended up in his pile of messages, Spicer-Simson had no way of knowing, but he opened it anyway.

In the long, rambling, disjointed, and nearly irrational message, Hope ranted that Lee's route was an idiot's dream and that Spicer-Simson was little wiser in following it. He also claimed that the commander had sent him on ahead but then had cut off all communications and left him completely unsupported.

By the time he'd finished Hope's obviously drunken rant, Spicer-Simson was boiling inside. He immediately penned a spicily worded message to the Admiralty, berating Hope and declaring he would arrest him for incompetence, disregard of authority, and public drunkenness at the earliest opportunity, most probably in Kabalo. He would send the man on to Cape Town, and if the Admiralty were wise, it would lock him in the same cell as John Lee.

Now that he'd mentioned Lee, Spicer-Simson took one more opportunity to criticize his former sub-commander for gross negligence in surveying and preparing the overland route.

"Before my arrival, it was nothing more than a cleared track not more than six feet wide,"Spicer-Simson scribbled. "By the time we passed, even that had narrowed considerably and had rapidly become thick bush again, except for the narrow footpath used by the natives." He also criticized the director of Naval Victualing for failing to deliver the proper supplies to Lukuga. "I shall arrive at Lukuga on schedule," he concluded, "but without additional stores, I fear I shall not be able to complete my mission."

He called in Dudley to take the message to the telegraph at the railhead, then feeling much better, he went outside to watch the preparations for the river journey.

So far, all the expedition gear and supplies, except for the boats, had been unloaded from the train, and a chain of African workers was passing the boxes, crates, tins, and drums from the depot to the base of the wharf.

Wainwright was on the wharf, directing the unloading, while Eastwood stood by, taking inventory. Hanschell was there, too, making sure his precious medical stores were accounted for and intact. Spicer-Simson joined them, but after a time, he lost interest in the proceedings. Lighting a cigarette and absently inserting it into the cigarette holder, he leaned on his cane and stared for some time at the broad but shallow Lualaba as it sat sluggishly around the wharf's pilings. Even if the water was low, the river was a lush and lovely sight after all that damn desert bush. He wished Amy were here to see it.

His reverie was broken by a voice calling urgently from the bank. "Commander, message for you, sir!"

It was Dudley, and he ran across the wharf and handed a paper to the commander, who grabbed it like a frustrated terrier going for the throat of some hapless rat.

"Hah!" he grunted as he read the message. "Hah!"

"What's it say, Commander?"

"It's from Admiral Gamble. He says he is pleased with our headway, and he commends my resourcefulness and perseverance." With a sudden movement, Spicer-Simson crumpled the paper and threw it into the water, where it simply soaked, unfolded, and floated on the surface of the water without moving an inch downstream. "Look at that bloody water!" Spicer-Simson shouted. "It's supposed to be a goddamn river, and instead it's more stagnant than a fucking gutter in Dublin!"

"Perhaps it will rain," Hanschell said hopefully. "After all, the rainy season is due to start...."

"Look at the goddamn sky, Doctor," Spicer-Simson growled without himself looking upward. "Does that look like impending rain?"

"I can't say it does, but...."

"But nothing. This damn joke of a river has—what would you say, Wainwright—perhaps two feet of water?"

"That's about the depth of it, sir."

"It'll take a month of steady rain to fill this up enough for the steamer to come upstream."

There was silence for several long minutes as Spicer-Simson brooded. Hanschell kept quiet for fear of inciting another outburst, Dudley stood by embarrassedly, and Wainwright wrinkled his brow.

"You know, sir," Wainwright said at last. "There might be a way."

"I'll entertain any suggestion, Lieutenant, even stupidity."

"I first thought that we might have to put the boats on the trailers again and haul them down the river as far as necessary. But I've a better idea. You know what they say about Mohammed, sir. If the mountain won't come to him, he's got to go to the mountain."

"Meaning?" the commander tapped his cane impatiently against the wooden dock.

"If the steamer can't come to us, let's go to the steamer."

"As I so patiently explained to the good doctor, Lieutenant, the water is too damn low. Our boats need four feet of draft and that," he jabbed his stick at the water, "is only two damn feet."

"We need four feet if we use the engines, sir, but not if we pole and row."

"Go on, Lieutenant." Spicer-Simson's clouded brow took on a hint of sunshine.

"We've got plenty of empty petrol tins and oil drums. We've been using them to carry water for the traction engines, and now we can use them to carry air. We can lash them all around the boats and help lift the hulls out of the water enough to make our way downstream. And downstream, we're bound to hit deeper water where we can run the engines."

"What about snags?" asked Dudley.

"Damn the snags," Wainwright said sharply. "We've seen worse on this trip and pissed on it. Piss on the snags. With your permission, Commander."

"What about the stores?" asked Hanschell, a little surprised at Wainwright's vehemence but posing a question to a problem that couldn't be pissed on.

"You speak Bantu, don't you, Dudley?" the commander asked.

"I understand several dialects, sir."

"Go 'round abouts here and requisition as many native canoes as you can. Get some names from Captain Mauritzen. Eastwood can augment our stores by purchasing what we need from the locals. Then load the stores in the canoes and proceed downstream to Musanga. When you arrive there, load the stores on the steamer, and await our arrival."

"Very good, sir."

"And Dudley, while you're at it, get some native paddlers to help with *Mimi* and *Toutou*."

"How many would you like, sir?"

"A couple of dozen, I'd say."

"Very good, sir." Dudley hurried off.

"Wainwright, let's go have a chat with Captain Mauritzen."

Spicer-Simson let Wainwright outline his plan, then he asked, "Can it be done? Will you help pilot?"

"Piloting is what I'm here for," Mauritzen replied, squinting across the river. "As for the rest, it's not a bad plan. Certainly you have a better chance getting down to the *Constantin* than she does making it this far upstream. When do you propose to leave?"

"I think we can be ready by Wednesday morning," Wainwright said.

"I will be ready."

"And so will we," Spicer-Simson said pointedly to Wainwright. "Lieutenant, get the boats off their cradles and into the water, but before you do, prepare them for shallow water travel."

Wainwright headed toward the boats, which were still parked on their flatcars on a short siding. He found Waterhouse and told him of the plans.

"We're going to put them in the river tomorrow?" the chief asked, a note of doubt in his voice. "Do you think it's a good idea, sir? The boats have been through some pretty tough times."

"You don't think them seaworthy?"

The chief shook his head. "Couldn't say, sir. But I'd be happier to test them ashore than in the water."

"Look at that water, chief." The chief looked.

"How deep is it?"

"Two feet or so, sir."

"Then the boats won't have far to sink."

"I suppose not, sir. Er, with the locos gone, how are we going to get them off the rail cars and into the water?"

"We can drag them off if necessary," Wainwright said, "but I think I have a better idea. Let's go have a talk with the stationmaster.

3

The next morning, the stationmaster produced a small gang of track layers, who proceeded to run a short spur from the railhead to the river and down the bank. By noon, the tracks were complete, and a station donkey engine backed each trailer down the spur to the water. Using a method similar to the one he'd devised to get the boats onto the tracks back at the covered bridge, Wainwright had the boats in their cradles lowered onto sleepers. Then the cradles were slid down the tracks into the water, where the boats floated free.

Wainwright had taken the precaution of having their engines removed, and it was a good thing because the boats didn't float for long. All the dryness and heat had opened seams in both hulls, and *Mimi* had sustained some damage as well when it careened down the Mitumba escarpment. Both hulls shipped enough water to take them to the bottom within minutes of their launching.

"No matter," Waterhouse said, staring at the gunwales sticking two feet above the surface. "Overnight in the river is probably the best thing for them. Tighten 'em right up."

And it did. The wood swelled, and both boats were bailed out and afloat by noon the next day. Wainwright ordered the engines reinstalled. He also had the boats' cradles dismantled and put in with the other gear. "We'll want them again when we get to Kabalo," he told Cross.

By the afternoon of the following day, the boats' engines were in place, and the next morning, Spicer-Simson boarded them for inspection. He wore his full-dress uniform and ordered the rest of the expedition to dress up, as well.

"It is the official launching of the craft," the commander explained to the doctor. "We must strive to adhere to protocol even if we are in the jungle."

The Africans who'd been hired to paddle were duly impressed, and so was the commander, though for different reasons.

"Lieutenant, is that water I see in the hold?" They were aboard *Mimi*.

"Yes, sir," Wainwright said. "One of the seams is still loose from that crash in the mountains. But once she's swelled some more, she'll be tight."

"Very good, Lieutenant. It seems we've taken proper care of our darlings despite the heat and other slight mishaps."

"Yes, sir."

Toutou was in even better shape, and the interior was completely dry.

"Very good," remarked the commander. "Now where the hell is Dudley with my canoes?"

Dudley arrived by late afternoon.

"Did you find canoes and paddlers?" Spicer-Simson demanded.

"I did, sir," Dudley replied. "I've contracted with about fifty natives who own among them several large canoes."

"Excellent," Spicer-Simson nodded. "But we'll need more to carry our stores."

"There's more, sir, but not canoes."

"More?" the commander said expectantly. "Get to it, man! Don't have me guessing!"

"Two barges, sir. They're small and simple, but they look large enough to haul most of our gear and supplies."

"Capital work, Sublieutenant."

"Will you want me to take command of the barges as well as the canoes?"

Dudley was obviously hoping for a chance to command his very own flotilla.

"Certainly. I can think of no better use for your talents. Take Tait with you."

"Tait?" The name came from both Dudley and Wainwright.

"Certainly, gentlemen," Spicer-Simson said. "Tait's a competent enough sailor for this mission. You don't agree, Wainwright?"

Wainwright caught his surprised expression and erased it from his face, even if it was too late.

"If I recall, sir, Tait was with the London Scottish Regiment before volunteering for the expedition."

"So he was. He and Mollison were serving at an officer's training camp."

"I didn't think he had any experience as a sailor, that's all."

"Of course he doesn't. Good God, man, they're going down a river. Hard to get lost doing that, eh?" The commander laughed. "Anyway, the natives will be doing the paddling and river navigation under Dudley's command. I'm sending Tait because he's a huge specimen. I don't want these natives to get the idea that they can make off with our stores. Tait will keep them in line. Give them some of that rugby heave-ho, what?"

"Yes, sir," Wainwright said. "Put that way, it makes perfect sense."

4

On October 4, Dudley and Tait set out with the majority of the remaining supplies, including sacks and casks of food—mostly rice and flour—and much of the petrol. The two modest river barges took about half, and the rest was loaded into the twelve huge dugout canoes that would pull the barges. Three ratings were assigned to go along to help guard the boats and their cargo.

Two days later, *Mimi* and *Toutou* followed. Wainwright had supervised the lashing of the empty tins and drums to the hulls of the boats. There was absolutely no thought of running the engines for fear of damaging the propellers and shafts, and some of the floats were tucked under the bilges to reduce the boats' draft and protect the shaft brackets and propellers. A team of eighteen paddlers was assigned to each boat and armed with long paddles to propel the boats and heavy poles to help work the hulls through the shallows they were certain to encounter.

They started just after noon mess, Mauritzen taking the lead in a canoe and *Mimi* and *Toutou* coming next, followed by several more canoes carrying the remainder of the gear, supplies, guns, and ammunition. At the commander's insistence, both boats were mounted with their stern Maxim guns.

"We're on the water, now," he said. "It is our duty to arm our craft. Besides, no telling what we might encounter, and we'll want to be prepared for any eventuality."

Spicer-Simson was aboard *Mimi*, as were Hanschell, Waterhouse, and a couple of the ratings. Wainwright was in charge of *Toutou*, and several expedition members were with him, too. The rest of the expedition was riding in the canoes along with the stores.

The men were in a fine mood. The Africans were doing all the work with their long-bladed paddles, so the expedition members could simply sit back and enjoy the ride.

"This is the way to travel," Hanschell heard Waterhouse comment in a satisfied voice. "All we need now is a pint and a pipe."

Yes it is, Hanschell agreed silently, listening to the Africans chanting together to keep time as their paddles dipped, surged, raised, and dipped again. He glanced up at Spicer-Simson, who was standing proudly next to the helmsman, staring straight ahead. For once, the commander seemed at peace with himself, and Hanschell was glad. The past two months had been hard, harder than any of them had imagined they might be, but now all that was past and they would soon be on Tanganyika. Hanschell glanced back, watching the wharves of Bukama shrink in the growing distance, then he turned to observe the passing scenery and the few visible kingfishers and hornbills.

Suddenly, *Mimi* gave a lurch that jerked Hanschell out of his seat and nearly threw Spicer-Simson onto the foredeck.

They'd run aground.

The commander didn't look pleased, but he held his temper and started to direct the Africans to free the boat from the mud bar, but he needn't have bothered. About half of them jumped overboard and began rocking the boat back and forth to loosen and lift it while the men on the boat shoved their poles into the bottom, gradually urging the boat across the bar. All the while, Mauritzen stood in his canoe, rifle at the ready, eyes searching steadily for crocodiles.

"Do you expect they'll attack?" Hanschell called out to the Dane, who shook his head.

"Not really. Too many people and too much splashing. They're really very sneaky creatures and not nearly as brave as they are fearsome to look at. They'd rather creep up on a solitary man and drag him into the water with as little fuss as possible."

Nonetheless, Mauritzen's eyes never once stopped scanning the tepid flow for any disturbance that might signal the movement of a large body under the surface. No crocs appeared, however, and in relatively short order, *Mimi* was afloat again, and when *Toutou* had been treated the same, the paddling resumed.

But just around the next bend another mud bar waited. And another around the following bend. Some of the bars were so shallow that all the gear had to be removed to the rear boat to make the lead one light enough to lift over the mud. Once the lead boat was past the bar, all the gear was transferred to it while the second boat was hauled over. The expedition's free ride was at an end, and everyone had to pitch in to keep the boats moving downstream.

At last, after only three grueling miles, Spicer-Simson called a halt for the night.

"This isn't any better than the desert," Waterhouse started to complain, but he quit when Wainwright reminded him that here, at least, they could take a bath and drink their fill.

But Wainwright hadn't bargained on the insects that descended on the camp as soon as dusk settled. The grunt, cry, and scream of the wildlife in the river and the jungle along the banks was soon overlaid with the constant swish and swat of fly whisks trying to beat back the six-legged hordes.

The next morning they set out early, but the going was, if anything, worse than the day before. There were mud bars at nearly every bend, and there were a lot of bends as the shallow river twisted across the landscape, meandering its gradual way downhill into the Congo Basin. And they had to watch constantly for snags that might tear into the hulls. That day, they ran aground fourteen times in twelve miles. Each time, some of the workers stood in the water to lift the boats, while others onboard poled them forward, occasionally portaging the boats for hundreds of yards before there was water enough to float them again.

"Good God, Mauritzen!" Spicer-Simson complained toward the end of the day as the boats were dragged across yet another stretch of mud. "Isn't there some way to avoid this bloody slag?" He slapped at a mosquito the size of a housefly.

"This is nothing," Mauritzen laughed from his canoe. "Count yourself lucky that the riverbed here *is* muddy. Farther downstream are many hidden rocks that will sink you if you are not careful. And the snags get worse. You would need an experienced pilot to navigate certain stretches, but you won't have to worry, because by then your expedition will be aboard the steamer."

The promise of the steamer was all that kept the men going. Even though they were on the river and there was plenty of water, the merciless tropical sun still blasted the air and baked their flesh, and mosquitoes, biting flies, ants, and other pests were a constant irritation. When they stopped ashore to camp for the night, they quickly learned to eat before nightfall because if they didn't, their plates of food and bowls of stew almost immediately would be inundated with insects struggling to breathe as much as to eat.

By mid morning the next day, the river became wider and deeper, and the boats were paddled for long distances without incident. Tall grasses, reeds, and ferns lined the banks, and beyond them, dense forest alternated with prairie thick with herds of antelope, waterbuck, and other creatures. Here and there, elephants, rhinoceroses, and even lions stood on the bank, watching with suspicion as the boats glided by.

Everyone was glad that, by now, the mud bars were a thing of the past because whole stretches of the river here were infested with crocodiles. For a time, the men took potshots at them, causing many to become meals for their fellows, but they soon tired of the sport.

Hippos were seen, too, but Mauritzen warned against antagonizing them.

"Make one angry," he said, "and it'll come right up under your canoe and chew it in half. And anyone in it, too."

The one creature they didn't see much of was man. Although an occasional village cropped up on the bank of this sad land through which they passed, for the most part, the river margins were simply empty of humans. More than once, the men remarked on the eeriness of passing through so much lush and sudden beauty without evidence of a single person. Hanschell, consulting his map, thought he knew why. Tsetse infestation was rampant in the area, and the pest and the dreaded disease it carried would have made human habitation of any duration problematic at best. The truth of his assumption soon became all too obvious to everyone.

"We're making good time," Spicer-Simson commented. "Have we reached a deeper part of the river?"

"We are approaching Lake Upemba," Mauritzen replied. It is deeper along here, but also very swampy. I advise your men to wear long sleeves and hats. Have you any mosquito netting?"

"Do we need it? Are mosquitoes especially bad here?"

"Not mosquitoes," Mauritzen said. "Tsetse flies."

Hanschell was immediately attentive.

"We must be careful, Commander," he said. "They carry African sleeping sickness."

"I know that, Doctor," Spicer-Simson snapped. "You've reminded me often enough. But you've taken care of all that, haven't you? You've inoculated us against just about everything except lust, and I sometimes wonder about that."

"We may be inoculated against the disease," Hanschell said, "but not against the flies' bites."

"Very well, Doctor. Proceed to inform the men to change uniform."

Grumbling that the temperature was already hot enough without having to bundle up, the men lackadaisically began to change. Some were not quick enough. One instant the air was perfectly clear, the next the boats were enveloped in a cloud of hungry insects.

Even the Africans squawked and slapped at the hordes of mini-mouths tearing tiny chunks of flesh from their bodies. Progress suddenly went from sluggish to rapid, and the paddlers and polers worked so furiously that *Mimi* and *Toutou* actually left wakes. In several minutes the cloud of insects was left behind.

An hour later, Lake Upemba came into sight around a bend. It was really little more than a several-square-mile depression into which the Lualaba spread out among trees and hummocks of mud sprouting brush and weeds. Suddenly the smooth water ahead was obscured as a dark, cloud-like mass drifted across the river.

"Will you look at that?" said Mauritzen.

"Is that what I think it is?" asked the commander.

"Tsetse," affirmed the Dane. "I've been here nearly twenty years, and I've never seen such a swarm."

"Commander," said Leighton, who was manning the aft Maxim gun. "I don't think I can make it through that. Not after that last bunch." His eyes were white with fear.

"What do you think, Mauritzen. Can we risk running the channel under our own power?"

"We haven't had any problems all morning," the pilot said. "And to tell the truth, I'm not anxious to spend any more time than I have to in that mess."

"Wainwright!" Spicer-Simson yelled across the water to *Toutou*. "Unlash your flotation and start your engines. We're going to run the channel through that!" He waved forward toward the cloud of insects. The cloud had stopped drifting, as if sensing the vulnerable flesh that the currents wafted toward it.

"What about the canoes?" asked Wainwright.

"We'll take the larger ones in tow," Spicer-Simson said. "The rest will have to fend for themselves."

The preparations were made quickly, and in minutes, the engines of both boats were roaring in the heavy air. The Africans aboard cowered at the sound, but as soon as the craft began to move, they cried out in delight and stood up to catch the wind.

They didn't stand up for long. In less than two minutes, everyone aboard was pelted with angry, ravenous flies that clung and bit, slipped in the wind, and bit again. Some of the men went so mad with slapping that they struck their mates, who either did not notice or appreciated the assistance. The cloud was so dense, deep, and vicious that it seemed as if they were splitting the scud of Hades.

"Goddamn!" Leighton cried over and over with rising hysteria, slapping and raking at his face and arms. "Goddamn bloody bastards!" Suddenly he stopped his wild movements and sat transfixed for an instant in the stern of the boat. Then he was in motion again, this time bending over his weapon. "Bastards!" he screamed as he cocked the Maxim gun. "You bloody goddamn bastards!"

The roar of the boats' engines was loud enough in the preternatural stillness of the heat on the water, but when Leighton's Maxim gun vented his rage on the tsetse, it truly seemed as if Hell's pandemonium had cracked through into reality on Lake Upemba. Every bird—whole flocks —within earshot suddenly took to the sky, nearly blotting out the sun and adding their own screeching to the din.

"Bastards!" Leighton screamed through a welted, pelted, tear-swollen face, but his words were lost in the shattered air. Spent shell casings from the Maxim hissed into the water and clattered into the boat's cockpit.

Chief Waterhouse managed to lunge through the clot of terrified Africans who shrank back from the demon in the stern who sought to blast the Kaiser's entire air force out of the sky with his one Maxim. He tackled the maddened rating and brought him to the deck. The Maxim gun, suddenly stilled, bobbed and twisted crazily on its mounting, and the Africans yelled in panic and surged forward, nearly ramming Spicer-Simson and Hanschell into the helmsman.

Then, miraculously, they were through the tsetse, and out on the calm, swampy waters of the lake.

Waterhouse dragged Leighton forward and deposited him in front of the commander, who looked down at the blubbering rating.

"I'm sorry, sir," Leighton gasped out between sobs. "I couldn't help myself."

Suddenly Spicer-Simson laughed, and the men looked at him as if he, too, might go mad.

"Get up, Mr. Leighton. No harm done. Just promise me one thing."

"Anything, sir. Anything."

"Promise that when you fight the Germans, you'll score more hits than you did against the tsetse."

5

The flight from the tsetse and Leighton's maddened machine gun frenzy lightened the mood after three days of slogging across mud bars, but everyone sobered when they landed that evening to camp. Within minutes, the ratings sent to scout the immediate area around the camp discovered an abandoned village surrounded by an overgrown boma, and they hurried back to fetch Spicer-Simson.

Spicer-Simson followed the ratings to the village, accompanied by the other officers and men, Hanschell included. They smelled the place before they could see it buried in the jungle. It stank of despair, death, and decay. All the huts were caved in or completely collapsed, and brush and tall grass overgrew the whole area. As they made their way through the eerie scene, their feet crunched on broken household goods and calabashes, and hordes of bott flies, disturbed by their passage, lofted from the foliage and filled the air with a dark buzz.

"It stinks," Berry said, slapping at the flies. Then he jumped and cursed as a large rat scurried over the toe of his boot. Several other men also flinched or kicked out as more rats, panicked by the men's approach, ran for safety. The rats were all fat.

They quickly discovered that the village hadn't been abandoned—it had died. Many of the potsherds scattered in the grass were actually human bones, broken and gnawed by scavengers.

"No wonder we haven't seen many natives since we started," Wainwright said. "What do you think caused this?"

"The thing I've been warning against," Hanschell said. "Sleeping sickness carried by tsetse flies."

"I hope Leighton got some of the bastards this afternoon," Berry quipped, but nobody laughed.

"Now you know why I've demanded that we take such stringent precautions," Hanschell said.

"And we shall continue to take them," Spicer-Simson replied. "I've seen this sort of thing before on the Gambia. Entire villages wiped out. Tonight we'd best surround the camp with fires. If there are many villages like this around here, the predators in the area will have learned a taste for human flesh, and we don't want to give them any openings. Wainwright, double the guard."

6

After a restless night, the expedition was up at daybreak, anxious to leave this haunted place. Spicer-Simson had the men fire up the engines and take all the canoes in tow, much to the delight of the African paddlers. By noon, they'd caught up with Dudley's flotilla, and after a short conference, they continued on down river, leaving Dudley's craft to follow at its own pace.

That night, they camped early and were up again at dawn and on the river within an hour. As the murky green water slid past the boats' hulls and the sun rose in the sky, the toil of the past few days melted in the heat and washed lazily away. It almost seemed to Hanschell that they were on some sort of grand holiday cruise, and as the day lengthened, so did his contentment.

Before noon, as *Mimi* purred around a long, slow bend under Waterhouse's gentle hand, Mauritzen, who was seated on the edge of the foredeck next to the chief called out, "The *Constantin de Burlay* coming up on the starboard bow!"

Spicer-Simson, who had been lounging in the rear of the cockpit, came forward. A stubby smokestack could be seen through the treetops at the end of the long bend, and within a few minutes, the hull of the steamer came into view.

"Remember what I told you about Captain Blaes," Mauritzen warned Spicer-Simson as Waterhouse pulled up at a dock near the steamer. "He's a pig, and he'll treat you the same."

"I believe so, if his ship is any indication," Spicer-Simson said scornfully.

They climbed onto the dock, and as soon as *Toutou* landed and Wainwright joined them, they approached the steamer. It *was* a sorry sight—dirty and sorely in need of paint—but Wainwright thought it looked seaworthy enough.

They were met at the gangway by a stubble-faced white man in slovenly khakis who told them to wait before disappearing into the cabin.

A moment later, a hoarse voice exploded, "Gotverdomme!" and the cabin door burst open. Out bustled Captain Blaes, followed by the man in khaki, and in half a dozen flat-footed, heavy strides, he was at the head of the gangway, surveying the expedition officers. They surveyed him back. He was large and thick-waisted, his red complexion mostly hidden by a week's growth of scruffy black beard. The bulbous, heavily veined nose that jutted above the beard was plain to see, as was his clear disdain.

"Mauritzen," he snorted, hooking dirty thumbs underneath dirtier suspenders that held up baggy trousers that might once have been khaki but now were a mottled gray. And that was the last thing the Englishmen understood, though Blaes went on for some time, punctuating his sentences with hawked globs of phlegm that he spat into the river or onto the deck depending on the direction he happened to be facing at the time.

Finally, he ran down, and Mauritzen turned to Spicer-Simson, a carefully blank look on his face.

"Blaes only speaks Flemish," he began but was immediately interrupted by a shout from the captain.

"Very well," Mauritzen began again. "He prefers to be called Capitaine de Steamer. As you can see, the Capitaine de Steamer doesn't like me very much. He thinks I'm a pompous know-it-all because he knows so little. He's been captain of the *Constantin* for eight years, and still he does not know the channel he pilots, and he hates me because I do. He thinks I want his job." Mauritzen shrugged. "Soon we will be at my home, and I will let you judge for yourself if I would lower myself to pilot this scow for longer than is absolutely necessary."

He paused to catch his breath and restore the bland look to his features. "At any rate, he will brook no interference with his command. He says that if you do interfere, he will shoot you, as is the right of any captain aboard his own vessel."

"Pompous swine, isn't he?" Spicer-Simson said with a sneer, taking the cigarette holder out of his mouth and deliberately flicking ash onto the gangway. "Well, if he gives us any trouble, we have enough ordnance to sink his little steamer before he has a chance to hitch up those filthy trousers. But tell him that we have no intention of interfering with his command, only that we wish a ride downstream." The commander replaced the cigarette holder between his teeth, baring them in what was as much a snarl as a smile.

Mauritzen translated, and Blaes, looking satisfied if no less wary, spoke again.

"He says he has already been paid by the Belgians to transport you, so if he must, he must. But he will have no further dealings with us personally. He's leaving all that up to his purser, Mr. Holmquist." Mauritzen waved to the man in khaki standing just behind Blaes. "He is Belgian, but he understands English."

"Very well," Spicer-Simson replied. "Tell him it's agreed. I'll have our quartermaster make arrangements with his purser."

Blaes merely grunted in reply to Mauritzen's translation before he turned on his heel and went back into the cabin, leaving Holmquist alone on deck.

"Come on, man," Spicer-Simson waved impatiently to the purser. "Mr. Wainwright here will introduce you to Mr. Eastwood."

7

It took Eastwood a couple of hours to dicker with Holmquist and for Holmquist to relay the expedition's wants to the Capitaine de Steamer and for Blaes to deign a reply, but at last it was done. Blaes would hold his ship until Dudley's supply flotilla arrived, and after taking the supplies aboard, he would steam down river to Kadja where the rest of the expedition would be waiting with *Mimi* and *Toutou*. They also paid off most of the paddlers, purchasing in the bargain two of the large canoes to haul their supplies.

With the negotiations complete, the expedition was off again, though they left Eastwood behind to deal with Dudley's flotilla when it arrived—and the blustery Capitaine de Steamer. Now that the water was deeper and the boats could travel under their own power and were pulling only one canoe each instead of several, they made excellent time. Mauritzen sat next to Spicer-Simson in *Mimi*'s cockpit, keeping his eye on the channel ahead, occasionally giving instructions to Waterhouse at the wheel.

They traveled a total of twenty-four miles before Spicer-Simson called a halt. The only entertainment was a hippopotamus that followed them for nearly three hours, poking its head above water every once in a while. Despite Mauritzen's warnings against antagonizing the huge animals, several of the ratings took potshots at it until Hanschell complained, and the commander ordered them to cease fire.

They were off to an early start the next morning, and although Spicer-Simson was hoping they'd make as good a time as they had the day before, Mauritzen warned him that they would have to go slowly once they reached Lake Kisali.

If Spicer-Simson had wondered why a lake should slow them down, he found out soon enough. Lake Kisali, at first, seemed no more than a widening of the river that soon spread out even further into swamp. But if the river was wide, again it was shallow, and Mauritzen stationed himself out on *Mimi*'s foredeck where he could direct Waterhouse through the single, narrow channel that wound its way across the lake.

At first, the boats, almost inching forward, had to muscle their way through thickets of reeds and dodge hammocks of mud topped with rough grass. Before long, however, the reeds were behind them, though their forward speed had to remain muted as they followed the channel.

But nobody paid much attention to the channel or its navigation once the boats were clear of the reeds because now the men could see the incredible proliferation of waterfowl that the reeds had masked from their sight. Everything from cranes, storks, and marabou to cormorants, ducks, geese, and even osprey had gathered at Kisali's shallow waters to stalk the insects and fish and be, in turn, stalked by crocodiles and other predators. Hanschell thought he'd never seen such a display of color and flurry of motion, especially when the boats passed near a hammock or half-submerged island and hundreds, or sometimes even thousands, of the birds would take to the air in dazzling waves of color.

"Quite as if we're leaving a wake in the heavens," Spicer-Simson commented, blowing a puff of cigarette smoke from between his teeth without removing the cigarette holder. Hanschell thought he was right. For hundreds of yards behind them, the air was filled with screaming, displaced flocks seeking to land again in the choicest spots.

"Look," Mauritzen pointed. "That's where another river, the Lufira, enters the lake. The water gets deeper after that, and we'll be able to run at a higher speed. But if you don't mind, Commander, I'd like to stop there for a short time."

"There" was a houseboat that seemed but a small speck as it floated out in the middle of the lake.

"It is my home," Mauritzen explained. "The flies aren't so bad this far from shore, and my wife is safe from unwelcome visitors—man and beast alike."

When *Mimi* and *Toutou* had been brought alongside the houseboat, Hanschell could see that it was a good-sized lodging build on a barge. A small engine was at one end, and a small motorboat and a couple of canoes were moored to the other.

"Perhaps you and Doctor Hanschell would care to join my wife and me for a civilized lunch?" Mauritzen said. "I'm sorry, but there isn't enough for the rest of your men."

"No matter," Spicer-Simson said. "Wainwright, we'll be here for a time. Have the men eat, but remain prepared for travel." Then he and Hanschell followed Mauritzen aboard the houseboat.

A servant met them at the door to take their hats, and the commander and the doctor found themselves in a room that was positively sumptuous considering its unusual construction and location. There were paintings and bookshelves and real furniture, and from another room wafted the notes of a Chopin piano concerto.

Mauritzen spoke with the servant, who went through a door, and a second later, the piano twinkled to a halt. A young blond woman came into the room. She was dressed in a white blouse and skirt, and she gave Mauritzen a hug and a kiss before turning to the guests.

"My dear, let me introduce Commander Spicer-Simson and Doctor Hanschell. Gentlemen, this is my wife."

"How lovely to have visitors," Mrs. Mauritzen said in perfect English that bore only a lilting trace of Danish.

Spicer-Simson took their hostess's hand and gave it a quick and chaste kiss, and Hanschell bowed.

The servant reappeared with drinks, and the four sat and chatted until lunch was ready.

"I imagine you get lonely out here," Hanschell ventured.

"Lonely for Europe and all those chilly, dirty cities?" She laughed. "I'd have to go to the zoo to get a glimpse of all the glorious creatures I see here daily. Believe me, gentlemen, I'm quite content."

"I'm afraid I've completely corrupted my wife," Mauritzen said. "She loves it here as much as I do. I don't think I'll ever get her to leave."

Lunch was soon served and, to Hanschell's taste, all too soon concluded. Afterward, as they boarded *Mimi*, Spicer-Simson and Hanschell thanked Mrs. Mauritzen.

"I feel rejuvenated," the commander said. "I expect it won't be long until Mrs. Mauritzen has civilized all of central Africa."

"I'm not sure I could," she replied. "Or that I'd want to. But thank you, anyway." After she kissed her husband, the expedition was off again, the strains of Chopin trailing them across the water.

"I'm quite astonished at your home and wife," Hanschell said to Mauritzen. "It seems as if a touch of European elegance lies in the heart of the Dark Continent."

"Thank you," the Dane nodded.

"I don't know why you should be surprised, old man," Spicer-Simson said. "European women have a peculiar propensity for being European under all circumstances. I remember how they were in Peking during the Boxer Rising, when the country was overrun by armed bands of fanatics shouting 'Death to the foreigners!' and looting, burning, and murdering. Of course, things should never have been allowed to come to such a pass. The situation had been badly handled. By the time I arrived on the scene, the nine hundred European residents, including two-hundred-and-fifty women and children, had taken refuge in the British Legation, and an international force was trying to get through to them. We had to go up the Pei-ho River to Tien-tsin and, from there, fight our way to Peking, ninety miles farther on. The ancient Chinese city was surrounded by a wall thirty feet high. Although I had been gravely wounded, I easily outdistanced my men and was the first to scale the wall. Fighting off the Boxers guarding the Shaow-men Gate, I threw it open from the inside, and the bluejackets rushed in. When we reached the legation by the sluice-gate entrance, I had the biggest surprise of my life. We were on a green, sunny lawn where beautifully dressed women were strolling up and down with parasols. They were half-starved, and they had been living for the past two months in the constant expectation of being massacred, but they looked as if they were at an English garden party. The contrast with the filth and horror surrounding them on all sides could not have been greater. I was reminded of this when we entered the Mauritzen's barge."

"There was, however, one important difference, sir," Hanschell said with a straight face. This was the first time he'd heard that Spicer-Simson was the hero of the Boxer Rebellion. "Mrs. Mauritzen shows not the least desire to be rescued."

Both the commander and Mauritzen laughed.

"And excellent commendation of Mr. Mauritzen," Spicer-Simson said. "And of Africa, too."

Within an hour, they were off the lake and back on the river, and by late afternoon, they arrived at Kadja, where they tied up to a pair of flimsy docks to wait for the *Constantine* to arrive with the rest of the expedition.

8

"Look, sir!" shouted Lamont, who was helping Cross tinker with *Toutou*'s engine. "The steamer!"

Wainwright heard the shout and looked upstream to see smoke from the *Constantin de Burlay*'s stack smoking over the trees that masked the first bend. The lieutenant was down at the docks trying to figure out just how he was going to get the boats hoisted aboard the *Constantin* when it arrived since the steamer itself had no crane, nor was there one at Kadja.

"Run up and get the commander," Wainwright told Lamont, then he and Cross watched as the steamer came into view, zigzagged downstream, and sloshed to a halt at the largest wharf, its hull crashing against the pilings in the process.

"That captain hasn't got a light touch, does he?" Cross commented.

Wainwright agreed, but he didn't say anything as he left *Toutou* and went over to the wharf, where a couple of African deckhands were tying off the steamer. He got there just as Dudley hopped off the ship, followed by Tait and the other ratings.

"Hello," Dudley said casually, then he threw a snappy salute when he saw Spicer-Simson approach, trailed by Mauritzen.

"Ah, Mr. Dudley," the commander said. "I trust everything went well."

"I don't know if it did, sir. That Captain Blaes can't steer worth a shit...I mean, sir, the trip seemed more awkward than was necessary. I think I could have done better myself."

Tubby Eastwood appeared from the cabin.

"Everything's in order, Commander," Eastwood said. "As soon as we get your supplies and the boats loaded, we're as good as arrived in Kabalo."

"Excellent," the commander said with a big grin. He turned to Wainwright "Let's get the boats on board."

"There is one slight problem," Wainwright said.

"And what might that be?" Spicer-Simson asked, his grin souring to grimace.

"As you see," Wainwright gestured around the tiny harbor, taking in the *Constantin* as well, "they haven't any cranes."

"I'm sure you'll see to it, Wainwright," Spicer-Simson said, and he turned and went back to the hut he'd picked for his headquarters.

"Yes, sir," Wainwright said to the commander's back. "I'll see to it."

Wainwright saw to it. Without cranes, the boats couldn't simply be lifted to the steamer's deck, but after pondering the problem, Wainwright saw it as two problems, really. The first was to get the boats out of the water and to the proper height, and the second was to get them onto the deck. Once he'd broken it down like that, a solution became apparent.

He had Eastwood hire a crew of locals to help, and under his own supervision, the crew cut two trees from the nearby jungle and hewed them both to twenty feet. These were dragged to a clear section of shore in the harbor and slid down the bank small ends first, about fifteen feet apart, to serve as a crude ramp. Meanwhile, half a dozen ratings under Cross's supervision rebuilt the cradles that had come downstream with the boats. By the time all this had been accomplished, the remainder of the expedition's supplies and gear had been loaded onto the steamer.

Then Wainwright had *Toutou* floated near the sunken ends of the tree trunks while the cradle was slid down the tree trunks into the water. In a few minutes, *Toutou* was maneuvered over the cradle and fastened to it, then, using the African crew and his trusty block-and-tackle, Wainwright had the boat and cradle dragged sideways up the crude ramp. The surly Captain Blaes was on deck to watch the operation, interested despite himself.

He wasn't the only one. The entire village had come down to the water to watch and cheer the proceeding. While the African crew and many of the ratings were in the water, helping shove the cradle up the inclined rails, Spicer-Simson noticed a minor commotion rising from a group of women on the nearby bank. He went over to see what they were oohing and aahing over so excitedly, and he saw the huge fisherman from Donegal in the water with the rest of the men. The one important difference was that the rating was buck naked, his red hair standing out against his pale skin like a flame. But it wasn't his hair that had ignited the women's interest: It was his huge size. And his huge size.

"Seaman!" Spicer-Simson shouted. "Where the hell is your uniform?"

"Sir?" the Irishman asked. "I didn't want it to get wet. And these people don't dress, so I didn't think it mattered."

"Not if you're civilized," the commander snapped. "For anyone else it's disgusting! Go get some clothes on! Where do you think you are, back in Donegal?"

Flushing almost as red as his hair, the rating climbed up the bank, eliciting greater excitement from the women gathered there. He brushed through them and went over to a tree where he'd piled his clothes and began putting them on.

As soon as *Toutou* was near the top of the bank, the ropes were tied off, and Wainwright signaled to Blaes to draw the *Constantin* close. When it was near enough, Wainwright's crew, again using the block-and-tackle, now tied to the tops of tall trees on the bank, hauled the ends of the two ramp trees out of the water until they wavered just above deck height. Again Blaes eased the steamer closer, beneath the trunks, and finally, the ends were lowered onto its deck. After that, it was a simple process for the steamer's donkey engine to drag the boat and cradle along the log rails and onto the deck.

They used the same method the following day to get *Mimi* aboard, and on the morning of October 16, the steamer set off down river. Though Spicer-Simson, claiming eminent domain, tried to commandeer the vessel for his exclusive use, Blaes refused and even took to carrying a huge but obsolete and rusty revolver tucked into the waistband of his worn and greasy trousers to augment his authority. Not that the commander's desires or the captain's refusal had much to do with the steamer's ultimate passenger list and cargo. As the ship left the dock, it was sagging to the waterline not only with the expedition's men, boats, and supplies but with African civilians long used to the convenience of the riverboat, East European traders, prisoners in chains, and, thankfully, guards for those same prisoners.

And everybody seemed to have plenty of baggage. Boxes, crates, bags, and baskets were piled high on the deck, and in a large space aft, elephant tusks, the property of the East European traders, lay in a huge mound. Wood to fuel the steamer's furnace was stuffed into every space not otherwise occupied, and more wood was heaped on top of the piles or anything flat enough to hold it. Chickens and goats wandered in the rare, tortuous isles, pecking and butting anyone who walked by and leaving such frequent droppings that, after a while, no one bothered to scrape their shoes, sandals, or bare feet.

With that sort of load, the steamer responded sluggishly to the helm, and so no one was really surprised when, at the first bend in the river, the ship didn't make the turn but ran straight into the far bank. Water churned aft as the engines went into full reverse against the current and began to swing the steamer around in midstream. The engines continued to pirouette the ship in reverse until, at last, the bows were pointed in the correct direction. Then, with the engines in forward, the steamer was off again.

"That's the most peculiar maneuver I've ever witnessed," Spicer-Simson said with a look of exasperated amazement. "I do believe the Capitaine de Steamer is a bit steamed himself."

"Very likely he is, Commander," Mauritzen said blandly. "But that's not the problem. If it was merely a drunken mistake, it might be acceptable. Unfortunately, it is the Capitaine de Steamer's modus operandi for any bend in the river. I'm afraid he never learned to steer very well. He's much safer to travel with going upriver since the current is enough to help him, but going downriver, he simply cannot maintain control."

They were prophetic words. At every bend, Blaes rammed the far bank, turned around backward until he was pointing downstream, and then proceeded as if nothing had happened.

But of course, something did happen. A mere two hours after leaving Kadja, the steamer ran aground so thoroughly that its engines couldn't pull it back into the current.

9

Getting the gunboats over the mud bars had been difficult enough, but freeing the heavily loaded steamer seemed impossible. After churning enough muck off the bottom to create a fresh mud bar downstream, all the while bellowing a steady stream of gotverdommes, Captain Blaes gave up and retreated to his cabin to drink.

"The Capitaine de Steamer says we will have to wait until the rains come and make the water deeper," his purser, Holmquist, relayed to Spicer-Simson.

"Goddamn it!" the commander bellowed in turn. "You mean we're stuck here indefinitely in the middle of the river?"

Holmquist, who was, by now, completely inured to that particular bellowed curse, simply shrugged. "What else can we do?"

"If you will permit me," Mauritzen broke in. "I think I might at last be of some service."

"You think you can free this scow?" Spicer-Simson asked.

"I will try. But as soon as Blaes hears the engines, he will want me out of the pilothouse."

"We'll see to it that you're undisturbed," the commander said. "Mr. Dudley, fetch Tait and Mollison."

While the two huge rugby players guarded the pilothouse door, Mauritzen had the crew bring up a head of steam. A group of passengers were enlisted to jump over the side with shovels and dig away as much of the muck clutching at the hull as they could. Then, using an alternating forward and back motion, Mauritzen gradually worked the steamer free from the bar. By this time, it was nearly dark, and Spicer-Simson had Mauritzen anchor for the night.

It turned out that the ship had anchored within a stone's throw of a village called Mulango. The ship was such a crowded mess that the expedition went ashore to camp, and because the jungle was thick along this stretch of river, they set up in the village itself. As the men made preparations for the night, Hanschell noticed that several of the local young men were approaching the ratings. He could tell that some sort of bargain was being struck, and it wasn't long before he saw a handful of young women emerge from their huts, and the nature of the bargains became obvious. Not only were the women a good-looking lot, they wore nothing but g-strings made out of bark.

When he realized what was going on, he went over to instruct the men to exercise restraint. Despite his harangue on the dangers of venereal disease, which, he claimed, was rampant in the area thanks to centuries of incursions of Arab slave traders, he was met variously with amazement, bald insubordination, and acquiescence that was glib and obviously insincere. After several such inconclusive encounters, he sought out the commander.

"Do you realize what is going on out there, sir? The men are flocking to the native women like bees to honey."

"Is that so?" Spicer-Simson drawled. "Well, I suspect that things will work themselves out."

"But, sir!"

"Men will be men," the commander said. "If you don't like it, I suggest that you do as I plan, and turn in before you witness something you think you should not. And I warn you, Doctor, no fires."

Hanschell, at a loss for words, went to his tent, but while Spicer-Simson's injunction to turn in early did prevent him from seeing what he did not wish to see, the tent's canvas did little to block the additional sounds that were added to the usual wild cacophony of the jungle. At last, he simply stuffed some wads of cotton in his ears and finally fell asleep.

10

The presence of Tait and Mollison on guard had proved unnecessary. Blaes did not budge from his cabin, even when the ship moved again in the current the next morning. He did not emerge until nearly noon, when he reassumed command as if nothing had changed.

And nothing had, really. At the first available bend, Blaes rammed the far bank, did his strange reverse pirouette in the middle of the river, and proceeded on downstream. But after half a dozen more such incidents, the *Constantin de Burlay* finally ran aground again.

"The Capitaine de Steamer is a very determined man," Mauritzen said after he had inspected the hull's perimeter. "I am afraid that this time he has finally succeeded." He looked at Spicer-Simson. "The *Constantin* will not leave this spot until the river rises."

The commander seemed fresh out of goddamns, but he wasn't at a loss.

"I suppose we can off-load the boats here and go the rest of the distance under our own power."

"What about all the supplies?" Eastwood asked. "We can't pull that many canoes, even if we had them."

"We may just have to hire more native paddlers. Or maybe a barge. Mr. Mauritzen, are there any barges within reach?"

"Possibly several," Mauritzen said. "But there might be something better. A steamer somewhat smaller than this one is somewhere not far downstream. The *Baron Jansenn*. The captain isn't much nicer than Blaes, but he's a far better pilot. And with a shallower draft and a deepening river, it shouldn't have any trouble taking you a far as Kabalo."

"I say, Wainwright!" Spicer-Simson shouted. "Grab those men!"

He pointed to three men preparing to drop a canoe over the side. With angry looks, they returned it to the deck as Wainwright and Dudley spoke to them.

"They want to know why are preventing them from returning to their families and their villages," Wainwright translated the men's sharp words.

"Tell them I simply do not want them to return empty-handed," the commander said. "Ask them what they want to transport Dudley and a couple of other men to the *Baron Jansenn*."

A deal was quickly made, and while Eastwood went to get three bottles of whiskey, Dudley asked, "What do you want me to do, Commander?"

"Scout, Dudley. Find this *Baron Jansenn* and make an arrangement with her captain. Take Lamont and Berry with you."

Dudley and the ratings quickly put together kits and were soon in the canoe riding around the first bend downstream and out of sight.

Mid afternoon brought them to a pitiful huddle of huts hugging a wide place in the river. The only remarkable thing about the place was the two-hundred-foot stern-wheel steamer that was moored against the bank just downstream from the village's four spindly docks.

"You can let us out here," Dudley told them, and the Africans, glad to be rid of their charge, stroked to one of the rickety excuses for a dock.

"What's the name of this place?" Dudley asked as he and the ratings clambered onto the shaky planking.

"Muyumba," one of the paddlers replied, then the canoe shot back out onto the river, headed downstream.

Dudley led the ratings ashore, down the bank, and across the gangplank that stretched from the deck of the steamer to the muddy shore.

"Hello, there!" Dudley called. "Anyone aboard."

There was, but Dudley had to ferret him out. The African, who appeared to be in his late fifties, said he was the caretaker and that the crew was away on leave.

"But I want to use the ship," Dudley said.

"No, no," the caretaker said, his eyes rolling in exasperation. "You may not. I have orders. Keep ship here until crew arrives."

"When will that be?"

"Don't know," said the caretaker. "Gone to Kisangani for women and good time. River low. Maybe be gone one moon or two."

"I see." Dudley left the lighter, trailed by the ratings, and returned to the docks. There he sat with his back against a piling and perused the quiet flow of the river.

"What are we going to do, Mr. Dudley?" Lamont asked at last.

"Hush. I'm trying to think like Mr. Wainwright."

So they sat, and from time to time, curious villagers peered at them from the perimeter of the hamlet, but they were bothered only by mosquitoes, flies, and the sun in their eyes. After about half an hour, Dudley suddenly stood up.

"Did you finish thinking like Mr. Wainwright, sir?" Lamont asked.

"I have. Come on," Dudley gestured, leaving the dock.

"Where are we going, sir?" Berry asked.

"Back to the *Baron Jansenn*," Dudley replied. "Gentlemen, we're about to become pirates."

11

"Ahoy," Dudley called out as he stepped off the gangplank onto the *Baron Jansenn*'s deck. "Is anybody here?"

"I tell you no sail until crew returns," the caretaker said, poking his head out of a doorway. "Crew not be back for long, long time."

"Yes, you said that," Dudley replied, walking over. "I'm not arguing."

"Why you here, then?"

"It seems my men and I are stuck here in Muyumba," Dudley said. "And as you probably know, the villagers aren't much company for sailing men like ourselves."

"Sailing men?" the caretaker asked, looking puzzled.

"That's right. Sailing men like you and me."

"I not sail steamer." The caretaker shook his head emphatically.

"I understand that," Dudley said. "And I'm not asking you to."

"What you here for?"

"Company," Dudley said. "And being a sailing man, I've come to the only other sailing man I know of in these parts."

"Me?" The caretaker poked himself in his chest with his forefinger.

"You," Dudley said.

"What we talk about?" the caretaker asked. "I not sailing man."

"You're in charge of this ship, aren't you?"

"That not make me sailing man."

"I suppose not," Dudley said sadly. "But you're the closest thing to one here. What say you show me around your ship."

"Not sail ship!"

"Not sail ship," Dudley agreed. "Drink this." He reached into his musette bag and pulled out a bottle of rum. He'd brought a couple of bottles along to pay for canoe passage back to the *Constantine*. "Fine stuff, this," he said, holding it up so the sun shone through it, sending amber refractions all across the deck. "What say we take a snort together?"

The caretaker looked dubious, but he also looked at the bottle and wet his lips.

"Not sail ship?"

"No. Drink this."

"Two sailing men?"

"Two sailing men and a sailing man's drink."

"We drink." The caretaker reached for the bottle, but Dudley drew it back.

"Not here," Dudley said.

"Where?"

"How about the captain's cabin?"

"Not captain's cabin," the caretaker said, looking worried. "He not like."

"He left you in charge, did he not?"

The caretaker nodded reluctantly.

"Then that makes you captain pro tem."

"What captain pro tem?"

"The captain when the captain is away."

"Yes. I must be captain pro tem."

"Then you should occupy the captain's cabin."

"He not like."

"I suppose not," Dudley said. "But then he's not here to not like, is he?"

"No," came the reluctant reply.

"And he's probably got the most comfortable cabin, has he not?"

"Yes."

"Then by God, man, let's go get comfortable and drink this stuff!" He waved the bottle, and the amber refractions shimmered and shook.

"Captain's cabin this way," the caretaker said.

"That's a good fellow," Dudley said.

Half an hour later the bottle was half drained. Dudley, lounging on the captain's bunk, put its mouth to his own and upended it. The amber liquid sloshed most convincingly, though he actually did little more than wet his lips. Then he passed the bottle to the caretaker, who likewise upended it but, unlike Dudley, took a huge swallow. Since they'd started on the bottle, Dudley had sipped little more than half a dram while the caretaker had emptied the rest, and he was quite drunk.

"Give the poor bastard a little something to repay him for the dirty trick I'm about the play," the sublieutenant thought, then he shook his head as the man tried to pass the bottle back. Aloud he said, "You hold on to her for a spell, mate. I've got to go bleed my lizard."

"Bleed a lizard?" the caretaker looked blearily puzzled.

"Yeah, you know...." Dudley shook an imaginary penis, and the caretaker brightened and laughed.

"You go bleed lizard. I stay here. I captain pro tem. I drink in my cabin."

"Yes, to be sure," Dudley said as he headed out the door. Once in the corridor, he shut the door and locked it with the key he had conveniently removed from the desk beside the captain's bunk.

Dudley listened at the door, but the caretaker didn't seem to notice that he'd been made a prisoner pro tem. With a chuckle, Dudley hurried to the deck and called out. Lamont and Berry instantly appeared.

"Ready, sir?" asked Lamont.

"You and Berry get up a head of steam. I'll check to make sure everything else is shipshape. I want to get back upriver by this time tomorrow."

On his way to the wheel house, Dudley passed by the captain's cabin. A thumping was coming from the door.

"English sailing man, you let me out. Must bleed lizard."

"I'm afraid you'll have to use the porthole," Dudley called back cheerily. "I seem to have lost the key."

12

"Splendid work, Dudley," Spicer-Simson commended as the *Baron Jansenn* drew alongside the grounded *Constantin de Burlay*. Then he turned and shouted, "You there, rating! Keep those men off the deck!"

The rating complied, pressing back the African passengers who tried to crowd onto the *Baron Jansenn* in hopes of actually making it down the river sometime within the month.

It took most of a laborious day to transfer the boats and supplies from ship to ship, and the men had a devil of a time keeping the passengers of the grounded steamer from boarding the lighter one. The passengers didn't seem to understand the commander's insistence that the *Baron Jansenn* had been commandeered for the expedition's exclusive use, and the East European traders even attempted bribery. When Spicer-Simson refused, one of them made the mistake of brandishing his rifle. With surprising speed and dexterity, Spicer-Simson disarmed him, knocked him to the deck, then pitched him overboard.

"I told you I'd learned some of the Chinese fighting arts when I was there, didn't I, Doctor?" the commander asked when he noticed Hanschell staring at him, mouth agape. "They didn't call them Boxers for nothing. Dudley, post extra ratings to prevent any more of these refugees from sneaking aboard."

When preparations were nearly complete, the ratings on guard dislodged the last desperate Africans from the rail and cast off the lines. The *Jansenn*, with Mauritzen at the wheel, drifted ever so slowly down stream in the sluggish current until it was a dozen yards from the grounded steamer. Then the engines were engaged, and the wide paddlewheel at the stern of the ship churned sludgy water. Captain Blaes watched from the *Constantin*'s deck as the smaller ship pulled away.

"I don't believe he looks any too happy," Wainwright remarked to Hanschell. "I'd have thought he'd be glad to get rid of us."

It was true. The Capitaine de Steamer appeared, if possible, even more disgruntled than when they'd been his guests.

"Misery loves company," Hanschell said. "He's a man in torment, and now that he no longer has us to share his suffering, he's more miserable than before."

In five minutes, the angry expression on Blaes's face no longer was distinguishable, and in twenty minutes, the *Jansenn* was around the first bend and out of sight of the grounded *Constantine*.

Two miles later, it ran aground.

The lucky three dozen Africans who had been offered a free ride in exchange for their labor, leapt overboard with shovels and ropes, and after much digging and hauling at lines and churning of the wheel and shouting all around, the boat was dragged free.

Three more days of cautious sailing—and no further groundings—found the craft approaching Kabalo. It was late afternoon on October 22.

13

As soon as they docked, Wainwright went to the rail yard to arrange the final rail trip of one hundred and eighty miles to the lake. Kabalo wasn't as miserable as most of the river villages they'd seen. Its port and rail yard were augmented with several mud brick buildings and corrugated sheet-metal sheds, and the village itself had a fair number of huts, most of which seemed well kept. But off to one side was a nondescript cluster of dismal huts surrounded by a flimsy thorn bush boma reinforced with wire. Several armed askaris lounged near the gate, watching curiously as Wainwright passed on his way to the rail yard.

Just as he entered the rail yard, he met Hope, who obviously was drunk despite the relatively early hour. Even taking the drunkenness into account, Hope didn't look at all fit. His cheeks were sunken, sallow, and unshaved, and he walked with a heavy-shouldered stoop.

"Ah, so you've finally arrived," Hope slurred. "Took you long enough."

Wainwright's temper flared, but he held his tongue in check.

"You must be Hope. Where is Tyrer?"

"That prig? He's in Lukuga with the Belgians. Where's Spicer-Simson?"

"The commander is down at the docks," Wainwright answered. "I think you should go see him."

"Bloody right I'll go see him," Hope said, and he staggered off in that direction.

In the rail yard, Wainwright noticed several very large piles of metal plate and girders. All the pieces had numbers painted on them, though the paint was peeling from the rust that coated everything. He found the stationmaster asleep at his desk, but the man roused quickly enough.

"Yes, I expected you," the man said. "The train will be ready soon. I'll take you to the engineer."

Meanwhile, Hope met with Spicer-Simson, who was standing with Dudley on the dock where the *Baron Jansenn* was tied.

"Hello, Commander," he said, executing a sloppy salute. "Reporting for duty."

"Ah, Lieutenant Hope," Spicer-Simson said, perusing the new arrival with distaste. "Just crawled in from the bush, did you?"

"Not exactly, sir," Hope said, puzzled by Spicer-Simson's aloofness. "I was in Lukuga, conferring with Commandant Stinghlamber, letting him know of your imminent arrival. I might add that the major has an excellent wine cellar."

"By what authority did you so confer?"

"I simply thought I might pave the way for you."

"Let me make something perfectly clear, Lieutenant. I have crossed half a continent and made my own road in the process. I do not need a junior officer to make advances for me, particularly with commanding officers of lesser rank than myself. Is that perfectly clear?"

"I only...."

"Never fail to address me as sir, Lieutenant," Spicer-Simson snapped.

"Yes, sir. Perfectly, sir." Suddenly Hope's clouded brow smoothed in surprise. "What the hell's that? Sir."

The commander followed Hope's pointing finger. Somewhere beyond the first several rows of buildings and huts, a pillar of black smoke was rising above the thatched roofs.

"Tell me, Lieutenant, is there a government rest house or brothel in that neighborhood?"

"Yes, sir," Hope replied, reddening. "I believe there is."

"Then I imagine it has just been paid a visit by our good Doctor Hanschell. He has a decided predilection for excising infection wherever he discovers it. Dudley, perhaps you should go see to a fire brigade."

"Yes, sir." Dudley hurried off toward the thickening cloud, disappointed at not being able to remain to hear the rest of the conversation.

No one, in fact, heard what was said, but after the commander had spoken for several minutes, Hope was seen to remove his sword and hand it to Spicer-Simson. When he left the dock at last, Hope looked even worse than before, although he seemed considerably more sober. He went straight to Hanschell's tent, which had been set up on the edge of the little shipyard. The doctor was in, smelling faintly of smoke and looking satisfied.

"Hello, old man," Hope said as he entered the tent.

His voice had the ring of familiarity, but Hanschell's first reaction was, oh, no, not another Spyridon. The visitor must have seen the lack of recognition in Hanschell's eyes.

"It's Hope," he said.

God, I trust not, Hanschell thought, looking at the gaunt man with his sunken cheeks and filmy eyes. Then he realized with a shock that it was *Douglas* Hope. That brought two more quick thoughts: one that Africa was not kind to Europeans and the other that his care in protecting the men of the expedition really had had a beneficial effect.

"You look ill," was all he could say, though the man's breath clearly indicated he'd been drinking as well.

"No, I'm fine. I just need some quinine. The commander is sending me back to Cape Town. The bastard has sacked me. Says I've been drunk, insulting the Belgians, and disobeying his orders."

"Sounds like what he said about Lee," Hanschell observed, suspecting something of Hope's culpability in that affair. But when he saw Hope's flush turn pale, he felt badly about the dig. "None of it true, I suppose," he amended, trying to sound sympathetic.

"Not the part about insulting the Belgians," Hope said. "I like 'em fine, and they've been a great help. And I don't think I disobeyed any orders. I just did what I thought the commander would want me to do."

"A dangerous mode of behavior, don't you agree?"

"The bastard even took my sword," Hope said, regaining a tinge of color. "Told me he's giving it to someone named Dudley."

"Please," Hanschell said, clearing a spot on his cot. "Sit down and let me examine you."

"No time. I'm to leave immediately. I've just come for quinine. Can you spare enough to get me to Cape Town?"

Hanschell dug out a couple of packets of quinine and passed them over. "That should last you."

"Thanks, old man."

With that, Hope departed, leaving Hanschell with a frown and a bad taste in his mouth. He decided to speak to Spicer-Simson about sending such an obviously ill man off on a twenty-five-hundred-mile journey alone, and he emerged from his tent and soon found the commander standing on the river bank with Mauritzen, surveying Kabalo. The closer he got, however, the more he realized that he'd better stay out of the matter. He'd heard the commander cursing Hope as he'd cursed Lee, and he feared being tarred with the same brush if he showed sympathy for the ousted lieutenant.

When he arrived, Spicer-Simson and Mauritzen were talking about the grubby group of huts surrounded by the fence and guards. It was, it seemed, a local prison.

"It looks like a bad spot for a prison," the doctor remarked. "Surely this ground turns marshy during the rainy season."

"Yes," Mauritzen affirmed. "But I'm told that no prisoners escape then. They can't get through the wall of mosquitoes that surround it!"

"Then we'd best be well away from here before then," Hanschell said.

"What are those?" Spicer-Simson asked, pointing to the piles of rusty plates and girders that Wainwright had noticed earlier.

"Those are the steamship *Baron Dhanis*," Mauritzen replied. "Or most of it. As soon as you take Tanganyika, the Belgians will ship the pieces to Lukuga and assemble them in safety."

"Well, then," the commander said, "we should not delay any longer than necessary."

A few minutes later, Wainwright joined them, accompanied by a man dressed in oil-stained ducks and a grimy billed cap.

"This is Monsieur Onfroi," Wainwright explained. "He's the engineer for our train to Lukuga."

"Well, I hope we will be on our way soon," Spicer-Simson sniffed, looking at Onfroi with nearly as much distaste as he'd surveyed Hope.

"We'll load the boats first thing in the morning, Commander," Wainwright assured him. "I'll have Eastwood begin transferring the supplies right now."

That night, as Spicer-Simson composed his latest dispatch to the Admiralty, he added as a footnote, "Lieutenant Hope has exhibited flagrant disregard for orders and for military propriety. He was dispatched ahead to scout and ordered to return immediately with news; instead he proceeded to go where he was ordered not to go, to imbibe himself into a drunken state, and to insult Major Stinghlamber and other of the Belgian officers at Lukuga. His conduct not only is unbecoming of an officer but has potentially placed the entire expedition in the gravest jeopardy with our allies and with the enemy. I have, therefore, discharged him and sent him back to Cape Town where, I trust, the Admiralty will deal with him as harshly as it has with Lee. We leave on the morrow for the final leg of the journey to Lukuga."

14

Simple though it was, Kabalo was enough of a river port to boast a steam crane, for which everyone was grateful, and in the morning, *Mimi* and *Toutou* were easily hoisted aboard a pair of goods wagons that Onfroi backed down a spur leading to the docks. Meanwhile, Eastwood supervised the final transfer of supplies, which had been considerably depleted. However, a telegram arrived from the Admiralty saying that the firm of Lever Brothers had plenty of supplies at Kinasha. The company had learned the expedition needed them and was volunteering to send them along.

"Who do you suppose told them we need supplies?" Eastwood wondered after Wainwright passed him the message.

"I don't know," the lieutenant replied, "but I'm sure the commander will take the credit."

"What should I tell them?" Eastwood asked.

"Have them ship everything to Lukuga," Wainwright instructed. "We'll be able to use them no matter how well-stocked the Belgians might be."

By noon, the train was ready, with the small passenger coach located immediately behind the engine and wood car, followed by several goods wagons piled high with gear and trailed by the cars carrying the boats.

Just before they departed, Mauritzen took his leave, wishing them well. He would take the *Baron Jansenn* back to Muyumba with the help of the disgruntled caretaker and the handful of African who had come to Kabalo with the expedition.

"Thank you for your valuable assistance," Spicer-Simson told him. "And give my fondest regards to Mrs. Mauritzen. Her presence brightens even darkest Africa."

And then they were off on the final leg to the lake. The first few dozen miles seemed little different than the depths of the upper Lualaba Valley, but soon after, the tracks began to climb into higher, rougher country reminiscent of the terrain they'd encountered on their descent from the Mitumba Plateau. Occasionally, the rail dipped down into a river valley, but the general trend was upward. About thirty miles beyond one particularly deep river valley that Onfroi told them was a portion of the Lukuga River—Tanganyika's sole outlet—the train pulled into a tiny station at a village named Kilu.

"We will have to wait here until Onfroi changes to a smaller locomotive," Wainwright informed Spicer-Simson.

"Why does he have to change?" the commander asked shortly. We've been making excellent time."

"He tells me that there are a great many frail wooden bridges between here and the lake, and they can't take the weight of the bigger engine."

"Very well," Spicer-Simson groused. "Do what is necessary, but just get on with it."

After Wainwright left, the commander and Hanschell descended from the passenger coach to stretch the numbness out of their rears and survey their surroundings. But Spicer-Simson's attention was soon arrested by activity at the end of the train. As it became apparent what was happening, his face reddened with anger, and he stalked off in that direction.

"Wainwright!" he bellowed.

"Yes, sir?" the lieutenant asked, hurrying up.

"Will you tell me what the hell is going on here?"

"They're unhooking *Toutou*'s car."

"I can see that, Lieutenant. On whose authorization?"

"I told them it was all right."

"Did you? And who authorized you to give that particular order?"

"You did, sir."

"I gave you no such order."

"You told me to do what was necessary and to get on with it, sir," Wainwright said as delicately as he could.

"I didn't mean that you were to leave one of our boats behind."

"It's the only way, sir. Onfroi says the trestles aren't strong enough to take both boats. Actually, he's worried that even one may be too heavy. He won't permit both boats at once."

"I don't give a damn what he will or will not permit. This is my expedition, and I give the orders."

"I think we should listen to him, Commander. Remember what happened at our first bridge back in Fungurume. Better to walk lightly than plunge everything into a gorge. He'll come back for *Toutou* right away."

"Very well, Wainwright," Spicer-Simson said in a tone that indicated bored exasperation and supercilious condescension. He looked askance of the heavens then shrugged melodramatically and gestured impatiently toward the train. "Proceed."

Toutou's car was shunted onto a siding along with half the flat cars bearing the supplies and a guard posted under Dudley. By then, the boiler of the new, smaller engine was charged with a full head of steam.

"Is our engineer finally satisfied, Wainwright?" Spicer-Simson asked snidely.

"I don't think so, sir. He says he thinks the train is still too heavy, but I told him we'll just have to chance it."

"As we have been doing since the beginning, Lieutenant."

"Yes, sir."

"Let's be off, then," Spicer-Simson ordered, and Wainwright signaled the engineer to pull out.

Rickety trestles or no, it was just as well that the engineer had refused to haul both boats at once, for the narrow-gauge engine only managed about fifteen miles per hour. Had both boats and all the supplies been in tow, their progress probably would have slowed to a walking pace. And when Spicer-Simson got a look at the first of the bridges, he was secretly thankful at Onfroi's insistence on lightening the train. He didn't pay close attention, but toward the end of the journey, Wainwright informed him that the train crossed at least fifty flimsy trestles in as many miles.

"It's like being back in the desert, sir."

"A damned sight easier, though, eh, Lieutenant, since you didn't have to build all the bridges."

"Right enough, Commander."

Spicer-Simson was getting into a jolly good mood, even if he was chaffing at the delay that making two trips had brought.

As the tracks ascended the Lukuga Valley, which cut its way between the Higher Marangu Mountains and the Lower Marangu, the countryside grew as beautifully strange as it was wild. Tremendous gorges debouched into the valley, many carrying lively streams that helped feed the roaring cataracts of the Lukuga as it fell toward the Lualaba. And here and there, the valley was sparsely littered with blocks of mountain fallen from the escarpments on either side. The blocks were so huge that trees grew from their upper faces.

Onfroi would invariably slow to a crawl and then stop whenever the tracks crossed one of the trestles, and everyone watched with intense and nervous curiosity as he emerged from his tiny locomotive to inspect the structure and tracks laid across it. As each one passed muster and was crossed—often with the passengers holding their breaths—Spicer-Simson grew positively gregarious. Their goal was in sight.

At about three in the afternoon of the third day, the train slowed, then ground to a halt.

"What is it?" Spicer-Simson asked, roused from a light doze. "Another bridge? Where are we?"

"I'll find out, sir." Wainwright left the passenger car and hurried up the tracks. He was back in several minutes, and the look on his face did not bode well.

"My God, Wainwright," an exasperated Spicer-Simson ground out. "What could possibly be the matter *now*?"

"We're in some village, sir. Makala, I think Onfroi said. We've stopped because the tracks have stopped."

"Stopped?" the commander sputtered. "Stopped? How in blazes could they have stopped?"

"The engineer says that the war came before the tracks were completed, and since then, there have been no new deliveries of rails to finish the line."

"Goddamn it!" Spicer-Simson shouted. "How the hell can I fight this war when everything stands in my way?" He spun and paced the length of the car then back, shoulders hunched, features taut.

"Why the hell didn't anyone tell us before?" he wanted to know.

"I don't know, sir," was all Wainwright could answer.

"Well, damn it, here we are."

"Yes, sir."

"How far from the lake are we?"

"Thirteen miles," Wainwright reported reluctantly.

"Thirteen miles." The commander said it softly. "Thirteen bloody goddamn miles." He wiped his face with his hand and looked a Wainwright with a jaundiced eye.

"You'd better have a plan, Lieutenant."

"I do, sir."

"Good man. I knew you would. What is it?"

"There's a roundabout, sir. For turning the train. First we pull *Mimi* onto the roundabout and send the engineer back for *Toutou*. Then we...."

"That's enough explanation, Wainwright," the commander said. "Carry on."

"Yes, sir," Wainwright said, and he turned and hurried off to give the orders.

Behind him, Spicer-Simson sank back in his seat, brow turned stormy.

The engineer and his toy train returned on the second day, hauling *Toutou* and the rest of the supply cars. Then, while Tait and Mollison prevented the Belgian engineer from doing more than voice his protest, Wainwright commandeered the engine and hauled a squad of men and track-laying tools three miles back down the track, where he had them set about dismantling the rails and ties and loading them aboard a flatcar.

In relatively short order, they'd worked their way back to Makala, and the requisitioned track was laid out in front. This wasn't difficult to do because the right-of-way already had been prepared and was only a little overgrown with scrub and weeds. And this close to the lake, the land, though rugged, required no new bridges or trestles.

And so, by continually tearing up the track behind, laying it ahead, driving on, and tearing up the track they'd just crossed, they leapfrogged the final distance to the lake.

But Spicer-Simson couldn't wait. The morning of October 28, as the last section of track was being torn up and re-laid, the commander, accompanied by Hanschell and Eastwood, set off on foot down the prepared grade. By now, the Lukuga River had broken from the flanking escarpments into a shallow valley where the water spread almost a mile between weedy banks.

The sky suddenly darkened, and Hanschell was reminded by a rising wind, stilts of jagged lightning, and cracks and booms of thunder that Tanganyika could be as temperamental as the commander. But the quick violence brought only a spattering of rain before the wind died and the skies began to clear.

A couple of miles later, they encountered a Belgian askari patrol along the right-of-way, and after the patrol learned who they were, the corporal directed them to a path that wound up a rocky hill. They went over the hill, winding through a thin growth of scrub and a number of huge termite hills until, a short distance beyond, the land seemed to end abruptly.

At long last, four and a half months after leaving England, Geoffrey Spicer-Simpson stepped from the scrub and hardship through which he had led his rugged band, and there, on a high bluff, he gazed out over Tanganyika's deep, blue-gray, choppy waters.

"I hope it was worth all the trouble it took getting here," Eastwood whispered to Hanschell.

The doctor couldn't bring himself to answer, but whether that was because of trepidation or mere reluctance, even he couldn't tell.

The final remnants of the sudden storm were blowing off the lake, and as the sun struck the water, the surface gleamed a deep, rich blue. From this height, the far shore could be seen, and beyond that, the huge massif of Kungwe surged above the lesser peaks of its range.

One final rumble of thunder from that direction reminded them all that there the enemy lay, and that all too soon, such sounds on the lake would be from cannon fire.

Chapter XI: The Belgians

"I SAY, WHO IS THAT, do you think?" Hanschell asked as they saw a slender man with white hair and beard approach the expedition's current location, which was a couple of miles shy of Lukuga.

"I don't know," Spicer-Simson said suspiciously. "But he's wearing one of our uniforms, and he'd better damn well have a good reason for it."

"I do believe it's Tyrer," Wainwright said in amazement. "What happened to his hair?"

Dudley, who had barely met Tyrer, remained silent, but they all stared, remembering the sublieutenant's canary yellow locks.

"He must have finished off his supply of hair dye," Eastwood said.

It *was* Tyrer, somewhat worse for the wear. As he came within earshot, they all quieted except for the commander, who said a bit ominously, "I hope you have things well in hand, Sublieutenant."

They all knew what his tone implied. Hope had helped send Lee packing, and now that Tyrer had played a role in ousting Hope, he'd better pray he wasn't next. But if Tyrer had any suspicions about what had happened to his erstwhile companion, he kept his expression free of the knowledge as he replied with a jaunty salute.

"Jolly well in hand, Commander, sir. Would you care to inspect the camp?"

"You have the camp prepared?"

"On the bluff above the lake," Tyrer explained. "We've built the huts on the typical native design, so if the Germans notice anything, they'll think it's just another village."

"Very good, Tyrer. Lead the way."

Not only was the camp large and spacious, the relatively flat bluff afforded enough area for a parade ground.

"Excellent, Sublieutenant," Spicer-Simson commented as Tyrer showed them around. "Have you contracted with the locals for labor?"

"Yes, sir. The native camp is over that way, through the trees." He pointed northwest of the bluff. "There are about three hundred of them right now, but we can hire more if we need them."

"Splendid. How far is the camp from Lukuga?"

"About two miles."

"I suppose that's no difficulty. We can establish a telephone link."

"Remember, sir, that we used our telephone wire making the dead men back on the Mitumba Mountains," Wainwright reminded him. "We still have it, but it isn't much better than light-duty cable now."

"We'll think of something, Lieutenant," the commander said. "We'll commandeer it from the Belgians, if nothing else. Tyrer, what about a harbor?"

"Major Stinghlamber has some ideas on that subject," Tyrer said, "but I've done some scouting and have noted a place more suitable for our use—and certainly more convenient—than anything the major has suggested. We can tour it any time you like, sir."

"We'll do that after morning mess. Right now, I suppose we should occupy our new camp."

"Shall we detrain the boats, sir?" Wainwright asked.

"Not without a harbor," Spicer-Simson said. "Extend the rail line to those trees," he pointed to the tree line just southwest of where they stood, "and keep them hidden there. As soon as we've constructed the harbor, we'll launch them." He looked around the camp. "Eastwood, have my headquarters set up in that large hut over there. The one with the flagpole in front. Dudley, let the men know that we will have a full-dress parade first thing in the morning. They've gone slack during the trek, and I've tolerated it since the going was so rough. But we're at Tanganyika now, facing German guns instead of bad terrain, and I want discipline restored. And Dudley, have them shave. Now gentlemen, to your posts!"

That evening, after mess, Hanschell sought out Spicer-Simson, who was ensconced in his large hut.

"What is it, Doctor?"

Hanschell had a somewhat difficult time concentrating on what he wanted to say. Instead, he found himself engrossed in what the commander was doing.

"Yes, sir. It's the huts. They simply aren't satisfactory."

"Nonsense, they're perfectly satisfactory. And I like this location very well. It has an excellent view of the lake as well as the advantage that it's already built."

"But the likelihood of disease...."

"Is probably much less here than back along the river," Spicer-Simson said. "Remember, Doctor, that we're 2,500 feet above sea level, and unless I'm mistaken, there's a nice breeze off the lake, which should help keep the pests away."

"But many of them have been occupied by askaris."

"The huts are only a month old. If we build new ones, will you insist we rebuild again after only a month's occupation?"

"I guess not, sir."

"I should hope not, otherwise we'd be spending all our time and energy building huts instead of preparing to bring the Hun under control, which is why we're here. So I warn you." He gave Hanschell a stern look. "No fires."

"No, sir. Excuse me, but what is that you're sewing?" Hanschell had a sinking feeling it was another skirt—one more colorful than the leather one most of them had become accustomed to. Perhaps for the full-dress parade in the morning?

"It's a flag." Spicer-Simson unfurled it and held it up, careful not to drop his needle. "I made it from the cloth remnants left over after purchasing all that water from those women back in the desert. Do you like it?"

The banner was white, bearing a St. George's Cross in the center and a red circle in the top, inner quarter.

"Isn't that a vice admiral's flag?" Hanschell was so astonished that he forgot the "sir."

"Appropriate enough, don't you think? After all, I'm the principal naval officer on an entire sea. I might as well act the part."

"But vice admiral, sir?"

Spicer-Simson only laughed.

"My dear Hanschell. Since you're most probably the only doctor within a thousand miles—at least, the only doctor I'd care to have treat me—I dare say you could call yourself president of the African Medical Society and get away with it."

Hanschell wasn't sure he'd go that far, but he could see that Spicer-Simson would not be swayed, either away from the camp's present location or from his self-promotion.

At morning parade, despite the fact that Spicer-Simson lectured the men vigorously on their lack of military deportment and discipline, there was surprisingly little grumbling. Hanschell thought that it was probably due to their collective amazement that the commander had actually gotten them to the lake. Maybe that was why the men now seemed completely anxious to please Spicer-Simson and to follow wherever he might lead. Or maybe they hungered for the glory of the battle and victory as much as Spicer-Simson did and were willing to forgive even the commander's harshness and erratic behavior. After all, if he'd gotten them here despite the journey's impossibilities, was victory any less likely?

2

"He can't be serious," whispered Dudley.

"I've known the commander for many months, now," Wainwright replied, "and I've never known him not to be serious."

Spicer-Simson had just emerged from his hut and was talking down to the Belgian lieutenant who had been waiting for him. The expedition had been at Tanganyika for nearly three days before Major Stinghlamber had sent the lieutenant to request Spicer-Simson's presence at a dinner in his honor.

The commander was, as usual, wearing his leather skirt, boots, and cap, but to these he had added a uniform tunic. His swagger stick swung smartly from his right hand, and his left hand occasionally readjusted the jaunty angle of his long, slender cigarette holder as it jutted from his face.

"But it's undignified," Dudley protested.

"He's been dressed like that since Fungurume."

"But we were in the bush then. This is different."

"Is it?" asked Dr. Hanschell.

"Yes, it is. Major Stinghlamber is the senior Belgian officer at Lukuga. Besides, it's Sunday dinner, and there will be women present."

"You don't think they'll find the commander's leather skirt fetching?" asked the doctor. "Perhaps they'll want a copy of the pattern."

"It certainly shows off his legs, don't you think?" Wainwright winked, and Hanschell chuckled.

"I still can't believe he's got those bloody pictures all over him," Dudley said, repressing a shudder. "I wonder if they hurt getting put on."

"Our commander is a man of iron constitution," Hanschell said. "Or he'd better be. I've warned him repeatedly about recklessly exposing his skin to the tsetse and mosquito, but he won't listen, yet he seems little worse for the wear."

"I'll wager the bugs take one look at those pictures and are afraid to bite for fear of being eaten," Dudley said.

"Come on, now, Doctor," Spicer-Simson called, waving his stick at Hanschell. "The lieutenant says we shall be late for dinner if we don't hurry."

Hanschell joined the commander, and they set off toward Lukuga, Spicer-Simson strolling purposefully behind the Belgian lieutenant, Hanschell trailing close behind.

"Why weren't we invited?" Dudley asked, staring after the small procession as it disappeared into the bush at the edge of the camp.

"I don't know," Wainwright said, "but it's probably best we weren't."

The Belgian lieutenant led the commander and the doctor through the not-so-small African town that had grown up next to Lukuga and then into the fortress itself. Soon after, they were ushered into the officer's mess, which was a huge, heavy-timbered room. A dozen or so Belgian officers were standing about in the main hall or sitting in clusters when the British entered, the largest group around Stinghlamber himself. Three white women were visible across the room. The Belgians' quickly thrown salutes dragged on in minor confusion when neither Spicer-Simson nor Stinghlamber saluted, and nobody knew when to relax.

Spicer-Simson took his cigarette holder from between his teeth and waved it around, dropping ash onto the floor.

"Quite a nice fortress, you have, Major," he said in his nasal French.

That seemed to do the trick, and one by one, the salutes hesitantly dropped.

"It serves its purpose, Commandant," Stinghlamber said, trying not to stare at Spicer-Simson's skirt.

"In comparative terms, Major, my correct rank is colonel."

"Ah, yes." Stinghlamber's eyes narrowed. "I suppose you are the senior *naval* officer. On the lake." He smiled thinly. "Tell me, your uniform...it is unfamiliar to me."

"Designed by me especially for the expedition," Spicer-Simson said. "Do you like it?"

"It is very...unusual."

"May I introduce our surgeon, Dr. Hanschell?"

Stinghlamber shook hands with Hanschell, whose French wasn't as good as the commander's but was sufficient enough to follow the conversation.

"The good doctor has managed to keep all of us free of everything from malaria to the pox. You'll have to ask him about the time I directed him to burn a local whorehouse to keep the men out of it." Spicer-Simson guffawed, then looked at the Belgian officers gathered around. "And who are these gentlemen?"

"This is *our* senior naval officer, Captain Georges Goor." Stinghlamber gestured to a tall, handsome, straight-backed man. "And Captain Dender, my second in command. This is Captain Schimmer. He commands our detachment at Uvira, on the north shore. And this is our chief engineer, Commandant Jadot."

With a bustling of dresses, the women came over, followed by a dour looking man in a black frock coat, a black round-brimmed hat with a round crown, and the collar of a priest.

"My wife," Stinghlamber introduced the women, "Mrs. Dender, and Miss Cosette. And this is Father Jacques. His mission is in Mpala, but he visits here frequently."

"Father Jacques has promised to build a large grotto shrine like that at Lourdes," Mrs. Stinghlamber said, "should the mission survive the war undamaged and the Congo remain Belgian."

Spicer-Simson's upper lip gave just the slightest disdainful curl as he nodded to the priest. "A White Father missionary?"

"You are familiar with our work, my son?"

"Familiar enough. One missionary's like all the rest."

Father Jacques' eyes opened slightly then narrowed as his shoulders stiffened. "I'm not certain I take your meaning, sir."

"I've lived on four continents," Spicer-Simson said regally. "And I've yet to see the good European religion has done on any of them."

"Are you impugning our good works?"

"God helps those who help themselves, does he not, Father?"

"So we are taught."

"And so we're helping ourselves, with God's help."

"Sir, you blaspheme."

"Sir, I remind you of the rape of the Americas by the Roman faith. I remind you of the Inquisition. I remind you of the Crusades."

"God's work is often difficult."

"And it always seems to fill God's coffers at the expense of others." Spicer-Simson turned deliberately from the White Father to look at Mrs. Stinghlamber. "Ah, madam, forgive me for being so rude. You and the other ladies are lovely."

"Thank you, Commander," Mrs. Stinghlamber responded, obviously as shocked by Spicer-Simson's words to the priest as she was by his leather skirt and tattooed legs. "How very nice of you to visit. We weren't certain that you would make it."

Spicer-Simson laughed.

"We have come a rather long way, haven't we?"

"Some of the men had a wager going that you wouldn't get as far as Bukama," Schimmer said. "They were giving you no better than one-hundred-to-one odds."

"I hope you bet against them," Spicer-Simson told him. "My men did back at the Mitumba Mountains and made a pretty day's pay."

"So we heard," Dender responded dryly.

"Really, Captain," Spicer-Simson said. "I couldn't allow the expedition to fail. It's true that we've faced tremendous hardships and nearly insurmountable obstacles at every step of the way, but I managed to solve all the difficulties in due course. We had to build nearly two hundred bridges using a construction technique of my own design. And take the time we were pulling the boats over the Mitumba. Nothing worked properly until I devised a pulley system and had the locos and oxen pull down hill instead of up. That did the trick."

Hanschell frowned, then he carefully erased the expression from his face, glad only that Wainwright wasn't here.

"Some tasks simply take a firm hand," the commander continued. "When we reached the Lualaba, I had to commandeer a steamer practically single-handed."

Several young, pretty local women in brightly colored clothes came up with drinks, and soon after, the party sat down to eat. Spicer-Simson looked over the dishes as the girls served them.

"I notice we're having mutton," he said. "I'm surprised you don't have your people serve native game. It's quite tasty when properly prepared. Our cook has become quite a culinary genius thanks to the recipes I gave him."

"There was plenty of game before the war," Stinghlamber said dryly, "but we've been short of meat, and now there's little left to shoot."

"What a pity," the commander said, taking a bite of mutton. "The hunting of big game is a passion with me. It's given me some of the most exciting moments of my life. I once had a remarkable day's shooting in the Galapagos Islands—the place, you know, where Darwin did his work. They're teeming with wild pigs and goats, and especially with wild cattle. In most countries, you know, it's considered too dangerous to hunt these animals except on horseback, but although there were no horses, I was determined to try my luck. It seemed a most uninviting place, all scrub and cinders, but it was a perfect day, and when we had climbed to a thousand feet or so, we came to a beautiful plateau with hills and woodlands—lots of cover for the animals—and good pasture in the valleys. I always took my chief petty officer with me; the best gun-bearer I ever had. The wild cattle were really wild—not used to being hunted. You had to kill, not wound. The first to fall, I remember, was a huge black bull with wicked-looking horns. I dropped him at forty yards, but before I could go up and examine him, two more thundered down on us. I bagged them with a left and right. 'There's another coming up astern, sir,' my CPO told me, putting a loaded rifle into my hand. I turned and saw that the one I had thought dead was charging us, not ten yards distant, head down, tail in the air, foam flying from his blood-stained nostrils. I gave him a ball between the eyes, and he fell dead at my feet."

Spicer-Simson paused to stuff another forkful of mutton into his mouth, not noticing the slightly horrified expression on the women's faces. Hanschell, remembering the last 'bull' the commander had bagged at point-blank range while it was staked and tied like a lamb at the slaughter, wondered if it would eventually become a wild, dangerous, thundering behemoth in some future retelling.

Mrs. Stinghlamber didn't look quite as blanched as the other women, and she tried to steer the conversation in a different direction.

"Tell us, Commander, how did you come to lead your most extraordinary expedition?"

"Yes," Stinghlamber agreed. "Most of us have been stranded here since the war began, and we're hungry for news."

"Extraordinary is truly the word for it, Madam, and that's the only thing that could have torn me from the important work I was engaged in as one of the senior naval officers at The Downs. After all, for a true Englishman, the defense of Britain stands over all else. But Sir Henry Jackson, our Lord Admiral, wanted one of his most trusted senior officers to handle the matter, and as I have had a great deal of experience against the German fleet, he personally insisted that I accept. He knew I would do anything for King George, but I already had gleaned from secret intelligence reports that retaking Tanganyika from the Germans was of paramount importance to the outcome of the war."

"We are under the impression that the real battles are taking place in Europe," Dender said, "and that Africa is of little consequence."

If Spicer-Simson noticed the trace of irony in Dender's tone, he ignored it.

"I assure you that the Admiralty does not consider it so," he replied. "After all, you can see that we are here to succor you."

Hanschell winced as a silence descended on the table.

"Perhaps you would enlighten us regarding your great experience, Commander." Dender's tone had turned slightly aggressive, but Hanschell heard real curiosity there, and he didn't wonder. He'd known Spicer-Simson for years, yet he was equally curious. Had he ever *really* known the commander?

"Gladly," the commander smiled, and Hanschell listened as attentively as the Belgians, certain that he would be entertained by something as surprising as it was new.

"As soon as war was declared, I was ordered to take command of a flotilla of destroyers based at Harwich. Night after night, we put to sea, hoping to intercept the German forces that from time to time made lightning raids on our vital communications with France. It was not very spectacular, you know—days and nights of cold and wet discomfort, but always with the possibility that something might happen.

"I wonder if any of you know, or can imagine, what it is like to be on a night patrol off the enemy coast in winter? There is little protection on the bridge of a destroyer, and one feels damnably alone, surrounded by impenetrable darkness. There is no light but the dim glow of the binnacle, the gleam of the bow wave, and the receding wake. One is almost mesmerized by the hiss of the water and the even throbbing of the engines—then suddenly a far-off searchlight stabs the darkness and goes out again. The dense smoke rolling from the funnels makes a blacker cloud against the leaden surface of the sea. Two hours before dawn. Action stations!

"I always close up the guns' crews and have everything in readiness for the first gray light of morning, for who knows what it will disclose? Of course, one has done it a hundred times, but it is impossible to relax, for one day it will happen, as it happened to me, that the first streak of dawn discloses an enemy vessel. Then it's a question of getting off the first salvo—a matter of seconds between destroying and being destroyed! But this time, the enemy wasn't another destroyer; it was a cruiser!

"Our first salvo was right on the target—but what could our two-inch guns do against the six-inch guns of our huge antagonist? Columns of water sprang up all around us—she had opened fire and straddled us! The descending columns cascaded across our decks. I take no credit for what happened next, for we really had no alternative. It was impossible to retreat—we would have been blown out of the water long before we were out of range.

"I rang for full ahead and, zigzagging wildly, closed the enemy! At only 1,000 yards range, I loosed off all my torpedoes, and one of them struck her amidships! Then I turned and made off under the protection of a smokescreen. It was purely a matter of self-defense, you see. I had little hope, perhaps little intention, of sinking the cruiser. One's actions in such cases are purely instinctive, but we heard later that after a long tow by her escorting destroyers, she capsized and sank just off the Jade Estuary."

Hanschell, knowing that the only torpedoing Spicer-Simson had taken part in had been on the receiving end when the *Niger* had been sunk while anchored, kept a straight face that masked his pity. If the commander wished to be a hero in his own eyes as well as those of the rest of the world, why spoil it for him? Besides, if he believed it deeply enough to give him the confidence he needed now, Hanschell was willing to go along. He only hoped that Stinghlamber or one of his officers wouldn't think to verify Spicer-Simson's story with the Admiralty.

By now, dinner was over, and while the ladies retreated to a tight huddle across the room, the men clustered around the bar, smoking and drinking.

"Tell me, Major," Spicer-Simson said. "If you have a senior naval officer, that must mean you have a navy."

Spicer-Simson's tone was casual, but Hanschell thought he noticed a touch of covetousness behind the words.

"We have one ship," Stinghlamber said. "Captain Goor commands her."

"Not really a ship, I fear," Goor said, turning to the Englishmen. "She is a ten-ton river barge with fearfully inadequate engines."

"But she's very aptly named." Jadot said with a slightly sardonic tone.

"Yes?" Hanschell prompted. "What is her name?"

"*Dix Tonne.*"

Everyone laughed, glad to have something to break the tension that had developed between Spicer-Simson and Stinghlamber.

"Actually," Dender said, we generally refer to her as the *Moselbak*."

Everyone laughed again, except Hanschell, who didn't recognize the word.

"It means 'clam bucket,' Doctor," Spicer-Simson supplied.

"On a good day, she'll do all of six knots," Jadot said.

The women, hearing the men laugh, took the opportunity to rejoin them. "Don't downplay *Dix Tonne*'s importance," Stinghlamber said a trifle defensively as his wife came to his side. "She *is* armed."

"She's got two guns," Dender amplified. "A 57 millimeter and a 47 millimeter. Most frightening of all to the Germans is her huge flag. They can see it halfway across the lake, and when they do, they flee at top speed."

"Really, Dender," Jadot said. "They needn't go all that fast."

"Seriously, though," said Goor, who was taking the jibes good-naturedly, "she helps keep the Germans from pestering us since they never know just where we've got her hiding."

"From what I hear of your boats, Commander," Schimmer remarked, "You will need all the help you can get."

"Is that so?"

Schimmer shrugged. "You'll pardon me, but two small boats with insignificant armament and no armor—of what use can they possibly be against what the German float on Tanganyika? All the hardships you went through to get them here hardly seem worth the effort."

"I dare say that if you'd been the one dragging my boats across the desert and over the Mitumba Mountains, you would hesitate to call them small," Spicer-Simson said. "But at any rate, size is no matter. I'll remind you of my earlier story about my destroyer sinking a cruiser ten times our tonnage. It's not size that matters; it's success."

"I hope," put in Mrs. Stinghlamber, "that you will allow me to be the godmother of your prizes. If you capture them."

"Do I detect a touch of irony in madam's voice?" Spicer-Simson asked with a minute bow in the lady's direction.

"I believe that my wife is noting the fact that a destroyer is more sophisticated than a couple of open motorboats," Stinghlamber said. "And you face a formidable foe on Tanganyika."

"We have been well briefed on the *Hedwig von Wissmann* and the *Kingani*," Spicer-Simson said nonchalantly. "Our boats aren't large, but they're quick and mount enough firepower to do real damage to anything the Germans have."

"Anything?" Stinghlamber said with a trace of triumphant superiority. "What do you say to a twelve-hundred-ton, two-hundred-foot steel-hulled steamer that can travel at eight knots and is armed with two 105s and two 88s?"

Hanschell thought he saw the commander pale beneath his tan.

"Are you saying there is such a ship on Tanganyika?"

"There is. The *Graf von Götzen*."

"Why wasn't I informed? I should have been informed."

"But Commander, I thought you said you had been thoroughly briefed."

Spicer-Simson's eyes suddenly narrowed as he remembered the closing of his final conversation with John Lee. In English, he muttered, "That bastard. That's what he was talking about."

The Belgians looked puzzled, then Spicer-Simson said to them in French, "How long has the *von Götzen* been on the lake?"

"She was launched in June," Goor said. "The ninth, I believe. Her larger guns were brought overland from the wreck of the *Königsberg*, and the 88s came from the *Moewe*. She was named for G. A. von Götzen, the first white man to reach Lake Kivu."

"She's the biggest ship ever launched on any of the African lakes," Schimmer put in. "They haven't quite finished her, so we've yet to see her in action. But she has the possibility of becoming a real danger, and not only because of her firepower. She'll be able to can carry nine hundred fully armed askaris anywhere along the German coast in forty-eight hours."

"Where is she harbored?" Spicer-Simson asked.

"All the German fleet is at Kigoma, on the opposite side of the lake and nearly two hundred kilometers north," Dender replied. "You may have sunk a cruiser with a destroyer, but against the *von Götzen*, your two boats are tsetse flies. Are you ready now to revise your estimates of chances for success?"

"We shall be victorious," Spicer-Simson said with perfect aplomb. "Tell me, Major, aside from your two fortresses and this ten-ton barge, what is the allied armament on the lake?"

"There are scattered shore batteries, but none with any real firepower," Stinghlamber said. "Captain Goor, please tell the commander about our other boats."

"There are two of them," Goor said. "One is a little speedboat named *Vedette*, armed with a Lewis gun, and the other is an open hydroglisseur called *Netta*. She's quite fast—probably the fastest craft on the lake—but she's too light to be of much use, though she also carries a Lewis gun. We use her primarily for patrols. And there is *Vengeur*, but she's not yet ready."

"*Vengeur*?" Spicer-Simson's eyes flickered.

"You probably know her as the *Alexandre Delcommune*."

"I understood that the commander of the *Hedwig*—what's his name? Horn?—blasted her to pieces right under the noses of your shore batteries, not just once, but twice."

Stinghlamber stiffened.

"Horn was lucky."

"Horn was successful because he's resourceful," Spicer-Simson retorted. "But there's another factor. He's a man in his own element. Tanganyika is water, and water won't be won by a bunch of soldiers sitting behind shore batteries. You have to go out there and get the spray in your face. And that's why our joint allied commanders have sent me here—to get the spray in my face and to take this lake. You say this *Vengeur* is still afloat?"

"We've begun rebuilding her, but she's little more than a hull right now," Goor explained. "And that barely floating. We have her hidden in the mouth of the Lukuga River. We plan to move her to the shipyard on the lake shore as soon as you launch your boats."

"Well," Spicer-Simson said. "I'd say that evens things up a bit, don't you think. The Germans may have an armada, but we have an impressive flotilla. And if what you tell me about the *von Götzen* is true, then I'm going to need every available craft at my disposal."

"I have no intention of throwing away my few craft on a foolhardy venture," Stinghlamber said.

"Let me put it another way, then. Your rank of commandant is equivalent to major. My rank of commander is equivalent to colonel. Thus it is simple. I am not simply the senior naval officer on the lake, I am the ranking Allied officer in the region. We are at war, gentlemen. Let us behave with military decorum as men who are at war must."

"Are you saying you are assuming my command?" Stinghlamber looked a trifle flustered as well as angry. "I refuse to acknowledge that your rank is superior to mine."

"I couldn't care less about your shore command," Spicer-Simson said. "But where the lake and boats are concerned, I will have every resource at my disposal, and I will accept no argument. I am here to take Tanganyika in the name of the Allied war effort, and that is exactly what I intend to do. Sit ashore if you like, but I *will* have your boats."

"Mon commandant!" Dender said, anger shaking his voice. "Must we listen to such outrage?" He turned on Spicer-Simson. "You traipse across Africa with your foolish uniforms and your pitiful boats and think you can take what is ours and use it as your own?" His lip twisted in disdain. "I, for one, won't accept orders of any sort from a commandant in skirts."

Spicer-Simson lit a cigarette and carefully fitted it into his long holder before he looked at Dender. When he did look, his eyes were filled with ice.

"Are you aware of the penalties for treason in time of war, Captain? I hope I don't have to go to General Tombeur. I'm certain he would not personally have requested and sanctioned my traipse across Africa with my foolish uniforms and pitiful boats if he did not completely back the plan I've been sent here to implement. A plan envisioned by Lord Henry Jackson himself. How shall General Tombeur respond when he learns the petty pride of a recalcitrant captain stands in the way of African victory?"

When Dender did not respond beyond a look of rage on his face, Spicer-Simson blew a cloud of smoke into the tense air, and took in all the Belgian officers with a sweeping look.

"One more thing gentlemen. I do not object to your references to my skirt. I designed it myself, and my wife made it for me. But I will not be referred to as commandant. My rank is colonel, so address me in that fashion in the future." He turned to Hanschell. "Shall we return to camp, Doctor?"

Hanschell followed Spicer-Simson as he strode purposefully out of the mess hall and away from the fortress. In the darkness, the doctor couldn't see the commander's expression, though he expected he would be seething. But about halfway back to camp, he was surprised to hear Spicer-Simson give a laugh.

"Mrs. Stinghlamber is quite a woman," the commander said. "Can you imagine—she asked me to allow her to be the godmother of our prizes?"

When they arrived in camp, Spicer-Simson went straight to his own hut, but Hanschell found Wainwright, Dudley, and Eastwood and told them what had transpired.

"Sounds like one hell of a Sunday dinner," Wainwright chuckled.

"Can he do that?" Dudley asked, looking surprised.

"Usurp command from Major Stinghlamber?" Wainwright shrugged. "I don't know, but I'm beginning to think that Geoffrey Spicer-Simson can do just about anything he wants."

"As long as he offends somebody in the doing of it," Eastwood said, shaking his head.

"Best be ready for tough dealing with the Belgian chain of command, Tubby," Wainwright warned. "None of them will be anxious to lend a willing hand to someone who's insulted their commanding officer."

"Don't worry," Eastwood said reassuringly. "I've been getting a lot of practice picking up after our commander."

3

The next day, Spicer-Simson went alone to Lukuga.

"I'm going to have a chat with Major Stinghlamber," he told Wainwright. "Proceed with preparing the camp while I'm gone."

Watching the commander stride to the edge of the cleared area and disappear in to the bush in the general direction of Lukuga, the lieutenant was glad Spicer-Simson hadn't asked him to come along. After hearing about the "party" the night before, he didn't want to be present if it came to a showdown between Spicer-Simson and the Belgian commandant—he had no desire to have Allied blood on his hands.

Stinghlamber gave Spicer-Simson a cool reception that heated up as soon as Spicer-Simson pulled rank on the Belgian, which occurred immediately after their greeting. It grew hotter yet when Stinghlamber insisted that the Allied flotilla harbor in the outlet of the Lukuga River instead of the site Tyrer had chosen: the mouth of the Kalemie River, which emptied into Tanganyika just south of the bluff where the expedition was camped.

"I'm afraid that a harbor at Lukuga simply won't do, Major," Spicer-Simson said offhandedly. "You aren't a sailor, so I can forgive your ignorance of maritime matters, but I have studied the charts and had the advice of an experienced hydrologist—one of your own, in fact—who knows the area, and the only logical conclusion is to build the harbor at the Kalemie's mouth."

"My men already have begun dredging the Lukuga," Stinghlamber said, frowning. "I simply cannot permit all their work to go to waste. Besides, my site is right under the fort's guns and will give you better protection."

"Lukuga is two miles from our camp," Spicer-Simson pointed out. "I can't have my men running back and forth all day. The call to stations may come at any time, and I'll want my men and boats ready immediately. That will be impossible if our camp is two miles away."

"There is plenty of room for you here in the fort," Stinghlamber said. "Anyway, I've already made arrangements with your Lieutenant Hope."

"*Sublieutenant* Hope has been sent home in disgrace," Spicer-Simson said haughtily. "In any case, I don't take the recommendations of drunken fools seriously. On the Lualaba, I had many conversations with Monsieur Mauritzen, the very excellent and capable pilot and hydrologist there. He tells me that there is a shallow bar across the Lukuga's mouth and that, during the typhoons whipped up by lake, the waves breaking over it can become dangerous. He said the Kalemie was a much more suitable location and that, before the war, there were plans afoot to create a new harbor there, anyway."

"As I told Hope, we can dredge the bar and deepen the Lukuga enough to protect your two little cruiser automobiles. At any rate, your proposed harbor could not be made large enough in a short time to protect the *Vengeur* and the *Baron Dhanis*."

Spicer-Simson darkened at the disparaging reference to the boats.

"From what Commandant Goor said, the *Vengeur* is, at present, a half-sunken hull, and I saw the *Baron Dhanis* still in pieces in Kabalo. By the time you need a harbor for them, I will have finished with the Germans. No, Major," he said, stressing the rank to remind Stinghlamber that he was army and knew nothing of the needs of seafarers. "I'm afraid it won't do. That bar on the Lukuga isn't there by chance but by the natural actions of the waves and currents. Even if we managed to be successful in dredging it sufficiently, one good typhoon or a month or so of currents will simply rebuild it. I want a harbor I can trust that won't take a steady commitment of time and effort to keep it serviceable. I've been exploring the area, and the mouth of the Kalemie, with a little work, will be an ideal harbor."

"Nonetheless," Stinghlamber insisted in a tight voice, "Lukuga would be safer. We do, after all, have a ship to help give you protection. And since I am charged with your protection, the harbor must be where I can accomplish that."

"Your *Moselbak* will have a difficult time protecting us," Spicer-Simson said, his tone hardening, "with our boats harboring at Kitute."

"Kitute?" Stinghlamber looked surprised. "That's at the southern tip of the lake."

"It is a British port," Spicer-Simson said. "We are the Royal Navy, and the Royal Navy quarters in a British port. If you insist on thwarting my plans to construct a proper port under my dual authority as a British naval officer and commanding Allied naval officer on the lake, then I'm afraid I will simply have to move my base of operations to a more suitable location. Of course, I will have to inform the Admiralty and General Tombeur of your flat refusal to cooperate. Since my government has gone to great expense—and I and my men to great effort and danger—to bring our warships to destroy German domination of the lake, I'm sure that Sir Henry himself will personally have a word with the Belgian high command regarding your refusal."

Stinghlamber didn't look so much angry at the threat as surprised. Then ire began to edge into his eyes, but he held himself in check, realizing that his career might be at stake. He had no faith that Spicer-Simson's tiny, under-armed motorboats would be able to wrest control of Tanganyika from the Germans, but he was willing to cooperate just enough to help provide this arrogant prick of a buffoon with the means for his own self-destruction.

"All right, Commander," he said carefully. "Kalemie it is."

"Look at it this way," Spicer-Simson said with a grin that bore no trace of smugness at his victory. "We shall simply help you along by starting the project, just as we will take back the lake for you."

"What will you require of me?"

"Jadot, for one," Spicer-Simson said. "And a shore battery. Two guns should do."

"And where will I get these guns?"

"You were planning on arming the *Vengeur* when you re-floated her. And the *Baron Dhanis* when it is rebuilt? You must have something available, at least until the ships are ready for ordnance. Perhaps by then we won't need their protection."

"Very well," Stinghlamber conceded. "A shore battery, but the men must remain under my command."

"Naturally," Spicer-Simson said. "They are, after all, manning a *shore* battery. And that brings me to my final requirement."

"Yes," Stinghlamber said, fearing what was to come.

"As senior naval officer, I must requisition all of your operational boats and their crews to serve under my direct command."

"Preposterous," Stinghlamber sputtered, knowing instantly that the preposterous was the order of the day.

"I *will* have them, Major," Spicer-Simson said sincerely. "But I assure you, I will return them as soon as I have taken back the lake."

4

While Spicer-Simson was wresting what he could from Stinghlamber, his officers supervised the men and local workers as they set up camp. Shortly after noon, while Wainwright, Cross, and Eastwood were discussing the best location for storing the munitions and supplies, Hanschell came over to them, his face serious.

"These huts, Wainwright," the doctor said. "They must be replaced."

"Replaced?" Wainwright looked genuinely puzzled. "Why, Doctor? I've spent a couple of nights in mine, and it seems sound."

"Sound enough, perhaps," Hanschell said, "but not clean. And I'll wager you have one of the best of the lot."

"That may be so," Wainwright admitted, "but I've been in many of the others, and I didn't think them terribly dirty. Not like some of the government rest houses you burned along the way. Say!" He eyed Hanschell suspiciously. "You aren't planning something like that, are you? The men already think you're something of a pyromaniac, and they'd certainly take offense if you set their quarters on fire."

"I promise nothing," Hanschell said archly, "except to keep the men safe from disease. Whether they like it or not, and by whatever means necessary."

"Don't act rashly just yet, Doctor," Eastwood put in. "When we've become established, I'll have a native work crew begin rebuilding all the huts one by one. Eventually, all will be fresh enough for even your discriminating taste."

"All right," Hanschell conceded. "But I'm holding you to it, Tubby, or your hut will be the subject of my first round of nocturnal visits."

With that, Hanschell left the two. He'd already set up his hospital in a large, airy hut, and Rupia was straightening up his quarters, so he had little to do for the time being but watch everybody bustle around. Not wanting to be in the way, he strolled through the camp, intending to walk to the edge of the bluff overlooking the lake.

On the way, he passed Dudley, who was wasting his energy scolding Tait and Mollison. Hanschell had to suppress a smile. The nervous and feisty Dudley looked like an antic monkey cavorting before two huge bears. Luckily for the monkey, the bears were as imperturbable as they were lumbering—unless the subject was rugby. That, however, seemed to incense Dudley all the more since he could not fathom how the two men got any work done.

But Hanschell knew that they did. They may have set about any task to which they were assigned at a deliberate and ponderous pace, but that pace never ceased until the task was finished. They never complained—not even when Dudley pressed them—and, taciturn as they seemed, they were unfailingly patient and pleasant in their own peculiar manner.

Hanschell skirted the trio, not wishing to disturb their tableau, and soon he was standing on the edge of the bluff, looking out over Tanganyika.

It was one of the most beautiful sights Hanschell had ever seen, with its deep blue water, brick-red cliffs, cerulean sky, and the hazy mountains rising beyond the far shore. Not homey beautiful like England but starkly majestic and almost frightening in its unpredictability. At any instant, the wind might whip the now-placid breakers washing the yellow-sand beach at the foot of the bluff into a foaming frenzy, while lightning shot and sizzled through swift dark clouds.

And that brought thoughts of the combat that, as surely as the next storm, would vent sudden violence upon the deceptive tranquility.

Would Spicer-Simson be up to it? Hanschell wondered, not for the first time. Or would all his arrogance and pretension and braggadocio evaporate as rapidly as a drop of water in the arid African breeze once he truly faced his enemy?

But Hanschell knew that Spicer-Simson had to confront another enemy as formidable as the Germans—himself. If he could conquer his own bizarre propensities—if only for a moment—he might actually win the day for England and redeem himself in his own eyes.

It might just be possible. Fantastic luck had carried the expedition this far. Luck with finding means to rebuild the trailers. Luck with bridges and brushfires and illness. Luck, even, with finding plentiful water in a parched landscape. And the passage on the Lualaba had its own share of good fortune, beginning with Mauritzen.

And there, perhaps, was the real core of Spicer-Simson's luck. All along, there had been good, strong, capable men to help shoulder the burden—men who seemed as determined to succeed as was Spicer-Simson. Wainwright. Dudley. Cross. Hell, even Tait and Mollison in their stolid, uncommunicative way. So it wasn't, perhaps, luck really but the wills of the men—their wills combined and channeled through Spicer-Simson—that was carrying them along. And, in some sense, that was

Spicer-Simon's greatest success. He may have ousted men who he felt threatened his ascendancy, but he also knew how to give a man a task and let him at it without interfering. Wainwright was the perfect example.

So even if Spicer-Simson all too often haughtily took credit for his men's successes, the men knew among themselves who had done what. That's why they affectionately called Wainwright the Old Loco Driver and obeyed instantly when he spoke, and why Waterhouse always had their attention. They even admired the high-strung Dudley for his bicycle ride and his brashness in commandeering the *Baron Jansenn*, saving them from the Capitaine de Steamer and his brutish touch on the *Constantine de Burlay*'s wheel.

Yes, Hanschell reflected sadly, staring over the lake. Good men. Brave men. How many of them, in the coming months, would be wounded, maimed, or disfigured? How many would die beneath his futile ministrations? How many would be lost to Tanganyika's unfathomable depths?

Melancholy came over him as suddenly as a storm on the lake. With heavy steps, he trudged back to his all-too-inadequate hospital.

That night, as he lay on his cot, listening to the drums and the cries and stamping of the people in the African camp as they drank pombe and caroused and danced, he was nearly overwhelmed with the menace, savagery, and loneliness of his surroundings. No Englishman, he thought, ought spend much time in Africa. It was too hot and too still and the air too oppressive. There was too much talk of witchcraft, and dark superstitions seemed to lie in ambush like a leopard waiting to drag its prey into the thick and unforgiving underbrush. The whole continent was too primitive and too full of flaring passion masking somber undercurrents. And into it all, Europeans had interjected their own uncouth war bristling with mechanized death.

It was enough to make Hanschell despair, but instead, he slept.

5

The morning of November 2 dawned peacefully. Above, the bright blue sky was strewn with wisps of cirrus clouds, and below, at the foot of the bluff, Tanganyika was a rich blue green. Both the air and water were still. As still as the men lined up in formal parade about a hundred feet from the edge of the bluff, facing the flagpole planted in front of Spicer-Simson's hut. It looked rather like a telegraph pole. Fifty yards behind the men, at the edge of the bush, several hundred Africans filled the margins of the parade ground, watching the proceedings with interest.

Of *Mimi* and *Toutou*, there was no sight. Both boats were hidden in the trees half a mile to the southwest, where the bluff was at its lowest near the mouth of the Kalemie.

"Men!" said Spicer-Simson as he swaggered down the line, appraising his navy with sharp, cool eyes. He was, as usual, wearing his skirt, and although he wore his shirt, the sleeves were rolled above his elbows in continuing flagrant contravention of Hanschell's medical recommendations.

"This morning I have received good news. I am now under direct orders of the commandant general in Rhodesia, our nearest British military commander. This means that Major Stinghlamber is clearly under my command. I have spoken with the major and commandeered the few watercraft he has to bolster our flotilla. As soon as we've constructed a harbor and launched *Mimi* and *Toutou*, they will be joining us. And now, to commemorate my elevation to His Majesty's naval commander on Lake Tanganyika, we will hoist our flag. Wainwright."

"Sir." Wainwright turned and shouted, "Chief."

A dour looking Chief Waterhouse, accompanied by two ratings, approached the flagpole, to which they attached a flag and hoisted it aloft. As it unfurled, a murmur sounded among the men.

"I know I'm just a dumb Afrikaner," Dudley whispered to Hanschell, "but I don't believe I've ever seen a flag like that before."

"It's a vice admiral's flag," Hanschell said. "Or near enough."

"Bit cheeky of him, isn't it, Doctor. Where'd he get it?"

"I believe he made it," Hanschell said.

"He sewed it himself?"

"Apparently our commander is quite handy with a needle and thread."

"Right you are, Doctor," Dudley said with a smile. "He is, at that."

A small group of tribesmen bearing a wooden chest separated from the rest of the Africans and, led by Eastwood, approached Spicer-Simson and Wainwright. "These are the headmen of the workers I hired," Eastwood explained.

"You are Bwana Chifunga Tumbo?" asked one of the men, who seemed to be chief among them and who wore a leopard skin over one shoulder.

Wainwright translated, suppressing the urge to burst into laughter at the name. Spicer-Simson, however, seemed quite pleased.

"Tell them I am Bwana Chifunga Tumbo," he said, puffing up a little. He puffed up even more as Wainwright relayed the headman's next statement.

"It is said you have captured all the mana of the Earth on your skin. It is said you command thunder and iron to defy the deserts, and you have carried canoes the size of houses across the mountains on your shoulders."

"You see the world's mana on my skin," Spicer-Simson replied grandly. "I harnessed thunder and iron to defy the deserts, and you have seen with your own eyes my large canoes resting in the trees. They wait there like chui waits to pounce on his prey."

He turned so that the tiger tattoo was displayed prominently to the men, flexing his muscle to make the beast glower.

"We have heard that the gaze of your eyes gives you power over man and beast," said another of the headmen, his own eyes shying away from the sight of the tiger.

"And women," another spoke. "It is said King Kasavubu's wives lamented when you left and that Kasavubu himself desired to know the source of Bwana Chifunga Tumbo's vitality."

"I tell you what I told King Kasavubu," Spicer-Simson said. "I am creating the future, and if you help me, I will give your villages and your people a place in it."

"And if we do not?"

"Go in peace, then. I will not enslave you like the Belgians do. I want your labor, but only if it is willingly given. If you stay, though, I promise that you will labor with all your might. But heed me well. I will destroy the Germans and any man who joins with them."

The headmen spoke briefly among themselves before turning back to Spicer-Simson.

"We will do your bidding, Bwana Chifunga Tumbo," the first headman said. "We wish a place in the future you will make without the Germans, who kill us for sport."

"Bwana Chifunga Tumbo is pleased," the commander said. "I hereby appoint you to the Royal Naval African Expedition."

He tapped each of the men once on the shoulder with his swagger stick, and in turn, they prostrated themselves and attempted to kiss his feet. The commander held his regal pose while Wainwright pulled the men to their feet and sent them off toward Spicer-Simson's tent with the chest.

"Wainwright, call the men to order," Spicer-Simson directed, since the ratings had relaxed during the exchange with the Africans.

"Ten, hut!"

"Right men. We've gotten here, and that was part of our mission, but we have a lot more to do before we can take on the real work we were sent here to accomplish. Tomorrow we begin work on a harbor where our boats will be safe. I have scouted a good location where the Kalemie River empties into the lake. It's not far from where *Mimi* and *Toutou* are now located. It will take some work, but I trust you will all pitch in just as you did on the trek overland."

A sudden gust of wind tried to lift the commander's skirt, and he clamped it down with his left hand while he pointed sporadically at his men with his swagger stick.

"And I'll have no laggards. It's the beginning of November, and I want to see *Mimi* and *Toutou* launched by the end of the year."

"Is that possible, sir?" asked Dudley.

"It is if I say it is, Sublieutenant."

Wainwright noticed that the new flag on the pole had begun whipping about as the wind abruptly picked up. Glancing over his left shoulder, he saw a darkly ominous line of clouds sweeping down the lake from the north.

"Commander," he called out, and gestured toward the rapidly advancing storm. "It looks like the rains have come at last."

"Quite right, Lieutenant," Spicer-Simson said. "We could have used them back on the Lualaba, but no matter. All right men, dismissed. Eat hearty tonight, for tomorrow we labor like animals."

"If not sooner, Commander," Wainwright said. "Look at those trees."

The trees he meant were nearly a mile up the lake, just at the edge of the storm. Everything behind the front was washed out in a seething gray haze of rain. Even as they watched, a swirling wind tore leaves from the branches, branches from trunks, and several trunks from the ground.

Without waiting for Spicer-Simson's orders, Wainwright yelled, "Take shelter, men!"

Everyone dashed for their newly occupied huts, and most reached them before the storm swept across the encampment. As he ran, Wainwright noticed that the Africans who had not long before flanked the proceedings at the parade ground already had disappeared from sight. He reached the hut he shared with Dudley just after the sublieutenant and, a second later, Magee hurried in after them.

"Sorry, fellows," Magee said. "Couldn't make my own hut in time."

"Stay as long as you like," Wainwright said, noticing as he spoke, the frown on Dudley's face. Ever since Lee had been bounced from the expedition, the men considered Magee a suspicious character at best. And they resented the fact that, when there was hard work to be done, Magee generally begged off, saying he had to keep up with his writing.

But Dudley had come along only at the tail end of the Lee affair, so that wasn't his gripe with Magee. Nor was it what others considered Magee's slackard behavior. Dudley, like Wainwright and Hanschell, was an educated man who knew that recording the events of the expedition was Magee's task and that words do not magically appear on paper but take time and their own sort of labor. Besides, Dudley had far-worse difficulties keeping his temper around Tait and Mollison, whose phlegmatically ponderous behavior continuously exasperated and inflamed the tightly wound sublieutenant to no end.

Instead, Dudley's particular resentment toward Magee had to do with the Queen's South African Medal that Magee wore on his uniform. The medal had been given to those who'd seen service in the Boer War, and despite the fact that Magee had a letter from Lord Northcliffe stating that Magee had been in more actions than any but the most seasoned officers, Dudley often voiced his doubts.

"If he did, he wrote his way through them," was one of Dudley's more frequent comments on the subject.

Dudley, himself, had seen considerable action—real fighting action—in the Boer War and could have worn a similar medal, but he simply didn't have one to wear.

But Wainwright really couldn't fault Magee. The journalist hadn't taken to wearing the medal until ordered to do so by Spicer-Simson, who had instructed everyone in the expedition to wear any and all decorations they had at all times. A lot of the men protested, like Dudley, that they hadn't brought their medals and ribbons with them, and the rest complained that if they wore what they'd brought, the medals would get lost and the ribbons dirty amid all the toil and labor.

The complaints had fallen on deaf ears, and Spicer-Simson stood firm. But what would you expect from a man who sewed his own vice admiral's flag? He was, he said, not only commander, but commander of Britain's finest, and he wanted everyone—especially the Belgians—to be completely aware of that fact.

The three of them stood in the doorway, watching Spicer-Simson—wearing all *his* medals and ribbons—complete his entirely leisurely stroll to his own hut. His right hand holding down his skirt and left hand his cap were his only concessions to the now-stiff gusts.

And then the rain came in such drenching sheets that it seemed they were standing beneath Victoria Falls themselves. Water began dribbling in through the roof thatch, and the whole roof began shaking like it was going to be lifted like a lid off a pot. Their hut wasn't unique. Through the doorway, Wainwright could see the same thing happening to many of the huts in the camp, and several roofs actually blew right off, leaving the men inside to scurry through the thrashing downpour to their nearest neighbors.

"Here," Wainwright ordered Dudley and Magee. Let's tie the roof down, or we're going to lose it."

After they were done, Wainwright and Dudley returned to the door to watch the fury of the storm, while Magee settled onto Dudley's cot with his notebook and pen and began scribbling. Curious, Wainwright approached and looked over his shoulder.

"The storm was ferocious," Magee had written. "It broke over our camp in a hurricane of wind accompanied by ear-splitting bursts of thunder and vivid lightning, which illuminated the country for miles around. The lake itself became a raging sea, enormous breakers rolling up and crashing on the shore, uprooting trees, and demolishing native huts...."

Magee glanced up. "Am I telling it right?"

"Right as rain," Wainwright told him.

"I like that," Magee laughed. "Right as rain." He went back to his writing, and Wainwright returned to the door.

Outside, Spicer-Simson's vice admiral's flag was standing nearly straight out in the wind.

"I wonder if the commander's flag will hold up to the punishment?" Wainwright mused.

"When those native chaps came up to the commander," Dudley asked, "what were they about?"

"They were pledging their fealty."

"Did they say what was in the chest?"

"Each family had donated an item of value to Bwana Chifunga Tumbo"

"Bwana Chifunga Tumbo?" Dudley repeated, eyes widening, then both men burst into laughter. Magee looked up from his scribbling, curious.

"All right," he said. "You chaps, speak the lingo, but I don't. Let me in on the joke. What's Bwana Chifunga Tumbo mean?"

"You tell him," Dudley said, still chortling. "I don't think I can."

"It means," Wainwright explained, controlling himself, "Lord and Master of the Loincloth."

6

The wind's ferocity abated within an hour, and the rain was gone in two, but the storm left considerable damage in its wake. Wainwright and Dudley joined the commander in surveying the ruined camp as the rest of the men, looking like half drowned rats, began setting things in order.

"My God, Wainwright," Spicer-Simson said. "Did you see that cyclone?"

"I did, sir."

"And those waves?"

"Yes, sir."

Spicer-Simson paced back and forth with a quick energy that several months earlier Wainwright would have called nervous but that he now recognized as ill-concealed excitement.

"We're definitely going to need a breakwater if we expect our harbor to be safe. Are the boats safe?"

"They are, sir. I just checked on them. They're well back in the trees and downhill of us. They're wet, but that's what happens to boats."

"And the supplies?"

"We were lucky. Tomorrow, I'll have Chief Waterhouse work out something more secure."

"Good, good. But don't assign him too many men. We must put our main effort into the harbor. The natives tell me that the prevailing winds are from the southeast, so we'll have to concentrate our initial efforts on the southern breakwater."

"Right, sir. I'll have the chief post the duty roster at the end of evening mess."

"Bring the natives in on the construction." Spicer-Simson paused, looking at Wainwright. "Did you hear what they called me?"

"I heard," Wainwright admitted, unsure of how to respond.

"Bwana Chifunga Tumbo," Spicer-Simson intoned, then he grinned. "Has a dandy ring, don't you think?"

"As long as they believe the 'lord and master' part, we should have no trouble getting the harbor built, Commander."

"I trust so. How goes the work rebuilding the huts?"

"They'll be done by nightfall. I...."

He was interrupted when Petty Officer Flynn came hurrying over to them. "Commander, sir! It's one of them, sir!"

"One of them?" Spicer-Simson looked down his nose at the petty officer.

"One, precisely, of whom?"

"The Germans, sir. One of their ships."

"Indeed?" Spicer-Simson looked barely more than interested, but he turned to Wainwright and said, "Make sure the men don't display themselves too openly. We don't want the enemy to know we're here. Then join me at my headquarters."

"Yes, sir," Wainwright said, feeling a thrill of excitement as he hurried to obey. Meanwhile, Spicer-Simson returned to his hut, went inside, and emerged with a pair of binoculars.

The German ship was just coming around the headland to the north of the bluff where the camp sat. It was cruising along at an unhurried pace about a mile off shore, though the angle of its wake indicated that it had been farther out when it had passed the fortress at Lukuga—no doubt to avoid inviting a shelling from the four-inch guns now stationed there. Spicer-Simson trained the field glasses on the ship, and he could clearly see the crew, especially those behind the circular shield of the single forward gun, which was aimed obliquely in the direction of the shore.

"Which one do you think it is, sir?" Wainwright asked. He'd come under the commander's veranda, and the other officers were close behind.

"One six pounder forward," Spicer-Simson said half musingly. "Wooden deckhouse and one funnel." He passed the glasses to Wainwright, who took a closer look for himself.

"She's got to be the *Kingani*," the lieutenant said, peering through the binoculars. "Looks like the reports on her gun were right—it doesn't have full sweep. Maybe just over a hundred degrees."

He started to return the binoculars to the commander, but Spicer-Simson waved them off, so he handed them to Dudley instead. Spicer-Simson lit a cigarette, stuck it into his holder, and puffed reflectively, watching the German ship angle off toward the south.

"She's not so big," Dudley said confidently, lowering the binoculars.

"I dare say she'd look big enough if you were staring down the bore of that six-pounder," Spicer-Simson said, but there was no bite to the words.

"You don't think we can take her, sir?" Dudley asked.

"Of course we can, Sublieutenant. That's what we came to do, and we shall. But we shall do it properly."

In another few minutes, the *Kingani* had passed out of sight down the coast, and the officers were about to quit Spicer-Simson's veranda when Rickson exclaimed, "Look! What's that?"

The others followed his gesture and saw a second craft, flat and lumbering, come chugging around the headland to the north, pursuing the *Kingani*'s wake.

"Can it be?" Spicer-Simson's voice dripped with false surprise as he retrieved his binoculars from Dudley and stared through them. "Indeed, it is," he answered himself. "I see Captain Goor standing by the helmsman." He lowered the glasses and gestured with them toward the object of their attention. "Gentlemen, witness the best the Belgians have to offer—the *Moselbak*. That translates as 'clam bucket.'"

Everyone laughed.

"Not her real name, of course. That's *Dix Tonne*. But you can see that *Moselbak* aptly suits her. She's probably doing top speed right now, which, I'm told, is about five knots. And she's only lightly armed."

"Is that one of the boats you're commandeering, sir?" Wainwright asked.

"It is. Do you have some objection, Lieutenant?"

"Only if you order me to command it, sir."

"You, Wainwright? Certainly not. We'll let Captain Goor keep his scow. I'm giving you and Dudley *Toutou*."

"Both of us, sir?" Wainwright looked puzzled. "I don't understand."

"Would you have me stay ashore while you and Dudley gallivant around in *Mimi* and *Toutou* both?"

"I guess not, sir. But we can't both command *Toutou* at the same time."

"Then you'll have to do it one at a time, won't you?"

"Yes, sir. But which one first?"

"Oh, you want *me* to decide?"

"That seems proper, sir."

"All right," Spicer-Simson dug into his pocket, produced a sovereign, and flipped it into the air. "Call it now, Lieutenant. Quickly."

"Tails," Wainwright said.

The coin fell into the dirt, and the two junior officers crowded over it to see which way it had landed.

"Heads!" crowed Dudley, picking up the coin and handing it back to the commander. "Does this mean I'm to command *Toutou*?"

"For our first engagement," Spicer-Simson said, pocketing the coin.

"I don't know what to say, sir," Dudley said.

"Say, thank you, sir."

"Yes, sir. Thank you, sir."

"What about me?" Wainwright asked, looking crestfallen.

"Do you want the *Clam Bucket*, Lieutenant?"

"Not exactly, sir."

"Then you will just have to wait your turn, won't you? You may command *Toutou* for our second engagement, should we have one."

By now, *Dix Tonne* had come abreast of the camp and was starting to make a wide, sluggish turn to head back the way it had come. Goor had given up the obviously futile chase.

With nothing left to distract them, the officers returned to their duties.

7

An hour later, one of the ratings on guard knocked at Spicer-Simson's door.

"A Belgian officer here to see you, sir."

"See him in," the commander said. "Ah, Major Jadot. So, Major Stinghlamber has finally relented and sent you?"

"Oui, Commander," Jadot replied with a quirky smile and a twinkle in his eye. "I have been trying to convince him for a year that his plan to dredge the Lukuga is folly. It is the lake's only outlet, and not only does its depth vary greatly, it is constantly clogged with sand and debris. I think your threat to relocate to Kitute is what did it."

"Frightened him with it, did I?"

"Don't underestimate Major Stinghlamber, Commander. He may be cautious and short-sighted, but he's a brave man. He's held on here despite the loss of his steamers and repeated bombardment from offshore. But he is anxious to avoid conflict between our governments."

"We are at war with a common enemy. It would behoove Major Stinghlamber to let experts deal with matters beyond his understanding. Kalemie is a much more rational choice."

"Yes, I agree, Commander," Jadot said, not noticing, or ignoring, Spicer-Simson's cutting tone. "But Major Stinghlamber still has other ideas. May I show you?" He held up a roll of papers that he was carrying.

"You may."

Jadot spread the papers on the table that Spicer-Simson was using as a desk. One was a chart of the shoreline for several miles on either side of the fortress.

"This is Kavalo Island." Jadot pointed to a small island just off a point of land north of Lukuga. "He thinks that the bay behind it would be an ideal spot for you to anchor your boats."

"That idiot," Spicer-Simson said with his best nasal snort. "That must be twenty-five miles away, and there are no shore defenses. And how are we to be supplied—by that scow the *Moselbak*?"

"I can't answer your questions, Commander. I am merely the messenger, and I do not claim to agree with the major's assessment that the bay is in your best interests."

"But it would be in *his* best interests?"

"How does your English saying go?" Jadot asked with a sly smile. "Out of sight, out of mind?"

"Blown out of the damn water's more like it," Spicer-Simson replied. "I'm sure the major would like nothing better."

"It is a perfect trap," Jadot agreed. "And the Germans have been known to anchor there. But we aren't going to give them or the major, the satisfaction, are we, Commander? Kalemie is the place, and you're just the man to make it happen. Look."

Jadot presented a set of detailed drawings of the proposed Kalemie harbor.

"The Kalemie enters the lake at a most fortuitous spot," Jadot went on. "There already is a small bay, and you can see there is the beginning of a natural breakwater formed by this spit of land here." He pointed to the southern edge of the bay. "Out in the water, the spit continues for some distance as a reef. The cliffs on either side will help shield the harbor from winds and from German eyes."

"What about the depth?"

"Fortunately, the river has carved a natural basin that should be sufficient for your boats except in the event of a storm. But we have many of those, and I think it would be wise to deepen it. We can easily blast these submerged rocks here and here, which should take care of matters. We can use the debris from the blasting to help build up the southern breakwater and one from the northern side as well. If that's not enough, there is a lot of loose rock left over when they cut the rail line through the Mitwanze Gorge a few miles back up the track. In fact, the blasting and breakwaters will make the basin large enough to hold the *Vengeur* and the *Baron Dhanis* once they are afloat. After the war, Kalemie could become a real harbor." He looked up and smiled. "That's what finally convinced Major Stinghlamber. I told him you would do the work, and he could take the credit for improving Belgian commercial interests in Africa."

"Excellent," Spicer-Simson nodded. "I see you are a man of vision. I'm tempted to think you have some British blood in you."

"No, Commander, I am of pure French extraction. But don't judge all of us by the example of Major Stinghlamber. Many of us think he's a stick-in-the-mud, as I believe your expression is said."

"So I gathered."

"Goor is a good man, and he'll follow your lead as long as it doesn't seem that he's directly disobeying the major. He does what he can with that tub he's got, considering the only thing he captained before the war was a fishery."

"A fishery?" Spicer-Simson asked, looking amused. "Hardly naval, but related, I suppose."

"Yes. He was stationed here originally to determine if Tanganyika could lend itself to commercial fishing to help feed the native population. The lake has an incredible abundance and variety of fish, many edible."

"Indeed?" Spicer-Simson perked up. "Perhaps I shall have my machinists make me a rod and reel."

"I wouldn't do that, if I were you, Commander."

"And why not?"

"By international law, Tanganyika is the private game preserve of King Leopold—a condition even your Admiralty would have to acknowledge. Any breach of that proviso might give Major Stinghlamber unwanted authority over violations."

"Diplomatically stated, Jadot," Spicer-Simson nodded. "I shall inform the men, especially the cook. I presume we may purchase fish."

"To your cook's content. As for the other officers, I'd watch out for Dender. He's a prickly sort, and he doesn't like what he considers interference. Besides," Jadot chuckled, "he lost heavily in the wagering that you would not make it to the lake."

"Goor is the only one who interests me. I couldn't care less what Dender thinks or does, as long as he does it on dry land." Spicer-Simson shook Jadot's hand. "Your help is appreciated."

"I'd do anything to get the Germans off the lake," Jadot said. "But don't thank me now. Wait until the harbor is finished."

Jadot left the charts and plans on the table when he departed, and Spicer-Simson had a rating fetch Wainwright. When the lieutenant arrived, the commander was in a buoyant mood as he pointed out the Kalemie's natural features, though his contempt was ill concealed when he related Stinghlamber's idea of using the backside of Kavalo Island as an anchorage.

"The fool has absolutely no conception of naval warfare," he sneered. "If it was left to him, we'd be sitting there in a perfect trap."

"Yes, sir," Wainwright replied, glancing at his superior. But the commander seemed not to hear him as his eyes went suddenly distant. Abruptly, a satisfied smile twisted his upper lip, and he looked at Wainwright.

"Yes, Lieutenant, a perfect trap. Now, where were we?"

The two of them went over everything for an hour, then Wainwright left to look for Chief Waterhouse to pass out assignments.

Eastwood told him that he thought he'd seen Waterhouse heading for the mess hut, so Wainwright went that way. He didn't find Waterhouse in the hut, which was large, windowless, and sparsely furnished with a simple long wooden table and a few chairs, but as soon as he entered, he knew something suspicious was going on. The cook and several ratings were clustered in the middle of the hut, laughing and jabbering excitedly about something in their midst.

"What the hell is that?" Wainwright demanded.

"What's what, sir?" asked the cook as the half dozen ratings in the mess hut nervously closed ranks, jostling shoulder to shoulder.

"You," Wainwright pointed to Atkins, who was in the center of the group and seemed to be struggling to keep something concealed behind him. "Break ranks."

"Me, sir?" Atkins squeaked. But his lost concentration made it unnecessary for him to comply with Wainwright's command. A small hairy face appeared over his shoulder. A second later, the rest of the creature followed. It was a baby chimpanzee.

"That's Josephine, Lieutenant," the cook said. "The men found her in the compound an hour ago. The storm must have driven her out of the jungle."

"She was hungry, no doubt," said Wainwright, the stern look on his face belying the humor he felt at seeing the young chimp's funny, forlorn face as she clung to Atkins's shoulder and neck.

"Yes, sir."

"And you fed her?"

"Yes, sir."

"Well, get her out of here before mess. And you men, get back to work."

"Yes, sir," came the chorused reply as Wainwright left the mess hut to continue his search for Waterhouse.

"Where'd the commander go?" Waterhouse asked when Wainwright found him.

"I didn't know he went anywhere. He was in his headquarters the last I saw him. I hope he didn't go out hunting. He's liable to get lost again."

"I don't think it was that," the chief said. "He was pointed toward Lukuga, and it looked like he was in a hurry."

That evening, Spicer-Simson didn't show up at mess, and Wainwright was on the verge of calling up a search party when one of the ratings came around to tell him the commander was in his hut and he wanted the officers to join him.

They found him red-faced and agitated, and Wainwright wondered what had transpired to so quickly alter his exuberance of this afternoon. It didn't take long to find out.

"That goddamn Stinghlamber is a goddamn idiot," the commander raged. "An insufferable goddamn idiot. I took that fool a perfectly sound proposal that could have solved our problems quite simply. If he'd thought of it himself, we wouldn't even be needed. But he has no incentive. No incentive whatsoever. No wonder the Belgians are losing the war here."

"What was the idea, Commander?" Wainwright asked.

"You remember when I showed you that absurd anchorage he wanted us to use on the backside of Kavalo Island?"

"Yes, sir."

"And you remember that I remarked that it was useless to us because it was a perfect trap?"

"I remember, sir."

"Well, that set me to thinking. If it would be a perfect trap for us, then it could be for the Germans, as well. All we have to do is have Stinghlamber use some of his askaris to haul one of his guns to the island and set it up in easy range of the anchorage. Any German ship that tried to use it would be sitting pretty. The charts show that the water is shallow there, and we could easily raise her for repairs and arm her with a larger gun than she already carries. That would give us kind of firepower we need to take on the *Graf von Götzen*."

"But, sir," Wainwright asked. "Do we really need it? *Mimi* and *Toutou* can swim circles around anything they've got."

There was a murmur of agreement from the others, but Spicer-Simson silenced them with a chop of his hand.

"Please be realistic, gentlemen," he said scathingly. "We may be able to outmaneuver them, but the *Graf von Götzen* is a twelve-hundred-ton steamer, and she's armed with two four-inch guns and two three-inch guns, and she must have several Maxims as well. She can carry nine hundred men. Great gods, can you imagine the small-arms fire alone."

"But we'd be out of small-arm range," Dudley said.

"I can see you've never been in a naval battle, Sublieutenant. The fact is, there is likely to be considerable small-arms fire. And while I trust our small cannon against the *Kingani* and the *Hedwig von Wissmann*, they only have armor-plated hulls. The *Graf von Götzen* is steel, and those old three-pounders of ours would barely dent her. No, gentlemen, we have to get a larger boat that we can arm with a larger gun than anything we now have. The Kavalo anchorage provides the perfect opportunity, but that pig-headed Stinghlamber can't see it."

"It is an ingenious idea," Wainwright began lamely, but a snort from Spicer-Simson stopped him.

"Of course it's an ingenious idea! All my life I've had ingenious ideas; the difficulty is to get people to accept them. I remember in the Channel Maneuvers of 1905—it was to test the efficiency of submarines—we had been told that the exercise was to be as realistic as possible. I had the idea of passing a line to another destroyer and dragging for periscopes. The submariners complained that I nearly sank one of their boats, but that doesn't matter. The point is that the idea worked!

"Of course, in peacetime, there's little scope in the Home Fleet for an officer with ingenious ideas—that's why I went to China. It was the result, as it happened, of another exercise, this time to test the defenses of Portsmouth Harbor. The destroyer I had the honor to command was part of the attacking force. We approached under cover of night and it was, I thought, a wonderful opportunity to distinguish myself. I went full ahead and made straight for the harbor entrance! Unfortunately, the coxswain was dazzled for a moment by searchlights, and instead of crashing into the harbor boom as I had intended, I piled my ship up on the beach at Southsea.

"I was court-martialed for hazarding my ship—and what was the result? I was able to demonstrate that mine had been the most dashing attack of all, and that I had got nearer to our objective than any other captain. So I was exonerated; even highly commended! But it so happened that there wasn't another destroyer available for me at the time, so I was given command of a gunboat on the Yangtze River. I was tired of fleet maneuvers, anyway. I wanted some real action, and that was the place for it! You will find that the best officers of the Royal Navy, the men to whom the country turns in its hour of need, all had their training on the China Station, and in gunboats."

"Excuse me, sir," Wainwright said.

"Yes, Lieutenant?" Spicer-Simson glared at him.

"We still have the mines, sir. We could lay them behind Kavalo Island. Maybe they'll do the trick, even if the Belgians won't cooperate."

"Excellent suggestion. See to it tomorrow. Now, gentlemen, please leave me. I have to inform the Admiralty of Stinghlamber's stuporous incompetence."

8

Much to Wainwright's secret delight, Josephine showed up for morning parade. Atkins was one of the men assigned to raise Spicer-Simson's vice admiral's standard, and he kept having to shoo the chimp away as he and his companion unfurled the flag and hoisted it aloft. As the flag rose, its colorful fluttering attracted Josephine's eye, and she scampered up the pole after it.

A jaundiced look from the commander's eye cut off the men's laughter as quickly as it had begun.

"What is that creature doing, Wainwright?"

"I think she's taken with your handiwork, Commander."

"Well, get her down off that flag pole immediately."

"You," Wainwright called to Atkins. "Get her down."

"How, sir?" the rating asked lamely.

"Perhaps you can offer her your breakfast."

"Josephine," Atkins called. "Come on down, now, that's a good girl."

"Josephine?" Spicer-Simson cut his eyes at Wainwright.

"Her name, Commander," Wainwright explained as the rating continued to coax the chimp. "It seems she's lately joined the expedition."

"So," Spicer-Simson said loudly. "A new recruit. I suppose next she'll be wanting a commission."

That drew laughter from the men.

"You," Spicer-Simson said to Atkins. "Stand back."

The rating obeyed, and Spicer-Simson strode up to the flagpole, rapped smartly on it with his swagger stick, and said in an authoritative tone, "You up there. There's no crow's nest on this mast, so come down this instant."

Josephine let go of the fistful of flag she'd gathered and stared down at Spicer-Simson. Then she shinnied down the pole and hopped into the commander's shoulder.

"Filthy beast," the commander said, smiling. "At least she knows who her commanding officer is. Cook! Have you fed this creature?"

"Yes, sir," the cook said hesitantly.

"Well if you plan on continuing to feed her, she'll have to eat at regular mess. I won't have anyone eating out of turn."

"Yes, sir!"

The men gave a laughing cheer that Spicer-Simson cut short.

"All right men, off to mess. Then we'll have at the breakwater. What do you say?"

The men gave another cheer and barely waited to be dismissed before they headed off to eat, Josephine in tow.

"That was generous of you, sir," Hanschell told the commander.

"Letting them keep the chimp?" Spicer-Simson shrugged. "Machiavelli said, let the people have their way in the small things and they'll more readily follow you in the large. Besides, I've always thought pets do humans a world of good. That chimp—Josephine, you say?—she'll give the men a spot of comic relief, and they'll need all they can get the next few weeks, as I plan on working them to death."

Chapter XII: Kalemie

WORK THEM NEARLY TO DEATH Spicer-Simson did. Starting that morning, every man in the British force labored fourteen hours a day, six days a week, assisted by four hundred African workmen. They blasted, cleared the shoreline, and transported stones from the Mitwanze Gorge to Kalemie and carried them out onto the breakwater. The Africans workers delighted in the explosions used to clear the rocky protrusions from the bottom of the basin and the subsequent waterspouts that fountained into the air, cheering every time one boomed out. Even the commander put in his fair share of time, carrying stones and doing other manual labor with the common sailors, and the Africans never tired of observing his tattoos writhe over his muscles. He became so popular that several askaris under a corporal had to be detailed to keep the crowd of adoring women and children at bay.

During the first few days, Waterhouse and a crew of three ratings were exempted from harbor-building duties to construct a magazine under Wainwright's direction. After seeing the ferocity of the storm and learning from the locals that it was by no means an unusually violent one, Wainwright selected a dry, rocky area of hillside on the west side of the camp from which even torrential rains would be likely to drain. There, he had the ratings burn off all the grass then dig a broad pit about eight feet deep. Into this, they put the ammunition and petrol and layered those with the expedition's foodstuffs and other stores. Atop that, they built a low-roofed shed of palm-leaf thatch and covered the whole with tarpaulins to keep it all cool.

Then they joined their comrades at the lake. When Wainwright went down for the first time, he found Spicer-Simson, his tattoos glistening with sweat, lugging a stone that must have weighted eighty pounds. The commander dropped the stone into the water at the end of the growing breakwater before returning to the beach where Wainwright waited.

"Good, God, man!" Spicer-Simson yelled, a huge grin on his face. He slapped the lieutenant on the back. "Is there anyplace you'd rather be!"

"No place in the world, Commander."

"Well, get to it, then!" Spicer-Simson gestured toward a pile of stones at the foot of the cliff. To make his point clear, he grabbed another himself and headed toward the breakwater. Wainwright, grinning himself, followed suit.

He wasn't smiling for long. Shortly before dark, a runner arrived with a message for Spicer-Simson, and a few minutes later, Eastwood came to call Wainwright to the commander's hut.

"The Lever Brothers supplies have arrived," Spicer-Simson said.

"They're here, sir? Already?"

"Not here, Wainwright. They are in Fungurume."

"Fungurume! What are they doing in Fungurume?"

"They are waiting for you to pick them up, Lieutenant."

"Me, sir?" Wainwright felt his heart sink. "Why me, sir?"

"Because you are the only one I can trust to get the job done properly."

"Thank you, sir," Wainwright said, not feeling particularly honored. The prospect of retracing their route all the way to the head of the rail line from Cape Town wasn't especially enticing.

"What about the harbor, sir?" Who's to supervise?"

"Cross and Dudley can manage well enough while you're away. Take a few ratings with you to act as guards. You can hire native bearers in Fungurume. I'm sure there are plenty who will want the work now that we've left the territory."

"Yes, sir. I'll leave first thing in the morning."

"Very good, Lieutenant. Have a good journey. It shouldn't prove nearly as arduous as the one that brought us here. I've learned, incidentally, that the Belgians have finally completed the rail line to Kabalo, which should make matters even simpler." He laughed. "I suppose they want Onfroi and their little train returned."

Eastwood followed Wainwright outside and over to the mess hut.

"There is one bright side, old chap," the quartermaster said.

"Pray, tell me, sir," Wainwright said glumly.

"I sneaked a peek at the telegram and now have a solution to the mystery of how Lever Brothers knew we needed their supplies."

"I'm all ears." Wainwright took off his cap and mopped his brow.

"Hope discovered they had them and asked their agent to turn them over. Apparently, the agent got in touch with the Lever Brothers home office, and the home office got in touch with the Admiralty."

"Hope." Wainwright couldn't help but chuckle as he donned his hat. "Well, the commander will certainly take the credit, then."

Both men laughed and went into the mess hut to eat.

2

Wainwright left, as he promised, at daybreak, so he did not get a chance to see Spicer-Simson's prediction about Josephine prove correct. By the end of the first week, she became the light of the men's lives, easing the burden of their incessant heavy labor. No matter how tired and cross they might become, she could raise a smile. They taught her to stand at attention during morning parade, and she would join the ranks at the end of the file, clutching at the leg of the rating next to her. And at mess, she ate with the rest of them, even learning how to use a spoon.

But the wildlife wasn't always so pleasant or amusing as Josephine. Now that the rains had begun, Hanschell began compulsively looking for signs of mosquitoes. He didn't see any, but the threat was enough to have him order everyone to rig mosquito netting over their cots and to take a quinine pill every day at evening mess.

But minute pests were nothing compared to the gristly discovery they made on the rock beach a few days after they'd begun work on the harbor. Hanschell hadn't even opened the medical hut for business when one of the ratings rushed in, pale and panting.

"Come quick, Doctor," he gasped. "Down at the harbor."

"What is it?"

"One of the native women, sir. She's dead."

"Dead?"

"She ain't got no legs, sir," the rating said, eyes goggling.

"No legs?" Hanschell struggled to understand even as he gathered up his medical kit.

"I seen it myself. Come quick."

"If she's dead, there's no need to hurry," Hanschell said, but he came quickly anyway.

There had been no need for haste. The woman, who looked to be in her thirties, had been dead for hours. Both legs were bitten off just above the knees.

"Crocodile," Hanschell pronounced to the gathering crowd. "She probably died from hemorrhage and shock. Get something to cover her up. Find out who she is so her family can take care of her." He looked at the men standing around, expression severe. "I can protect you from disease but not from your own stupidity if you attempt to come down here alone or in the dark. Let this poor woman's fate be lesson enough. Tanganyika's full of crocs, and I'll warrant they'd find an Englishman as tasty as one of the natives."

Despite an inquiry, no kin of the woman could be found, and in the end, they buried her in a cleared area of the forest just inland from the camp. Spicer-Simson said a few words over her then ordered that armed guards stand watch over the harbor construction to prevent further incidents. Every once in a while, one of the guards would shoot into shadows in the water, but if a crocodile was hit, it never surfaced to tell the tale.

"Most likely the explosions are keeping them away during the daytime," Cross told the commander. "That woman might not even have been killed here. Her body could have washed up from elsewhere."

"Quite right, Lieutenant," Spicer-Simson replied. "But I want the guard maintained. I don't want that cemetery of one to find itself the resting place of a single man of my expedition."

The good weather continued, the construction moved steadily ahead, and even the Belgians were more cooperative. As Stinghlamber grudgingly promised, they set up a couple of twelve-pounders on the bluff above the harbor. The guns had been intended for the *Baron Dhanis*, but as Spicer-Simson had pointed out to the Belgian commandant, that ship was still in pieces in Kabalo. The gunners were Belgians and remained under Stinghlamber's command.

The only spot of trouble came not from the building of the breakwater but from a brush fire that swept across the bluff. The cook and the few men in camp managed to divert it from the huts but could not prevent the wind from carrying the flames straight toward the magazine and petrol depot. Luckily, Wainwright, who remembered all too well the brush fires during the trek, had ordered the men to burn off the surrounding grasses before digging the pit, and the flames never got close enough to ignite the munitions or fuel.

3

"Dudley!" The voice was faint but no less throaty for the distance.

"Is that the commander?" Cross asked, straightening from his labor on the breakwater and pointing at the rim of the bluff above Kalemie Harbor.

"Do you think he needs me?" Dudley replied with a wink. He hopped from rock to rock along the breakwater until he was on the beach, then he trotted toward the path that led up to the bluff. As he reached the top, he paused to stare down. Although they'd barely begun construction, the lengthening breakwater already made Kalemie truly look like a harbor.

Dudley smiled as he recalled hearing a group of tribesmen discussing the breakwater. The general consensus among them was that the British were going to build a bridge across the lake. There, opinion split almost evenly between those who thought the bridge was so soldiers could march over to attack the Germans and those who believed the bridge was for the rail spur, which had been extended down to the water to carry rocks from Mitwanze Gorge.

The placid water within the growing arm of breakwater gleamed like a mirror under the sun. There was no dock. The rocky beach shelved off so gradually that any dock that they actually could use would be too long and flimsy, and they'd seen what Tanganyika's storms could do. Instead, they'd bought several canoes to serve as shore boats, and *Mimi* and *Toutou* would be anchored out in the harbor along with *Dix Tonne* and the two Belgian motorboats, *Vedette* and *Netta*. The latter was flat-bottomed, overpowered for its size, and fast as hell but too light for anything more than a couple of men and its Lewis gun.

Dudley smiled again, thinking how Spicer-Simson, backed up by himself, Wainwright, Chief Waterhouse, and a dozen armed ratings, practically confiscated the two motorboats from the reluctant Belgians the week before. Stinghlamber had finally consented to let Spicer-Simson take the two craft, but he'd refused to relinquish *Dix Tonne* without a guarantee that Goor would remain in command of the barge.

"Goor can keep it," Spicer-Simson told the Belgian as he and his men boarded the two smaller boats to take them back to Kalemie. "None of us want it."

They agreed to let *Dix Tonne* remain at Lukuga until Kalemie was finished. Then the British took the motorboats to the fledgling harbor. They weren't much, but they were a start, and counting *Mimi* and *Toutou*, which were still hidden in the trees half a mile from the water, the commander had the beginnings of a real flotilla.

Dudley's breast filled unexpectedly with something akin to pride as he stood there. The commander was a daft bird, but he'd inspired the men to do great work. If they never accomplished anything more than what they already had, they could be proud—even if the Germans blasted the tiny flotilla to pieces in its makeshift harbor. And damn, if he wasn't part of it. A big part....

"Dudley, blast it!" Spicer-Simson strode into view, whiskers bristling. "What the blazes are you doing?"

"The harbor, sir," Dudley said, jumping. How Wainwright managed to put up with the commander so well never ceased to amaze him. "It caught my eye. It looks lovely, and see," he pointed. "There is the beginning of your flotilla."

"Come on, man! There's no time for mooning with work to be done."

"Yes, sir."

"I've been up at Lukuga talking to that insufferable snot Stinghlamber. I asked him for some telephone gear so we can keep in touch more readily than sending a runner. After all, it's more than two miles, and this is war. We need the best communications possible, but he said he had none to spare."

"He wouldn't supply you, sir?"

"I don't know if he doesn't have the wire or if he's still miffed that I took his boats." Spicer-Simson gave a sudden guffaw and poked Dudley with his elbow. "Probably the boats, eh, Sublieutenant?"

"He seemed pretty attached to them, sir, when you took them away," Dudley said with a smile.

"No matter. Let him sulk. We've got a war to win, with or without Stinghlamber, though I confess I do not comprehend his attitude. When I was on the Gambia, even the governor acknowledged the superiority of my rank to his."

"I understood that the governor of a colony took precedence over everybody," Dudley said.

"Not in this case," Spicer-Simson answered. "The governor of Gambia, before he joined the colonial service, had been a major in the army. A major does not forfeit his rank on resigning; therefore, he was still a major, and as such, had to acknowledge the seniority of my rank of lieutenant-commander. We served together on the Boundary Commission. But these Belgians don't even know the difference between a commandant and a commander."

Spicer-Simson lit a cigarette, inserted it into his holder, and took a few thoughtful puffs.

"But I did learn something interesting in my visit," he said. "Captain Dender let slip that the Germans have telephone lines strung down the entire eastern shore of the lake so they can keep in touch between Kigoma and the German fort at Bismarckburg at the southern end of the lake and all their military posts in between."

"Are you thinking what I'm thinking, sir?"

"Absolutely not, Dudley. You're thinking what I'm thinking. Tell Waterhouse to make *Vedette* and *Netta* ready to sail, and have them stocked with provisions for a couple of days. Have him divide the supplies into six packs. He's also to select four askaris to accompany us. Who's that signalman who knows about telephones and telegraph?"

"Tasker, sir."

"Find him, and tell him to fall out of whatever duty he's in. Everyone should eat and doss out. We'll be leaving at two am."

"That's rather early, isn't it, sir?"

"Of course it's early. I'd much prefer arriving on the German shore before daylight in case lookouts are posted."

"Yes, sir. That's right. You're going with us?"

"Of course."

"But, sir. Are you sure? It might be dangerous."

"Hang the danger. I want to piss on the enemy's soil."
"Yes, sir." Dudley turned to go.
"And Dudley."
"Sir?"
"Everyone is to be armed, and bring ammunition for the Lewis guns."
"Yes, sir!" Dudley hurried off toward the top of the bluff, the view of the harbor below forgotten.

Spicer-Simson did not immediately follow Dudley back to the camp. Instead, he, too, paused at the cliff's edge and took in the view, though it was impossible to tell from his expression whether or not a feeling of pride swelled his own breast.

4

Just before 2 AM, Dudley, Tasker, and four askaris worked their laborious way down the dark path to the harbor. The askaris were carrying, in addition to their rifles, two canvas bags of tools and a pair of contraptions fabricated out of pipe. Spicer-Simson was waiting next to the boats, which Waterhouse had drawn up on the beach and made ready. For the trip, the commander was wearing trousers instead of his skirt.

"What are those?" Spicer-Simson pointed to the contraptions.

"Wire rollers, sir," Dudley said. "Mr. Cross had them worked up. We'll need them if we're going for wire."

"Right you are. All right men, let's be off!"

In five minutes, everything was stowed, and the men were aboard. The chief helmed *Vedette,* and Tasker *Netta.* Askaris manned the Lewis guns, while the commander and Dudley took positions amidships in *Vedette.* In short order, the motorboats pulled out of Kalemie Harbor on the first official sortie of the Tanganyika flotilla.

Dudley watched the few lantern lights of Kalemie Harbor gradually recede as the boats edged through the moderate waves raised by an easy night breeze. Between Kalemie and the eastern shore, he knew, Tanganyika was fifty miles wide. After about half a mile, he could see a few lights on top of the bluff where the camp stood, but within twenty minutes, the lights dimmed to little more than the twinkle of the stars above. Ahead was only darkness. And Germans.

The enemy had to know by now that the British expedition had joined the Belgians at Lukuga, and the construction of a harbor only could mean that boats of some sort soon would occupy it. It was only a matter of time before they started nosing around, and they probably already were on their guard.

He peered nervously into the darkness that surrounded them but could see nothing, so he stole a glance at the commander, whose features were dimly lit by the half moon. Spicer-Simson was staring intently into the darkness ahead, too, as if his vision could penetrate the ebon atmosphere all the way to the eastern shore, seemingly oblivious to the danger that an approaching German vessel might pose.

Dudley tried to relax. Even if the Germans were on the lake in their vicinity, they'd be moored for the night, not sailing. Besides, the fastest German ship was *Hedwig von Wissmann*, which could do only ten knots on a good day. The motorboats were quick enough to pace *Hedwig* if not outrun her and nimble enough to make shelling at a distance an unlikely possibility, even in full daylight.

Nonetheless, Dudley couldn't help but cast around for some glimmer in the blackness. He wasn't sure if he was keeping an eye out to ensure that he saw the Germans before the Germans saw them, or if he just needed to spot something tangible in this murky insubstantiality through which *Vedette* and *Netta* ploughed. It was eerie out here with nothing on which the senses could attach themselves except the distant and cold stars and the pallid crescent moon. But Dudley found solace in the sound of the petrol engine, which Waterhouse kept humming at a good clip, and the chief's sure hand on the tiller as he steered by those same stars toward the enemy coast.

"Imagine it, Dudley," Spicer-Simson said after a time, waving expansively and staring about at the seemingly endless darkness. "Tanganyika. Tanganyika, man! I've done it! I've actually gotten us here."

"Yes, sir. You have."

"And here we are in the middle of it. Did you know, Dudley, that Tanganyika is the second deepest lake in the world?"

"What's the first, sir?"

Spicer-Simson turned his head in Dudley's direction and laughed.

"Did I say something funny, sir?"

"We are sailing—no, flying—through a veritable void of mystery and enchantment, with eternity not just above but all around—yes, buoying our very craft—and you ask some sort of technical question."

Spicer-Simson fell silent as he lit a cigarette with effort in the wind of their passage, inserted it into his long cigarette holder, and stuck the holder between his teeth. The coal glowed fiercely in the wind.

"Sir?"

"Yes?"

"What *is* the deepest lake?"

"Baikal," the commander said, taking the cigarette holder from between his teeth. Dudley caught a brief scent of smoke that instantly dissipated. "In Russia. The hearts of Mother Russia and darkest Africa both tap deep springs."

"Have you ever been there, sir? Lake Baikal, I mean."

"To my regret, no. I was on the way there when Archduke Ferdinand was assassinated and I was called home to my command. But no matter. I'm on Tanganyika now." He glanced back at Waterhouse. "Steady on course, Chief."

"Right, sir. Steady as she goes."

Gradually, after four hours on the water, the eastern shore edged up out of the horizon. Dudley could see the line of bluffs standing dark against the now-brightening eastern sky.

"Throttle down, Chief," Spicer-Simson ordered. "If there are Germans on the shore, we don't want them to hear our engines, and sound is bound to carry a long way over the water."

"What if the Germans are waiting for us, Commander?" Dudley ventured.

"We'll have our Lewis guns send them packing, don't you think?"

The boats landed without incident, however, in a little rocky cove that would keep them hidden from all but direct scrutiny. The first thing that Spicer-Simson did after hopping out of *Vedette* was to urinate copiously on the narrow, pebbled beach.

"Been saving that up the whole way over here," he said, buttoning his fly. "All right, men. Shoulder your packs, take up your arms, and let's go find some telephone wire."

"How shall we find it, sir," asked Tasker. "It could be out there, anywhere." His gesture took in half of Africa.

"I believe we can assume that it won't be trying to hide, Tasker. We're not hunting snark, here. If there's a line, it'll be strung up and down from the north to the south. All we have to do is bear directly east, and we'll eventually cross it, don't you think?"

"Yes, sir. I guess that's so."

"Then take hold of your tool kit, and let's get hunting."

They found the line not two miles inland. It was strung through the bush on low poles that did little more than give enough clearance for most four-legged beasts.

"Here's where we split up, men," the commander said. "Dudley, you take Tasker and two askaris north, and I will head south with everyone else."

"How far should we go, sir?" Dudley asked.

"We want plenty of line while it's here for the taking. After the Germans discover it gone, we won't have so easy a time getting more. Make it ten or twelve miles, then start rolling. Meet back here tomorrow, and don't dawdle. As soon as the Germans lose contact, they'll send a patrol to make repairs, and we'll want to be gone before they arrive."

"Yes, sir."

"And remember," Spicer-Simson admonished Dudley. "If you see any Germans, do not engage them. Don't even let them see you if you can help it. We're here to appropriate wire, not become captives."

The two groups separated and, within minutes, were out of sight of each other.

5

"Do you think there are any out here, sir?"

"Any what?" Dudley stared at Tasker.

"Germans."

"His Majesty wouldn't have sent us out here if he didn't expect us to encounter them sooner or later," Dudley said.

"Yes, sir." Tasker took a determined grip on his rifle. "I expect you're right, sir."

They followed the wire, moving cautiously through the brush for most of the morning without incident except for a brief encounter with a family of elephants, which they skirted hastily. Suddenly Dudley stopped just at the edge of a clearing, and Tasker bumped into him. The askaris crouched behind them.

"What, sir?"

"Look." Dudley pointed across the clearing at a concrete bunker sunk between two boulders and partially masked by scrubby bushes.

"Don't move," Wainwright warned as he reached into his hip pocket and pulled out a handkerchief. Its color was more ecru from being washed in African streams than white, but Dudley sincerely hoped that the Germans in the bunker wouldn't be overly picky. Tentatively, he waved the makeshift flag.

Nothing happened.

Dudley peered into the darkness of the gun slit. No weapons protruded their deadly snouts. He stuffed the handkerchief back into its pocket.

"We're a bunch of lucky bastards," he told Tasker. "It's empty. We could easily have been mowed down. Come on, but keep watch. The tenants might be close by."

Dudley hurried over to the bunker, the others following. The outpost's interior was spartan—a couple of bunks, a table, and some cooking pans and utensils, including a single-burner Primus stove. And a telephone. A thin layer of dust lay over everything, but that meant nothing. After they left, dust would cover their traces in a matter of days, or maybe hours.

"Let's go on up the line," Dudley said. "We'll be passing back this way." They continued on for another few miles before Dudley judged they'd gone far enough and called a halt. He had the askaris unship the wire roller, then one shinnied up the short pole and snipped the wire.

"That's the alarm," Dudley said. "As soon as the Germans realize their line is broken, they'll send someone along to repair it. We'd best get cracking."

"You think they're that close, sir?" Tasker asked.

"The outpost we found was deserted," Dudley said, "But that doesn't mean the next one up or down the line is. Keep your weapons handy."

"But the commander said not to...."

"Do you fancy spending the entire war in a German prison camp in the middle of Africa? Or getting shot at and not shooting back?"

"No, sir."

"Then I suggest we engage the enemy if he shows his face, no matter what the commander said. In the meantime, let's get to work."

Tasker showed the askaris how to feed the end of the clipped line into the wire roller and cranking the spindle. As they neared the first pole, the askari who wasn't cranking went up and detached the wire from the insulators. As an afterthought, Dudley had him use his bayonet to pry the insulators loose, and the sublieutenant popped them into his pack.

Thus began the laborious task of moving south again, winding the line onto the roller as they went. After half a mile, the coil of wire had filled the roller, so they pulled it free and started a fresh roll. By late afternoon, they'd reached the bunker. They spent the remaining couple of hours before dusk returning north to collect the coils they'd left strewn along their route.

"Damn, sir, but I'm bushed," Tasker said as they returned to the bunker with the last two coils.

"Me, too. But we'll feel better after a hot meal."

"That's right, sir. The Jerries left us a Primus. Think they'll mind if we use it?"

"I shan't mind. Fire it up, Tasker, and I'll lay out our meal."

"Very good, sir."

An hour later, the four men sat outside the door of the bunker, listening as Africa woke up for the night.

"Should we post a guard, Mr. Dudley?" Tasker asked.

"I think not. We have the bunker for protection if simba's out and about. And if the Germans are coming, they're not expecting anything but a broken line, so they won't be risking travel after dark. And we'll need our rest. We've got a long day ahead of us."

And it *was* a long day. They actually arrived at the rendezvous point by noon, but they had to return along their route as far as the bunker to collect the nearly twenty-five coils of wire they'd rolled since the day before. They didn't meet up with Spicer-Simson's party immediately, but there was ample evidence of the commander's progress in the growing number of coils left in the interim between trips down the line.

As late afternoon approached, Dudley told Tasker and the askaris to collect the last few coils of wire, then he returned to the bunker to pilfer its contents. He stuffed everything into his knapsack and, chuckling beneath its weight, trotted back to the rendezvous point.

When he got there, he found Waterhouse, Tasker, and two of the askaris loading themselves up with coils from the pile and Spicer-Simson walking up from the direction of the boat, trailed by the other two askaris.

"Look, Commander," Dudley panted as he drew near. He unshouldered the knapsack and opened it to display its contents to Spicer-Simson.

"A telephone! Excellent!" the commander said. "We came across one, too."

"And I took their Primus, sir. Imagine when they come to fix the line all hungry and tired and upset at the missing line and then finding nothing to cook with." He laughed, and Spicer-Simson joined in.

"All right," the commander said. "Put that back on, and let's get this wire down to the boats."

The two of them loaded up with coils and carried them down to the beach. By dark, all the wire and the purloined gear were stowed aboard *Vedette* and *Netta*, and the pilots were steering the launches out of the little cove.

Exhaustion took all conversation out of the men, but not Dudley's vigilance for German craft. The boats, burdened with several hundred pounds of extra weight, were laboring sluggishly through the water, and the sublieutenant no longer was convinced they could outrun any of the German ships if one happened to give chase. But it was unlikely that one of the enemy craft would be running in the dark, and the return trip remained as uneventful as the initial journey. By the time they crossed the center of the lake, twinkling lights were just barely visible on the western bluff. The lights gradually resolved themselves as those of the Belgian fortress, so it was a simple task for Waterhouse to bear slightly south.

And then they saw a beacon—dim to be sure, but definite. It was a kerosene lantern hoisted on a tall pole stuck in the rocks at the end of the southern breakwater.

"Damn thoughtful," Dudley commented tiredly but thankfully.

"Keep back!" Spicer-Simson called to Tasker. As *Netta* fell astern of *Vedette*, Spicer-Simson stood in *Vedette*'s bow like a conquering hero.

More lanterns danced on shore, and a sudden cheer rose from that direction as *Vedette* eased through the globe of light cast by the beacon lantern and headed for the beach. *Netta* came close behind.

"We thought you were lost, sir," Cross said breathlessly as Dudley tossed him the painter and he tied it off.

"I'm never lost, Cross," Spicer-Simson said, ragged exhaustion lending hoarseness to his imperious tone. "Now, would you be so good as to have our cargo taken up to the camp? And have someone run ahead and tell cook that famished men who've done an honest day's work are in need of comestibles and libation."

"Comestibles and libation, sir?"

"Precisely. Cook's best. Nothing plebeian for the heroes of the hour."

As a handful of ratings began unloading *Vedette*, Cross hurried up the path. The commando group, led by Spicer-Simson, began to trudge wearily in his wake. Suddenly the commander halted and gave a braying laugh.

"Sir?" Dudley prompted.

"Gustav Zimmer."

"The German commander, sir?"

"Precisely. Wouldn't you love to see his face when he learns he's just been stripped of twenty-five miles of wire and two telephones."

"Perhaps he'll think the natives took it."

"Ah, but *I* took it, Dudley. *I* took it."

6

Zimmer didn't know what to think, and he didn't know who took his telephone wire, but he was damned sure it had something to do with the mysterious construction that was taking place just south of Lukuga, around the mouth of the Kalemie River.

The spot was fairly well protected, however, and the Belgians had mounted a couple of guns there large enough to keep the *Kingani* and *Hedwig* at bay, and he was reluctant to send out the *Graf von Götzen* because it was, as yet, barely more than a floating hull and deck, and only the two smaller *Moewe* guns were in place. The big ones would have to wait for the shipbuilders to strengthen the deck. Without any weapons but the two 88s, he didn't want to risk taking the ship anywhere within range of Lukuga. Not, at least, until the two 105s were mounted. Then he'd do what he liked.

But the fact that his ships' crews couldn't make out the purpose of the activity on the western shore didn't mean that the work had gone entirely unobserved. His Ba-HoloHolo informants claimed that a new harbor was being built but that it only had two small motorboats. The natives said they were the canoes that the mysterious expedition with the strange blue uniforms had brought overland, but from the descriptions he'd heard, they sounded more like those two little Belgian boats, *Vedette* and *Netta*.

Unfortunately, any supposition was further complicated by the fact that some of his informants told him that hundreds of laborers were carting stone down to the lake in an effort to build a bridge across to German territory. Some even reported that the rail line had been run right down to the water and that the bridge was to carry trains of troops across the lake.

That, of course, was patently absurd, but the harbor continued to trouble him. Why would the Belgians be building a new harbor for two motorboats? Or even for that lead-bottomed barge with the ridiculously large flag. He would have laughed at the thought, but various Ba-Holo-Holo also reported that the Belgians were constructing another steamer. The problem with those reports, however, was that one group said the new steamer was in pieces at Kabalo, while others claimed that the Belgians were repairing the *Alexandre Delcommune* at the mouth of the Lukuga. If so, why the new harbor? Maybe they were rebuilding the ship at Lukuga to protect it from his boats and planned to move it later to the new harbor. Or was it being rebuilt by whoever was inhabiting the camp on the bluff above the new harbor? And if they weren't Belgians, who were they? The English? But if it was the English, all they had were those two large canoes they'd supposedly brought overland, and Zimmer couldn't believe that those could be large enough to threaten his ships.

Whatever the case was, there were too many maybes and too many unknowns. Worse, the information he'd been getting from the Ba-HoloHolo had been less than dependable of late even though he'd personally sent tribesmen of his most trusted informant, Chief Kabwema of Tembwe, back into the region. He suspected that his ships had bombarded the native villages once too often, and they'd finally turned against him. Not that he cared what the filthy beasts did, but he did require more information, and it had to be reliable. Zimmer called his aide and had him send for Lieutenant Job Odebrecht, the current commander of the *Hedwig von Wissmann*.

At dusk, Odebrecht sailed south from Kigoma, moving slowly and keeping well out on the lake until the *Hedwig* came abreast of the Lukuga River. He ordered the helmsman come around and begin inching toward the Belgian fortress, maintaining as slow a speed as possible to keep the steady chuk-chuk of the engines down and to eliminate sparks from the stacks. By 1 AM, the *Hedwig* was as close to Lukuga as Odebrecht dared approach, but the darkness on the shore masked whatever might be there as effectively as it hid the *Hedwig*.

He had his men lower the *Hedwig*'s dinghy, and he quietly rowed close to shore. What he saw in the moonlight stunned him. It was a shipyard—crude, but quite large enough to accommodate a ship of 250 feet. That was bigger than the *Graf von Götzen*. The fact that no ship was present meant nothing. The Belgians wouldn't have prepared the shipyard if they didn't expect to build a ship there.

Odebrecht rowed back to the *Hedwig*, and the German craft sneaked back out onto open water before sailing at full speed for Kigoma. Zimmer must be informed about this new development immediately.

The German captain wasn't pleased. First, his valuable shore communications had been stolen, and now the Belgians were building something large enough to challenge his flagship. Or were planning to. And even if the shipyard was presently empty, the Ba-HoloHolo couldn't be entirely wrong about either the *Alexandre Delcommune* or the unnamed vessel in pieces at Kabalo. He had a sinking feeling in his gut when he realized that both rumors might be true. Two ships! What would von Lettow-Vorbeck think? No, what would he expect?

If there was a new Belgian ship on the lake, Zimmer had to destroy it, but to do that, he had to find it. He knew, though, that Odebrecht couldn't do all the surveillance alone, so he sent for the *Kingani*'s skipper, Lieutenant Job Rosenthal, late of the cruiser *Königsberg*. It pleased Zimmer to have two Jobs captaining his two ships. It made things balance nicely.

For the next couple of weeks, Odebrecht and Rosenthal took turns at reconnaissance, each round lasting several days at a stretch. At night, they sneaked their boats beneath Lukuga's guns and went ashore. To protect themselves, they smeared their faces with burned cork and wrapped themselves in long blankets in an effort to look like just two more Africans. The disguises worked so well that several times they actually bluffed their way through the Belgian picket lines.

But they didn't learn much. Neither of the two reported ships were anywhere around the shipyard, and they dared not ask direct questions. They didn't even learn the name of the *Baron Dhanis*. But at last, one night at the end of November, the *Kingani* happened on a small patrol skiff manned by Belgian askaris. After the *Kingani* captured the craft, the prisoners told Rosenthal that a vessel now was under construction in the yard.

Rosenthal knew he had to take its photograph. Zimmer would expect no less. Just before dawn on November 29, he sailed the *Kingani* in toward the shipyard. At two hundred yards, he set the camera at its longest exposure, and although he couldn't see a ship, he began shooting anyway. By now, the German boat had been spotted by the Belgians, and the shore batteries began blasting away, but the dawn light spoiled their aim, and before they could score a hit, the *Kingani* pulled out of range. Unfortunately, the dimness of the light ruined the photographs, as well, but Rosenthal didn't know that at the time. Loading another roll of film in the camera, he ordered his helmsman to head for the Kalemie River. He might as well get shots of the construction there, too.

It took only a few minutes to reach the spot. It was, as yet, too early for construction, and the only men visible were a few askari sentries. They saw the *Kingani* immediately, of course, but Rosenthal didn't care. He snapped the camera's shutter again and again, though little was obvious except the growing line of a breakwater.

By now, the askari sentries were popping away at them with rifles, but to little effect.

"Toss a couple at them," Rosenthal ordered his gunner. "See if you can hit that." He pointed to a rail flatcar parked near the shore end of the breakwater, loaded with rock.

His gunner complied, and his second shot hit the car, which fell over on its side, tumbling its contents right on top of the breakwater, accidentally adding to its bulk.

"Damn!" Rosenthal could have kicked himself. He didn't know about Spicer-Simson's phenomenal luck, but he did realize he'd just helped the enemy's construction project with the ill-conceived hit. Determined to salvage something from the fiasco, he told his gunner to lob a few shots onto the bluff.

The first two distant booms of the German gun awakened most of the expedition, but it wasn't until the third shot whined overhead and detonated in a clump of trees that most of them realized that they were being shelled. In seconds, the entire expedition was dashing, bewildered, blinking, and half-dressed into the morning air. Another gun blast sounded, exploding more trees much nearer the camp.

"It's the Germans!" Dudley yelled to Spicer-Simson, who had donned his skirt and boots but nothing else, not even his hat.

They rushed to the edge of the bluff just as another shell screamed overhead and disintegrated a hut.

The *Kingani* lay about two hundred feet out from the breakwater, and her single gun belched again. The shell gouged a hole in the parade ground, but by now, the Belgians were firing back with both the twelve-pounders. Water gouted just outside the breakwater, a good hundred and fifty feet from the German vessel.

"God!" exclaimed Cross. "A blind man could do better than that!"

More shots from the Belgians did fall closer to the *Kingani*, but not quickly enough. Smoke billowing from its stack, the boat churned toward open water, and none of the Belgian shells landed within fifty feet of it.

"What incompetence," Waterhouse said, shaking his head. "I can't believe they didn't hit her."

But they hadn't, and in minutes, the *Kingani* was a dancing, distant speck on the lake.

"I'll have to let Major Stinghlamber borrow our gunners until we need them," Spicer-Simson commented dryly. "Teach his men how to shoot."

"You'd better, sir," Dudley agreed. "If the Germans keep sneaking up on us like that, they're bound to get lucky eventually and hit something worth hitting."

"Have the sentries keep a sharp eye," the commander ordered. "If the Germans return, I want those Belgian guns blazing."

And continue sneaking up the Germans did. The *Kingani* was spotted two days later, but the Belgian guns remained silent because the boat kept well out of range. And they saw it again the following day, almost at dusk, this time still out of range, but just barely.

"What's she doing?" Dudley asked as he and Spicer-Simson watched the vessel through field glasses.

"They can tell we're building a harbor, and undoubtedly the Ba-HoloHolo have told them as much," the commander replied. "They're trying to find out what mischief we're up to."

"I can't believe we're still a secret," Dudley chuckled. "After all the trouble getting the boats here and the attention we've drawn from the Africans, half the continent must know about us by now."

7

Rosenthal was undeterred by his failure to photograph the shipyard. The day after he'd fired at the camp on the bluff, he again brought the *Kingani* close to shore under cover of darkness. The wind was up that night, and the surf was heavy, which he knew would keep the crocodiles off the beach. That didn't mean he wasn't afraid, but he slipped into the water and swam ashore anyway. His feet found the rocky beach half a mile north of the shipyard, and once on land, he trotted through the darkness, hoping he was right about the crocodiles. He wasn't half the way to the shipyard, though, when an askari spotted him, shouted out, shot at him a couple of times, and gave chase. Plunging back into the water, Rosenthal swam for the *Kingani*. To his intense relief, he found it without mishap.

The next night, he tried again and actually made it to the shipyard. No ship was there, but lights and noises from the new harbor at Kalemie attracted his attention, and he set off in that direction. The trip took him well over an hour as he picked his way over the rocky coast and frequently hid from sentries, but what he saw made the effort well worthwhile. In the forest at the edge of the Kalemie shore, a lot of work was going on, with the sound of steel striking steel. He crept over to have a look. What he found was equally as interesting as a ship—two large gunboats sat on railway spur that led almost to the water.

Neither boat showed a gun, but when he crept up to one for a closer inspection, he discovered that it had a mount for a gun large enough to do real damage to either the *Kingani* or the *Hedwig*. And the boat's lines indicated it was probably considerably faster than either of the German vessels.

His reconnaissance had taken far longer than he'd planned, but no matter. This was definitely something that would interest Zimmer. Rosenthal sneaked back up the beach and, only an hour before dawn, went into the water, and struck out for the *Kingani*.

But the *Kingani* was gone. His crew, expecting him to return hours earlier, had decided he'd either been captured or fallen prey to a crocodile, and they'd abandoned him.

Left with little choice, Rosenthal tiredly swam ashore. Just as dawn lightened the sky, he found a thicket to hide in and wait until dark so he could make his way past Lukuga. He was a long way from East Africa, but what else could he do? Maybe he could find an old blanket and some charcoal and don the disguise that had served him so well in the past and eventually make his way around to German territory. But Rosenthal's luck had run out. An askari on patrol discovered him before the morning was out and noticed his ill-covered German naval insignia. By noon, Rosenthal was locked in a cell in Lukuga.

Spicer-Simson didn't learn of Rosenthal's capture from the Belgians. His relations with Stinghlamber had, by now, completely deteriorated, especially since the Belgian commandant apparently had rethought his position and once more was refusing to release *Dix Tonne*. The two men weren't speaking, not even enough for Stinghlamber to tell Spicer-Simson that the *Kingani*'s commander had been captured right on the edge of Kalemie Harbor.

But that didn't mean that Spicer-Simson was unaware of Rosenthal's presence in a Belgian cell. The expedition's own Africans heard of it from the askaris, and they told the British officers by the time the day was done. At first, Spicer-Simson expected word from the Belgian commander, but when three days passed without a message, he called Goor to his hut and told him to tell Stinghlamber that he needed to interrogate Rosenthal at the earliest opportunity.

Goor returned two hours later with bad news.

"They've sent him away," he reported.

"Sent him away?" Spicer-Simson asked, his jaw clenching. "And just where did they send him away to?"

"A prisoner of war camp. Major Stinghlamber said to tell you that since Lieutenant Rosenthal was captured on the land, he was under Belgian authority."

"That goddamn fool!" Spicer-Simson spit. "Rosenthal might have vital information. You can tell that pissant Stinghlamber that I will lodge a formal complaint against him. If my mission is jeopardized through his idiocy, I'll see to it that he is court-martialed."

The situation could have been worse. Before Rosenthal was shipped off to the Belgian prison camp, his captors permitted him to inform Captain Zimmer of his POW status. He was told to write nothing else, and when Stinghlamber read the message, it seemed that the German lieutenant had followed instructions. But there was something that Stinghlamber couldn't see. Using his own urine, Rosenthal had written an invisible message on the back of his letter. The secret message warned Zimmer that the British had two armed motor launches on the lake. The only thing that saved the expedition from immediate exposure was further Belgian incompetence. The letter sat in Lukuga for a month before taking another full month to make it around the lake. By the time it reached Kigoma, Zimmer already was aware of *Mimi* and *Toutou*.

8

About the same time that the Germans began their nightly reconnoiters, Wainwright was finishing his up-river ride aboard the *Constantin de Burlay*. Blaes, the Capitaine de Steamer, was just as much a drunken lout as on the down-river journey—so much so that he didn't even recognize Wainwright, for which the lieutenant was grateful. But Blaes's purser, Holmquist, did, and he seemed happy to have a European beside his captain to talk to. Wainwright, imagining himself stuck alone with Spicer-Simson in the middle of Africa, sympathized perfectly.

Since the rains had started, the river was up enough for the *Constantin* to make sporadic runs as far as Bukama, so Wainwright arranged with Holmquist to take him that far and return in three weeks to pick him up again. That would save him the trouble of hiring a canoe to paddle him upstream to Bukama and dozens to bring him back down again.

The *Constantin* arrived in Bukama in relatively short order and with relatively few groundings. Apparently, the water was too high or Blaes was too drunk to spot the mud bars so that he could aim for them. The speed of the journey left Wainwright with only one regret, and that was that he'd missed Mauritzen, who was occupied farther downstream. At Bukama, he boarded the train for the short hop to Sankisia, then he and his men set off back along the track they'd carved across the landscape and that had taken so long to traverse.

Wainwright had thought that the only real difference would be that he'd make better time walking than dragging the boats, and in that he was right, but there were other differences as well. The rains had begun, and the dust that had coated everything on the outward trip was now muck that frequently made walking difficult. But greenery was beginning to overtake the drabness of the earth, and water was, if not plentiful, then at least not in desperately short supply.

Many of the makeshift bridges they'd piled in gullies and streams appeared to have washed away, but during one stop at a village, he learned that the locals had quickly removed most of the felled timber before the water had a chance to destroy the bridges. The wood had gone into a building frenzy that resulted in new bomas, huts, and communal buildings in villages all along the track.

In general, everyone they met seemed happier and less oppressed than they'd been when the expedition first passed this way. He even saw a lot of women wearing dresses made from the gaudy cloth that Spicer-Simson had traded for water.

One of the most surprising elements, though, was the track itself. After the expedition arrived in Sankisia, Wainwright really hadn't given much thought to the road they'd cut, but now, going back over it, he was astonished to see that it had become a major thoroughfare for the region. In addition to people walking on foot between villages, there was even a little oxcart traffic along some stretches. This continued, though irregularly— only a few people were using the steep, rough grades over the Mitumba Mountains, for instance—but it was obvious that the expedition had left a permanent route on the land and opened the region in a way that might have taken decades longer had the expedition not come through.

After five days, Wainwright and his men reached Fungurume. The railhead had not advanced one whit since he'd last seen it. Down at the river, most of the first bridge had washed away, and that suited him fine —he desired no remnant to remind him of that nearly disastrous failure.

The supplies were stacked by the railhead, waiting for him and guarded by a squad of askaris under a Lieutenant James. James agreed to stay until Wainwright found bearers and began the return trek, but it was an easy promise since the next train wasn't due for another fortnight.

As Spicer-Simson predicted, Wainwright had no trouble finding bearers. Many of the men who'd worked for him the first time were waiting at the depot and welcomed him like a long-lost brother as much as an employer. They all remembered the amusing failures, the ingenious solutions to difficult problems, and the ultimate triumph of actually reaching Sankisia more than they did the hardships. And most were not surprised to learn that the boats actually had reached their destination intact. Could the outcome have been any different with Bwana Chifunga Tumbo in charge?

By the end of the first day in Fungurume, Wainwright had a full crew of bearers, and they set out two days later. The going remained easy, and they arrived in Sankisia in slightly more than a week. Rather than waste time loading the supplies onto the slow train, Wainwright had the bearers carry everything on to Bukama. The bearers could move along nearly as fast as the narrow-gauge locomotive, anyway. Then there was nothing to do but wait for the *Constantin* to return to pick them up.

As the British sat beside the now-flowing Lualaba, Wainwright chuckled at how easy this trip had been compared to the first. Then he stopped laughing.

No need to tempt Fate, he reminded himself. Every time they had, Fate had bitten back.

9

So far, Tanganyika had presented a relatively calm face to the expedition, but the day after Spicer-Simson learned that Rosenthal had been summarily shipped off to a POW camp, the lake seemed to follow his mood and grow blustery. Work continued through the afternoon, but as the next dawn grayed a torn sky filled with rain, one look at the harbor made it obvious that no work would get done that day. The seas were high, though it looked like the unfinished breakwater was doing a good job of holding back the breakers.

But it partially collapsed the next day as gale force winds slashed waves right over the rocks and whipped the harbor water into its own minor frenzy. All day, the expedition members crouched in their huts, trying to keep dry as rain dribbled continually through the thatching. And those were the lucky ones. Several huts again had their roofs blown off, sending the men inside scattering for other shelter.

Amazingly, Wainwright returned from his journey to Fungurume just as the storm reached a fever pitch. Leaving the handful of ratings he'd taken with him to guard the stores on Onfroi's little train, he reported immediately to Spicer-Simson, who looked relieved to see him, though the commander quickly hid it.

"Hell of a situation out there, Wainwright. Wasn't sure I'd see you again. How was your journey?"

"Not too bad, sir. The Lualaba's up, so we made good time. Even the Capitaine de Steamer was hard-pressed to find anything to run aground on, though his method of maneuvering around the bends hasn't changed."

Spicer-Simson chuckled.

"There's something else that might interest you, sir. That road we made from Fungurume to Sankisia? Already it's become a regular route. I even saw some oxcarts on it."

"You don't say." Spicer-Simson stroked his goatee. "So I've left my mark on Africa, have I?"

"It seems that you have, sir. Well, I'd better get back to the train before Onfroi decides the storm is too much for him and tries to go back to Kabalo."

"Fine, Lieutenant," Spicer-Simson replied. "We'll survey the harbor tomorrow if this bloody storm lets up."

The next day began badly, and it didn't look like anybody would be inspecting anything, but by midday, the rains stopped and the winds began to abate. Evening brought a stillness that was almost eerie after the storm's ferocity. Even before the winds and rain entirely had ceased, Wainwright, trailed by Dudley, ventured out onto the bluff. Spicer-Simson already was there, and together they surveyed the harbor.

"Damn, sir," said Wainwright. "Look at that mess."

"It's a bloody sorry sight, indeed," Spicer-Simson affirmed.

The harbor was in shambles. Eighty yards of breakwater, so laboriously constructed the month before, had been smashed, and *Vedette* and *Netta* were washed up on the rocky beach, scarred but sound.

"Crazy damn lake," Spicer-Simson snorted, squinting his eyes and staring out over the water, where the light chop was glistening with deceptive gaiety under the reborn sun. "One minute smooth as a lady's cheek and the next throwing itself at you like a harpy in heat."

"Good luck that the railway is safe," Dudley ventured.

He was referring to the rail spur that ran from the brush, across the rocky margin, and down to the edge of the water, which they'd been using to ferry stones from the quarry to the harbor.

"Don't worry," Wainwright said sourly. "There's time enough for that to be smashed, too.

"We must rebuild with all haste," Spicer-Simson said, turning to the two officers. "I want the boats launched by Christmas."

"Christ," Dudley began, then caught himself. "I mean, sir, that's less than two weeks away. Can we do it?"

"The breakwater hasn't vanished, Sublieutenant; it's merely toppled. Look." The commander pointed to a shadowy crescent that hovered just beneath the rippling waves. "It's still there, which simply gives us a broader foundation on which to build. The next storm won't be so lucky. All we have to do is build it above the water again. Wainwright?"

"We can do it, sir."

Early the next morning, Wainwright assembled the African crews and had them, under Dudley and Cross's supervision, start loading more rock onto the flat car they were using for hauling between the quarry and the harbor. Meanwhile, he took charge of the two hundred men he had carrying the stones from the car out onto the breakwater.

But work halted about noon when Waterhouse came down from the bluff above looking worried.

"Better get up there, Mr. Wainwright," the chief said. "Trouble's a brewing."

Hurrying up the path, Wainwright found Spicer-Simson in a heated argument with Captain Dender.

"You can tell your major he's a blundering ass!" the commander shouted. "He can't do this! I'll have him up on charges!"

Dender was red-faced himself, but he managed to keep his voice level, though it was filled with fury. "Major Stinghlamber believes the storm damage here proves he was right about the drawbacks of using the Kalemie as a harbor. He is going to resume dredging the Lukuga, and he needs these men."

"Well he can't have them."

"He will have them," Dender said, his voice rising a notch at last. "You will remember that he has dominion over the land, and these natives are subjects of the King of Belgium." He put a hand suggestively on the butt of his pistol.

"Your pissant king can kiss my royal ass," Spicer-Simson raged.

Dender's face purpled, and for a moment, it seemed that he actually was going to draw his gun. But as he and his handful of askaris were completely surrounded, he thought better of it. He turned to his chief askari and spoke a few words. Within minutes, the other askaris had communicated with the tribal leaders, and in half an hour, only a dozen Africans remained at the construction site. Dender tried to take them, too, but Spicer-Simson stepped in front of the Belgian captain.

"Those men are not Belgian subjects," he snarled. "They came with me from Rhodesia. If you attempt to conscript them, I will have you shot."

Several of the armed ratings cocked their weapons to drive home the point.

"And if you or any goddamn landlubber Belgian officer enters this camp again," the commander warned, "I will have him clapped in irons for the duration."

"You would not dare."

Spicer-Simson leapt forward, his nose nearly touching Dender's.

"I will dare anything for the Royal Navy, my country, and King George."

Dender held his ground for a moment, but he could not long endure either the anger on Spicer-Simson's face or the madness in his eyes. He turned, waved to his askaris, and stalked out of the camp.

As soon as he'd been swallowed by the brush, Spicer-Simson walked stiffly to his hut and disappeared inside.

"What are we going to do?" Waterhouse asked.

"I'm not sure," Wainwright said.

"Are you going to talk to him?" the chief asked, nodding toward Spicer-Simson's hut.

"Are you insane?" Wainwright sighed and passed a palm over his face. "Sorry, Chief. I will talk to him, of course, but I think it's probably safer if I wait a while."

"Much safer, sir."

10

Early the next morning, Wainwright dashed under Spicer-Simson's veranda and pounded on the door.

"What the hell is it?" came a gruff snarl from inside the building. "It's Wainwright, sir. I have some good news."

"Stinghlamber is dead?"

"No, sir. Better than that."

"Nothing could be better than that," the commander said, but Wainwright heard his footsteps coming across the floor, then the door opened.

"This better be good, Lieutenant. I'm in the middle of writing a dispatch demanding Stinghlamber's head on a platter for thwarting the designs of the British Admiralty and endangering the entire African theater of war."

"Look, sir," Wainwright said, stepping back to give Spicer-Simson a better view and waving toward the parade ground. It was thronged with hundreds of Africans, all staring at Spicer-Simson.

"They say they've come to work for Bwana Chifunga Tumbo. They say Bwana Chifunga Tumbo has greater musunga than the Belgians. That's white man's magic power, sir. If the askaris come to take them away, they will kill them. In short, Commander, they say they are your people and you are their chief."

"Is that what they say?" A slow smile spread across Spicer-Simson's face, and he stepped to the edge of the veranda. As the sunlight hit his tattooed torso, a great cheer went up from the crowd. "Perhaps I should visit with my people, then. What do you think, Wainwright?"

"I think you should do no less, sir."

"Fetch my hat and sword."

Wainwright did, and Spicer-Simson donned them before emerging from the veranda.

The edge of the crowd lay less than a hundred feet from Spicer-Simson's hut, and he strode across the space with a measured stride that even Wainwright thought leonine. When he reached the leading edge, the commander strode back and forth in front of them, hands on his hips, a serious expression on his face.

"Wainwright, is this man still the chief headman? The one wearing the leopard skin?"

"Yes, sir, that's him."

The commander approached the man Wainwright indicated, stood before him, and drew his sword. A great gasp went up from those close enough to see, and the headman's eyes grew wide. Spicer-Simson simply smiled and tapped the man lightly on each shoulder with the flat of the blade.

"Tell him I have just formally knighted him," Spicer-Simson said without taking his eyes off the African. "Tell him he is my number one man among his tribe."

Wainwright spoke quickly in Swahili, and as his words registered, the man's chest swelled with pride. Then suddenly he fell at Spicer-Simson's feet and began showering his own head with fistfuls of dirt.

"What's he doing, Wainwright."

"It's a sign of fealty, sir. You named him your man; he is demonstrating his obedience."

Almost instantly, all the Africans went prostrate, following the headman's example.

"How long will this go on, Lieutenant?" Spicer-Simson asked, sheathing his sword. "At this rate, they'll be digging a pit in our parade ground."

"Not long, sir. May I suggest that you walk among them?"

"Certainly." Spicer-Simson began striding through the supplicants, casting his glance here and there, eliciting moans wherever he went. Finally, he returned to the front and had Wainwright raise up the headman.

"Tell him his people already have accomplished great things for Bwana Chifunga Tumbo, but that I will require even more effort. We must rebuild the harbor by Christmas, and if they work hard, they will be justly rewarded."

Wainwright translated the message, and the headman nodded, eyes bright.

"Oh, yes, Wainwright, you can tell them that you're my number one man here, and that I command them to obey your every order."

Wainwright relayed the message then said to Spicer-Simson, "Thank you, sir."

"Don't thank me. It's on your head if the harbor isn't finished by Christmas."

"Yes, sir."

"All right, Lieutenant, you know what to do. I'm returning to my command post to finish my dispatch to the Admiralty. These honorable people have given me fresh hope, but I shall not spare Stinghlamber."

The Africans chanted exuberantly while the commander returned to his hut, then the headman looked at Wainwright.

"Yes," the lieutenant muttered to himself. "On with it."

That night, the commander called his officers together after evening mess.

"I'm concerned about the safety of the boats when we launch them," he said, looking around at them.

"We should have plenty of warning in the event of another storm," Eastwood said. "We can just keep the boats back in the forest until it blows over."

"It's not the storms I'm worried about," Spicer-Simson said. "It's the German patrols. Launching the boats could take hours, and if they come snooping around, they'll see us launching *Mimi* and *Toutou* and will blast them out of the water and go home laughing."

"I do have a possible solution, sir," Wainwright said.

"Of course you do, Lieutenant. Let's hear it."

"We've already run the rail spur down to the harbor to haul rock from the quarry. It would be a simple matter to extend the rails out into water deep enough to float the boats. Then we just have to slide their trucks down the rails and float them right out of their cradles. We can probably launch each one in half an hour. We did something similar back in Bukama, if you recall."

"How long will it take to extend the rails?"

"Three or four days, sir. As soon as the breakwater is complete, we can get started."

"So ordained," Spicer-Simson said. In his next regular dispatch to the Admiralty, he took full credit for the idea.

11

Their luck held, as it always had. Wainwright had larger blocks blasted from the cliffs along the rail line to make the breakwater difficult for even a heavy sea to move, and by the end of the week, the damage was repaired and fresh progress made. Finally, on December 20, the breakwater was complete. Spicer-Simson called for a formal parade to mark the occasion, and he even sent an invitation to Stinghlamber and the other Belgian officers.

"Is that wise?" Hanschell asked him.

"Oh, they'll never come, of course," the commander said jauntily. "I just wanted to rub our success in his face."

Stinghlamber already had plenty rubbed in his face. The day after the Africans returned to the English camp, he sent Goor with a demand that Spicer-Simson return them, and Spicer-Simson sent Goor back with a flat refusal.

"They say I am their chief now," he'd told Goor. "Who am I to argue? There are simply too many of them, and I would hesitate to make them angry."

Goor came back to say that Stinghlamber was furious and that he was sending his askaris to round up the Africans. But the askaris, knowing the danger to themselves if they tried to intervene, had balked, further enraging the Belgian commandant.

All the men of the expedition, done up in their best uniforms, lined up at the base of the breakwater. Josephine was with them, and the beach was crowded with the Africans who'd helped, dressed in their most colorful finery. Everyone was silently expectant, but suddenly, the Africans, who were looking out from the shore, began shouting and pointing, and a chugging sound came from around the headland to the north. *Dix Tonne* appeared, moving slowly but steadily toward the harbor entrance.

"Look at that, sir! The colors!" Dudley exclaimed.

In place of the huge Belgian flag that had once flapped from the vessel's single mast there now hung a smaller Belgian flag, and above it fluttered a British one.

The men of the expedition cheered as *Dix Tonne* plowed into the harbor and settled in a wash.

"Permission to join you, Commander!" Goor shouted from the deck.

"Permission granted!" Spicer-Simson called back, a huge grin on his face. "Waterhouse! Send out canoes to bring our new comrades ashore!"

He jammed his cigarette holder between his teeth as the Belgian crew anchored *Dix Tonne* then boarded the canoes. When they were on the beach, Goor stepped up and saluted.

"I believe you have command of the water, sir," he said. "As you can see, not all Belgians are landlubbers."

"And not all are irresponsible fools," Spicer-Simson said, returning the salute. "Glad to have you with us. I promise that I'll never again call your vessel the *Clam Bucket,* and if I hear one of my men do so, I'll have him keelhauled. She's the *Dix Tonne*. Hear that men? Now, give them a cheer!"

And they did, pounding the handful of Belgian sailors on their backs and pulling them into the ranks.

"I have only one request, Commander," Goor said.

"Yes?"

"Do you have any extra huts in your camp? I'm afraid that Major Stinghlamber and the others might make things difficult for my men."

"Eastwood will see to it," Spicer-Simson said. "And now, Captain, you're just in time for the dedication. Wainwright, call the men to attention. Dudley, go fetch the natives' headman and bring him out here with us."

The men snapped to attention, and the headman, looking equally uncertain and proud, joined the officers and stood at attention along with them. Spicer-Simson drew his sword, held it dramatically aloft, then slashed it down, and a dozen ratings with rifles fired out over the water. As soon as the firearms' salute quit echoing from the cliffs, Spicer-Simson raised a bottle of vermouth and dashed it across the breakwater.

"I hereby name this harbor Kalemie," he called out in resounding tones. "May she safeguard our boats and our lives."

The cheers of the British force were drowned by the roar of the tribal throng when they learned that the harbor was named after their own river. The rest of the day was taken with relaxation, and the commander allowed the men at what little whiskey remained.

12

The next morning, Wainwright's crew began running the rail line out into the water. The task wasn't easy. The lake bottom was uneven, and he had to construct a wooden plow, weighed down with rock, to smooth the way. Then he had a crew plant two parallel lines of posts in the mud, as far apart as the rails needed to be. The rails were lashed to the posts, then ties were nailed beneath the rails, and within a week, he had a crude rail line built a little more than a hundred feet across the water, suspended between the posts.

"I confess I'm nervous about the next step," Wainwright told Dudley and Cross. "I've had the workers practice, but if just one of them has his timing off, everything might be spoiled."

But there was nothing to do except face the moment bravely. He yelled the order, and a man standing beside each post hacked a machete or large knife or hatchet against the ropes tying the rails to the post. Wainwright needn't have worried. Insane miracles had dogged the expedition every step of the way, and they weren't about to stop now. All the ropes severed in the same instant, and the rails vanished into the water intact and perfectly aligned, their nether ends about eight feet deep. When he was thoroughly satisfied that another miracle had occurred on schedule, Wainwright had the workers remove the posts, and the underwater railway was ready to help launch the boats.

At parade the following morning, Spicer-Simson announced that the boats were to be launched immediately, and the entire British complement hurried down to the harbor, accompanied by the best of the local workers. The remainder of the Africans crowded along the bluff above, many so close to the edge that Hanschell, watching among them, feared some would fall or be pushed off.

But he soon lost interest in what was happening on the bluff around him as the process of launching the boats got under way. The locomotive, pushing the truck bearing *Toutou*, crept out of the bush and down the incline toward the shore. When it reached the water's edge, it stopped, and the truck was disengaged. Then, with Spicer-Simson, Wainwright, and Waterhouse aboard *Toutou*, one end of a long, loose cable was made fast between the flat car and the locomotive to pull the car out of the water once the boat was launched. Then the locomotive gave a gentle push, and in a gradual, stately roll, *Toutou*'s flat car dipped into the water and kept going, pulled by gravity down the long, gentle incline.

Before the flat car had gone deep enough to have its wheels half covered, though, it bumped to a halt.

"Blazes!" Spicer-Simson snarled. "What's wrong, Wainwright? I thought you had this business worked out."

"I did, too, sir. I think we just need a push."

"You!" Spicer-Simson bellowed to the Africans on the rocky beach. "Get down here and give us a push." He waved his swagger stick like a baton, directing his troops.

Whether they understood his words or not, his intention was plain enough. Within seconds, men were all around the flat car, jostling for position and handholds.

"Careful!" Wainwright yelled as the flat car began to roll deeper into the water. "Don't get caught under the wheels! Don't...."

Suddenly his words were drowned out by a massive cheer from the men in the harbor and the people on the bluff, and Wainwright realized with a shock that *Toutou* was floating free of its cradle. Elation surging in his breast, he whirled to face Spicer-Simson.

"Commander! We're afloat! We did it!"

"Indeed."

Wainwright wasn't sure what response he expected of his mercurial commander, but what he saw was a steely reserve that seemed, for the moment at least, to have wiped out even Spicer-Simson's braggadocio. In a second, it flashed on the lieutenant that everything until now—planning the expedition, transporting the boats across half of Africa, and building the harbor—had been a game. Even stealing the German's telephone wire from the other side of the lake had been as much prank as practical necessity. But now, everything was different. With the boats on the lake would come real confrontation, and their German foes were not only better armed but more experienced in both the nature of Tanganyika and the realities of warfare on its waters.

"Sir," he said. "You will prevail."

"Thank you, Wainwright," Spicer-Simson said, looking at the lieutenant. A mischievous twinkle lightened the seriousness of his eyes. "I have so far, haven't I?"

"Smashingly, sir."

"Then let's dawdle no longer. Chief, take *Toutou* in to anchor. Wainwright, let's get the gun mounted. We can't leave our little babe unprotected."

Mimi followed her sister into the harbor that afternoon. Each boat had taken only about twenty minutes to launch.

13

The next morning, with Spicer-Simson commanding *Mimi* and Wainwright *Toutou*, the motorboats roared out of Kalemie Harbor and onto the open waters of Lake Tanganyika. It was Christmas Eve.

The noise of the engines brought exclamations of awe from the assembled Africans, but even the sound of the hundreds of cheering voices was quickly left behind. Both boats were fully armed and their fuel tanks topped up. Wainwright stood amidships, thrilling at the feel of the wind on his face and the sight of the open water disappearing in their wake, glad that the commander had let him take command for the trial run, even if Dudley would have her for the first engagement. He glanced ahead at *Mimi*, some twenty yards off *Toutou*'s port bow. Spicer-Simson was standing in the cockpit just behind the three-pounder and Waterhouse, who was serving as gunner. His cigarette holder was jauntily clenched in his teeth.

Lamont had been assigned to *Mimi*'s Maxim gun, and for some unknown reason, the commander had chosen Tait and Mollison to pilot *Mimi* and *Toutou* respectively, despite their inexperience with boats. Wainwright knew that Spicer-Simson had taken a liking to the two taciturn men, and Wainwright suspected it was because they'd so quickly sought out the commander back in England to volunteer for the expedition. But liking a man wasn't necessarily reason enough to trust him to operate a machine with which he had no experience. Wainwright had seen the truth of that in his own past when he'd let a tyro locomotive driver have the controls with results that, blessedly, fell short of disastrous.

Flynn was at *Toutou*'s three-pounder, and Berry served double duty as mechanic and gunner for the Maxim. *Toutou* also carried the signalman, Tasker, so the two crews could communicate with one another. Spicer-Simson had assured Wainwright that *Mimi* didn't need its own signalman since he, himself, was quite versed in semaphore.

"See if you can catch up," Wainwright yelled to Mollison, who opened up the powerful engine to a full-throated roar. In seconds, they drew near *Mimi*. As soon as the commander saw them closing, he shouted something to Tait that Wainwright couldn't hear. But he knew what it had been because the sound of *Mimi*'s engine deepened, and Spicer-Simson's boat pulled away from *Toutou* in a spume of wake.

Wainwright urged Mollison on, and soon the two boats were bow to bow, racing across the water.

Wainwright leaned over Mollison and yelled, "How fast?"

"Better than thirteen knots, sir," came the shouted reply.

Wainwright clambered back amidships and shot a glance at the commander, who was still staring straight ahead, the implacable look Wainwright had seen during *Mimi*'s launch back on his face. Then the commander turned toward *Toutou* and began waving with his arms.

"What's he saying?" Wainwright asked Tasker.

"I'm sorry, sir," the signalman replied. "I can't tell. His semaphore doesn't make any sense."

"Try telling him you don't understand."

Tasker signaled several times then stopped and said, "It's no use, sir. I don't think he can read my signals any more than I can read his."

"You *have* been trained in semaphore, haven't you, Tasker?"

"Yes, sir. I've never had any problems before."

"Try him again."

Tasker did, but his movements just made the commander's arms wave more wildly. After a few more minutes, Spicer-Simson abruptly chopped his hand down and shouted something to Tait, who throttled back.

"I think that's clear enough," Wainwright said. "Full stop, Mr. Mollison."

As the two craft came to a drifting halt, Spicer-Simson hailed Wainwright through cupped hands.

"I think it's time we test-fired the guns," he called. "One shell each. Aim toward the German shore."

The three-pounders had a range of only a few hundred yards, and the shells would never reach the distant shore, but Wainwright appreciated the symbolic gesture. When the prows of both boats were pointed east, toward the mountains whose distant peaks were dimly visible just above the waterline, the gunners loaded the Hotchkisses and stood by waiting for the commander's signal.

With an almost casual gesture, Spicer-Simson called out, "Fire!"

The guns barked, and both boats jumped back half a foot with the recoil.

"Splendid!" Spicer-Simson yelled, his earlier pensiveness vanished. "Bring them around and let's try the Maxims."

For several seconds, the Maxims tore the air with lead and fury, then the commander yelled, "Let's have another turn or two before we return to port!"

Wainwright was glad to oblige. The wind and bow spray were invigorating, and he could almost imagine they were chasing an enemy vessel. When the boats finally reached Kalemie, Spicer-Simson confronted Tasker on the shore.

"What's the matter with you?" the commander demanded. "Aren't you a signalman? Aren't you qualified?"

"Yes, sir," Tasker said, looking abashed. "It's only your semaphore I can't read, sir."

Spicer-Simson's jaw clenched, and he turned and walked across the pebbled beach to Wainwright. The lieutenant was supervising the men loading petrol tins onto *Vedette* to take them out to refuel the gunboats.

"That man Tasker is a disgrace," the commander snapped.

"The signalman, sir?"

"Signalman, my bloody ass. He couldn't tell semaphore from semiconscious. I don't want him in the boats again. Understood?"

"But how are we to signal?"

"I'll do all the signaling that is necessary, Lieutenant." Spicer-Simson turned to go, then faced Wainwright again. "And don't let him near the command hut, either." With that, he stalked off, leaving Wainwright staring after him.

Mess that evening was a rough time for Tasker. No one would speak to him, instead giving incomprehensible hand signals then laughing uproariously. After the fifth or sixth time, just after Lamont had waved with frantic seriousness, Tasker, a big grin on his face, signaled back. Dudley, who *could* understand semaphore, burst out laughing.

Turning red, Lamont asked, "What did he say, sir?"

"You don't want to know, Mr. Lamont," Dudley said, still chuckling. After that, everyone once again spoke civilly to the disgraced signalman.

14

Spicer-Simson wasn't there to witness the humiliation he'd done to Tasker. Instead, he and Hanschell were taking Christmas Eve dinner at the Belgian fort. Spicer-Simson hadn't been invited, but that didn't stop him from crashing the party, anyway, and dragging Hanschell along to witness his victory.

Now that he'd launched the boats, the commander was even more pompous than ever. Hanschell was prepared to cringe, but then he realized that Stinghlamber was nearly as inflated as Spicer-Simson. The Belgian officer sent back his own volley, bragging about his shipyard and the progress that was being made on the *Vengeur*, which would be hauled to the shipyard now that the gunboats were in the water.

During the exchange, Hanschell found himself surreptitiously watching Mrs. Stinghlamber. He thought he'd caught an interesting gleam in her eyes as the two officers argued. If he could have read her mind, he would have been even more intrigued. Mrs. Stinghlamber certainly recognized Spicer-Simson for an eccentric, class-hopping cad, but she also saw something else in him that caused her to feel a bit fluttery and—dare she admit it, even to herself?—excited.

"So you think your little motorboats actually will accomplish anything worth accomplishing?" Stinghlamber asked, sneer not in place but certainly implied by his military stiffness.

"I own the lake," Spicer-Simson replied blandly. "I have only to convince the German commander."

"He will take some convincing," Stinghlamber said. "I know the man."

"I know the arrogance of commanders," Spicer-Simson said, not thinking of himself. "I will be able to use his predilections against him. You ask what I hope to accomplish? What I was sent here to accomplish, at the very least. And Zimmer will make it more than possible—he will make it easy because he believes nothing probable but his own narrow perspective, taken from a very comfortable chair. Sadly, that all too often happens in the modern military."

"How can you decry the military viewpoint," Dender asked, "and at the same time, lead your men into mortal danger and expect them to follow? Isn't that a contradiction?"

"Your statement is a non sequitur," Spicer-Simson said, removing the cigarette butt from his holder and crushing it in an ashtray. "Military code and behavior are the best of what we offer and are completely divorced from the vision—or lack thereof—of our leaders. Those of us on the front lines whose day-to-day decisions mean success or failure are the ones whose actions count. In any case, I understand my men and their capabilities, and I do not ascribe to a policy that automatically considers them chattel to be exploited."

"But haven't you yourself exploited the natives?" Dender pressed. "They slave for you, but you give them nothing in return."

"Perhaps you simply do not comprehend what it is that I give them," Spicer-Simson said. "The Belgians, for certain reasons, are not greatly esteemed in the Congo, but these people have learnt that we English are a different sort of white man, and they place great store in me. As do the British Admiralty and, not coincidentally, your own government."

The statement left a hostile silence in its wake that even Spicer-Simson could not ignore.

"Come, Doctor," he said. "Let us return to the camp. We have more important matters to attend to than idle chatter."

With that, he swaggered out, Hanschell hurrying after him.

15

Christmas Day dawned, quiet and cheerless. All the liquor was gone, the problems with the Belgians had disrupted the mail, and the commander refused to allow any sort of Christmas celebration.

"Remember, men," he told them at morning parade. "We are now on the brink of battle. Our jolly vacation is over. I want everyone ready at a moment's notice when the time comes to engage the enemy."

Later, privately, he railed to Hanschell about the lack of intelligence on the *Graf von Götzen* and that the *Vengeur* wouldn't be seaworthy for months.

"That ship is our only hope against the *von Götzen*," he said. "They've bloody well sent us into an impossible situation. Two gunboats against an armored ship that size!"

"I don't suppose they knew," Hanschell ventured.

"They bloody well must have known!" the commander snarled. "That prick Stinghlamber made sure to show me a copy of the report forwarded to the Admiralty. Gloated about it, the pompous ass. And look at this."

He passed a sheet of paper to the doctor, who saw that it was part of a letter from General Edwards. Apparently, Hope had stopped in Salisbury to air his complaints about Spicer-Simson's handling of the expedition, and Edwards was reluctant to act against Hope in light of the fact that Spicer-Simson had used Hope's testimony to destroy Lee's credibility.

"I have examined all of Hope's reports," a passage read, "and can find nothing as objectionable in them as you suggest. I believe this whole matter is a case of misunderstanding and that we can resolve it without difficulty."

"Can you believe it?" the commander snarled. "I give Hope every chance, and the bastard practically ruins us, then he has the temerity to say *I'm* to blame. I tell you, I shall demand a full inquiry."

When Hanschell finally left Spicer-Simson's hut, he wanted to go around the entire camp and warn all the men of the foul mood into which the commander had sunk, but he had a group of local patients lined up at the medical hut, so he went to work. It was too bad he hadn't. Not long after Hanschell departed, Spicer-Simson set out on his daily inspection of the camp. He called several ratings to task for untidiness, but when he entered the hut shared by Tait and Mollison, his already harsh expression grew red hot.

Tait, it seemed, had not taken to heart the commander's injunction against celebrating the holiday, and he'd festively decorated the hut with foliage he'd gathered in the forest.

"Looks nice, don't it?" he'd asked his hut-mate not ten minutes before the commander came through the door.

"Hmm." Mollison had replied.

"I like Christmas, don't you?"

"Hmm."

Spicer-Simson's response wasn't so elementary, but it was no less terse. "What's this?" he sneered. "A whorehouse? Take all that down and burn it."

News of the commander's aroused temper spread quickly through the camp, but even with everyone on alert, none of them were prepared for what happened at evening mess. Halfway through the meal, Spicer-Simson lurched to his feet and pounded the table with his fist. There wasn't much talk going on, but what little there was instantly choked off.

"It's not fair!" the commander proclaimed loudly. "The Admiralty never intended that I should go afloat. I was supposed to command the base, to be responsible for discipline, for naval routine, for arranging supplies. Lee and Hope were supposed to command the launches. Now, owing to their disgraceful defection, I have to go out in the launches and take all the risks of a naval action, because there isn't another worthy seaman among you!"

Even the breathing in the room stopped, and everyone stared at him —some aghast, some grim-faced, and some as if they thought he'd finally gone completely mad.

Giving a bitter, derisive chuckle and dismissing them all with a disdainful wave of his hand, Spicer-Simson stalked out into the night.

Chapter XII: *Kingani*

"Sir, Captain Goor is here to see the commander."

Bit early for visitors, Wainwright thought as he glanced up from his breakfast. Even Goor. Tommy Atkins, the sentry who'd brought the news, shifted his rifle from one arm to the other.

"Why tell me if he wants to see the commander?"

"Well, sir. I ain't seen the commander this morning. I think he's still in his hut."

"And you didn't want to disturb his beauty rest, is that it, Atkins?" the lieutenant asked with a wry twist of his lips.

"If you'll pardon me, sir, it seemed like work for an officer."

Yes, Wainwright thought. After the commander's outbursts the day before, none of the men wanted to be the next target of his wrath.

"Right you are," Wainwright said. "Bring Captain Goor over. I'll take him to the commander."

Atkins hurried out of the mess hut, and after taking a last swig of tea, Wainwright followed into the early morning air. Not only was it the day after Christmas, it was Sunday, to boot, and he could have slept in just as the commander was doing. But after nearly half a year of rising early, working like a slave all day, and going to sleep soon after nightfall, he'd come to love waking in that ungodly hour before dawn when the air itself seemed tensely poised between the nocturnal clamor of the bush and the molten heat of day.

Atkins returned with Goor, and Wainwright greeted the Belgian captain then dismissed Atkins, who hurried back to his post.

"What brings you out so early?" Wainwright asked.

"A message for the commander," Goor replied. His face looked equally serious and feverishly bright.

"I don't think he's up, yet," Wainwright said. "Would you care to have breakfast with me while we wait for him to rise?"

"My news is important."

"Perhaps if you told me, I could relay the message when he's up."

"I think the commander will want to hear this directly."

"Very well," Wainwright said. "Let's wake him, shall we?"

Spicer-Simson woke easily enough, though he wasn't particularly happy about it.

"Blast it, Wainwright!" he said, sitting up in his bunk, mustache twisted out of kilter. "What time is it?"

"A few minutes after seven, sir."

"Good God, Lieutenant. It's Boxing Day. What in blazes do you want?"

"It's not me, sir. It's Captain Goor. He's brought a message. He wouldn't tell me. Said he has to tell you himself."

"All right, then," Spicer-Simson said, rolling off his cot and reaching for his skirt.

As he did, Wainwright caught a glimpse of the commander completely in the buff and knew that he and Hanschell had won the pool regarding the extent of the tattoos. But he couldn't say anything yet because no one would believe him without corroborating witnesses. Until then, he'd just have to be patient.

"I just hope it isn't more nonsense from that idiotic Major Stinghlamber," the commander said, belting on the skirt. "Send him in."

"Will you want me here."

"To do what? Translate?"

"Of course not, sir."

"Get on with you. I'll handle Goor."

"Very well, sir." Wainwright saluted, stepped out of the hut, and gestured for Goor to enter. "He's ready for you," he said, and as Goor went inside, he muttered under his breath, "God help you if it isn't important."

Instead of leaving immediately, though, Wainwright moved around the wall of the hut and listened through the thin mud thatching. Unfortunately, Goor began speaking quickly in a barrage of French, and Wainwright couldn't understand a word he said. The commander said a few things back, some of which sounded like questions, which Goor answered, then the Belgian captain came out of the hut and trotted off in the direction of Lukuga.

Wainwright may not have understood Goor's words, but the Belgian captain's excitement was plain enough. Something was afoot, Wainwright just didn't know exactly what. Not for the first time, he found himself resenting Spicer-Simson's secretiveness as much as his snotty arrogance. It all seemed so unnecessary—even counterproductive. But there was little Wainwright could do but musingly return to the mess hut for a second cup of tea, mentally counting his winnings from the pool.

Wainwright wouldn't have been so calm if he'd known that Goor had just reported to Spicer-Simson that one of the German vessels had been spotted off Mtoa. It was steaming down the coast at a relatively slow speed and would undoubtedly skirt the Belgian batteries at Lukuga. Goor estimated that the vessel should arrive off Kalemie Harbor in about two hours.

"Do you intend to attack, Commander?" Goor had asked.

"I'm not certain. Did the lookouts identify which craft it is?"

"They didn't say. It was early, though, and the light wasn't the best. Perhaps they could not tell."

"I'll wait, then, until I know more," Spicer-Simson replied laconically. "There might be more than one. Or it might be the *Graf von Götzen*. In any case, she may not come all the way to Kalemie."

"If you attack, sir, I pledge myself fully to you," Goor said huskily, placing his hand on his heart. "My ship and my crew are at your command."

"Of course they are, Captain. I shall rely on you."

"Excellent, Commander. I will prepare *Dix Tonne* immediately."

"Ah, Goor."

The captain halted in mid turn and looked back at Spicer-Simson. "Commander?"

"It would be better if you stood to for the time being. And don't mention this conversation to any of my men. No sense in getting up their dander over nothing. Go back to Lukuga, and let me know the instant the German passes. A positive identification would be most beneficial."

"Yes, sir. Anything else?"

"Do your duty, Captain Goor. That's all I ask."

"I shall not fail you." Goor snapped a salute and headed out the door. Spicer-Simson finished dressing and left for the mess hut.

Wainwright was still there when Spicer-Simson entered and sat down at the head of the table. A rating came over, set a cup of tea in front of the commander, then went back for his breakfast.

"Did Captain Goor bring news, sir?" Wainwright asked after the rating left.

"What? Oh, that. It was nothing." Spicer-Simson raised the cup of tea, blew steam off the top, and took a sip.

Bloody hell it was nothing, Wainwright thought, but all he said was, "Quite right, sir. What time would you like the men to fall in for morning parade and church service?"

Spicer-Simson consulted his watch, seemingly lost in thought, as if he were going through a rapid set of calculations.

"Nine-thirty should do it nicely," the commander said.

"Bit late, isn't it, sir? We still have to make repairs to the breakwater, and you wanted that telephone wire strung...."

"Wainwright."

"Sir?"

"I' believe you asked me what time I wanted you to convene the men for parade?"

"Yes, sir."

"And what did I say?"

"You said nine-thirty, sir."

"Very good. An excellent time. See to it, there's a good fellow."

Wainwright left the mess tent, nearly bumping into Dudley coming in for breakfast.

"Have the men fall in for morning parade at nine-thirty," Wainwright said as he passed the sublieutenant.

"Isn't that a bit late?"

"Perhaps, but the commander seems to think it appropriate. Better watch him. He's acting strangely." As if he wasn't always, he thought.

2

At nine-thirty sharp, the bugler blew, and Chief Waterhouse ordered the men to fall in. As soon as they had, with Josephine in her usual place at the end of the front line, Spicer-Simson looked them over then strode up and down their ranks. The inspection was brief as the men were dressed in their Sunday best and everything was in order. Inspection completed, Spicer-Simson returned to the front of his men, where Eastwood handed him a prayer book, already opened.

"Let us pray."

The assembled men lowered their heads and the commander began to read, his nasal and affected voice making the ancient words seem insincere and almost comical. But no one was laughing—they'd all heard the same droning delivery too often the past nine months to think much of it one way or another.

Abruptly, a African, barely a teenager but wearing the uniform of a Belgian askari, broke from the crowd of men, women, and children that habitually arrayed themselves at the edge of the parade ground each morning and rushed over to Spicer-Simson. The commander stopped reading and looked at the young man. Wainwright thought he detected an unusual light in the commander's eyes but attributed it to the glare from the morning sun, which was full in the commander's face since the men were ranked with their backs to the lake.

The young askari handed Spicer-Simson a folded piece of paper, which the commander opened and read without visible interest. Then he nodded to the askari and, as the youth hurried off the parade ground, pocketed the note and went on droning the Sunday service.

Hanschell nudged Wainwright.

"What was that all about?" the doctor whispered in the lieutenant's ear. Wainwright shrugged, wishing he knew. All he did know was that something was different about the commander, some peculiar set to his rounded shoulders that....

"Do you see that?" Dudley hissed loudly in Wainwright's other ear.

"Mr. Dudley, did your mother train you to interrupt Sunday service?" Spicer-Simson asked, voice dripping acidly.

"No, sir, Commander, but do you see...?"

"I'll see you in the brig if you say another word," the commander said.

Dudley started to say, "Yes, sir," but thought better of it.

The ranked men gave each other quizzical looks at the exchange, then again bowed their heads as Spicer-Simson continued with the morning prayer. But not Wainwright or the other officers, who could see what the men couldn't—the *Kingani*, trailing a thin plume of smoke from its stack, was rounding the point of land just to the north of Kalemie. It was about a mile out on the lake, moving fairly slowly through the light chop.

Dudley nudged Wainwright

"Do you see it?" he hissed again, but much more quietly.

"I see it," Wainwright whispered back.

"What are we going to do? We need to tell the commander."

"I believe he already knows," Wainwright replied, thinking, *the bugger's known since Goor came into camp more than two hours ago.*

At last, Spicer-Simson completed the prayer, snapped the prayer book shut, and looked over the ranked men with narrowed eyes.

"Men," he said. "This morning Belgian lookouts twenty miles to the north reported seeing the *Kingani* heading south down the lake."

Murmuring broke out among the men, but Wainwright quickly quieted it.

"Obviously, the Germans are on a reconnaissance of Lukuga and Kalemie. They are at this very moment, passing our position."

With that statement, military order instantly broke as all the men spun. By now, the ship was almost even with the bluff. It was all the officers could do to reorganize the men and quiet them so the commander could continue.

"Chief Waterhouse," Spicer-Simson said with unexpected blasé. "Dismiss the men. All hands clean into fighting rig." That was the order for the battle crews to put on clean clothes to reduce the danger of infections in case of wounds. It was an odd instruction since everyone already was dressed in their Sunday finest. "Oh, yes," Spicer-Simson added. "Man the launches for immediate action."

There was a stunned silence, then the chief yelled, "You heard the commander! Dismissed!"

The now loosely ranked men broke for their tents in a mad scramble. Wainwright grabbed a pair of field glasses and trained them on the steamer. It was the *Kingani*, all right, cruising, as near as Wainwright could judge, at an easy five knots. The fifty-five-foot ship's forward gun, a Hotchkiss six-pounder, was mounted in front of the wheel, and just aft of the wheel was the funnel. Immediately behind that lay the boiler-room skylight, and back of that sat a small deck cabin. With Rosenthal in a Belgian stockade, there was no telling who now was in charge.

In five minutes, the men were all out of their huts again, racing down the path toward the harbor, dressed, to a man, in their dirty and tattered old uniforms. Wainwright grabbed one, Berry, who barely seemed to recognize the lieutenant.

"Didn't the commander order you into clean clothes?"

"And mess up my Sunday best? No, thank ye, sir."

Then the man was gone with his fellows, leaving Wainwright on the parade ground with Spicer-Simson, Dudley, Hanschell, Eastwood, Tyrer, and Magee. Around them, the crowd of Africans, who were extremely curious at the obvious excitement of the British, had grown much larger than might be expected from the expedition's workforce. The were edging deferentially across the parade ground toward the edge of the bluff.

"Shouldn't we be going, sir?" Dudley asked anxiously.

"No hurry, Sublieutenant."

"But the *Kingani*, sir. She's almost out of sight."

It was true. The German steamer was disappearing around the point of land to the south of Kalemie.

"How obliging," Spicer-Simson said.

"I don't understand, sir."

"No matter. Let's go on down, shall we?"

"Couldn't I go?" Hanschell asked, feeling completely left out.

"Nonsense!" Spicer-Simson said down his nose. "You're much too valuable to risk afloat! But you'll be prepared, I trust, in the event we need your services upon our return."

"Take care, Commander, and you shan't need me."

Spicer-Simson did not reply but proceeded with stately grace down the path to the harbor, Wainwright and Dudley in his wake, the latter bouncing like an excited terrier.

The other officers watched them go then went to the edge of the cliff, where they could view the activity below. Soon, they were surrounded by a huge crowd of Africans who wanted to see what was happening, too.

"I hope all goes well," Hanschell muttered.

"It will be all right," Eastwood assured him. "I've felt all along that the hand of God is over this expedition."

"Why shouldn't it be equally over the Germans?" Hanschell snorted.

The quartermaster flushed and looked taken aback, but he quickly recovered. "You'll see," he said. "We'll all get home safely, every one of us."

3

"What time is it, Wainwright?" "Ten-ten, sir."

Spicer-Simson and the lieutenant stood on the rocky shore, trying to keep out of the way. Kalemie harbor was a storm of activity. The commander had ordered the entire flotilla into action, including the confiscated Belgian craft, and everyone was frantically trying to get the boats ready.

"*Kingani*'s been out of sight for nearly an hour, sir," Wainwright reminded the commander.

"Correct, Lieutenant. I wish to see her well down the lake before we start."

"Is that wise, sir?"

"Perhaps you'll have noticed that *Kingani*'s only threat to us is her bow gun. It may be a six-pounder, but reports give it only a one-hundred-twenty-degree arc of fire forward. I want her captain to see us coming and, hoping to outrun us, turn tail for home. The farther down the lake he is, the more desperate he'll be to do so and the farther he'll have to go. In addition, he is certain to turn to port because a turn to starboard will bring him too close to Lukuga as he flees back up the lake. And all the while, he'll not only be unable to shell us, but he'll leave his backsides exposed."

"You're certain he'll run, sir?"

"Most assuredly. *Kingani*'s was Rosenthal's, and he might have faced us down, but he's been shipped off to prison. This will be a new man, probably untried at command and anxious to preserve his ship. He will run."

Wainwright wasn't as sure as Spicer-Simson of the proclivities of *Kingani*'s new captain, but the commander would have his way whether Wainwright agreed or not. He glanced up at the bluffs surrounding the harbor. They were lined with Africans, who watched the activity below with great interest. It looked to Wainwright like there were at least a thousand of them—nearly double the number that normally occupied the margins of the British camp. Obviously the word had gone out quickly that something important was afoot, drawing in more locals from the surrounding region. Wainwright thought he could spot a good number of Ba-HoloHolo, but he wasn't entirely sure.

"You'll have quite an audience, Commander," he said.

"What?" Spicer-Simson glanced up. "Oh, yes. I suppose I shall. Lieutenant, go tell the men we depart in half an hour." He gestured with his swagger stick toward the boats.

Wainwright left, and Spicer-Simson stared at *Dix Tonne*, which was waiting in the middle of the harbor. He could see Goor staring back. Spicer-Simson pointed his swagger stick at the barge then slashed the stick toward the harbor mouth. Goor threw a jaunty salute and shouted an order to his crew. With its severely underpowered engine sputtering, the ungainly barge began blundering toward the harbor mouth, its twin flags waving in the light breeze. The time was 10:15.

Spicer-Simson watched the hubbub in the harbor for a few more minutes, then he waved at Dudley, who left the canoe that was taking the last of *Toutou*'s crew out to the boat and hurried over.

"As soon as the *Kingani* spots us, she'll turn to port and try to run for Kigoma," he told the excited young sublieutenant. "After she's turned, I want you to come at her from the port quarter, while I have at her from the starboard. Whatever you do, don't let *Toutou* get between *Mimi* and *Kingani*."

"No, sir. I saw *Dix Tonne* leave. When do we sail, sir?"

"It won't be long, Sublieutenant. Don't worry, we shall overtake *Dix Tonne* well before it has a chance to fire on Kingani. Go finish your preparations."

"Yes, sir!"

Another hour passed, and by now, Dix Tonne had long since lumbered out of sight around the southern headland. Except for the commander, the men were aboard their crafts and obviously getting anxious, but Wainwright spoke to them from *Vedette*, reassuring them, saying that the wait was part of the commander's plan.

At last, Spicer-Simson strode to a canoe and was paddled out to *Mimi*. As he boarded, all eyes watched him, though he affected not to notice. After several seconds, he addressed the crews.

"Men," he called out. "This is our first engagement, the purpose for which we have traveled thousands of miles and struggled and suffered. Acquit yourselves well."

There was a momentary silence followed by a unanimous cheer from the crews.

At 11:25, Spicer-Simson gave a command to Tait, and *Mimi*, roaring and spitting a wake of foam, dashed toward the harbor mouth. Waterhouse was on the Hotchkiss and the huge, red-headed seaman from Donegal was just behind him to feed him shells. Next went *Toutou*, with Dudley in command, Mollison at the helm, Flynn as gunner, and Cobb to serve him ammunition. Each boat carried a rating on the aft Maxim gun, though neither expected to fire a shot. *Vedette* followed, with Wainwright in charge, and last out was *Netta*, loaded with cans of petrol.

"We don't want to run out of fuel if the chase is prolonged," Spicer-Simson had told Wainwright, as glad not to be on such a floating bomb as he was disappointed that Dudley had won the toss for *Toutou*. *Vedette*, armed only with its Lewis gun, would be little more than a showpiece in a battle between boats with cannon. Wainwright was there primarily to rescue survivors if *Mimi* or *Toutou* was hit. Frankly, he didn't think there'd be survivors if one of *Kingani*'s six-pound shells landed a good one on either of the gunboats.

His disappointment at not being in the thick of the battle vanished, however, as *Vedette* surged out onto the waves, replaced by an emotion that hovered somewhere between elation and hysteria. He turned to look back at the vanishing mouth of the harbor and was astonished at what he saw. The British and Africans on the bluff above Kalemie had been joined by a large number of Belgians, easily distinguishable in their bright uniforms. But the crowd on the bluff was just a small part of the audience. Now that the boats were out onto the lake, he could see that several nearby hills had their own gatherings of dark bodies. There must have been another two thousand, and even at half a mile, he could hear chants rippling through the throngs.

"Almost like they've come to watch a rugby match, eh, sir?" the helmsman commented, noting Wainwright's gaze and glancing back himself.

"Nice day for it," Wainwright commented. The sky was a cloudless metallic blue, a light breeze freshened the air, and the blazing sun lent a dazzling crown to each of the wave tops.

Wainwright turned again to look ahead, where *Mimi* and *Toutou* already were overtaking the sluggish *Dix Tonne*. *Vedette* could not keep up with the gunboats, but he wanted to be as close to the action as possible without interfering with them or drawing German fire, so he ordered the helmsman to pass the Belgian barge.

For action there would be this day. The smell of it was strong in the air. As if by some unknown agency of excitement, the breeze suddenly picked up, swelling the low waves into a moderate chop.

Up ahead, *Mimi* and *Toutou* now were dancing abreast across the water, and nearly three miles ahead of them, the *Kingani* ambled forward as if unaware of the pursuit.

In fact, *Kingani*'s new captain, Lieutenant Jung, didn't yet realize he was being hunted. As Spicer-Simson surmised, Jung was new to his command, having been promoted when Rosenthal failed to return from his shore reconnaissance a few weeks earlier. Whether Rosenthal had been captured or eaten by crocodiles was a moot point to Jung. He'd liked Rosenthal well enough, but he liked having his own command even more, and he was relishing *Kingani*'s steady thrust through the waves, the breeze, and the passing shoreline of Upper Tembwe Bay.

Jung was only half looking for the *Alexandre Delcommune*, which he been sent to spot—and shell if he could—after the Ba-HoloHolo had reported that the Belgians were trying to refloat her. It was nonsense, of course. The ship wasn't at Lukuga or any of the other likely spots, and he was of the mind that it was a fiction, anyway. So now it was to be an easy journey south for another few hours before anchoring for the night and steaming farther south the next day to inspect some new batteries the English were building on the Northern Rhodesian shore.

Jung bent and patted the goat they'd brought along for fresh meat. It was tethered just in front of the gun mount, and whether or not it realized that soon it was destined to become food, it was somewhat skittish.

Probably just frightened of being on the rolling, vibrating deck, Jung thought, as he turned lazily to watch Tanganyika unfold behind *Kingani*. It was 11:40. In an instant, he spotted several boats pursuing his wake. They were a long way back, perhaps 5,000 yards.

At first, he chuckled to think that the Belgian's Captain Goor and his miserable little flotilla would have the temerity even to attempt a chase and confrontation. Then, as Jung's eyes scanned the pursuers, he realized that there were five of them. He lifted his binoculars and dialed in the focus. There was, some distance back, a barge that had to be *Dix Tonne*, and next were the two little motorboats, *Vedette* and *Netta*. But the two powerful-looking motor launches in the lead were a surprise, and his shock flared into alarm when he saw the British White Ensign flying over both of them. His fright deepened to distress as he realized that both boats were armed with what looked like three-pounders.

"Die Engländer sind hier!" he yelled.

4

Jung shouted at the helmsman to turn *Kingani* ninety degrees to port and bring her up to full speed of seven knots. Smoke billowing from her stack, *Kingani* swung about, and Jung ordered his gunner, Petty Officer Schwarz, to open fire on the nearest of the two attackers. Almost instantly, *Kingani*'s six-pounder bellowed.

Jung expected the gun to easily dispatch the pursuers, but the first shot fell short, and the next three went over as the boat, more agile than Jung suspected, bounded this way and that in the increasing chop. Then the boat—which was *Mimi*—ducked out of *Kingani*'s cramped arc of fire and began steadily rounding astern. The German gunners managed to get off three shots at the second pursuer—*Toutou*—all of which missed before it could be ranged, and it, too, swept too far astern. For the first time, Jung cursed his ship as he realized just how limited his weaponry was against moving targets, especially when they were behind him.

And then, as *Mimi* moved to *Kingani*'s starboard and *Toutou* to port, Jung knew that all he could do was run. And he cursed his ship again, for it was readily apparent that *Kingani*'s seven knots couldn't match the speed of the nimble gunboats. His only hope was his original 5,000-yard lead, though by now that had diminished by more than 1,000 yards. If he could make it far enough up the coast without taking a hit, he might survive his first command.

He ordered every available hand, including his navigator, Penning, to arm themselves with rifles and fire at the pursuers. Penning joined Schwarz in the gun housing, hoping the extra elevation might give him some advantage.

In seconds, the water around the two gunboats was spattered with bullets from the rifles and *Kingani*'s aft Maxim gun, but the gunfire did little to slow them. And to Jung's amazement, there was a madman standing bolt upright in the boat to starboard, completely oblivious to the German fire. Jung raised his binoculars and saw the man, wearing a peculiar blue tunic and...was that a skirt?...staring back through his own pair of field glasses. A long cigarette holder jutted from a fierce grin, and he was waving something long and shiny in his right hand. Mien gott! It was a sword!

That madman was Geoffrey Spicer-Simson, who stood precariously on the foredeck just behind Waterhouse, completely exposed to enemy fire. Waterhouse kept on his knees by his gun since, with its truncated mounting, that was the only way he could fire it. If the commander noticed the lead sizzling in the air around him, he gave no sign but stood completely erect, field glasses to his eyes, furiously puffing on his cigarette holder and waving his sword. The positions were difficult for both him and Waterhouse as the boats bucked and dropped in a motion that would have been sickening if all aboard hadn't been so keyed up.

"Get down, sir!" Tait yelled. "You'll be shot!"

Spicer-Simson either didn't hear him or chose to ignore the warning, so Tait yelled the same thing at the red-headed seaman, who was standing in the cockpit but, because of his towering stature, presented nearly as much of a target as the commander.

"No thank you, sir!" the seaman yelled back. "I can see better standing up!"

It wasn't long before the bucking of the boat over the waves forced Spicer-Simson to quit waving his sword, much to Waterhouse's intense relief since he'd almost been slashed several times as he knelt beside the Hotchkiss. The commander simply jabbed the point into the wood of the foredeck and used the planted weapon to brace himself against the boat's bounding.

By now, *Kingani* had completed her wide swing to port and was obviously fleeing. Spicer-Simson dropped the binoculars and shouted something that Tait could just barely hear over the full-bellied roar of the engines and rush of wind. But even if he hadn't heard, he would have understood what the commander wanted. With a jerk, he jacked the throttle all the way up to its last notch.

As he did, Spicer-Simson turned to face *Toutou* and began waving wildly.

In *Toutou*, Dudley kept himself behind the relative safety of the gunwales. He clutched his hat to keep it from being blown from his head and watched the commander's peculiar arm signals, a puzzled expression on his face. Then he clambered up beside his own gunner.

"Do you understand semaphore?" he shouted to Flynn.

"Some, sir," the gunner nodded.

"I thought I did, too. What's he saying?" Dudley gestured toward Spicer-Simson's gesticulating figure.

Flynn watched for a several long seconds before shrugging.

"Can't tell, sir. The commander must have his own personal sort of code."

"I'm sure he does," Dudley snorted as he scrambled back into the cockpit behind the gunner.

At 11:47, seven minutes after the chase had begun in earnest, *Mimi* and *Toutou* rounded within two thousand yards of *Kingani*, and Spicer-Simson bellowed at Waterhouse, "Fire!"

As *Mimi*'s three-pounder barked, Dudley nodded at Flynn, and *Toutou*'s gun followed suit. With the increasing seas bucking the two boats like wild horses, both shots went completely wild. The gunners reloaded with difficulty and fired again. And again. And again. The waves were pitching the boats so badly that neither gunner could fire more than one round a minute. And for those many minutes, water geysered all around *Kingani* as the water heaving beneath the gunboats' bows threw the shells everywhere but at their target. All the while, Spicer-Simson stood on the plunging deck yelling commands and invectives that nobody understood.

Despite the nasty chop, the two gunboats' superior speed brought them closer and closer to their prey, and at 11:52 one of *Mimi*'s shells carried away *Kingani*'s mast.

"Lyddite!" Spicer-Simson screamed into Waterhouse's ear as he saw the mast fall. It was the first coherent word Waterhouse had heard the commander utter since the engagement had begun. Until now, he and Flynn on *Toutou* had been firing common shells loaded with TNT, but each boat also carried shells with the more explosive lyddite charge, and Waterhouse rammed one into the breech of his gun.

As the range reduced to just over 1,000 yards, Waterhouse's first lyddite shell fell into the gun shield of *Kingani*'s Hotchkiss. The blast threw a halo of fire over the foredeck, followed by a billow of oily black smoke. A man—*Kingani*'s warrant officer—staggered from the tiny cabin aft of the stack and rushed toward the gun, where he was enveloped in the smoke. And at that instant, Waterhouse, with his range down, quickly dropped another shell into the middle of the smoke. As it exploded, *Kingani* shuddered but miraculously kept moving through the water.

Suddenly two dark bodies dived off the boat into the lake just in time to avoid a rapid series of explosions as shells from both *Mimi* and *Toutou* tore into *Kingani*. By now, both gunboats were leaking, their hulls suffering the effects of being dry for too long then battered by the waves and their keels and decks tortured by the shocks of repeated recoils from the three-pounders, but no one aboard either vessel seemed to notice.

"Cease fire!" Spicer-Simson shouted.

Since both gunboats had now slowed to a crawl, even Dudley could hear the shouted order at two hundred feet.

While the two Africans swam away from *Kingani*, the ship began a slow, aimless turn to starboard until it pointed at the German shore, as if giving a last, forlorn look at home. Then suddenly she went dead in the water. A man blackened by soot and oil lurched out of the cloud of smoke spread over the deck, waving a dirty white handkerchief. He was followed a second later by another man. It was 11:58—only eleven minutes since the British had opened fire.

In *Toutou*, Dudley saw the commander again giving some of his incomprehensible semaphore signals. At least that's what he at first thought. As his own boat came up closer, though, he realized that Spicer-Simson was not signaling but dancing and jerking with utter joy. He'd even dropped his precious cigarette holder. Then the commander gave Dudley a dark glare and a hand signal whose meaning was utterly clear.

"Stand to," Dudley yelled at Mollison. "The commander wants to board his prize."

The boarding maneuver was well intended but not well executed. Tait was a novice helmsman and heavy-handed on the throttle. With a jerk, *Mimi* bounded toward *Kingani*, and Tait cut the throttle too late. With a splintering crash, *Mimi* struck the *Kingani* amidships, punching a hole in the larger ship and splitting *Mimi*'s own seams. Spicer-Simson, still standing on the foredeck next to Waterhouse, was thrown to the planks.

Back on the bluff above Kalemie, where the clear day made for marvelous viewing, the shore-bound British and the Belgians had been watching the action. Captain Dender lowered his binoculars and began roaring with laughter.

"What is it?" Hanschell asked.

"Your gallant commander has rammed his prize!" the Belgian replied.

5

Mimi was taking on water, but Spicer-Simson, completely enraptured with his victory, didn't seem to notice that any more than he did his bruised knees. He scrambled to his feet, laughing, pressed past the simmering barrel of *Mimi*'s Hotchkiss, and leapt aboard *Kingani*, snatching his sword from *Mimi*'s deck along the way. Waterhouse and Tait followed him, while the red-headed seaman and the Maxim gunner frantically began bailing.

Kingani's foredeck was awash with blood and littered with chunks of meat—all that was left of Schwarz and Penning, who'd been in the gun shield when Waterhouse's first hit had landed, and the warrant officer, who'd walked right into Waterhouse's second. Lieutenant Jung was leaning over the edge of the gun shield, and at first, it appeared as if he was trying to rise. But as Spicer-Simson and Waterhouse approached, they saw that his movement was caused by the rocking of the ship, for Jung was as dead as the others, his torso shredded and one leg blown away at the hip. Right next to him, a tethered white goat bleated in panic and clattered its hooves on the deck. Not only had it come through the shelling unscathed, it was as clean as if freshly bathed despite the vast quantities of blood splashed all around it.

While Waterhouse stared around wide-eyed at his handiwork, the ship lurched, and Jung's corpse bent forward as if acknowledging defeat then fell to the deck, splashing blood onto Waterhouse's trousers. The chief moaned, staggered to the rail, and vomited over the side.

Meanwhile *Toutou* had come alongside, and its crew clambered aboard *Kingani*. Quickly, they took custody of the two Germans. One seemed reasonably collected, but the one with the handkerchief was glazed-eyed and shell-shocked. They gave up trying to pull the white handkerchief from his paralyzed grip but simply put him into *Toutou* and left him under the guard of the Maxim gunner before returning to *Kingani*. The other German spoke a little English, and the commander quickly learned that he was the chief engineer.

"Check below," Spicer-Simson ordered, and *Toutou*'s crew disappeared down the ladder.

After they'd gone, Spicer-Simson went to Jung's corpse, bent, and rolled it over. He shut the lieutenant's eyes then noticed a signet ring on Jung's finger.

He tried to tug it off, but the blood made skin and gold equally slippery, and he couldn't get it loose. He closed Jung's hand into a fist with the ring finger extended, drew his sword, and bent down once again. In a moment, he straightened, resheathed his sword, dropped Jung's severed finger to the deck beside the corpse, and slipped the bloody ring onto his own hand.

Dudley, Mollison, and Flynn returned to the deck, accompanied by one more German and eight African stokers and deckhands, all liberally covered with soot and grease.

"Sir," Dudley said urgently. "At least one shell went through the engine room skylight, and there's another dead one down there. Also, she's got a hole in her port bunker just below the water line. We haven't much time."

Kingani was beginning to list to starboard.

"Sir," said Tait. "*Mimi*'s taking on water fast, too. Maybe we should...."

"Quite right," Spicer-Simson snapped. "Here's *Vedette* and *Dix Tonne*." The two craft were just coming alongside. Wainwright already had picked up one of the Africans who'd leapt overboard. The other had vanished, presumably beneath Tanganyika's waves. Goor looked flushed, but if he was disappointed at not participating in the battle, he didn't show it.

"Put two prisoners into *Mimi* to bail and the rest into *Dix Tonne*," Spicer-Simson ordered. "Dudley! Do you think we can get *Kingani* back to Kalemie?"

"If we hurry, sir."

"Good! Flynn, you take her in. Keep one of the gun-layer assistants to guard the prisoners. You and you," the commander pointed to the German chief engineer and the sailor who'd been brought up from the engine room. "Get back below, and get the engines moving."

"We will need a stoker," the German chief engineer said stoically.

"Take one."

The engineer gestured to one of the Africans, who followed the two Germans as they hurried below. Inside two minutes, *Kingani* shuddered, and smoke belched from the funnel.

By now, the rest of the prisoners were aboard *Dix Tonne*, and while the commander was taking one last look at his prize, Dudley ordered Flynn and the red-headed seaman to clear the worst of the body parts from the wheel house, then he went to the commander.

"We'd better hurry, sir," he said anxiously.

"Yes, we'd better. Tait, back to *Mimi*!"

Spicer-Simson and his crew quickly boarded *Mimi* and cast off. Dudley and *Toutou* weren't far behind, and within a minute, all the vessels were plowing through the waves toward Kalemie, *Mimi* and *Kingani* leaking like sieves.

Mimi, more lightly loaded despite the water sloshing in her bottom, ran out ahead, accompanied by *Toutou*, and sped toward the harbor, the two African captives bailing furiously. Meanwhile, *Vedette* hovered in *Kingani*'s wake, prepared to rescue the men aboard if the ship went down.

Within ten minutes, *Mimi*, half full of water, rounded the breakwater and wallowed into the harbor.

"Better not stop," Spicer-Simson advised Tait. "Run her aground."

Tait, experienced at running the boat into things to stop it, complied, and *Mimi*'s bow plowed up onto the beach. As Spicer-Simson hopped onto the pebbles, the boat, with a gurgling sigh, settled to its keel in the water.

Spicer-Simson watched the boat's last gasp, then dazed, silent, and unsmiling, turned to face the ecstatic crowd of British, Belgians, and Africans alike who were rushing down from the bluff toward him.

6

"Your commander doesn't appear to be too pleased with his victory," Dender yelled to Hanschell as they were jostled this way and that by the flowing, jubilant African mob as it clogged the path to the harbor.

Hanschell was saved from answering by the blare of Belgian and African bugles and the roaring cheers that shook the air. But Hanschell did have an opinion, even if he didn't voice it. After a career of failure and disappointment, Spicer-Simson truly had done something heroic, and the doctor wasn't sure the commander actually was prepared for that fact. Until now, everything had been a sport, but the prize at hand proved more than sport—especially now that *Kingani*, riding low and listing badly to the right side, turned into the harbor, and Hanschell could see, even at this distance, the crimson splatter on its deck. The sport was finished.

Hanschell sighed and thought, half in prayer, "At least let him have his glory. He's had the responsibility, now let him have his rewards."

Just then, the Belgian shore batteries at Lukuga, having been apprised of the victory, added the booming muscle of its four-inch guns to the already cacophonous din. Hanschell paused and gave one last look at the tide of men surrounding the commander before watching the *Kingani* make it halfway from the harbor mouth to the beach, where it foundered in about seven feet of water. Shouldering his way out of the crowd, he hurried toward the hospital hut. There would be wounded, and he needed his bag.

Vedette pulled alongside *Kingani* and took off the crew and the goat. Flynn, white faced and shaken from the smell of gore, powder, and smoke, promptly fainted the second his task of piloting the small ship ended, but he revived before Wainwright could have him taken aboard *Vedette*. No sooner had the launch come ashore than, almost sedately, as if it were reclining to take a nap, the *Kingani* capsized on her starboard side.

Flynn and Wainwright were hauled out of the boat, pummeled, and dragged into the ecstatic crowd, which almost immediately opened in awe and respect as Spicer-Simson strode up to the two men.

"I see you brought her in for me, Mr. Flynn."

"Yes, sir!"

"Well done."

The crowd roared again then fell silent as the commander raised his arms for attention. Lieutenant Jung's bloody ring glittered in the sunlight.

"Men," he shouted. "We have won a naval action in miniature!"

With that, the frenzied crowd bellowed again, surged, and practically carried the victors, commander and rating alike, up the path to the British camp. Hanschell, holding his medical bag, stood at the top of the cliff, waiting for the dancing river of people to flow by before going down to see what medical service he could render. But Wainwright, passing, clutched his arm and pulled him along with the rest.

"Don't bother, Doctor," the lieutenant shouted. "There are no wounded—only the dead and the living! The dead don't need you, so you might as well join the living for the victory celebration!"

7

The sun beat down on the parade ground. Spicer-Simson sat just in front of the flagpole on a carved wooden throne. Flanking him were two beautiful young Ba-HoloHolo women suspending palm-frond sunshades over his head to protect him from the sun. A Ba-HoloHolo boy stood at hand, bearing a tray with a glass of vermouth.

Even the commander himself had tried to shoo them away, but they would have none of it. They wanted to be close to the him because he had the most powerful musunga the Ba-HoloHolo had ever known. For more than a year, now, the Germans and their ships had shelled fear into their hearts, but Spicer-Simson had, in one morning, chased and beat one of them into submission, capturing it in the bargain. The two women hoped to increase their fertility, and the boy was the son of the local chief, who wanted Spicer-Simson's potency to rub off on his progeny. The throne was the chief's, too, and there for the same reason.

Before Spicer-Simson were arrayed the captives from the *Kingani*— eleven all told: the chief engineer, his assistant engineer, the shell-shocked seaman who'd waved the handkerchief, and eight African deck hands and stokers. Ringing this tableau was the entire British contingent, the Belgians mixing in, and around them spread several thousand Africans, muttered translations of the Europeans' transactions rippling through the throngs.

"Is this the lot?" Spicer-Simson asked.

"All except one of the natives who jumped off right after the first explosion, sir," Wainwright said.

"We didn't pick him up?"

"No, sir. I lost him in all the excitement."

"Oh, well, we can't start looking for him now. I suppose the crocs found him soon enough." Spicer-Simson pointed with his swagger stick at the German chief engineer. "Did your captain know about our gunboats?"

"We knew nothing of them," the chief engineer said with a voluble gesture. "We hear from the Ba-HoloHolo that the British have canoes they try to drag over the Mitumba Mountains. It is a distraction, ya? A big joke." He laughed then caught himself and looked around at his captors. "Ya, a big joke on us."

"What were you looking for?" the commander asked.

"The *Alexandre Delcommune*. We hear reports the Belgians try to assemble and launch her."

The commander looked at Dender. "I assume you have adequate quarters for our captors?"

"We can accommodate them," the Belgian captain assured him.

"Excuse me, sir," said Dudley.

"Yes, Sublieutenant," Spicer-Simson said with an imperious tilt of his head.

"The goat, sir. The men would like to keep her, if that's all right with you, sir. As a mascot."

Spicer-Simson looked back at Dender. "You may take the prisoners, but leave the goat."

"As you wish, Commander." Dender spoke to his soldiers, who took the captives in hand and marched them away.

"Wainwright."

"Sir?"

"The dead. We will accord them a full military funeral."

"Yes, sir. Chief Waterhouse is attending to the graves. He tells me they should be ready before evening mess."

"Waterhouse? He's one of the heroes of the hour. He shouldn't have pulled burial duty."

"He told me he wanted it, sir. Said he killed them, so he should be the one to see them safely in the ground."

"Very well. Have the men convene in full dress for the service."

The funeral was a solemn and formal affair, and for the occasion, the commander wore knee socks in addition to his boots, skirt, and tunic. After he read the service and called for a prayer, fifty askaris lifted rifles to their shoulders and fired a thunderous salute.

As the men dispersed to eat, Wainwright motioned to Cross.

"Assign a detail of men to guard the graves," he said.

"A guard? You aren't afraid they're going to get up for a bit of revenge, are you?"

"It's not the dead I'm worried about," Wainwright said, nodding toward a cluster of Ba-HoloHolo men who had stayed behind. The men smiled, showing pointed teeth. "They'll dig them up as soon as our backs are turned. It's bad enough we killed the blokes without permitting our new-found allies to eat them."

"Quite right." A queasy look crossed Cross's features.

"Keep the guard posted day and night. A week should do it. If they still want them after that, maybe we should give them the pleasure."

"I'll assign a guard." Cross hurried off.

While he waited for the guard to come, Wainwright sat on a rock, staring at the German graves, arrayed near the one belonging to the woman who'd been killed by the crocodile, wondering how fast this little city of the dead would grow here in the jungle. Spicer-Simson had taken the day today, and Wainwright had to admit that the commander's strategy, as off-handed and strange as it had been, had worked in trapping the *Kingani* and running her down. But he also remembered the almost haphazard way the battle had gone, and he knew they couldn't be so lucky the next time out. They would then face a larger, better-equipped foe—the *Hedwig von Wissmann,* or maybe even the *Graf von Götzen.* Since it would be his turn to command one of the gunboats, he fully expected to come back to the camp in as many bloody chunks as the Germans they'd just buried. If he managed to come back at all.

A gloom darker than the falling evening settled on his shoulders, and he glanced up at the group of Ba-HoloHolo still milling about at the edge of the bush. It was now too dark to see their pointed smiles, but it felt all too much as if those smiles were appraising the tenderness of his own flesh. At last the guard arrived, and Wainwright hurried off to mess, trying not to think about the BaHoloHolo's plans for their next meal.

Mess did go a long way toward dispelling his gloom, however. Spicer-Simson had not yet arrived, and the men were full of animated chatter. It was obvious to Wainwright that none of them had considered what might befall them at the next battle, and he wasn't about to shake them from their almost childish excitement. As he ate, he caught snippets of conversation, most about Spicer-Simson.

"Aye, he was a cool one, I'll tell you," Flynn said. "He didn't bat an eyelash while the Germans sailed right on by."

"Little did they know what was in store," Magee said.

"But he had 'em pegged, for sure," Mullen put in. "Let 'em get right past, pretty as you please, then goin' after 'em when all they could do was run."

"I guess they didn't run fast enough!" said Lamont, and everyone laughed.

"And did you see him standing there the whole time?" Dudley asked when the laughter had quieted. "The bullets were flying all around us. He must have nerves of iron. I was crouching behind the gunwale the whole time."

"He was quite unmoved," Tait agreed sonorously. "He stood there in full view of the enemy, his long cigarette holder in his mouth and his eyes glued to his binoculars. The only thing that shifted him was the collision when I rammed the prize. He fell flat, and when he got up, he was laughing!"

That brought another round of cheers from the assembly, as much at Tait's unexpected loquaciousness as the imagery he called up. Then everyone quieted at an exchange over who had actually fired the fatal shot.

"I know the first two hits were mine," Waterhouse proclaimed.

"That's so," Flynn agreed. "But the one that went into the engine room could've been either of us."

"Whoever it was," exclaimed Dudley, "it was fine shooting."

"And it wasn't easy, sir," Waterhouse said. "The way we was pitchin' and rollin', sometimes the gun was pointin' down at the sea, sometimes up at the sky. And the commander worried me, sir. He was shoutin' all the time, and I couldn't hear a word he said, what with the wind and the spray and the roar of the motors—and him with that cigarette holder between his teeth. He must have thought I put up a very poor show."

"Nonsense, Chief Waterhouse."

Everyone turned to see Spicer-Simson enter the room. The commander wore a big smile, and his mood obviously had lightened since the funeral service. "Tell me, Chief, how many rounds did we fire?"

"Thirteen, sir. But I think I only hit *Kingani* three times."

"Not by my counting. I was just down in the harbor inspecting our prize, and I counted twelve distinct hits."

Twelve shrapnel marks, maybe, Wainwright thought, frowning. But only four or five real hits. But he kept his mouth shut.

"Twelve hits out of thirteen, Chief. I call that wonderful shooting considering you were on your knees doing it."

"I was down there praying as much as firing, sir."

As the men laughed and cheered, Hanschell noticed that Spicer-Simson was playing with something that gleamed on the ring finger of his right hand. Leaning over to Wainwright, he whispered, "Is that a ring on the commander's finger."

"It is," Wainwright answered tersely.

Odd, Hanschell thought. I don't remember him wearing a ring before today. He wondered somewhat mordantly if Spicer-Simson had it made before leaving England to commemorate his victory and now was finally able to wear it. That would be like him. But then Wainwright said something that turned his heart cold.

"I believe he cut it from Captain Jung's finger after the engagement."

Feeling bewildered, and not a little uneasy, Hanschell remembered that Jung had been missing his right ring finger.

"We were just trying to figure out, sir, who it was that dropped the one into the engine room," Waterhouse was saying. "I think Flynn...."

"Flynn?" The commander snorted contemptuously. "Certainly not. Petty Officer Flynn couldn't have possibly scored a hit without a qualified naval officer to spot for him and give him the ranges, as I did for you."

The boisterous mood in the room suddenly curdled, and all eyes turned to Flynn, who reddened.

"Brum Waterhouse can have all the hits, sir," Flynn said glumly. "I'm happy so long as none of theirs hit us."

"I've always had great luck as a gunnery officer," Spicer-Simson went on, and he began to regale them with one of his tales, not noticing that the life had suddenly gone out of the party.

Outside, however, the frenzy of the African celebration continued unabated, accompanied by singing and chanting and dancing to the rhythms of hundreds of drums. It was a sorry contrast to the mood in the mess hut.

8

Hanschell was awakened at dawn by someone knocking and softly calling his name from just outside his door. He groaned. The African party had stopped only an hour or so earlier, and he hadn't had more than a couple of hours of sleep all night. And he'd made the mistake of being the last to leave the cheerless party. Before he knew it, he'd been obliged to hear more of the commander's braggadocio until Spicer-Simson himself finally ran down.

It had been nerves, really, Hanschell finally realized as he trudged back to his hut. The commander's long string of failures so suddenly, decisively, and violently ended, that all of his pent-up energies had to find some release. Hanschell just wished he hadn't been the one to bear its brunt. But, then, who else could it have been? Who else could Spicer-Simson really trust enough to call friend but the one man in the expedition who knew him longest and best?

The knock came again.

"Who is it?" Hanschell muttered groggily.

"It's Flynn," said a subdued voice.

Hanschell swung his legs off his cot, drew on his trousers, and stuffed his feet into his boots. Without bothering to tie the laces, he shuffled to the hut door and threw it open. With Flynn was the red-headed seaman from Donegal.

"We was wonderin', sir, if you have any sort of preservative," Flynn said.

"Preservative?"

"Yes, sir. Something to keep meat fresh."

"There's salt in the mess."

"Not like that, sir," the red-headed seaman said. "More like them, whatyoucall'em...you know, species in a bottle."

"Species in a bottle?" Hanschell shuffled back to his bunk, sat, and rubbed a hand across his face, feeling the dry rasp of whiskers beneath his palm. "You mean specimens in a bottle?"

"That's it," Flynn said. "Specimens. Do you have anything like that liquid they keep 'em in?"

"I suppose I do," Hanschell said after a pause. "What do you need it for?"

"Specimens, sir," the red-head said. "We want to keep 'em fresh."

"What kind of specimens?"

"Well, sir...." Flynn shuffled uneasily, and the red-head looked diligently at his own feet.

"Get on with it," Hanschell said testily. "If you wake me up at the crack of dawn, you'd better damn well be prepared to explain yourselves."

"Yes, sir," agreed Flynn.

He hesitantly produced a small jar in which rested a small, pale sausage. That's what Hanschell thought when his bleary eyes first settled on Flynn's specimen. Then he took the jar and looked closer.

"This is a finger."

"It is, sir."

"Where did you get it?"

"Found it lyin' on *Kingani*'s deck."

"This is Captain Jung's finger. Why didn't you have it buried with him?"

"I don't know, sir. Seemed like a good idea to keep a souvenir, I guess."

"I see." Hanschell turned to the red-headed seaman. "And I suppose you have a finger, too."

"I wasn't so lucky, sir. All I got is this." He pulled out a bottle half filled with blood, already separating and clotting.

Hanschell shook his head and rubbed his face again, wondering if he was dreaming. First there'd been the grueling, impossible trek across Africa, then the naval battle in miniature, and the commander sporting the dead German's ring, not to mention having to post guards over the graves to keep cannibals from enjoying a gruesome feast. And now there were these gristly souvenirs.

What the hell am I doing here? he wondered. I'm supposed to prevent tropical disease and patch up the wounded not supply preservative to indulge ghoulish eccentricities. I shouldn't be here at all, he realized. I should be back in London, teaching in the medical school.

The school where human specimens reigned supreme, and where he'd bottled not a few of them himself.

With a sigh, he got up and went over to his medical trunks.

"Thymol will do," he said. "It'll have to; it's all I've got."

After he'd decanted enough thymol to cover the specimens and the seamen hurried off gloating, Hanschell went back to his bunk and lay down, but sleep wouldn't return. He kept seeing the pallid sausage floating in Flynn's bottle of thymol, wondering how many more specimens he'd have to pickle before he was again safe in England.

Chapter XIV: January

"Heave!" yelled Wainwright. Early in the morning on the day after the engagement, several of the cables that had been used to haul *Mimi* and *Toutou* up and over the Mitumba Mountains were passed beneath *Kingani*'s hull, parbuckling her, and gradually, she was hauled ashore by gangs of laborers more anxious than ever to serve Bwana Chifunga Tumbo. At last, the small ship lay beached broadside, water streaming out of hatchways and a ragged hole in its side.

"This looks like damage from one of our guns," said Cross as he inspected the hull, "but what made this triangular hole?"

"I believe that's where the commander's boat rammed her," Wainwright said.

The chief engineer turned a jaundiced eye on the lieutenant, who gave back a wry grin and asked, "How long will it take to patch her?"

"Not long. We'll build a forge right here on the beach, and we can get enough steel plating from the Belgians. Her hull will be ready in a couple of days. The engine might take a little longer, though. I don't know what damage was done, and everything will have to be taken apart, regreased, and put back together again."

"Do it as fast as you can. And while you're at it, remove the forward gun."

"The forward gun? That's a six-pounder. We're not putting a three-pounder in its place?"

"No. The Belgians have a twelve-pounder semi-automatic gun they meant to mount on the *Vengeur*, but since they haven't got her repaired yet, the commander's convinced them to give it to us. They say it can fire twenty-eight rounds a minute."

"Damn!" Cross breathed. "That's a big gun. Maybe too big for *Kingani*. It'll be like those three-pounders on *Mimi* and *Toutou*, liable to go over the side if you fire it any direction but straight ahead."

"The *Kingani* only has a limited arc or fire, anyway, so that shouldn't be much of a problem. Besides, it's what the commander wants. It's to be his flagship, and he wants firepower equal to his rank."

"We can reinforce *Kingani*'s bow, but I'm making no guarantees."

"Do your best. And mount our spare three-pounder aft, next to the Maxim. We don't want to be fighting blind to the rear like the Germans had to do. Not that we'll be running *from* the Germans, but better to be safe than sorry."

"I'll get her floated and fixed," the chief engineer promised. "She'll be as ready for battle as she'll ever be."

2

With nothing to do for the moment, Wainwright went to visit Hanschell. The doctor regaled him again about the need for clean new huts, and the lieutenant had just agreed to mention the matter again to the commander when Dudley popped into the medical hut.

"Have a look at this, fellows," the sublieutenant said, setting an eight-inch tall statue molded in clay onto the table.

It was facing Wainwright, so he was the first to notice the features. He started laughing.

"What's so funny?" Hanschell demanded. He turned the statue then joined in the laughter. "My God! He's become a fetish!"

"Splendid likeness, isn't it?" Dudley asked.

It was. Everything was there, from the goatee to the rounded shoulders to the skirt. There were even bas-relief hints of tattooed snakes and birds on the bare chest, arms, and legs.

"Where'd you get it?" Wainwright wanted to know.

"From one of the natives. I must've seen half a dozen this morning."

"Bwana Chifunga Tumbo is powerful juju in these parts," Hanschell said.

"I'm going to set it up in our hut," Dudley said. "Maybe it'll bring us more good luck."

"Better not let the commander see it," Hanschell advised.

"You don't think he'd like it?" Dudley asked.

"He'd like it too much," Wainwright supplied, and they all laughed again.

At that moment, Spicer-Simson blustered into the hut, and Wainwright quickly swept the statue off the table and hid it behind his back. He needn't have bothered, for the commander set its twin in its place.

"What do you think of that?" he demanded jovially.

"It's a splendid likeness," Wainwright said, winking at Dudley.

"Isn't it, though. I'd know myself anywhere."

"There are quite a few of them among the natives, sir," Dudley ventured.

"You're quite right. Can't say I'm surprised. This sort of thing happens all the time when a more primitive race gets a glimpse of what a real Englishman can do." Spicer-Simson stroked his goatee.

"What are you going to do with it, sir?" Wainwright asked.

"I suppose I'll send it home to my wife."

"It'll be better than a photograph, won't it, sir?" Dudley asked.

"Indeed it is, Sublieutenant," the commander said, a twinkle in his eyes. "Not quite like having me home in the flesh, though, do you think?"

"No, sir."

"On the other hand, perhaps you chaps would like to keep this one. I can easily get another. Well, I must be going. Seems I've become something of a divinity to these people. Their women are constantly bringing me their babies to kiss and whatnot—the duties of a demigod, you know."

With that, Spicer-Simson strode out of the tent, leaving the statue of himself on the table. Wainwright pulled the first statue from behind his back and set it on the table by its mate.

"Bookends," Dudley said.

"Only the book of our commander's exploits would fit comfortably between those fellows," Wainwright commented dryly.

"It would be a library by the time he was done with it," Hanschell amplified, and the three men burst into fresh laughter.

But no amount of laughter could erase the bald and uncomfortable fact that they now took orders from an African demigod.

3

The next morning, Dudley strolled down the path from the bluff toward Kalemie Harbor. It was a beautiful day, as they all were when terrible storms weren't ripping down the lake, and he whistled an air he'd last heard in a music hall the week before he'd hopped aboard his bicycle to chase down the Tanganyika expedition. God, that seemed years ago. A lifetime, he amended, thinking of the battle just two days before. It seemed his life before then had been but a shallow dream from which the thunder of the gunboats' three-pounders and the stench of gun smoke and blood had abruptly awakened him.

Not that he was upset in any way at the memory of the charnel house that the fire from his own command had helped create. He'd seen battle before, and had the German commander, Jung, been able to bring his own gun properly to bear, Dudley himself might have been rotting in one of the graves on the bluff. No, this was war, and Dudley could accept that slaughter was part and parcel of the whole affair.

So, Dudley was not upset. Quite the contrary, he was elated, and the world around him seemed to shimmer with a reality he'd never before experienced. He'd tasted the heady draught of victory, and he desired nothing more than to drink again at its fount.

In this pleasant mood, he approached the rocky beach where Lieutenant Cross was laboring with his small crew of ratings and a somewhat larger crew of Africans to repair the *Kingani* and the gunboats.

"Good morning, Mr. Cross," Dudley said. "How fare the repairs on the *Kingani*?"

"We'll have an easy time of it. The holes are clean, and there's been no damage to her engines, just a soaking, and that's nothing for a steam engine. We'll have her floated in another day or two, just as soon as we fashion and fit the patches. Now that we've got a good forge, that won't take long. Want to see it? The forge, I mean."

"I'm at your disposal."

"This way." The chief engineer started off toward a group of men and equipment clustered near the base of the bluff, well back from the water. "By the way, that was a marvelously good battle you and the commander put on. To tell the truth, I wasn't sure but that we'd dragged the boats all across Africa just to see them sunk on the threshold of glory."

"Thank God that didn't happen, or I wouldn't be here to witness your new forge."

"Right you are," Cross said, then he called up ahead. "Clear the way, you blokes. We've got a hero among us!"

Dudley suffered the barrage of the ratings' quick, bright cheer without visible harm, then the chief engineer waved proudly toward his newest construction. Dudley walked up to the forge and circled it, taking in its sturdy stone firebox and blistering fire pit.

"Is that firebox double-walled?"

"It is. Makes the most of the heat."

Dudley nodded appreciatively. "And the bellows—I don't believe I've ever seen one quite like it."

"Quite a darling, isn't it?"

"An ingenious design, if somewhat unorthodox. I congratulate you."

"Bless me if I wouldn't like the credit for myself, but it has to go to someone else."

"One of your men?"

"One of the Africans."

"How extraordinary. Which one?"

"Why...." Cross craned his neck, peering at the local workers clustered nearby. "He was here only a minute ago. I hope he's not gone. He's quite clever with his hands. You," he pointed to one of the Africans in the front rank, the headman of the group. "Where is that fellow who made this?" He gestured to the forge.

The man assumed a stoical expression and took a step back. So did the rest of the group.

"Here, you boys. What's wrong with you? Where's that fellow. Fundi. Yes, that's his name. Where is Fundi?"

The Africans looked like they were going to take another step backwards, but before they could, Cross waded into them.

"I know he's here. Blast. I saw him not two minutes before you came. Yes! There you are!"

The chief engineer disappeared among the Africans then reemerged, dragging a man by the arm.

"He was hiding back there."

"He doesn't seem too proud of his good work."

Dudley was right. The forge designer's head was hung and his face averted as Cross pulled him in front of Dudley.

"Does he speak English?"

"He knows a few words, I'll warrant."

"Fundi. Understand?" Dudley tried.

The African remained unresponsive except for trying to pull out of the chief engineer's grip.

"No matter," Dudley said, and he tried a few words of Bantu, but they only made Fundi pull away more.

"Hold still, you," Cross said, jerking on the man's arm. "Look at the sublieutenant when he's talking to you." He forced Fundi to raise his head.

"Well, I'll be," Dudley said with a smile. "No wonder he doesn't want me to see him."

"You know this boy?" Cross asked.

"I've only seen him once, but that was enough and only two days ago. He's one of the two fellows who jumped off *Kingani* after the first few hits. The one who disappeared."

"Is that true?" Cross demanded of the headman as the rest of the African workers began chattering nervously.

"Him good friend," the headman said, averting his eyes.

"Him good German friend," Dudley snorted.

"No, bwana. Him build furnace. Him friend to you."

"Him must go with rest of prisoners," Dudley said.

Abruptly, Fundi fell to his knees in front of Dudley and began speaking rapidly in an imploring voice.

"What's he say?" Cross wanted to know after Fundi was forced to pause for breath.

"He admits he was working for the Germans as the chief stoker on the Kingani," Dudley relayed, "but he says they forced him into labor. He's begging me not turn him over to the Bula Matari. That's the Belgians. He says the Germans are terrible masters, but the Belgians are demons."

"He's right enough there. Have you seen how they treat even the best of their servants? Makes me ashamed to say we're brother Europeans."

Fundi spoke again, but Dudley silenced him with a couple of terse words.

"What are we going to do?" Cross asked.

"We'll have to take him to the commander. Only he can decide."

They took Fundi to camp, followed by everyone who'd been on the beach.

"What do we have here?" Spicer-Simson demanded as they led the prisoner through the habitual throng of Africans that gathered on the parade ground whenever the commander was in evidence. He had been going over plans for gradually rebuilding the camp's huts with Wainwright and Eastwood.

"A new prisoner, commander," Dudley said. "He's the one we thought was missing after he jumped overboard during the battle. He swam to shore before we could pick him up."

"A strong swimmer, eh?" Spicer-Simson, running a judicious eye over Fundi. "What's his name?"

"Fundi, sir. It seems the other natives have been hiding him since he came ashore."

"Best send him packing with the other prisoners, then."

"Begging your pardon, sir, but he says he wants to work for us."

"In what capacity could he serve?"

"He was the chief stoker on *Kingani*, sir. He knows the ship, and that could be to our advantage."

"He's deucedly clever with his hands, sir," Cross put in quickly. "He built us a forge to repair *Kingani*'s damage just like that." He snapped his fingers. "We're days ahead of schedule thanks to him."

"He's afraid of the Belgians, sir," Dudley said.

"As well he should be, the imperialist bastards." Spicer-Simson turned to Wainwright. "What do you think? Shall we keep him?"

"He seems like an asset, sir, but if we do, Major Stinghlamber will not be pleased. We're supposed to turn all prisoners over to the Belgians."

"Capital idea, then! We'll keep him. Serve that bloody arrogant bugger right." Beaming, the commander turned to Fundi and drew his sword. The stoker's eyes nearly bugged out of his head as the commander stepped forward with the bared blade to tap him on the shoulders.

"Rise, Sir Fundi," Spicer-Simson intoned, stepping back. "As commanding officer of His Majesty's fleet on Tanganyika, I hereby commission you an official Royal Naval stoker on the prize of our first engagement, the former *Kingani*, soon to be my own flagship."

Speechless, Fundi threw himself down at the commander's feet, scooped up a double handful of dusty earth, and dumped it over his own head.

The gathered Africans roared their approval.

4

In another two days, *Kingani*'s hull was completely patched, and Cross had her dragged back into the water and set to work refurbishing her engines. That took another four days. On the second of those, an oxcart hauled the new gun—a twelve-pounder—from Lukuga to the edge of the bluff above Kalemie, where Wainwright had constructed a crane to lower it to the beach below. Once it was there, he let Cross take over. While Cross directed a crew of workers in wrestling the twelve-pounder first onto *Vedette* and then aboard the *Kingani*, Wainwright toured the small ship.

When he stepped onto the deck, the Old Loco Driver couldn't help but feel in his blood the call of *Kingani*'s steam engine. He descended into the engine room to see if there was anything he could do to help. Lamont and Berry were down there, sweating and bruising themselves as they struggled to tear down the engine.

The place was a shambles, but not because of the effects of the shelling—that debris had been cleared out to give the men room to work. Engine parts lay strewn all about the confined space, and Lamont and Berry kept bumping into each other and the bulkheads and smacking their heads on the beams of the low ceiling. They cursed continually, not seeming to care that an officer was present. In fact, they totally ignored Wainwright as they twisted at a large and particularly stubborn nut with an impressively sized wrench.

As Wainwright's eyes adjusted to the dim light, he saw that the engine hadn't been in the greatest shape even before the shelling. It seemed as much patched together as riveted. A makeshift apparatus attached to the steam valve caught his eye. It was a lever arrangement with a cable that snaked away from it and up through a hole in the deck. He poked his head through the hatch and saw that the cable ran to the wheelhouse. He ducked back down and stared curiously at the apparatus again, then his eyes lit.

"Good grief," he muttered, realizing that the ungainly device was a way for whoever was piloting to control the steam pressure from the wheelhouse. It looked primitively effective but not at all safe. He turned to the sweating, grunting engine room artificers.

"Why don't you...?"

That was as far as he got. Lamont sat bolt upright, red-faced, at the first word, and by the third, he'd drowned out Wainwright with a string of epithets that would have had even the most profane expedition member squirming.

If they could have understood him.

Wainwright caught a few words, but Lamont's Glasgow accent had thickened in direct proportion to his frustration and fury, rendering him almost unintelligible. But even if Wainwright couldn't understand everything the ERA shouted, he got the distinct impression that Lamont didn't need some old landlubber locomotive driver telling him a damn thing about seafaring steam engines.

When Lamont ran down, sputtering, and returned to the wrench. Wainwright calmly and silently climbed out of the engine room. Up on deck, everyone was staring, and Cross immediately came over.

"Was that Lamont?" he asked.

"It was," Wainwright blinked. "I didn't know he had it in him."

"I'll have a word with him later."

"No need," Wainwright said, chuckling. "He was probably right, and I admire him for saying so."

5

Spicer-Simson showed up soon after to watch the mounting of the gun, which was considerably larger than the six-pounder that had previously armed the ship.

"Damn priapic, eh, Wainwright?" the commander called from the canoe that was bringing him out as the gun was being swung into place.

Kingani measured fifty-five feet, and the barrel of the 12-foot gun jutting more than half its length over the bow gave the boat an air of warlike excitement.

"A real flagship for you, Commander," Wainwright said, leaning over the rail to catch the canoe's painter. "By the way, how did you manage to convince the Belgians to give up the gun?" He was surprised that Stinghlamber had given anything at all to Spicer-Simson.

"I traded them for it," Spicer-Simson said. "It appears that the French had a pair of six-inch guns back on the Lualaba. The French had lost track of them, the Belgians didn't know they were there, and no one on the river realized what they had. I got hold of them, and Stinghlamber couldn't say no when I dangled them in front of his greedy eyes."

"How did you know about the guns, sir?"

"I have my ways, Lieutenant," was all the Spicer-Simson said as he stepped onto *Kingani*'s deck.

He wasn't about to tell Wainwright that Hope was actually the one to discover the guns, stored in a shed, during his preliminary survey of the river. Their presence was noted in his final report to the commander, filed just before Spicer-Simson sacked him.

"Welcome aboard your prize, Commander," Cross proclaimed. "She's as sound as she ever was."

"Show me your work, Lieutenant."

"Why, yes, sir. Shall we take a look at your new gun, first?"

"Lead the way."

"She's quite lovely," Cross said when they arrived at the bow. "A twelve-pounder semiautomatic that can fire twenty-eight rounds a minute. She's a mighty weapon, sir." Cross ran a palm along the big barrel. "You'll have firepower to spare."

"I notice you've removed the gun shield."

"Did the Germans more harm than good, sir," Cross said. "We're also going to take off the canopy so it won't catch shrapnel and such if she's hit."

"Excellent idea."

"I'm sure the commander also has noticed that the gun is somewhat large for her craft. We're going to have to move the wheelhouse back a few feet. And there is another problem."

"Don't tell me that we can only fire it straight ahead."

"I'm afraid it's true, Commander. We had to mount it something like the three-pounders on the gunboats. We've reinforced the deck, ribs, and thwarts with scrap iron, so it's possible that if you did fire abeam the gun would not tear loose and throw itself overboard as those on the gunboats would. But even if the deck'll stand the strain of firing abeam, I'd strongly advise against it. The recoil would likely capsize the ship."

"Blast," Spicer-Simson grated, his brow darkening.

"Let me show you this, sir," Cross said quickly. "Maybe it'll lift your heart." He led the commander and Wainwright to the rear of the ship, where a rating was working on a second, smaller gun mount next to the Maxim gun. "We're putting a three-pounder here, and it'll have full sweep astern. You'll have some limitations, but all-in-all she's a fine warship with plenty of firepower and gun range."

"Good work, Lieutenant," Spicer-Simson said, somewhat mollified. "When will she be ready for her maiden voyage under the White Ensign?"

"A week more, sir, to finish alterations, get her engines in order, and bring up a head of steam."

"We'll be anxiously waiting your signal, Cross."

"Would the commander like to inspect *Mimi* while he's here?"

"You've put her in working order?"

"Let me show you, sir."

The gunboat was up on shore, a couple of ratings at work on her.

"We're repairing the sprung planks as best we can, sir, and we're putting in extra strengthening." Cross tapped the deck with his foot. "I'm afraid she'll never be quite as good as new, sir, but even if she makes a lot of water, she'll not sink."

"She's a good girl," the commander said, bending down and giving the bow an affectionate pat. "We'll get use out of her, yet."

6

About a week following the victory, a special dispatch was delivered to Spicer-Simson's hut. Within half an hour, he called for a full-dress parade of all the expedition's personnel, even those working in the harbor.

"I have received a telegram," the commander said in his most patrician tone. "It is a direct dispatch from Buckingham Palace."

A murmur rose among the men, and the commander let it play out and die down on its own, relishing the drama of the occasion.

"Let me read it to you. It says, 'His Majesty the King desires to express his appreciation of the wonderful work carried out by his most remote expedition.'"

The men broke into cheers, quickly followed by stamping and cries from the throngs of Africans as a translation circulated that their god's king had singled out Bwana Chifunga Tumbo for personal glory.

At last the noise died enough for the commander to continue.

"In addition, I am please to announce that accompanying His Majesty's message is a second from the Admiralty. In it, they announce two items of special interest. First, gentlemen, we have made history. *Kingani* is the first German warship to be captured and recommissioned in the Royal Navy."

The Britons hurrahed again, and the Africans joined in, but Spicer-Simson cut them short.

"As for the second piece of news, retroactive to the action with the *Kingani*, I have been promoted to full commander."

Again applause and cheers rolled across the parade ground, and Spicer-Simson, beaming with pride, held out his arms as if to embrace his adoring multitude.

"One more item," he said when the crowd quieted again. "Dudley, step forward."

Looking confused, Dudley went to his commander.

"Also retroactive to the date of the engagement, you have been promoted to full lieutenant. Congratulations." The commander took Dudley's hand and shook it vigorously, and suddenly the British broke ranks and rushed in to surround their commander yelling congratulations that were lost in the din of the milling Africans all around. Wainwright just managed to reach Dudley, who was grinning as manically as the commander, and slap him on the back.

From the margins of the crowd, Hanschell looked on as Commander Spicer-Simson basked in his glory. His men loved him, the Africans idolized him, the Admiralty had promoted him, and even King George found him worthy of personal praise. Not bad for a man who just the year before not only had been one of the oldest lieutenant-commanders in the Royal Navy but who, it seemed, would never command anything larger than a desk in some obscure and dusty office. Recognition had been long in coming for Spicer-Simson, the doctor thought. Let him revel in it.

Hanschell was relishing a victory of his own. The fact that the members of the British expedition remained healthy while the Belgians often were ill with a variety of tropical diseases and other ailments such as dysentery—occasionally resulting in death—had not gone unnoticed by the commander. Just the day before, he had congratulated the doctor on his foresight in all matters of hygiene, including the placement and design of the latrines and his insistence on the building of fresh, clean huts for all the men. The commander also—finally—recognized that Hanschell's attempt to keep the men from stealing out at night to visit women in the nearby villages was probably a prudent policy, and he formally ordered that that sort of activity cease immediately. He did not, however, place a guard on the camp, so the order had little real effect.

Hanschell had to content himself with the new huts and the general cleanliness of the British camp, even if he could not halt the men's nocturnal prowls. The rainy season, which had miraculously held off just long enough to give Spicer-Simson a chance to get the expedition safely to Tanganyika, was now showering the region almost daily. All the formerly dry gullies and ravines were rushing with water, low-lying areas away from the streams rapidly became pools and ponds, and the mouth of the Kalemie River turned into a slough more than a mile wide and filled with reeds and swamp grass. All that water, the doctor knew, spelled danger from the more nefarious of African sicknesses because it would breed hosts of malaria-carrying mosquitoes and tsetse flies infected with sleeping sickness.

Backed by Spicer-Simson's express command, he detailed a work party to clear all the brush and weeds from the camp and the immediate vicinity, all the way down to the edge of the Kalemie. In some ways, the effort hardly mattered because he couldn't touch the other side of the river, which was Belgian territory, and prowling flies and mosquitoes do not recognize political boundaries.

So, during his daily clinic for the Africans, he began examining the children, in particular. Of the first forty-three he examined, eleven had enlarged spleens—a sure sign of malaria. Any mosquito that bit these children and then bit another human would certainly cause an infection. With Rupia's help, he began catching mosquitoes around the camp and dissecting them under his precious microscope. One in twenty was carrying the malaria parasite.

He began recording every mosquito bite suffered by any member of the expedition. The men obviously thought him daft until several ratings came down with fever and showed malaria parasites in their blood. Hanschell immediate stepped up the everyone's daily quinine dose, wondering why only ratings were getting sick but not the officers. He eventually traced it to the mosquito nets that covered the cots. The officers' nets were three feet wide but the ratings' were only eighteen inches, which meant that their arms were more likely to lay right against the netting while they were asleep, allowing the mosquitoes to bite them in the night.

With the increased quinine dose, most of the men who came down ill were confined to their beds for only a week, and Hanschell often had a difficult time keeping them there that long. The victory over *Kingani* was fresh in their minds, and everyone was hungry for a try at one of the other enemy ships. Nobody wanted to be sick in bed when that chance came.

7

The flaps that covered the wide, mosquito-netted windows of the hospital hut were propped up, giving Dr. Hanschell and Father Jacques a capital view of the camp. Rupia puttered around, putting things in order, while the Europeans conversed.

The Belgian White Fathers missionary had come to the camp to beg quinine from Hanschell, as he regularly had since the British arrived. Hanschell had stepped up his supply orders for the drug to oblige the missionary, not out of sympathy for either the man or his beliefs but because Hanschell was on his own personal crusade against tropical diseases.

But quinine wasn't what presently engrossed Father Jacques. Instead, he was staring out the window at Spicer-Simson as the commander made his way across the parade ground, which was practically covered with Africans. As the commander passed with his stately, swaggering walk, whole ranks of Africans knelt, clapped their hands, and chanted, while many of those nearest prostrated themselves in his path and poured handfuls of dirt through their hair.

"It's sacrilege," Father Jacques spat. "The man is demoralizing Africa."

"I think you take the matter far too seriously, Father. I'll admit that the commander's fame has spread quickly, but his influence is neither as wicked nor as wide as you imply."

"I, on the contrary, believe that the commander would like to paint the entire region with a bestiary akin to his own menagerie," Father Jacques lamented. "Have you heard what they call him? Bwana Chifunga Tumbo."

"Lord and Master of the Loincloth," Hanschell chuckled. "A rather amusing appellation."

"It's no laughing matter, Doctor. That's all you hear from Uvira to Sumbu Bay. He's known halfway into the Congo. They attach magical significance to him, especially those bestial tattoos of his. Have you seen the way they chant and bow when he's out in one of the launches signaling? They think he's using magic to call the other German ships into his power."

Hanschell had to laugh again. "He may well be doing just that, for all we know. Our own signalmen can't understand a thing he means. If the natives grant magical power to the commander's tattoos or his semaphoring, maybe they understand the commander better than his own people do."

"Have you not seen the idols?"

"I have one right here," Hanschell admitted. He found it behind a medical trunk. "Rather a good likeness, don't you think? I believe he sent one home to his wife."

"It's bad enough that he's corrupting the heathen, but now he's infecting Europe. The man is ruining all my good work by misleading them and pretending to powers he doesn't have."

"You could lay the same charge against me," Hanschell said, politely refraining from including Father Jacques, as well.

"I don't see how. You are a trained medical man."

"Every day, the natives line up at my door to see the Great White Witchdoctor, and I can't do a damn thing more for them than their own witchdoctors. I examine them and make a diagnosis, but that is often more to my benefit than theirs because half the time, I can't treat them, either because I don't have the proper facilities or supplies of drugs or time or because they are untreatable. All I really do is shake my bottle of purgative pills over them a few times like a rattle and then give them a couple. The next day, they have a great bowel movement and so grant my magic great potency. But the ones with sleeping sickness do not get better, the ones with cancer will die all the same, and the children with malnutrition or parasites or malaria still suffer at their parents' knees."

"At least you give them hope," the missionary said.

"Empty hope," Hanschell said tiredly. "And for the same reason, you can hardly blame them for abandoning your church."

"I'm not sure I quite understand you," Father Jacques said a trifle hotly.

"You offer them something abstract," Hanschell explained, "while the commander has shown them something real that they can understand."

"Nothing is more real to the immortal soul than the Kingdom of God."

"These tribesmen have seen you, Father, come into their country spouting fire and brimstone, and still they are enslaved and maltreated by the Germans and even worse by your own countrymen, who have made the Congo its own version of Hell. Our commander has come in with locomotives spouting fire and brimstone and dragging two small warships over deserts, mountains, and swamps, and in one fell swoop, he has captured an enemy ship and painted its deck with their blood and taken them prisoner."

"It's the blood," the White Father lamented. "These heathen love blood."

"Not blood," Rupia said unexpectedly. "Bwana Chifunga Tumbo have powerful musunga."

"Musunga be damned," snorted the missionary. "What can *you* know about anything?"

"Come now, Father," Hanschell said. "You have to admit, he does have a peculiar sort of charisma."

"Idolaters will burn in hell, as will those they idolize," Father Jacques said, but his tone was less vehement than before, and he hung his head.

"Nevertheless, you accept his food and his quinine," Hanschell said, holding out a packet of the medicine to the missionary, who practically snatched it from his hand.

"And he makes me pay dearly for it with his ridicule as well as by turning my flock away from Jesus Christ," Father Jacques said.

"Cheer up, Father. Our commander may have played havoc with your congregation, but there have been positive side effects, as well."

"What, pray tell?"

"In a manner of speaking, the Ba-HoloHolo *have* been converted—to our side. They were working for the Germans only because the Germans treat them with slightly less indignity and violence than do your own countrymen. There was always the possibility that the Germans could have incited them to rebel, and now the commander has turned the tables. With their allegiance turned to him, your own people need not fear an uprising, and we can use the natives to our advantage."

"Small advantage to be served by a bunch of bestial cannibals," Father Jacques retorted.

"Not that you didn't try hard enough, Father," Hanschell snapped, patience lost at last. "You have your quinine, now, if you'll excuse me, I have work to attend to."

The White Fathers missionary started to sputter, then thought better of endangering his quinine supply and, in a huff, left the hospital.

"The Belgian Christian not know musunga," said Rupia after the missionary had gone. "That why he not loved like Bwana Chifunga Tumbo."

"I suspect you're right," Hanschell said, then he chuckled. "I dare say that Father Jacques would have been even more upset than he was if he'd witnessed our commander at his bath, eh?"

Rupia laughed, too, and with good reason. It was the commander's bi-weekly and very public bath, instituted immediately following the victory over *Kingani*, that finally had settled the long-standing pool regarding the extent of the commander's tattoos, although Wainwright earlier had confided to Hanschell that they were destined to share the pot. On the occasion of the first public bath, most of the expedition members were working down in the harbor, but within an hour, everyone knew that Hanschell and Wainwright were the winners. The commander was, as they predicted, completely illustrated—or at least as completely as was humanly endurable.

The two, though, didn't gloat over their victory. Instead, since all the liquor—except Spicer-Simson's personal supply of vermouth—had been exhausted before Christmas, they used the winnings to buy whiskey and rum and several cases of cigarettes for the men.

Within two weeks, the commander's bath had become a ritual that brought an audience that seldom numbered less than several hundred, though few could have been close enough to take in the spectacle in detail. On Wednesdays and Saturdays, precisely at four PM, Tom, set up a folding tub of green canvas in front of the commander's hut. He then poured cans of boiling-hot water into it while the crowd's jostling for position took on a reckless note that continued as a large grass bath mat was spread in front of the tub and a stool set up beside it holding a bottle of vermouth and a glass. Finally, the servant tested the water, and when it was the right temperature, he went to the door of the commander's hut and gave the word.

Then the god appeared at the door of his hut, a towel wrapped around his waist instead of his skirt, a pair of slippers on his feet, and his long cigarette holder gripped in his jaw. He stood there smoking, exhibiting his profusely illustrated skin, and taking in the throng while it voiced its approval with wails and stamping and clapping. When the applause died, he walked with regal aplomb to the side of the tub. There, he passed his cigarette holder and towel to Tom and, naked, commenced a peculiar exercise routine of several minutes length, flexing and waving his muscular arms and legs and twisting and bending his torso. The tattooed beasts, birds, reptiles, flowers, vines, and insects writhed and wriggled and danced over his bunching muscles, the Africans moaning and chanting and swaying all the while.

Finally, Spicer-Simson stepped into the tub, where he worked up a considerable lather with a heavily perfumed soap that filled the air with flowery scent. After some minutes of this, he stood and let Tom pour fresh, cold water over him. Completely rinsed, the commander then emerged from the tub and toweled dry while Tom stuck a fresh cigarette into the holder and decanted a glass of vermouth. Draping the towel around his waist, the commander took the cigarette holder, held the cigarette to a match presented by Tom, and inhaled several puffs before accepting the glass of vermouth, which he raised in salute to the attendant throng before downing it with obvious relish. His handing the empty glass to the servant signaled a fresh round of response, and with this, the commander returned to his hut. Immediately, the local women rushed forward bearing jars they hoped to be fortunate enough to fill with the bath water, which they took home to bathe their own children.

It appeared that Spicer-Simson's apotheosis was nearly complete.

8

The Belgians seem to have taken fresh heart at the British victory. The *Vengeur* was hauled to the shipyard on the day after, and the engines were repaired even while the repairs to the *Kingani* were completed. *Vengeur*'s boilers, though damaged, could manage enough steam pressure to shove the hulk at six knots. In accordance with the division of authority, the ship was handed over to Spicer-Simson, who had it brought to Kalemie. Cross mounted the six-pounder that had been taken off *Kingani* on its foredeck, and the two hundred rounds of ammunition for the gun that had been aboard the *Kingani* were brought aboard as well. Everyone seemed pleased that the gun, which had started its life on the *City of Winchester* before it had been captured and installed first on the *Königsberg* and later on the *Kingani*, once again would be trained on the Germans.

The presence of the gun further delayed *Vengeur*'s official launch because the deck plating had to be strengthened to withstand the recoil, and the extra weight proved too much for the tub's old engines, now reduced to a pitiful three knots. New, larger engines were requisitioned, but they wouldn't arrive for a month or more.

But there remained plenty to do. Although *Vengeur* was technically under Spicer-Simson's command, its crew was Belgian, and Spicer-Simson was skeptical about their competence.

"I've seen their gunners in action," the commander commented. "They can't hit anything from dry land. How are they going to hit a moving ship from the deck of a moving ship?" He ordered one of the expedition's gun-layers, Murphy, to train *Vengeur*'s gunners. "But be careful, PO," Spicer-Simson said firmly. "Don't let them practice too much. We only have two hundred rounds for that gun."

In the meantime, Stinghlamber ordered that the pieces of the 1,500-ton *Baron Dhanis* be brought from Kabalo. They would be reassembled under the new six-inch guns of the fort then brought to anchor in Kalemie Harbor, but the process was likely to take months to accomplish. Even so, in preparation, Jadot began to build a stone jetty parallel to Kalemie's northern breakwater to make a relatively storm-proof landing slip for the ship once it was completed.

Spicer-Simon's little flotilla was growing into a real navy.

But as preparation were made to resume hunting the enemy, there was, for a time, no enemy to hunt. So, to occupy the time on days when the rain wasn't too much of an impediment, the men frequently went out with guides to hunt the local African wildlife, which included elephants, antelope, and buffalo. Lions and leopards were the most popular game, however, not only because of the prestige of bringing in one of the ferocious creatures but because the big cats constantly roamed the camp at night. No one would consider going about alone or unarmed after nightfall, but the worst of it was the loss of the cattle, goats, and poultry that were kept around for fresh meat.

The one goat who was safe from both predator and camp cook was the one that had survived the explosions on *Kingani*. This fortunate animal, which was named for the ship from which it was taken, was looked on as a symbol of great fortune and had become a camp mascot to equal Josephine. The men took turns walking Kingani about camp on a rope, and it slept indoors to protect it from the cats.

No one was safe, though, from Spicer-Simson, not even himself, though he enjoyed such a reputation these days that his men seemed to forgive his every foible and autocratic pronouncement. One day, the men spotted half a dozen huge, dark pillars hovering over the lake.

"What are those things?" Lamont asked.

"Waterspouts," Berry answered.

"They don't look like waterspouts to me," Waterhouse said.

"They got to be, Chief," Berry said.

Everyone seemed to have an opinion one way or another, until Hanschell said, "They can't be waterspouts. They're not moving."

"Nonsense, Doctor," Spicer-Simson snorted. "These are waterspouts. Very dangerous things! Only local atmospheric pressures prevent them from moving. At sea, one learns to give them a wide berth. I've seen much bigger ones off the China Coast—tremendous vortices of thousands of tons of water. I remember one incident as captain of a gunboat on the Yangtze River. I'm expert, you know, at speaking Chinese, and I had warned a large junk to alter course to avoid one of these waterspouts. They're enormously strong craft, hundreds of years old, some of them— three masts, sails of matting. The gear's so heavy it takes a dozen men to steer them. The Chinese believe that the air is full of devils, and that particularly nasty ones run up and down the waterspouts, and so at the head of the mainmast, they fix a long painted bamboo. If a devil sits on it, it breaks, and he falls into the sea. Ingenious concept, but that junk would have done better to listen to my advice. The waterspout broke over it— my little gunboat was nearly capsized by the huge wave set up by the falling water. When the dense cloud of spray had blown away, there was nothing left of the junk but a few shattered timbers and a little painted bamboo stick bobbing up and down in the sea."

The commander's tale seemed to end all possibility of argument, but Cross called out hastily, "I believe they're moving now, and right toward us!"

Almost all of the dusky columns were drifting toward the shore, and in a panic, everyone prepared to dash for shelter. Suddenly, from the margins of the camp, hundreds of Africans rushed across the parade ground toward the bluff, carrying what looked like small fishing nets.

"Do they expect the waterspouts to bring fish ashore with them?" Dudley wondered aloud.

His question was soon answered—as was the debate over the true composition of the dark pillars. As the first reached land, it lifted over the edge of the bluff, flattening as it did, and the men found themselves engulfed in a swarm of insects.

"They're kungu flies!" Hanschell cried, grabbing a few from the air.

The tribesmen were running around, crying out happily as they scooped up large numbers of the flies in their nets.

"What are they going to do with them?" Cross asked, and his question was answered as the people grabbed handfuls of flies from the nets and squeezed them into patties, which they promptly began to munch.

"Good, God," Eastwood said. "They're eating them!"

"A good lesson for a quartermaster, don't you think, Tubby," Hanschell laughed.

Before long, the feast had been joined by birds, and even baboons and village dogs rushed out and began gobbling as many flies as they could.

"What a marvelous lesson in natural history," Hanschell said, batting flies from around his face.

He got an even better lesson that evening after mess. He was relaxing in his hut, when a savory odor wafted by his nostrils. Outside, Rupia squatted in front of a fire, grilling one of the kungu cakes. Hanschell watched with interest until Rupia popped one of the grilled cakes onto a plate and offered it to the doctor.

"No thank you, Rupia."

"Good, Bwana Doctor. You eat."

"But I've already eaten."

"You eat."

The look on Rupia's face was insistent, and Hanschell thought of how hard the African had worked for him in the hospital and in helping Hanschell gather the hundreds of insect and reptile specimens that sat in bottles and jars and tins in Hanschell's trunks. The least he could do would be to take a taste. After all, the Africans seemed to be eating them with relish and without harm.

Taking the plate, he ventured a nibble. Then another and another. It really *was* good—something like marrow. In short order, the cake was gone, and Hanschell surprised himself by looking to see if there was more. Rupia, smiling, scooped another onto his plate.

"Rupia tell you kungu good."

"Yes, Rupia," Hanschell agreed. "Kungu good."

9

The *Kingani* was fully repaired and armed by January 13, but its maiden voyage the following day was unexpected. Rains and squalls had disrupted activities several times since the beginning of the month, but the storm that now hit could only be termed a typhoon. Winds began to blow down the lake from the north in the wee hours of the morning, and by dawn, Tanganyika was whipped into a frenzy. Each of the larger boats maintained small watch crews, and this day, the task of guarding *Kingani* had fallen to Petty Officer Murphy, a pair of seamen, and Fundi, who had been keeping the boilers hot since the beginning of the storm.

For the most part, the men stayed inside, trying not to be beaten to death against the bulkheads as the hull heaved. Taking turns every fifteen minutes, one of them would stumble outside to check on things before returning to, if not the relative safety, then the definitely dryer region of the cabin.

"That damn breakwater ain't up to snuff, Mr. Murphy," one of the seamen, Leighton, commented after taking a turn on the deck. "The water's so rough, we're starting to drag our anchor. It won't likely hold out much longer in this weather, and if it gives, we'll be on the rocks in no time."

The second seaman, Vickars, who went out next, came back with even worse news.

"I think our anchor cable is tangled on the *Vengeur*."

Murphy hurried onto the bounding, wind-blown, rain-drenched deck, both seamen following him. Sure enough, the anchor cable was fouled with *Vengeur*'s bow mooring. And it also appeared that *Vengeur*'s stern moorings had come free, leaving the larger vessel swinging dangerously in the wind. The way the wind was blowing, the two vessels would be smashing against one another at any moment. *Vengeur*, being considerably larger than *Kingani*, would likely crush the smaller craft and sink it before its maiden voyage under the White Ensign. And if *Vengeur* didn't get *Kingani*, there always were the rocks of the breakwater, now dangerously close on the other side of the small ship.

"Not on my watch!"

"What, Mr. Murphy?" asked Vickars.

Murphy hadn't realized that he'd spoke aloud.

"Go below," he ordered Vickars. "Tell Fundi we're going to take her out of the harbor."

"We're going to what?"

"Snap to it, man. I'd rather lose *Kingani* trying to save her than just sit here and be battered to pieces."

The seaman disappeared into the cabin, and Murphy turned to the anchor windlass with Leighton.

"We can't pull the anchor in," he said. "When I give the order, I want you to run the cable out all the way and lose it."

He barely heard the "Right, sir" as he hurried back into the cabin. There he met Vickars coming up from the engine compartment.

"Fundi's ready, sir."

"Stand with me at the wheel," Murphy ordered. "I may need help steering."

"What about the anchor, sir?"

"Leighton's going to slip it. Signal Fundi for full reverse."

The steam engine hissed and chugged, and gradually, the *Kingani* began to back away from *Vengeur* as the windlass ran out the anchor cable. Then, with a little backward bound, the ship was free. As Leighton staggered into the wheelhouse, Murphy and Vickars fought to turn the prow toward the harbor mouth.

When he could glimpse it through the driving rain, Murphy shouted, "Full ahead!"

With a little better speed, the ship fought against the heavy waves sloshing between the arms of the breakwater. They barely cleared the tip of the southern line of rocks, and then they were in open water, the gale howling around their ears.

"We did it, sir," Vickars said. "What are we going to do, now?"

"We need to get farther out," Murphy answered. "Then we keep her bow to the wind and her steam up, and we pray."

They should have prayed first. Barely two hundred yards out of the harbor, the engines abruptly strained and clashed, and the ship lost momentum.

"What the hell is happening?" Murphy demanded as *Kingani* wallowed dangerously in the swells.

Vickars clambered below and was back a couple of minutes later.

"Fundi says it's not the engine. He thinks there may be something wrong with the propeller."

"Try to keep her headed into the wind," Murphy told the Leighton, then he and Vickars rushed outside.

As the swells came and went, exposing the floundering stern, Murphy could see a hank of rotted rope fouling the propeller.

"Shit!" He looked up, toward the shore, realizing that the storm was blowing the *Kingani* back toward the rocks of the breakwater. We need an anchor, he thought. But the ship's only anchor was back in the harbor, now attached to *Vengeur*'s cables. There must be something. He cast around frantically, seeing nothing that might help them. Quickly, he made a circuit of the deck, and as he passed the bow, his eyes fell on the anchor windlass.

"Get a wrench!" he shouted at Vickars, who was following him. "Come back and unbolt the windlass!"

While the seaman hurried to obey, Murphy went to a locker and found a length of rope strong enough to serve his purpose. He returned to the windlass and tied one end of the rope to a nearby stanchion. By now, Vickars was back with a wrench, and while he unbolted the windlass, Murphy tied the other end of the rope to it. When both had finished, they wrestled the heavy machine to the edge of the deck and heaved it overboard.

"I hope to God that rope is long enough," he said. And strong enough, he amended silently.

Apparently it was long enough, for immediately, the ship began to swing around, its bow to the wind, the makeshift anchor rope stretched taut from the bow. Murphy stared through the blinding rain until he could make out a feature on the all-too-close shoreline, and he watched it for several minutes.

Yes, *Kingani* was dragging its makeshift anchor, but not much. Only time would tell if their luck would outlast the storm.

And then everyone, the stoker included, huddled together in the tiny pitching cabin, watching through the windows as the rain-soaked air blustered around them and waves sent sheets of water over the bow.

Back on shore, Spicer-Simson, Wainwright, Cross, and Dudley, accompanied by about thirty Africans, had made their way through the winds and lashing rain to the harbor to see how the flotilla fared. They were just in time to witness *Kingani*'s narrow escape, but what might have happened to the flotilla's newest addition was amply demonstrated just moments later. *Vedette*, wallowing and pitching in the heavy seas, suddenly tore lose from its anchor and headed right for the rocky shore.

"Get down there!" Wainwright shouted to the Africans over the gale. He led them into the rough surf as the motorboat washed toward them, but their combined strength was unequal to Tanganyika's fury, and within minutes, the motorboat crashed against the rocks. In a flash, battered and upended, it slid off the rocks and sank with only its bow showing in the troughs of the waves.

Wainwright commanded the Africans to return to the safety of the shore then hurried back to Spicer-Simson.

"I'm sorry, sir," he yelled. "We just couldn't stop her."

Spicer-Simson merely nodded as he turned a concerned eye to *Mimi* and *Toutou* pitching dangerously in the surf.

Hours passed, and they seemed like days, especially to *Kingani*'s tiny crew. But at last, the tossing began to subside, and they could begin to discern a darker gray that was the shore.

"Blimey, sir," said Vickars. "We've blown down into Tembwe Bay."

"I believe you're right. The storm's weakening. Fundi, get below and build a fresh head of steam. We'll head for port as soon as it blows out. You two, help me clear the propeller."

By the time *Kingani* steamed back into Kalemie late in the afternoon, the storm had not only blown itself out but left beautiful skies and clear air in its wake. And it had left the harbor in shambles. Expedition members and African workers were scurrying around, checking the other boats for damage. The commander was waiting on the beach when *Kingani* rounded the point, and he immediately leapt into *Netta* with Wainwright and sped out to the ship.

"Sorry, sir," Murphy apologized as Spicer-Simson came aboard. "*Vengeur* had broken lose and our anchor was fouled with her forward mooring. I was afraid we'd be battered to pieces."

"Thank God she's safe," Spicer-Simson said, and it was the first time Wainwright had ever heard the commander utter words in such a tone of gratitude. "You did the right thing, Murphy. I'll see to it that you get a commendation for your quick thinking."

"It wasn't just me, sir. The men did their part. And Fundi, too."

"Fundi? The German stoker?"

"He kept the fires hot and ran the engine, too, sir. We'd never have made it out of the harbor without him."

"So," Spicer-Simson said, beckoning the stoker forward. "You've become a valuable asset to us, it seems."

"Fundi serve," the stoker said solemnly. "Fundi serve Bwana Chifunga Tumbo."

10

Surprisingly, considering the storm's ferocity, only *Vedette* and *Vengeur* suffered any damage—*Vedette* to its propeller shaft and *Vengeur* to the propeller itself. Both were soon repaired even as final preparations were made for *Kingani*'s official maiden voyage.

Before that, though, the Belgian lookouts at Kavalo spotted *Hedwig von Wissmann* steaming down the coast but keeping well off shore to avoid danger from Lukuga's new batteries.

Wainwright was with Spicer-Simson when a runner brought them the news.

"Too bad we can't engage them, now," Wainwright said wistfully.

"We're simply not prepared, Lieutenant," the commander said.

Wainwright knew it was true. *Kingani* was still undergoing its final outfitting, and *Mimi* and *Toutou* needed their decking reinforced after the strain of firing the guns during the first battle and the terrible jostling they'd taken during the typhoon.

"Go down to the harbor, Wainwright, and make sure *Kingani* is concealed. No need in letting the enemy know we've purloined their vessel."

Wainwright did as the commander bid, though the task wasn't difficult since most of *Kingani* was hidden by the breakwater. And anyway, the shore batteries both at Lukuga and Kalemie were sufficient to keep *Hedwig* far enough out on the lake that even a powerful set of field glasses would not be able to distinguish any details against the cliffs.

By January 15, all the boats and harbor were back in trim, and Spicer-Simson called a full dress parade in the harbor the following morning to christen his new flagship and take it on a test run.

"As you know, men," Spicer-Simson told the gathered expedition, "our new flagship is the first German vessel to be captured during the war, and likewise, it is the first to fly the White Ensign."

He signaled, and two ratings aboard the boat raised the flag of the Royal Navy. The rest of the men cheered, and the Africans followed suit.

"In joining our two motor launches," Spicer-Simson continued, "she gives us a dominating advantage on the lake. May she carry us to swift and decisive victory and bear honor to the flag she now flies. Of course, we cannot continue to call her *Kingani*—that would be an insult to the flag and country we love so well." He turned to Eastwood, who was standing by. "Eastwood, the bottle," Spicer-Simson commanded.

Eastwood handed him a clay jug filled with pombe—at this point, European spirits were too high in demand to be wasted on the launching of a boat or on the waters of Tanganyika.

"I hereby christen thee," Spicer-Simson loudly proclaimed, raising the jug, "HMS *Fifi*!"

The smashing of the jug on *Fifi*'s prow helped hide the dismayed grunts and disgusted sniggers that rose from the ranked men. Only the Africans cheered.

"Did you say *Fifi*, sir?" Wainwright managed to ask with a straight face. He wished he had a drink of the native beer, nasty as it was, that was running down the steel plating of the former *Kingani*'s bow.

"*Fifi*," Spicer-Simson affirmed. "It means 'tweet-tweet,' like the sound of a bird. A suitable name for the companion of *Mimi* and *Toutou*, don't you think, Lieutenant?"

"Certainly similar, sir. What made you think of it?"

"Wasn't me at all. Mrs. Stinghlamber suggested it. It's after her pet bird. You weren't there, but one night at dinner, she asked to be the godmother of our prizes, so I allowed her the honor of naming our first."

Hanschell, who had been present that night, seemed to remember that Mrs. Stinghlamber's remark had been ironic, then he realized that she'd continued the irony with *Kingani*'s new name. He chuckled to himself, not because the major's wife had pulled one over on Spicer-Simson but because the commander could probably read her attempt at a jest, carry it through, and be satisfied on his own terms. *Fifi* did go well with *Mimi* and *Toutou*. It went well, in fact, with the whole tenor of this crazy undertaking.

Next to him, Waterhouse shook his head sadly and muttered, "I never thought I'd be working warships with the names the likes of these."

"Cheer up, Chief," the doctor said cheerily. "It may have an odd name, but it's armored and well mounted. She's a warship, whatever she's called."

"Crew!" the commander bellowed. "To your stations!"

HMS *Fifi*'s official maiden voyage under the White Ensign was about to begin. Aboard were the commander, Wainwright, Dudley, Cross, five ratings, Fundi and his assistant stoker, and Waterhouse, who, since the engagement that had captured their prize, had become the commander's favorite gun layer. Also aboard for the ride was Captain Goor.

"Full speed ahead!" Spicer-Simson yelled around his cigarette holder as *Fifi* churned past the end of the breakwater and onto open water. Cross complied, and Spicer-Simson called out, "Mr. Tait, bring her to port."

"We're to head up lake, then, sir?" Wainwright asked. "Going to show off our prize to the Belgians?"

"Precisely, Lieutenant." Spicer-Simson flashed a toothy grin. Then he strode forward, placed a hand on the shaft of the gun's 12-foot barrel, and took a broad, regal stance. In just a few minutes, they were passing the buttresses of the fort at Lukuga, and the commander gave the fortifications an exaggerated salute. Then he dropped his hand and swaggered back to the wheelhouse.

"That's all the buggers deserve," he said. "Bring her about, Wainwright."

"Down-lake, sir?" the lieutenant asked.

"Some of our own infantry chaps have fortifications near Baudouinville. They've been here nigh on two years and haven't had the success we've had in our first week on the water. I'm going to show them what stuff the Royal Navy is made of."

"Jolly good, sir. We'll show those buggers, too."

"That we will, Wainwright. That we will."

The day was a beautiful one for a voyage, and gradually the lake slipped beneath *Fifi*'s bow and boiled astern as the ship sailed on past Tembwe Bay and into waters where the Tanganyika expedition had not yet ventured. The ship managed a steady seven knots, even with the added weight of the new gun.

After an hour or so, the commander said, "Time to test our new gun. Wainwright, bring her to port and keep her at full speed and steady as she goes."

As the ship turned to point at the German shore, the commander went out to the gun, where he instructed Waterhouse to load a round. In seconds the breech was opened, the round loaded, and the breech closed.

"Ready, sir!" Waterhouse called, proud to serve.

Spicer-Simson drew his sword, held it high, then slashed downward. The gun roared and belched a lateral pillar of smoke.

Everything that was loose on the deck and in the cabin and engine room—including the men—suddenly flew forward four feet. Or seemed to. The powerful recoil from the big gun had brought the boat to a complete standstill from seven knots.

Gradually, screws furiously churning the water aft, *Fifi* began moving again and picked up speed.

The commander, who'd staggered and nearly fallen as the ship stopped dead in the water, recovered, and Waterhouse gave him such a sheepish look that Wainwright almost burst out laughing. Then the lieutenant had an even funnier thought. The gun could fire twenty-eight rounds a minute, but if it even approached that rate in an action, the repeated recoil would actually propel the ship away from the enemy!

"Not only priapic," Spicer-Simson said to Wainwright, "but explosively so, eh, Lieutenant?"

"Deafeningly so, sir," Wainwright replied, keeping a straight face.

"We were up to full speed?"

"Yes, sir."

"I think we should try it again. Load her up, Waterhouse."

Waterhouse loaded, the commander signaled, the gun roared, and the ship stopped dead in the water.

"Mr. Cross," the commander called out as the ship began moving again. "You were right." Then to Waterhouse, he said, "Mr. Waterhouse, never dare fire that gun abeam."

"No, sir. I won't."

Satisfied with *Fifi*'s firepower, the commander strolled the decks and went below, inspecting the extent of his new domain. Wainwright went over to keep Tait company in the wheelhouse, and after half an hour, Spicer-Simson joined them, apparently satisfied that all was well.

"Damn fine ship," he said.

Wainwright didn't bother to reply to this obviously rhetorical statement, but he agreed that *Fifi* certainly was a cut above the gunboats.

Before long, the British fortifications at Baudouinville came into view. Spicer-Simson hurried out of the wheelhouse and took his regal stance beside the gun. As *Fifi* drew abreast of the fortifications, the commander faced them and gave his broad salute.

At that instant, three plumes of smoke issued from the shore batteries, followed by dull booms. Almost immediately, three geysers fountained around *Fifi*.

"Goddamn it!" Spicer-Simson bellowed, and more cannon shot splashed the deck with water. "Tait! They're firing on us! Take evasive action!"

The pilot twisted the wheel and angled the ship farther out onto the lake.

"What the blazes do they think they're doing?" the commander yelled as he rushed into the wheelhouse, his skirt and uniform tunic dripping, his cigarette damp and dangling pitifully from its holder.

"They must think we're Germans," Wainwright said. "Did you tell them we'd captured the *Kingani*, sir?"

"Tell them?" Spicer-Simson snarled, shaking the disintegrating cigarette onto the deck. "Why in blazes should *I* have to tell them? Stinghlamber knew. *He* should have told them."

"It seems he didn't, Commander," Goor said. "You know how communications sometimes are. I will send a messenger down as soon as we return to port."

"You do that. Until then, we'd better keep to the middle of the lake." Spicer-Simson straightened his tunic, which was already drying in the heat. "Fired on by our own shore batteries. Now that's a jolly good turn, eh, Wainwright."

"It is, sir."

Suddenly Spicer-Simson laughed. "Miserable shots, aren't they? No better than the Belgians at Lukuga. We couldn't have been more than two hundred yards offshore. That's a land crab for you."

11

The mercurial weather continued through the rest of January, ferocious storms sweeping the lake in successive waves separated by troughs of relative calm. The storms' timing left much to be desired, for twice during that time, *Hedwig von Wissmann* was spotted steaming down the coast, but the heavy seas made leaving the safety of Kalemie and giving chase an impossibility.

The *Hedwig*'s crew, commanded by Lieutenant Odebrecht, was looking for some sign of the *Kingani*—either wreckage or survivors. Most of the German officers on the lake had decided that the *Kingani* had either gone down in a storm or been sunk by the Belgian shore batteries. They were tending toward the latter, however, since there'd been no severe weather during the scheduled several-day jaunt that had ended with the *Kingani*'s failure to return to Kigoma.

It galled Captain Zimmer to think that the *Kingani* might have inadvertently wandered into the Belgians' gun range, but he didn't discount the possibility. Jung had been new at command, having replaced Rosenthal when the lieutenant had disappeared during one of his midnight forays to the Belgian shipyard. But there also was the possibility that something was going on with the Belgians of which he was unaware, and that galled him even more than the possible loss of the *Kingani*. Winning a war required strategy, and strategy required intelligence, and Zimmer's one sure source of intelligence on Belgian activities seemed to have completely dried up and blown away on the hot African wind.

That source had been the Ba-HoloHolo, who hated the Belgians even more than they hated his own countrymen. But ever since the *Kingani* had vanished, so had the Ba-HoloHolo. He'd heard reports, as vague as they were unsettling, that the Ba-HoloHolo had discovered some new god that they worshipped, whom they called Bwana Chifunga Tumbo, which his translators informed him meant Lord of the Bellycloth.

Zimmer couldn't imagine what *that* signified, but whatever it was, he would be forced to keep sending out the *Hedwig* until it found the *Kingani* or its survivors or he was assured by the passage of time that they'd never be found. It was an expensive, time-consuming, and dangerous way of conducting a search, but it was all he had. If only the *Graf von Götzen* was ready, he'd simply shell Lukuga to rubble. But it wasn't ready, and all he could do was order Odebrecht to keep well off shore and return to port immediately if the weather became too severe. No sense in losing the *Hedwig* as he had the *Kingani*. So far, though, Odebrecht had returned empty-handed.

Zimmer sighed and resigned himself to sending another negative report to Colonel von Lettow-Vorbeck. He hated to tell his demanding superior about losing the Ba-HoloHolo almost as much as about losing the *Kingani*. The colonel had hoped to foment the tribe into creating an uprising in the Belgian's own territory. Such an event would not have been decisive, but Zimmer would have liked to see the Belgians squirm.

Too bad he wasn't in Lukuga, because Stinghlamber was squirming a lot these days, but for reasons other than native uprisings or German predations. Those were nothing compared to the private war between himself and Spicer-Simson. And it was getting worse since the commander's victory. That damn vulgar British upstart had made good against all odds, and Stinghlamber worried that Spicer-Simson was presumptive enough to try to use the victory as leverage to gain permanent control of Stinghlamber's own command at Lukuga.

Stinghlamber didn't put it like that to the Belgian high command. He simply reiterated in every dispatch that the ranking army commander should continue to control Lukuga even if the ranking naval commander had all of the lake he could manage to acquire. The apparent agreement of the Belgian high command, however, did little to soothe him. They weren't, after all, in direct and constant contact with the bizarre British officer, as Stinghlamber was.

Stinghlamber's initial annoyance with Spicer-Simson and his uncouth ways had grown to active resentment and dislike. Mrs. Stinghlamber frequently tried to console him, pointing out that eventually the war would end and the British would leave, but her words had no more effect on her stressed husband than those of his own high command.

Privately, though, the Belgian major's wife had a certain fascination with the British commander. True, he was uncouth, overbearing, and a braggart, and that leather skirt he affected was more than laughable, but he was never less than gentlemanly with her and the other ladies. He'd even used her suggestion in naming his prize. And in only a few months, the man had produced more positive effects on the morale of the Allied troops in the region than her husband had in years. She didn't like him, exactly, but she had to admire his panache. In fact, she had, unknown to her husband, acquired one of the little clay effigies of Spicer-Simson, which she kept in the recesses of her trunk, and she occasionally removed it to gaze on its cleverly wrought features. She planned to take it home with her when she and her husband returned to Belgium.

Unfortunately—or perhaps fortunately—that came all too soon. The internecine bickering had caused some bad blood between the Belgian high command and the British Admiralty, and the Belgians decided that Stinghlamber wasn't the man to stand his ground against Spicer-Simson, even if the Admiralty *had* ordered the commander to confine himself to the lake. Stinghlamber was to be replaced by one Lieutenant Colonel Moulaert, whose rank and bearing they believed would provide more than a match for the British commander. Moulaert was on his way but wouldn't arrive for another few weeks. In the meantime, Major Stinghlamber and his wife found themselves prepared to leave for Europe by packing their belongings—the effigy of Spicer-Simson included In Mrs. Stinghlamber's trunk.

12

January 24 brought another storm, though not as fierce as some, and although the breakwater held and so did most of the boats, *Toutou* managed to break free. For a few minutes, it looked like the open hull might take on enough water to sink the boat and make it more sluggish in the currents and, perhaps, preserve it, but that proved not to be. Wave after wave lofted the boat toward the southern breakwater, finally smashing it bow first onto the rocks.

As soon as the waves subsided, *Toutou* was dragged ashore, but there it was destined to sit for some time. The crew numbered expert mechanics and engineers who could build, repair, and operate metal-hulled craft, but *Toutou* was wooden and, finally, was beyond their expertise to repair since there wasn't a real shipwright among them.

Spicer-Simson seemed unconcerned now that he had *Fifi*, but Dudley was upset because he'd looked forward to commanding *Toutou* while Wainwright took over *Mimi*, and now he believed his chance for further glory was gone.

Hanschell, who had been treating locals all morning, missed the activity around *Toutou*. After noon, tired of doling out purging pills, he got Rupia, and the two of them went into the bush to hunt specimens. Hanschell's accumulation of insects, reptiles, and small mammals was growing nicely, and he hoped to take it back to the London School of Tropical Medicine to augment its collection.

They were two or three miles from camp, foraging through the brush, when Rupia emerged from the foliage, his face excited.

"Navyman god!" Rupia exclaimed.

"Navyman no different than Rupia," Hanschell said, puzzled at Rupia's sudden appearance and odd statement. "You should know that by now. You've seen enough European blood to know that the differences between us are only on the surface."

"No, Bwana Doctor," Rupia insisted, shaking his head. "Navyman god. You come see?"

"See where?"

"You come, Bwana Doctor. This way."

Rupia took Hanschell by the arm, practically dragging him through the underbrush. In a couple of minutes, they entered a clearing, and Hanschell understood what Rupia was talking about. At the other side, beneath the branches of a large tree, stood a statue of the commander. Unlike the eight-inch clay figurines Hanschell had seen so far, this one was nearly three feet in height. Its size permitted a great deal of detail, and the likeness was worthy of any European sculptor. There was a carved goatee; tiny bas-relief representations of tattoos covered the arms, legs, and torso; and a hugely erect penis depending large testicles protruded from beneath the skirt. A clay pith helmet perched on the statue's head, and its hands held a pair of clay binoculars.

The statue stood atop an altar of carefully placed stones, adorned with feathers, scraps of cloth, and snake skins. Surrounding it all was a circle of round stones splashed with blood.

"Navyman god," Hanschell muttered to himself, for Rupia had vanished back into the bush, leaving his bwana to commune with the deity.

It looked to Hanschell as if much of the feathers and cloth and other material lying around on the altar had been put there by worshipers, and he had a sudden but overwhelming urge to add his own totem. All he had with him, though, was a half-empty bottle of purging pills. He shook the bottle like a rattle over the shrine then set the bottle beside the effigy in a mound of feathers. He felt a little foolish doing it but afterward was glad he had.

As he stood there, looking at the blood scattered over everything, he thought of what Father Jacques had said just a few weeks earlier.

"It's the blood," the White Father had lamented. "These heathen love blood."

"Not blood," Rupia had replied. "Bwana Chifunga Tumbo have powerful musunga."

"Here's both," Hanschell muttered. "And I expect we're in for more."

Chapter XV: *Hedwig von Wissmann*

THE STORMS THAT HAD PLAGUED the region most of the month of January finally abated in the first week of February. But the clearing skies were no palliative to Captain Zimmer, who had given up hope of finding the wreckage of *Kingani* or its survivors. Then early in the morning on February 8, Lieutenant Odebrecht came to Zimmer's headquarters trailing an African under armed guard.

"He's one of our spies," Odebrecht said to Zimmer. "And he tells an interesting tale. Go on," he prodded the African. "Speak about the *Kingani*."

"*Kingani*, him sunk," the tribesman said.

"Where?" Zimmer demanded. "How?"

"New big guns from Belgian," the African said. "In not deep water. Him chimney still above water."

"Does he have any idea where?" the captain asked Odebrecht.

"He has only a vague notion," the handsome young lieutenant replied. "Somewhere not far north of Lukuga. He says he didn't see it himself, only heard about it. I'd put the spot near Buena Dengwe Point. We've had reports that the Belgians have installed some new guns there, and maybe Jung was careless."

"I want you over there, Odebrecht," Zimmer said. "We'll settle this matter once and for all. Take the *Hedwig* and *Wami* and depart at once." Zimmer pointed to the wall chart of Tanganyika. "Sail down to Kunwego, leave *Wami* there, and proceed to the Belgian side of the lake. If you approach before dawn, you should be able to scout sufficiently before the sun is well up and the Belgians alerted. See if you can find the *Kingani*, but whether you do or not, return to Kunwego by noon. I'll follow in the *Graf von Götzen* as soon as we can get up steam and should be at the rendezvous by the time you arrive."

"Yes, sir." Odebrecht saluted and hurried from the room. By noon, the *Hedwig von Wissmann* had left Kigoma in the distance, *Wami* riding her wake.

The seventy-foot iron-plated wooden steamer displaced one hundred tons, though as Machinist Mate Mewes euphemistically put it, "She rolls in such a way that there is no pleasure voyaging in her." But pleasure or no, until the *von Götzen* had been built, the *Hedwig* had been queen of the lake. In addition to Odebrecht and Mewes, she carried eleven German seamen and nine African deckhands and stokers. Her bow was armed with two six-pounders with a range of two miles, and the aft deck mounted a one-pounder Hotchkiss. The gun had half a mile less range than the bow guns, but it was a quick-firer with five revolving barrels and could cast a deadly pall within its reach.

The twenty-five-foot *Wami* was a clunky, sluggish steam-powered boat that had seen better days. As of yet, it was unarmed and, because of its lack of speed, served primarily as a support vessel.

Their immediate destination, Kunwego, was the German base almost directly opposite Lukuga, in the bight of Kungwe Bay. The ship and boat arrived there in mid-afternoon, and Odebrecht ordered his men to rest so they would be fresh for the night crossing that would bring them within sight of the Belgian defenses by dawn.

At midnight, Odebrecht ordered the *Hedwig* out onto the dark lake. The air was unusually still. Not a breath stirred the air, and the water was almost flat. Knowing that sound, even a human voice, could carry for miles across such a calm surface, Odebrecht had the helmsman keep speed to five knots, half the speed of which the ship was capable. He ordered the men to maintain silence and forbade open lights.

Odebrecht had no trouble finding his way to Buena Dengwe Point, even in the darkness. The lights of Lukuga shone like beacons, and he simply told his helmsman to steer a bearing that would take them somewhat north of the point. The Ba-HoloHolo spy had not been clear on the *Kingani*'s precise location, and Odebrecht intended to make a single sweep down the coast. He certainly didn't want to spend a lot of time sailing up and down in front of the Belgian guns, looking for a wreck, especially if those guns had caused the wreck.

Dawn found the *Hedwig* a mile and a half off shore and about five miles north of Buena Dengwe Point, cruising slowly south. Odebrecht called for the lookouts to check the waterline near shore for sign of the *Kingani*'s funnel, and he raised his own binoculars to his eyes. He sincerely hoped they found the *Kingani* this trip, for he was tired of chasing ghosts.

2

After Spicer-Simson had been partially surprised by the approach of the *Kingani*, he'd had the telephone wire he'd stolen strung to several Belgian outposts up the lake from Kalemie. At 6:15, the telephone rang in the expedition's communications hut. Tasker took the message then ran to the commander's headquarters.

Spicer-Simson was up and dressed, and Tasker quickly spouted his news.

"The Belgians, sir," he panted. "At Mtoa. They report a German ship steaming south. She's just come round Buena Dengwe Point."

"Hmm," the commander mused, fingering his cigarette holder. "Twenty miles from here. Could they identify the ship?"

"Looks like the *Hedwig von Wissmann*, sir. Shall I roust the men? Call 'em to battle stations?"

"Roust the men, by all means," the commander said with a shrug. "Call them to breakfast."

"To breakfast, sir?" Tasker gave Spicer-Simson a dumbfounded look. "What about the *Hedwig*?"

"I suspect she'll continue down the coast. We'll see." He stuck a fresh cigarette in the holder and lit it with a nonchalant air. "Run along, now."

"To breakfast, sir?"

"To breakfast."

Tasker hurried off. Within fifteen minutes, all the men were in the mess line, but the general excitement kept most of them talking instead of eating. At last, the commander entered, and everyone stood expectantly at attention even before Chief Waterhouse could issue the order. But Spicer-Simson merely waved for the men to stand at ease and return to their meals before he sat at his own table and dug into his food with relish.

"What's he doing?" Dudley, fired up, whispered to Wainwright. "The *Hedwig*'s coming right into our arms, and he's just sitting there."

"It appears that he's eating," Wainwright answered.

"But what about the *Hedwig*?"

"Who can truly sound the depths of our commander?" Wainwright said. "He's brought us this far. Let's continue to have faith in him."

Dudley seemed partially placated, though he couldn't stop squirming.

"Damn, but I wish *Toutou* was in commission."

"You've had your go," Wainwright said with a grin. "I believe it's my turn, now."

A few minutes later, Spicer-Simson swallowed the last of his morning tea, stood, stretched, and said, "Excellent meal. My complements to Cook." He looked around as if expecting the gathered men to burst into vocal agreement, but he was met with dead silence.

"All right, men," he said lazily. "I suppose it's time for action. Clean into fighting rig."

The silent men sat for a stunned second longer, then pandemonium broke loose as they all rose and simultaneously tried to exit the hut. Even the imperturbable Wainwright was caught up in the excited rush, and a mad minute later, the hall was empty of everyone save the support staff, Hanschell, and Spicer-Simson, who was standing stark still, staring at the now vacant doorway.

The commander's face was a flat mask, but Hanschell thought he caught a peculiar look flicker briefly within his eyes. He tried to guess, and not for the first time, what was going on in that inexplicable mind. Was the emotion furtively flickering there fear? For an instant, Hanschell thought so, but even as it vanished, he cast that supposition aside for another—sadness. Before he could fully analyze it, though, the look was gone, washed away by a fierce glee that lit up the commander's entire face. With this look burning his characteristic grin into place, Spicer-Simson headed for the door.

"Commander?" Hanschell's call halted Spicer-Simson just shy of the door. "May I go this time? You might need me."

"If we need you, Doctor, we need you in one piece. I've told you that already."

"Yes, sir," Hanschell said, disappointed nonetheless. "Well, good luck, sir."

"Thank you, Hanschell. I'll keep that in mind." The commander turned to go but paused on the threshold. "Have the natives pray for me, will you old boy?"

"We pray for you constantly, Commander," the doctor assured him. "All of us."

Spicer-Simson nodded, the fierce look replaced for an instant by solemn gravity. Then he was gone, following his men to the harbor.

"What are our plans for the engagement, Commander?" Wainwright asked as soon as Spicer-Simson reached Kalemie.

Spicer-Simson casually removed the cigarette holder from between his teeth and thoughtfully stared up along the edge of the bluff. If it was possible, even more Africans lined the shore than had prior to the engagement with *Kingani*. There were thousands of them, and their ranks were getting deeper by the minute.

The commander shifted his gaze past the end of the breakwater. An eerie calm lay on the lake; its surface was like a sheet of glass, and the air was completely still. Above, a watery sun barely broke through a leaden haze that lay so close to Tanganyika that it was indistinguishable in color and texture from the water itself. The exact level of the horizon was impossible to ascertain; it could have been inches away as easily as miles. But visibility, though distended and uncertain, had not entirely vanished. The *Hedwig* could be seen rounding the headland north of the harbor, though the ship appeared to be suspended in a gray void.

It was not the calm before the storm, but complete limbo.

There were sounds, though. From *Fifi* came the muffled work song of Fundi and his assistant stoker as they fed wood to the fires, and *Mimi*'s engine burst into a throaty roar, sending out smoke that almost instantly blended into the uniform gray of the air and water.

The commander pulled out his watch, consulted it, then said, "We have the *Hedwig* outgunned and outnumbered. We'll close with her immediately, as soon as *Fifi* has steam up."

The crews hurried to their battle stations.

"Dudley, you're with me," Spicer-Simson continued. "Since *Vengeur* isn't ready for action, go ask Captain Goor if he would care to join me aboard *Fifi* for the engagement."

Dudley rushed off and came back a few minutes later with Goor, who was flushed with excitement. He thanked Spicer-Simson for the opportunity to be on hand, saying he'd left his second in command, Lieutenant Wautier, in charge of *Dix Tonne*. *Netta* was still to function as a fuel reserve for *Mimi*, with Sous-Officer Baptiste in charge.

Fifi had Tait at the helm, and Lamont and Berry were minding the engines. Waterhouse was on the forward gun and Murphy on the rear, and two ratings were on hand to feed them shells. A rating manned the Maxim. With Wainwright in *Mimi* were Mollison at the helm, Flynn on the gun, and two ratings—one on the Maxim and one to feed shells to Flynn.

At 7:45, Spicer-Simson's flotilla left Kalemie. *Fifi* and *Mimi* rapidly rounded the end of the breakwater and moved out onto the supernaturally smooth lake, while *Dix Tonne* and *Vedette* struggled along behind, vainly trying to keep up, and *Netta* purposefully held back.

3

Aboard the *Hedwig von Wissmann*, most of the crew were scanning the shallow water off shore for some sign of the *Kingani*. Despite the early hour, the air was muggy with a strange haze. Lieutenant Odebrecht, used to winters of Teutonic proportions, always found the African climate stifling, and he was dressed as usual when out on the lake—only in trousers, undershirt, and boots. Around his neck, though, was the Iron Cross he'd won in the Indian Ocean while aboard the *Königsberg*. That, he wouldn't do without.

They'd been looking for nearly three hours, now, and scanned more than twenty miles of coast without a sign of the *Kingani*. Odebrecht decided it was time to give up the search, particularly given the strange, hazy weather that had significantly reduced visibility. If the weird haze increased, he'd have to cruise closer to shore than was safe in order to spot anything of significance. Besides, they already were entering upper Tembwe Bay, much farther south than where the *Kingani* was reportedly sunk.

Then, at 8:30, the ghost he'd sought for nearly six weeks emerged from the haze. That was his first thought as he saw the ship materialize some 6,000 meters away and steam toward the *Hedwig*.

"It's the *Kingani*!" shouted two of the lookouts almost simultaneously. "We've found her!"

Odebrecht felt his heart leap with gratitude, and he called to the helmsman to head for the lost ship. As the *Hedwig* came about, Odebrecht twiddled the focus knob on his binoculars. It was the *Kingani*, all right—he'd recognize her configurations anywhere. But there was something peculiar about the way she looked, as if the lost weeks had wrought some arcane alteration to her. For one thing, her deckhouse seemed misplaced and her bow heavier....

That's when he saw a muscular gunboat racing out from behind *Kingani*, and his vague but growing suspicions were wiped out in an instant as the terrible truth struck him. The White Ensign was flying from *Kingani*'s mast! The British were on Tanganyika, and they had *Kingani*!

No wonder he had been unable to locate survivors or wreckage—the survivors were in a prison camp and the wreckage had been hiding in that clever little harbor at the mouth of the Kalemie. He didn't know how the British had gotten hold of the *Kingani*, but he realized that he and Zimmer had been wrong to attribute her loss to the Belgian batteries. That nasty little gunboat must have had something to do with it.

He focused more sharply on *Kingani*'s unusually heavy bow, and his heart sank. There was a new gun weighing down her foredeck—a 12-pounder, he judged, but he didn't wait to examine it more fully. He shouted for the helmsman to come about at full speed and flee for Kigoma. He could only hope that the *Kingani*'s deck would prove no match for the recoil from such a disproportionate weapon, for a direct hit would spell disaster for the *Hedwig*.

"Arm the guns!" he yelled, silently cursing the limited reach of the aft Hotchkiss. "Fire if they get within range! Get riflemen on deck!"

Now he could see several other boats emerging from the haze behind the *Kingani* and the gunboat, but his eyes quickly returned to the two lead craft. They were the danger, and the others didn't matter. Only those two.

Up on the bluff, Hanschell had been joined by the rest of the British, who'd rushed up from the harbor to get a better view, and by a large group of Belgians from Lukuga, led by Major Stinghlamber himself. They watched the British leave the harbor just as the *Hedwig von Wissmann*, three or more miles out on the lake, steamed past Kalemie.

And then the *Hedwig* turned north and ran.

"It seems your commander has entered the fray prematurely," Stinghlamber said with ill-concealed satisfaction. "The *Hedwig* has a three-mile head start and a clear run to Kigoma."

Hanschell feared Stinghlamber might be right, though he urged *Fifi* and *Mimi* on with every pulse of his racing heart.

"*Fifi* has a one-knot advantage and *Mimi* even more," he replied, surprised at both the tartness in his voice and the certainty of his words. "Commander Spicer-Simson will run them down."

And then, suddenly, the *Hedwig von Wissmann* vanished into the eerie haze that lay upon the water.

"Perhaps, Doctor," Stinghlamber responded. "But your commander must be able to see his prey to catch her."

Hanschell did not speak but simply watched as, within minutes, *Fifi* and *Mimi* also seemed to vaporize, leaving the sluggish *Dix Tonne*, accompanied by *Vedette* and *Netta*, to waddle after them. It was now 9:30.

The gathered multitude gave a collective sigh of disappointment and strained forward as if to penetrate the mist and distance, but even the muffled roar of *Mimi*'s engine now was lost. A frustrated mutter ran through the crowd, and Stinghlamber himself seemed let down, though Hanschell couldn't tell if that were a good thing or bad.

Suddenly, an electric tension galvanized the gathered Africans, who began shouting and waving and pressing toward the edge of the bluff as if to throw themselves over, like lemmings, to the rocky beach below. Hanschell looked around, bewildered, when the cook tugged at his arm.

"The sky, Doctor!" the cook yelled. "Look at the sky!"

Hanschell followed the man's jabbing gesture, and the electricity took hold of his own soul.

There in the firmament above the lake, thousands of feet up in that realm of haze and deceptive distances, raced three gigantic ships: the *Hedwig von Wissmann*, *Fifi*, and *Mimi*, all magnified ten times their actual size! Unnaturally thick black smoke began to billow from the stacks of both of the larger craft.

"They're pouring oil on the fires for more speed!" Stinghlamber cried in an excited voice as he gazed at the mirages surging across the firmament like clouds made corporeal.

The Africans were stamping and shouting and moaning, and the Europeans—Belgian and Briton alike—were drawn into the frenzy.

It's an illusion, the doctor told himself. Mirages like one sees on the desert, where tricks of light and distance and heat conspire to make cities of sand and castles of imagination. It's only an illusion. It's only an....

The illusions began firing at each other.

4

By 9:30, Odebrecht thought he might have a chance to escape. He knew that the *Kingani* had a one-knot advantage over the *Hedwig*, and that the gunboat probably was even faster, but his long lead and the obscuring haze might work to his advantage. If he couldn't see the ship and gunboat chasing him, they couldn't see him, either. He ordered the helmsman to make a slight adjustment to their heading—not enough to appreciably narrow the gap between the *Hedwig* and her pursuers but enough to throw them off her course.

"Have the stokers pour oil into the furnaces," he ordered. "We'll need every knot we can manage."

The gray smoke billowing from the funnel suddenly belched black with petrochemical soot as the stokers dumped cans of oil onto the wood and fed the wood to the fires. As the flaring oil raised the furnace temperature, the already heady thrumming of the steam engine took on a note of urgency, and there was a slight surge as the ship gained a knot or two.

After more than five minutes of full speed on the altered course, Odebrecht began to relax. Surely they would get away, now. Surely.

At 10:00, a single distant boom interrupted his hopes.

"They're firing!" Mewes exclaimed. "But at what?"

"They're firing blindly," Odebrecht laughed. "They can't see us, and they're...."

A heavy object splashed into the water a hundred yards off their bow and instantly erupted in dull concussion and a geyser of water.

Odebrecht's laugh died in his throat. The shot had been grossly overranged but the bearing was right. Somehow, the damned British had divined his maneuver. He didn't know how, but that hardly mattered. All that mattered now was escape.

"More oil!" he yelled.

Divine intervention on behalf of the British hadn't entered the picture. It didn't need to. As with the watchers on the shore, the crews of *Fifi* and *Mimi* could see a huge *Hedwig von Wissmann* sailing through the sky.

"God!" said Tait at the wheel. "She's bigger than we thought, sir. Are you sure she's not the *Graf von Götzen*?"

"It's some kind of optical illusion," Spicer-Simson snapped. "Look! She's changed course! Fifteen degrees to port!"

Tait, who didn't know a degree from his missing finger, simply steered for the huge image.

"She's poured on the oil!" yelled Dudley as black smoke tarred the sky behind the illusion.

"Lamont!" the commander shouted down the hatchway to the engine room. "Oil! Pour on the oil!"

In seconds, *Fifi* was polluting the air above Tanganyika, increasing the ship's speed by three knots.

Spicer-Simson rushed out of the cabin and up to Waterhouse, who stood staring at the *Hedwig*'s mirage, his mouth open, finger hovering over the trigger.

"I'm not sure I can shoot at that thing," he said lamely as the commander came up behind him. "Sir."

"You'll have your chance, Waterhouse. See, we're closing already." It was true. The illusion had dropped a degree or two toward the horizon and shrunk considerably. "We should be in range in less than half an hour, and maybe by then we'll have a decent visual sighting. Be ready when I give the order."

"Yes, sir."

Spicer-Simson returned to the wheelhouse, which was beginning to heat up from the ship's funnel just a few feet aft. But the commander didn't stay. Instead, he rushed onto the port deck and began waving his arms with wild and peculiar gestures at *Mimi*, cruising some fifty yards away.

At first, Wainwright and his crew failed to notice the commander's semaphoring, for they were busy watching the gigantic *Hedwig* sail away across the sky. The pall of black smoke from *Fifi*'s funnel drew Wainwright and Mollison's attention, and they saw the figure of their commander waving frantically.

"What d'ya think he means by them signals, sir?" asked the helmsman. "I can't make heads or tails of 'em."

"Don't worry, Mollison, no one can. I'm not sure even the commander knows for certain since they seem to change every time he tries them."

He stared hard at *Fifi*, and he could see Spicer-Simson's mouth working, so he yelled at the helmsman, "Best steer for her. I think he's got something to tell me."

"Keep back!" Spicer-Simson yelled to Wainwright when *Mimi* had pulled abreast of *Fifi*. "Hold station astern!"

Wainwright signaled that he understood, and told Mollison to drop back. In seconds, *Mimi* was bouncing in *Fifi*'s wake.

"I think the commander wants all the glory, sir," Mollison commented. "Beggin' yer pardon, sir."

"No offense taken," the lieutenant said. "I have a feeling you're right." Privately, though, he promised himself a piece of the action no matter what the commander wanted. He hadn't dragged the damn boats halfway across Africa and built all those bridges and solved all the other problems to cast away his chance at glory.

True to Spicer-Simson's prediction to Waterhouse, the chase went on for another half hour before the commander judged that the *Hedwig* was in range. It was a raw guess, really, because the German vessel was still a large, though reduced, mirage steaming away from them through the gray haze above the indeterminable horizon. But the commander had grown impatient, and at 10 o'clock, he gave Waterhouse the signal to fire.

Waterhouse jerked the trigger, the 12-pounder boomed, and *Fifi* stopped dead in the water. Behind the ship, Mollison twisted *Mimi*'s wheel, and the gunboat barely missed crashing into *Fifi*'s stern. Water boiled around *Fifi*'s screws as the engines struggled to get the ship moving again, and gradually the ship picked up speed, but by then, the *Hedwig* had gained a few hundred yards.

Spicer-Simson was right in thinking he was within range of the *Hedwig*, because by the time he had Waterhouse fire, *Fifi* had closed to 3,000 yards. Unfortunately, the mirage did not include a sign of the shell striking the lake, so Waterhouse had no idea that he'd over-shot Hedwig and thus no notion about how to correct his aim.

Spicer-Simson seemed oblivious to that fact.

"Fire!" he bellowed again and again, and Waterhouse had no choice but to aim blindly, pull the trigger, and stop *Fifi* dead in the water again and again. By the time he'd fired twenty times or so, the repeated stalls had given the *Hedwig* a chance to considerably widen the gap, and it was escaping.

"More wood!" Spicer-Simson screamed down the hatch at Lamont. "More oil!" He also seemed oblivious to the fact that *Fifi*'s funnel was almost red hot and that the wheel was only a few feet in front of it. Both Dudley and Goor were relieving Tait, now, and the three were trading off at five-minute intervals to keep from roasting alive.

But there was an even worse problem that wasn't readily apparent until *Fifi*'s speed decreased by a couple of knots despite the heat radiating off the stack and the thick black smoke boiling from it. The *Hedwig*'s mirage grew more distant, larger, and fainter.

"Blast it!" the commander raged. "What's happening? We're losing her!"

He ducked down the hatch to the engine room, where he met with a taste of hell. The heat in the engine room was so fierce that every metal surface burned to the touch, and even the air threatened to turn molten. Fundi and his assistant, stripped to loincloths and dancing in front of the furnace like fire demons, sweated as if they'd just emerged from a swim in the lake.

"The air's too still out there, sir!" Lamont yelled over the din of engine, fire, and stokers' work chant. "We've got no draft in the funnel, and she's choking on her own smoke!" He tapped a fogged gauge. "We're losing boiler pressure, and there's nothing we can do about it if we keep pouring on the oil!"

Spicer-Simson, unable to breathe, or even think, in the intense heat, scrambled back up to the deck, and stared at the diminishing image of the *Hedwig von Wissmann*.

"Goddamn it! She's getting away!" he raged, and he kicked the side of the wheelhouse. "Goddamn it to h...."

At that instant, his voice was drowned by a roar that thundered up from behind *Fifi*. The commander flung himself to the rail, staring wide-eyed as *Mimi* surged past, going full out across the glasslike water, chasing the vanishing enemy vessel.

"No!" screamed Spicer-Simson, frantically waving his arm in his incomprehensible semaphore. "Back to station, Wainwright! Get back!" His voice was lost in the thunder of the gunboat's engine, and Wainwright pointedly ignored his arm signals.

"Blast him!" Wainwright yelled to Mollison. "Full speed ahead, or she'll escape!"

Able to travel nearly twice the *Hedwig*'s speed, *Mimi* quickly closed the gap to a little more than 2,000 yards. At this range, the actual *Hedwig* could be seen steaming beneath its dissipating mirage.

"Fire!" Wainwright yelled to Flynn, and the three-pounder spoke repeatedly, stepping shells closer and closer to the German ship.

Odebrecht had watched the gunboat run up from behind, and as its fire crept toward his ship, he ordered his aft gunners to open up. The gunners cranked the Hotchkiss, and the five barrels spun and spit their shells, but the gunboat kept just out of its range, all the while firing shells of its own.

And then, one of those shells exploded against the *Hedwig*'s hull, jarring the ship though doing little damage. But Odebrecht knew that now that the British gunner had the range, he'd soon drop shells onto the deck, and the German wasn't about to let that happen without giving the gunboat a fight. He had to bring the two six-pounders on the bow into play.

"Forty-five degrees to port!" he yelled at the helmsman, then to the gunners on the bow, "Fire as soon as you can bring to bear!"

The forward gunners already had their guns loaded, and moments after the course change, one of them was aimed right at *Mimi*. The gunner fired.

On *Mimi*, Wainwright saw the muzzle flash of the port bow gun. Not being a nautical man, he ignored nautical terms.

"Turn right!" he shouted and was almost flung overboard as Mollison, no nautical man himself, wrenched the wheel.

The German shell geysered the water several yards from *Mimi*'s stern.

"Keep firing!" Wainwright yelled to Flynn as Mollison straightened out behind the *Hedwig*. Flynn heartily obliged now that he had the range.

His next shot crashed into the ship's starboard side, again doing little damage to the steel plating.

"She's turning right!" yelled Wainwright, and he saw the muzzle flash of the *Hedwig*'s starboard bow gun. "Turn left!" he barked, this time bracing himself for the jolting turn.

The shell splashed harmlessly into *Mimi*'s wake.

The *Hedwig* again turned to port and fired. *Mimi* danced to starboard, and the shot missed.

Wainwright laughed and turned to Mollison.

"I believe we can keep this up all morning if they like!"

The zigzag dance went on for half an hour, with the *Hedwig* turning this way and that, pitching shells and trying to blast the pesky gunboat out of the water, and the speedier and more agile *Mimi* zigging when Odebrecht's gunners thought it was zagging.

The constant shift in angle and range didn't allow Flynn to effect any real damage, but the two vessels' wavering course slowed the *Hedwig*'s relative forward progress, and gradually *Fifi*, draft now restored in its stack, began to come up from behind and close the range. When Spicer-Simson was about 3,000 yards out, he ordered Waterhouse to recommence firing.

Waterhouse still could not see his target clearly, and his shells tore up the water well beyond the *Hedwig*. Wainwright watched more than a dozen of them land without effect then signaled for Mollison to come about and head back to *Fifi* at full speed.

He was almost sorry he had, for as the gunboat came alongside the small ship, Spicer-Simson leaned over the rail, face crimson and mouth agape, issuing a barrage of cursing that made the lieutenant's toes curl.

"But Commander," Wainwright tried to break in. "You're a hundred yards over!"

"You bloody tosspot!" the commander fumed. "When I tell you to stay back, I mean for you to stay back!" Then he caught himself as Wainwright's words registered. "A hundred yards over?" He turned to Waterhouse, and yelled, "You're a hundred yards over, you bloody fool!"

Waterhouse, who had heard Wainwright the first time, already had lowered his aim, but he had a greater concern. "We only have three more shells, Commander."

"What! Well, you goddamn better make them count! If the *Hedwig* comes about, we'll be in trouble. And if she gets away, she'll come back with the *von Götzen*, and we'll have lost the element of surprise!"

Waterhouse, praying that he *would* make every shell count, pulled the trigger and braced himself for the shock of the ship coming to a dead stop.

No such thing happened. In his excitement, he hadn't seated the shell properly, and it had jammed in the breech.

"We'll have to wait, sir. If we open it now, it's liable to go off."

"How long, Waterhouse? How bloody goddamn long do we have to wait?"

"Half an hour would be best, sir."

"I'll give you twenty minutes. And if it goes off in your hands, you'll be a damn sight luckier than if you miss with the next one."

While the twenty tense minutes passed, *Mimi* returned within range of the *Hedwig*, harrying the larger ship so it wouldn't turn on the now-defenseless *Fifi*.

Then Waterhouse waved everyone back and gingerly opened the breech. There was the shell, live and scary as hell. He eased it out, warily carried it to the side, and dropped it overboard into Tanganyika's unnaturally placid water. Thanking God that it didn't explode when it hit the surface, he hurried back to his gun.

He inserted the next shell, his next to the last, as carefully as he could under the glaring eyes of his commander. When he'd closed and locked the breech, he looked up at Spicer-Simson, whose expression wondered why he was captain over a ship of fools.

"Fire at will, Mr. Waterhouse," the commander said dryly, and Waterhouse, holding his breath, aimed and fired.

Mimi was close enough to the *Hedwig* that Wainwright could see the shell crash right through the ship's deck and into her engine room. The explosion instantly killed everyone there, destroyed the engine, ruptured an oil tank, and ripped through the ship's side. Waterhouse's next—and final—shot fell right after the first. The second explosion ignited the spilled oil, sending out a gout of flame and billows of ebon smoke that blanketed the vessel.

The *Hedwig* listed sharply to starboard and began sinking at the bow.

German sailors and African deckhands leapt off the sinking ship and swam out into the lake. A few tried to lower the lifeboat, but one of *Mimi*'s shells had ripped it like a sieve, and it sank before anyone could get aboard. Odebrecht, observing naval formality, made sure he was the last to abandon ship, and after pulling off his boots, he followed his crew into the water and swam away from the sinking hull.

It was 11 o'clock, two and one-half hours since Odebrecht had spotted *Fifi* and *Mimi* and tried to run.

5

Fifi and *Mimi* rounded up the *Hedwig*'s survivors, which totaled twelve Germans and eight Africans, before any of them sank after their ship.

Lieutenant Odebrecht was one of those dragged aboard *Fifi*, and after he'd been set on stocking feet, he stared around himself in amazement.

"Welcome aboard HMS *Fifi*," Spicer-Simson said, swaggering across the deck to the prisoners. "I am Commander Geoffrey Spicer-Simson of His Majesty's Royal Navy. You commanded the *Hedwig*?"

"Yes," Odebrecht admitted ruefully. "Lieutenant Job Odebrecht and his crew, at your mercy." The Iron Cross on its chain around his neck looked incongruous against his dripping undershirt. He tried out of habit to click his heels together, not remembering that he'd taken off his boots before leaping off the deck of the burning ship. He glanced ruefully at his half-dressed state then back at Spicer-Simson. "Please forgive my unorthodox appearance."

"Certainly," Spicer-Simson waved it off. "You are surprised at our ship?"

"Very. I had not thought to see her again, especially under these circumstances. What did you call her?"

"*Fifi*."

"*Fifi*?" Odebrecht looked puzzled, then he laughed. "Yes, a joke."

"I never joke about my ships," Spicer-Simson said.

"But *Fifi*? That is the name of a Marseilles whore, not a warship."

"Well, she certainly brought the pox upon you, eh, Odebrecht?"

The German lieutenant, dripping water from his defeat, had no rejoinder for that, nor would he have had time for one, as a loud cry came from the rail. "There she goes!"

All eyes were riveted on the *Hedwig*, her hull wallowing and deck awash in flames. Suddenly, the ship gave a loud gasp of drenched boilers and a huge belch of air, and her bow went completely under, lifting her stern high. She poised there in a graceful dive, suspended between air and water, black smoke billowing like a pennant over the lake. Then with a gurgling sigh, the *Hedwig von Wissmann* slid straight down beneath the surface of Lake Tanganyika. For couple of minutes, bubbles of smoke and a rubble of broken and charred flotsam broke the vanishing swirl left by the sinking ship until all was still as the surface of the lake settled again into its supernatural calm.

Mimi came alongside *Fifi*, and Wainwright clambered aboard.

"Congratulations, Commander," Wainwright said. "Another victory for you."

"And one for you, Lieutenant."

"You're not angry with me, Commander?"

"Of course I am," Spicer-Simson said. "But pleased as well. We'll discuss the matter later. This is Lieutenant Odebrecht. Odebrecht, Lieutenant Wainwright, my second in command."

The two lieutenants exchanged salutes.

By now, *Dix Tonne*, *Vedette*, and *Netta* churned up, and Wainwright suggested quietly to Spicer-Simson that the prisoners be spread out among the several Allied vessels.

"There's as many of them as us, Commander. It'd be a shame if they took back *Fifi* by sheer force of numbers."

"Good thinking, Wainwright. See to it. But leave Lieutenant Odebrecht aboard."

In a few minutes, most of the prisoners had been transferred.

"Well," Spicer-Simson called out. "No sense dawdling. Let's take our victory back to Kalemie."

While those needed to operate the boats took their stations and the rest kept guard on the prisoners, the flotilla started for home.

"Hold it!" yelled Spicer-Simson before they'd gone two hundred feet. "Full stop!"

"What's the matter, Commander?" Wainwright called from *Mimi*'s deck.

The commander, who'd been standing in the *Fifi*'s bow like the figurehead of a returning hero, pointed to the water ahead of them.

"What's that?"

Wainwright directed Mollison to steer *Mimi* toward the flotsam. He fumbled over the side for a minute before drawing a polished wooden case from the water. When *Mimi* was back alongside *Fifi*, Wainwright passed the box up to Dudley, who took it to the commander.

"It's a flag locker," Spicer-Simson said. He opened the case, removed a flag, and shook it out.

"Do you realize what we have, men?" Spicer-Simson asked.

"It's a German flag, sir," Dudley answered.

"Not just any German flag, Lieutenant. "This is a German naval ensign." Spicer-Simson gazed lovingly at the unfurled flag. "Yours, Lieutenant Odebrecht?"

The German nodded.

"Do you realize that this is the first German ensign to be captured in a naval battle in this or any other war?"

"I do not relish that distinction, Commander," Odebrecht said glumly.

"No, I suppose it's natural that you wouldn't. But you won't begrudge me my moment of satisfaction."

"The moment is all yours, sir."

"Yes," the commander said, almost fiendish delight torn across his face. "It is." He jammed his cigarette holder between his teeth and signaled to his helmsman.

"Make for Kalemie!"

The Allied vessels hurried toward home.

Chapter XVI: Bwana Chifunga Tumbo

THE HAZE HAD LIFTED CONSIDERABLY by the time they approached Kalemie, but they could have found the harbor with their eyes closed.

"Is that thunder, sir?" Waterhouse asked looking up at the sky. "I hope we're not in for another blow."

It wasn't thunder, it was the sound of thousands of shouting voices and thousands of assegai spears battered against hide shields. It was the sound of complete and abject adulation that men give only to a god in human form.

What the crews of the returning victors could not know was that while they had fought the enemy on an uncertain horizon, the entire engagement, from chase to battle to sinking, had been continuously projected across the sky in titanic mirages. The Africans were beside themselves in a frenzy of ecstatic worship, and even the Europeans on the shore were caught up in the madness. As the ships entered the harbor, the askaris on the bluff who were armed fired volley after volley from their rifles, while the rest set fire to hastily gathered piles of wood. Meanwhile, a huge contingent of warriors, led by the local Ba-HoloHolo chief and a handful of his prettiest wives, streamed down the path to the harbor.

The boats anchored, and one of the canoes, with an anxious Hanschell aboard, was paddled out to *Fifi*. As they neared, Tait waved.

"Nothing for you, Doctor. We've no casualties, and there are no wounded survivors."

When the canoe was tied to *Fifi*, Hanschell clambered aboard the larger craft.

"My God, Commander!" he said, rushing over to Spicer-Simson. "You were magnificent!"

"We'd never have done it if it hadn't been for Mr. Wainwright," Waterhouse called out. "And a close thing it was, sir. It was our last two shells that got her!"

"Did you see the way he had me goin' this way and that?" Mollison asked animatedly from *Mimi*, anchored ten feet away. "'Turn right!' he says. 'Turn left! Turn right!' And all the time, Flynn pepperin' 'em, and them losin' headway."

Hanschell looked over at Wainwright and saw the lieutenant beaming, his face flushed and bright with excitement.

"The commander was quite angry with me at first," Wainwright laughed. "I hope you've forgiven me, sir."

"You are, indeed, the man of the hour," Spicer-Simson replied, and though Hanschell listened closely for some sign of sarcasm, bitterness, or envy, all he saw was a genuine smile on the commander's face. But Spicer-Simson also looked like he had after the victory over *Kingani*—glassy-eyed, detached, and a trifle unsteady.

"Don't forget Fundi!" Lamont sang out, presenting the stoker, whose smile gleamed from his sweat-streaked, soot-coated face. "I never saw such a head of steam. And when the engine was choking, it was him what figured out what to do!"

By now, *Dix Tonne*, with *Vedette* and *Netta* close behind, churned into the harbor. Within a few minutes, the smaller boats ferried the crews and prisoners ashore.

There, to everyone's complete surprise, Stinghlamber grabbed Spicer-Simson by the shoulders, embraced him, and planted kisses on both his cheeks. And then the Africans descended, shouldering the Europeans aside.

The chieftain threw himself at the commander's feet, and his wives poured sand over his head, symbolically giving his land over to the new deity. Spicer-Simson, always a quick study, acknowledged the gesture by grasping a handful of the sand heaping the chief's head and tossing it into the air before proceeding toward the path to the bluff, leading his prisoners and victorious crew. Before him, the awestruck multitude melted to the ground, moaning in obeisance, only to rise after he'd passed and press after him for one more glimpse of the man who could leave the shore in tiny ships and wage a huge and glorious battle of the gods across the sky.

If the commander was surprised by the adulation, he didn't let it show on his face or in his stately walk through his adoring subjects. Even Hanschell, who realized that freak optics on the lake had magnified the battle all out of proportion, felt the irrational urge to prostrate himself in Spicer-Simson's path and pour sand over his own head.

2

A call from Tait, who was handing the prisoners over to the Belgians, brought Hanschell back to reality. It seemed that one of the prisoners was wounded, after all.

Tait came over, leading Odebrecht, who was helping one of his seamen. The man, face pale, clutched a bloody hand to his chest. Both Germans saluted Hanschell, the seaman using his uninjured hand, and while Tait returned Odebrecht to the Belgian guards, Hanschell, accompanied by Eastwood, took the seaman to the hospital hut. After cleaning the man's hand, Hanschell saw that it had been severely damaged by shrapnel, and he knew that he'd have to amputate at least two of the fingers and very possibly a portion of a third.

"I'm going to have to operate immediately," he said to Eastwood. "But my German's not good enough to tell this fellow. All I can get is that his name is Kasemann. Go get Lieutenant Odebrecht. He can translate and calm the fellow's fears."

Hanschell made Kasemann as comfortable as possible, though it was obvious the man was in considerable pain. Luckily, Eastwood returned in just a few minutes, Odebrecht in tow.

Hanschell explained what he had to do, and while Odebrecht relayed the information to the seaman, Hanschell showed Rupia how to use the chloroform. As Hanschell opened the bottle, he realized that the seal was still intact. This was the first time during the entire expedition that he'd had to use it.

Soon the man was under, and Hanschell performed the surgery. When he was finished, Eastwood and Odebrecht put Kasemann into a camp bed, then Eastwood went to get an askari to post guard.

While Eastwood was gone, Hanschell turned his attention to Odebrecht and saw, for the first time, how drawn and dejected the lieutenant seemed. And no wonder.

"How do you feel?" Hanschell asked, sitting the German down.

"I am in much pain," the lieutenant said, touching his head.

A quick examination showed no sign of wound or concussion, but Hanschell could only imagine the kind of headache the man must have. He gave the lieutenant a packet of aspirin and a glass of brandy. Before the glass was empty, Eastwood returned with the askari guard and a piece of paper.

Although Odebrecht could speak some English, he could not read it, so Eastwood did the honors.

"Ah," Odebrecht said wistfully. "My parole."

"Sign here," Eastwood indicated, and when Odebrecht did, Eastwood told him, "Commander Spicer-Simson requests your presence at evening mess, if you feel up to it."

Odebrecht glanced at the wounded seaman on the bed.

"He'll sleep until morning," Hanschell assured him. "No sense in you staying here to watch over him. I'm sure you could use a hot meal."

Odebrecht still looked hesitant. "I am afraid I cannot." Odebrecht spread his hands, palms up. "I am not properly dressed. My shirt was lost in the ship and my boots in the water."

"That's no difficulty. You're about my size. I have a spare uniform tunic and boots. You're welcome to them."

"Your offer is accepted." Odebrecht smiled ruefully at the doctor. "Especially the boots."

The doctor fetched a shirt and an extra pair of boots he had yet to wear, and soon Odebrecht had cleaned up and donned the loaned clothing. Hanschell noticed that the German lieutenant made a point of wearing the Iron Cross on the outside of his tunic.

"It is the only mark of rank I still have," Odebrecht explained. "Is there time to visit my men?"

"Eastwood will take you," Hanschell told him. To Eastwood, he said, "Bring him back here after he sees his men. Another dose of aspirin before dinner wouldn't hurt him."

No sooner had they gone, than Father Jacques showed up in a high state of agitation.

"Can't you speak to him, Doctor? Make him see reason? The man is walking sacrilege. I've labored for years to bring Christ to these godforsaken people, and your commander has not only destroyed all my efforts, but he relishes his role."

"The commander and the others went out to battle," the doctor reminded the missionary. "They risked their lives, and their limbs." He gestured to the German's bandaged hand. "Can you who worship God or I who worship science say as much?"

"At least we look to higher powers," the missionary said. "But the commander worships only himself." He threw his arms into the air and looked toward the ceiling, or rather, through it. "The greater his hubris, the greater will be his downfall."

"Perhaps," Hanschell said coolly. "But did you not witness the battle in the sky?"

"I saw it. A trick of the light and atmosphere."

"No doubt," the doctor agreed, "but a most fortuitous one, wouldn't you say?"

"What is your meaning?"

"What power could have arranged such a display at such an exactly appropriate moment?"

"If you mean that God did that evil work, Doctor, you are more sadly gone in your worship of science than I realized. Any sane man could see in it the hand of Satan."

"I think that we are entering a world where sanity is changing its parameters," Hanschell said. "Our European notions of right and wrong do not apply here any more than does our sense of propriety."

"It matters little what you think," Father Jacques said, his lip curling. "Those who do the devil's work will rot in hell."

"I only know that all of us will eventually rot, even the God-fearing. And speaking of rot, you'll have to excuse me, Father. I have to dispose of these." He waved the shallow dish containing the remains of German seaman's fingers. "If I don't, Flynn will come around looking to add them to his collection."

Father Jacques was obviously puzzled by this last statement, but his indignation overcame his curiosity, and he huffed out of the hospital. After he'd gone, Hanschell took the digits outside to Rupia's little fire pit, where he spent the next hour burning them to ash.

3

Darkness was falling when Hanschell raked out the last ashes to be sure all the remains of the remains were burned. While he'd squatted there, he'd watched in amazement as the hillsides and bush for miles around lit with fires that cast wild and strange shadows of dancers. The air was so alive with drums and chants and excitement that the pervasive sounds of the bush was completely drowned out. The expedition's African contingent—already considerable after the victory over *Kingani*—seemed to have multiplied many times over in just a few short hours.

While Hanschell took it all in, Eastwood returned with Odebrecht, and the three of them walked over to the mess hut.

Spicer-Simson and the other officers were there, eating a simple meal of catfish, stew, and coffee. The commander was in one of his grandiloquent moods, but if Hanschell expected boisterous celebration from the officers and men, he was disappointed. The pendulum of elation brought by the battle and victory had swung the other way, leaving a quiet, almost exhausted air hanging in the room.

As Hanschell and Eastwood came in with Odebrecht, everyone except Spicer-Simson stood up.

"Commander Spicer-Simson, here is Senior Lieutenant Job Odebrecht, on your invitation, to join our mess."

Odebrecht bowed stiffly and clicked the heels of his borrowed boots together sharply.

Then the three new-comers sat and were served by the mess staff.

After a few minutes, Odebrecht turned to Wainwright and said, "I think your commander owes you a great debt." He smiled wanly. "Not that I regret not having blown you out of the water."

"It wasn't for lack of trying," Wainwright said, and they all laughed, even Odebrecht. "I can't believe it," Wainwright said, turning to the doctor. "They say the action was projected on the clouds."

"Yes," said Hanschell. "It was huge."

At Odebrecht's blank look, Hanschell described the optical effect that caused the action to be projected in the sky.

"A marvelous engagement," Spicer-Simson said when the doctor had finished.

"For you, Commander," Odebrecht said.

"Yes," Spicer-Simson replied, and there was stiff silence.

"I must congratulate you, Commander, for your daring in bringing your gunboats across Africa and taking *Kingani* and *Hedwig von Wissmann*," Odebrecht said, breaking the silence. "At this time yesterday, I wouldn't have thought it possible. We had no idea that the Allies had armed vessels on the lake."

"Surely you must have heard something of our expedition," Wainwright said. "It wasn't a well-kept secret among the natives."

"We heard that there was an expedition coming up from the south, yes, but we had no idea of who it was or its purpose. The color of your uniforms misled us. We only heard that you finally arrived at Lukuga with two large canoes, and later, our native spies told us that you were trying to build a railway bridge across the lake. It seemed absurd."

"It would have been, had it been true," Spicer-Simson amplified. "That was our launching device. We let the gunboats out into the lake via railway car."

"Ingenious," Odebrecht said.

"Thank you," Spicer-Simson replied. "My own idea."

Hanschell smiled to himself as he saw a resigned but tolerant expression flicker in Wainwright's eyes.

The meal was soon over, and the officers and ratings dispersed to their huts.

Hanschell took Odebrecht back to the hospital and gave him a bed near the still-sleeping seaman.

"How are you?" the doctor asked.

"Still there is much pain," Odebrecht said.

Hanschell prepared a light dose of morphine, and injected it into Odebrecht's arm.

"Sleep, Lieutenant. You'll feel better in the morning."

4

The evening was not so celebratory for Captain Zimmer. He, too, had retired early, but he could not sleep. What could have happened? Where was the *Hedwig von Wissmann*? All he knew was that the *Graf von Götzen* had arrived at Kunwego by noon, expecting to find the *Hedwig* with some news, good or bad, or the missing *Kingani*, but all he'd found was the *Wami* waiting alone.

The lieutenant in charge could only tell him that *Hedwig* had not yet returned, but the *Wami*'s crew had heard heavy gunfire from the direction of Lukuga between nine and eleven o'clock.

And then silence.

Zimmer ordered the *von Götzen* to return to Kigoma, and as the massive steamer ploughed through the water, he pondered how he could explain to von Lettow-Vorbeck that he had lost two ships and didn't even know what had happened to them. By morning, however, he did. One of the radiomen at Kigoma intercepted a Belgian transmission that said the *Hedwig* had been sunk in an engagement on the lake.

So, he thought. There had been a battle, and the *Hedwig* was gone. Undoubtedly, something similar must have happened to the *Kingani*. The Belgians must have resurrected the *Alexandre Delcommune* or secretly brought that other ship from Kabalo to the lake and assembled it.

Zimmer cursed the loss of the ships, he cursed the capture of Lieutenant Rosenthal, whose daring nighttime sorties could have supplied advance information, and he cursed this new god of the Ba-HoloHolo, this Bwana Chifunga Tumbo, who had so won over the cannibals that they no longer supplied Zimmer with accurate information.

If he had anything to be thankful for, it was that he still had the *von Götzen*, and it now was battle-ready, with the two *Königsberg*'s 105s in place along with the two 88s. And there also was the *Adjutant*, a 250-ton tug captured from the British nearly a year before at Rufiji. Once the British blockade of the East African coast went into effect, the *Adjutant* had slipped by the British and was presently anchored in Dar-es-Salaam, doing nothing. She could be broken down and shipped to Kigoma. Huebner, the naval architect who'd built the *von Götzen*, assured him the tug could be ready by early March, and Zimmer had a spare 37 to mount on her. He felt somewhat better at the prospect of rebuilding his small fleet, but he needed something more to assure victory. He'd already lost two ships, and when he went up against the Belgians, he intended to shell everything they had afloat to matchwood and scrap iron. He wondered if he could get hold of another big gun. Surely von Lettow-Vorbeck would understand the urgency.

Zimmer nervously resolved to wire his colonel with the request, but for now, he realized he had only one choice, and his mood darkened again. He'd have to send the *Graf von Götzen* under the noses of the Belgians in search of the survivors of *Hedwig von Wissmann*. This time, though, he would personally lead the search, and he would be ready. If the *Alexandre Delcommune* or *Baron Dhanis* dared to trifle with his dreadnaught, he welcomed the opportunity to send them to the deep, deep bottom of Lake Tanganyika.

He would begin the search at first light.

5

"Your seaman is resting. He should be all right if infection doesn't set in."

"I'm glad to see that your expedition has someone of professional caliber," Odebrecht said, his somewhat condescending tone masking an obvious sense of disgruntlement.

"I'm not sure what you mean," Hanschell said. "Are you implying that Commander Spicer-Simson and the others are less than professional?"

"Reserve officers," Odebrecht said. "It's a disgrace to be defeated by amateurs."

Hanschell laughed, remembering that's what Lieutenant Freiesleben had called them before he'd lost all his money. "You believe His Majesty would send amateurs on an expedition of this importance? Let me assure you, Lieutenant, that nothing could be farther from the truth. Commander Spicer-Simson is a career naval man and a commissioned officer in the Royal Navy. As for his men, nearly every one of them was handpicked for his expertise. You may have reason to regret defeat, but you have no cause for disgrace."

The sourness fell away from Odebrecht's face like sloughed skin, and he stood up.

"I had not realized. Thank you, Doctor. You have healed me as well as my injured seaman."

"No thanks are necessary," Hanschell said. "I defend my commander as easily as I do my patients or England."

Hanschell might have said more, but a sudden shouting drew the two men outside the medical hut.

"Ship ahoy!" the doctor heard as he stepped into the sunlight.

The boat crews, lead by Wainwright and Dudley, were making a mad dash for the path leading to the harbor. Hurrying to the cliff side, Hanschell and Odebrecht could see what the excitement was all about. Nearly four miles out on the lake, steaming south along the coast, was a ship.

A real ship.

"The *Graf von Götzen*?" Hanschell asked, looking at Odebrecht.

"The *von Götzen*," the German lieutenant affirmed soberly.

My God, Hanschell thought. She's come right into our grasp, just like the other two.

"She is looking for you," he said aloud.

"Yes, as we looked for *Kingani*."

A roar sounded from the harbor as Wainwright's crew fired up *Mimi*'s engine, and *Fifi*'s stack belched smoke as Fundi worked up a head of steam. Hanschell saw Odebrecht stiffen, and he wondered if the German was thinking of his own defeat or that his sister ship would be attacked as *Kingani* and *Hedwig* had been.

Hanschell seriously wondered, himself. The *von Götzen* looked as if it far out-classed Spicer-Simson's flotilla. It certainly presented an impressive image, even at four miles.

The excitement in the harbor reached a fever pitch, but gradually, the German warship worked its way south. And still, Spicer-Simson had not stirred from his hut.

A figure detached itself from the hubbub below and raced toward the path to the top of the bluff. It was Dudley, his expression both excited and worried. The lieutenant passed by the doctor and Odebrecht as if they weren't there, making straight for Spicer-Simson's hut, which he entered with barely a knock.

Spicer-Simson emerged, trailed by Dudley and carrying a pair of field glasses, which he trained on the ship. Dudley said something to him. The distance was too great for Hanschell to hear the words, but the lieutenant's gestures were voluble enough—he was obviously urging the commander to attack at once. Hanschell left Odebrecht and edged closer to the commander's hut.

"Sir?"

Spicer-Simson, still intently watching the German dreadnaught, appeared not to hear Dudley.

"Commander, sir?" the lieutenant said louder.

Spicer-Simson lowered the binoculars, but didn't take his eyes from the lake.

"It's the *von Götzen*, sir."

"How perceptive of you, Dudley."

"Wainwright's making the boats ready, sir."

"Is he, now?" Spicer-Simson turned at last to look at the lieutenant. His face was impassive, but his eyes were hard. "And who gave that order, Lieutenant?"

"Well...no one did, sir. We just thought...."

"I'll do the thinking, if you don't mind," the commander interrupted, though not loudly. He raised the binoculars and continued to watch the *von Götzen*. Several minutes passed.

"Aren't we going to attack, sir?"

"I think not, Dudley."

"But, sir...."

"Are you arguing with me, Lieutenant?"

"No, sir. I just don't understand."

"Then take a look." Spicer-Simson passed the binoculars to Dudley, who trained them on the distant ship.

"What do you see?"

"I see the *von Götzen*, sir."

"Look more closely, Dudley."

"It hasn't changed, sir," Dudley said after a long look. "It's still the *von Götzen*."

"Precisely."

Dudley lowered the binoculars and looked at Spicer-Simson, his brows wrinkling with puzzlement.

"I still don't understand, sir."

"Then let me enlighten you," Spicer-Simson said in a tone laden with sarcastic patience. "When you look at the *von Götzen*, you see only another victim ripe for the taking. But I see something different. I see two hundred feet of steel hull armed with four guns, two of them four-inch. Do you have any conception of the range and firepower those guns have?"

Dudley shook his head.

"That's because you're a land crab at heart, Dudley. Well, I will tell you: they range twice that of *Fifi*'s gun. Do you know what Napoleon said about guns, Lieutenant?"

"No, sir."

"He said, 'God fights on the side with the best artillery.' Furthermore, the *von Götzen* won't stall when she fires, and she's as fast as *Fifi*. We'd never even get close."

"But what about *Mimi*, sir? She's quick. Wainwright dodged *Hedwig*'s shells."

"*Hedwig*'s guns were both in her bow, and she could only bring one to bear at a time and had to make obvious maneuvers to do that much. *Von Götzen* won't have that difficulty with two 88s astern."

"We can do it, sir. I know we can. *Mimi* can get in close and...."

"And do what? The three-pounder did little more than dent *Hedwig*. Even if *Mimi* can get within point-blank range, it's doubtful her gun will do even that much against *von Götzen*'s armor. And then everyone aboard will be easy targets for small arms fire or *von Götzen*'s smaller guns firing in tandem."

"Isn't there anything I can do to persuade you, sir?"

Spicer-Simson stared Dudley straight in the eyes, and the lieutenant flinched at the anger he saw bristling there. But he also thought he caught a shadow of something else lurking in the commander's gaze that he could only define as a petulant stubbornness.

"I haven't come halfway across Africa and won two decisive battles to be persuaded by a tyro lieutenant to throw it all away in a vain attempt to magnify his or anyone else's glory. Now, proceed to the harbor and tell Wainwright to have the crews stand down."

Dudley remained rooted, too uncomprehending and shocked to move. Then abruptly, Spicer-Simson turned on his heel and entered his hut, closing the door behind him.

Dudley, face a red mask of scowling frustration, shoulders rigid with tension, stalked off the veranda. He went by Hanschell and Odebrecht, again without acknowledging them.

"...goddamn, pigheaded, skirt-wearing...," was all Hanschell heard as the lieutenant approached and passed them, going straight for the path to meet Wainwright and Goor, who were on their way up.

Hanschell looked at Odebrecht, and the German lieutenant looked back, his face set in an impassive mask, though contempt touched his eyes. Abruptly, Odebrecht turned and walked stiffly back to the medical hut.

"What's the news," Wainwright asked as Dudley approached. "The *von Götzen*'s already disappeared around the point. When's the commander coming?"

"He's not."

"What?" Wainwright waved his hands helplessly in the air.

"He says *von Götzen*'s too much for us. Too much steel, too much firepower. We're not going out after her."

"Shit!" Wainwright spat. "What the bloody hell's wrong with him?" Then he looked helplessly at the other man. "I guess we'd better give the bad news to the crews."

6

From the bluff, Hanschell watched the activity in the harbor quiet to a definite air of despondency. Soon, the men began ascending the path to the camp, and unable to do anything—unable even to understand what had happened—Hanschell hurried back to his hut. The look of contempt that lurked in Odebrecht's stare was hard enough to endure, and if a defeated foe had that in his heart, how had Spicer-Simson's refusal compromised him in the eyes of his own men?

Thankfully, Odebrecht remained in the room with the injured seaman because Hanschell didn't think he could face him. After pouring a stiff dose of medicinal brandy, the doctor sat in his chair and contemplated the eight-inch effigy of the commander that sat against one wall.

"Here's to you, sir," he said, raising the glass and taking a bitter swallow.

But even if he was still feeling disgruntled about the affair, he had to admit that maybe the commander was right. The *Graf von Götzen* was tremendously larger than anything Spicer-Simson had at his disposal. It was a real, honest-to-god ship, not a gunboat or steam yacht. *Fifi*'s gun might be able to inflict some damage if *Fifi* could get close enough, but *Mimi* and *Toutou*, even together, would be worthless—like children with popguns going up against an elephant. And while *Vengeur* eventually might be mounted with a gun big enough to damage the *von Götzen*, it remained pitifully underpowered. Until its new engines were installed, it couldn't engage the enemy.

Hanschell shook his head, unable to resolve his emotions and rationalizations. But he was saved from drowning them in a second glass of brandy by a knock at the door.

"Doctor Hanschell?"

It was Lamont and, behind him, Fundi.

"Look at him, Doctor," Lamont said, pushing a reluctant Fundi forward. "And we thought we had no wounded on our side. Didn't say nothing, the brave beggar. And him ready to stoke after the *von Götzen* without a word of complaint."

"Lie down here," Hanschell urged the stoker. Fundi grinned shyly, but any joy that might have been in the smile was leeched out by the obvious pain in his eyes.

The stoker's hands, forearms, shins, and feet were mats of raw blisters.

"It was hot in that engine room," Lamont said. "God-awful hot. I was burning, and I was six feet away from the furnace. But he was right on top of it, and he just kept on stokin' and pouring on the oil and not giving an inch the whole time we chased *Hedwig*. And he was going to do it all again without a whimper. I only just now saw, when we came out of the engine room. You got to do something for him, Doctor."

"These are bad burns, but I don't see any charred flesh," Hanschell said. "He'll be in pain, but as long as we can keep the infection down, he should be all right in a few weeks. He's a young lad and strong. He'll heal well. Here, stay by him while I prepare a salve."

"Thanks, Doctor. Hear that, Fundi? Doctor Hanschell will fix you up. He says you'll be fine."

Fundi smiled wistfully.

"Thank you, Bwana Lamont. Fundi sorry he cause trouble."

"You silly bastard." Lamont patted his arm. "I couldn't do without you."

7

Odebrecht and the rest of the prisoners were shipped off the next day, but not before the German lieutenant thanked Hanschell for treating Kasemann.

"Think nothing of it," Hanschell said. "I'm just glad it wasn't worse."

"You also have been most considerate in letting me stay here," Odebrecht said. "It has been far more comfortable than a Belgian cell."

"Don't think I did it for you, Lieutenant," Hanschell laughed. "I did it as much for me. Refined company is scarce in this part of the world."

"Yes," Odebrecht said, extending his hand. "Perhaps we shall meet again after the war and under more favorable circumstances."

"I would like that," Hanschell said, shaking with him. "Best of luck."

And then Odebrecht and Kasemann were gone, leaving Hanschell with Fundi as his only patient. But Hanschell soon discovered that wasn't a bad thing. The stoker was an ingenious man with a clever mechanical sense and considerable energy. Even suffering as he was from his burns, he spent the afternoon puttering about in the back of the medical hut while Hanschell treated a number of local patients. When the doctor finally took a break, he discovered that Fundi, under Rupia's direction, had fabricated a small stove with a flue to take fumes and smoke through the hut's back wall.

"Doctor need hot tools," Rupia explained, pointing to the makeshift autoclave. "Need good fire. Steady fire. Fundi fix."

"Yes," Hanschell nodded approvingly to the stoker. "Now Fundi need rest. Sit down, so I can treat your burns."

"Not feel so bad," Fundi said as Hanschell smeared burn ointment on his arms, hands, and legs. "Bwana Doctor good doctor. Not like doctor in village."

"You have a doctor in your village?"

"Yes. Kabaka. Him dance and make smoke and noise, but Fundi think burns still burn."

"Yes," Hanschell chuckled. "Very likely."

8

"We have again received accolades from home," a beaming Spicer-Simson informed the men at parade the next morning. "The message arrived last night from Lord Admiral Sir Henry Jackson, himself. I quote: 'I doubt whether any one tactical operation of such miniature proportions has exercised so important an influence on enemy operations.'"

The men cheered, and the commander let the sound linger before holding up his hand for silence.

"On myself," he continued, "Sir Henry bestows the Distinguished Service Order." More cheers that lingered a trifle less longer. Most of the men were still upset that they hadn't chased down the *von Götzen* when it put in an appearance the day after the battle with the *Hedwig*.

"But I am not the only one to be decorated. Upon Lieutenant Wainwright, for faithfully executing my orders to attack the *Hedwig von Wissmann* when *Fifi* stalled, is bestowed the Distinguished Service Cross. And last but not least, a Distinguished Service Medal goes to our gunners, CPO Waterhouse and PO Flynn, and to our two engineers, Lamont and Berry, who nearly burnt up in their own engine room attempting to put more fire in *Fifi*'s belly."

The men laughed as Berry grinned sheepishly. But Lamont just wrinkled his nose, a pained expression on his face. After Spicer-Simson dismissed the men, everyone hurried off to mess except Lamont. Instead, the seaman strode purposefully toward the path leading down to Kalemie.

No one seemed to notice until near the end of the meal, when Magee wanted to gather the Distinguished Service winners together for a group photo.

"Where's Lamont?" Wainwright asked.

"I think he's down in the harbor, sir," Berry answered. "He was grousing about something."

"How could he grouse about winning a bloomin' medal, is what I'd like to know," Waterhouse said.

"Let's go get him!" someone shouted, and within minutes the mess hall was empty as the entire British contingent, Spicer-Simson included, headed for the path to the harbor.

The found Lamont sweating over the forge on the beach.

"Lamont!" Spicer-Simson snapped as they came up to the forge. "You can't leave a victory celebration like that. It's not good form."

"Sorry, sir," Lamont said hastily, straightening from his work.

"What are you doing there that's so blasted important?"

"A medal, sir. I'm striking a medal."

"What?" The commander laughed. "You can't wait until you get the King's own delivered into your hands?"

"It's not for me, sir."

"For whom, then?" Spicer-Simson asked, surprised

"Fundi, sir."

"Fundi? The native stoker?"

"You said something about me and Berry withstanding the fires of hell, sir. Well Fundi was the man stokin' 'em. And you should see how he's burned for it. Ask Doctor Hanschell, who's treatin' him. If me and Berry deserves medals, Fundi does, too. I just now finished it."

"Let me see what you've done," Spicer-Simson ordered. Lamont held up a six-inch disk of brass with a hole in the top, and the commander read aloud what Lamont had engraved: "HMS Fifi, 9 February, 1916." He flipped the medal over and recited, "Fundi, RN" Holding the medal high, he declared, "It appears that even an engine room artificer can out-think an admiral on occasion." The gathered men laughed, and Spicer-Simson turned to Waterhouse. "Have the men fall in for formal parade. Lamont, you and Berry get Fundi from the medical hut and bring him to the parade ground."

And so it was that half an hour later, at formal parade with the entire British contingent standing at attention and saluting, Spicer-Simson bestowed the most unique medal in the history of the Royal Navy on an African stoker he'd captured from an enemy vessel and promoted to his own.

The gathered Africans shouted and clapped their spears against their shields as Fundi, grinning widely and seeming to forget his wounds, limped along the British ranks, shaking everybody's hand.

9

Once the celebrations ended, however, attitudes around camp began to register the frustration the men felt. Spicer-Simson might have heard the mutterings himself, but it was Hanschell who bore the brunt of them. Typical were the comments of PO Flynn when he came in for his regular examination and dose of quinine.

"We can't understand why the commander didn't do anything," Flynn said. "There she was, big as daylight. We could have gone after her. Her big guns don't scare anyone, sir. We wouldn't have been the ones to sink— we were on a winning run! But I suppose our luck will change now."

"Don't be too certain of that, Mr. Flynn. Think of all the times since we began that things looked hopeless."

"There *were* times like that, sir."

"Yes," Hanschell said. "Remember the time the boat and traction engine fell through the bridge?"

"And then the engine fell over trying to get up that far bank," Flynn laughed.

"Or the Mitumba Mountains. Steep as all hell, but we got across them."

"Thanks to the Old Loco Driver," Flynn said. "He's saved the day more than one time before he helped sink the *Hedwig*."

"Do you know what I think, Mr. Flynn? I think the commander probably made a wise decision not to go after *von Götzen* right away."

"How's that, sir?"

"Right now, all he's got are *Mimi* and *Fifi*. But in a week, *Toutou* will be repaired. The *Vengeur*'s new engine will be installed soon, and *Baron Dhanis* will be ready to sail in a month or so. She's 600 tons and armor plated, too, and she'll mount several large guns. With them, we'll be more than a match for the *von Götzen*."

Flynn brightened. "You may be right, sir. In a month, the commander will have more than a flotilla; he'll have a regular navy."

"That's right, Mr. Flynn. You wait and see. The commander will act soon enough."

"If only he didn't seem so distant, these days," Flynn said. "The men might have more confidence. But he's barely said a word to anyone since we sank the *Hedwig*. Even to Mr. Wainwright, though when he does, he's ever so polite. More than he used to be, yelling, 'Mr. Wainwright this,' and 'Mr. Wainwright that.'"

"I'm sure Lieutenant Wainwright knows how much the commander appreciates him for all he's done. As you said, we probably wouldn't be here without him."

"Yes, sir. That's a certainty. Well, I'd better get back to my duties. Thank you for the quinine."

"Quite right, Mr. Flynn. Oh," he stopped the petty officer as he was heading for the door. "You'll spread the word about the commander's new navy, won't you?"

"I'll do my best," Flynn said, "But it won't be easy. The men are feeling pretty low about him."

Then he was gone, leaving Hanschell to ponder the effigy of Spicer-Simson sitting against the wall.

10

"It's happening," Spicer-Simson said. He'd just come through the hospital door, waving a piece of paper.

"What, sir?" Hanschell asked.

"Now that we've cleared the lake of the *Kingani* and the *Hedwig*, the Allied high command has decided it's time for the land assault against German East Africa. The Belgians will invade around the north end of the lake, through Rwanda, and combined British forces will move in from Kenya and Rhodesia."

"That's wonderful news!"

"Is it?" Spicer-Simson looked glum. "What about the *von Götzen*? They just don't seem to understand. She's a formidable foe, but they know nothing of her, though I've tried to tell them often enough. You know what they said? They said that if I could take care of two enemy ships, surely I could take care of one more. One more! That one more is more than I've got put together, yet they insist we go after her."

"Well, then," Hanschell said. "That's what you should do."

"You, too, Hanschell?" the commander asked in the same tone that Julius Caesar must have used with Brutus. "Taking sides against me?"

"You know better than that. But you can't just sit and do nothing."

"As I have told the men a thousand times, the *von Götzen* is too powerful for our boats. If we'd managed to take *Hedwig* without sinking her, perhaps things would be different. But *Fifi* and the gunboats cannot possibly conquer a vessel the size of the *Graf von Götzen*. And matters will soon grow worse. The African prisoners revealed that the Germans are bringing in a second ship from the coast, though not as large as the *von Götzen*. That will give them two, when the first was too much."

"The men think we can do it."

"Bah! What do they know?"

"They helped you take the *Kingani* and *Hedwig*. That should count for something."

"I'm not sending out boys with slingshots against a Goliath just to let them satisfy some vaunted but stupid sense of honor and glory," Spicer-Simson snapped. Hanschell noticed that the commander nervously twisted at something on his finger. It was the dead Lieutenant Jung's ring.

"This is war, Commander."

"Now you're an expert on war, Doctor?"

"Expert enough to know that we are called on to take chances with our lives and that every victory is risky."

"I'll not risk my men futilely."

"That's not your decision to make."

"Isn't it?" The commander's voice was harsh.

"No. You're last in line. First is the Allied high command, and then it's the men themselves. They volunteered, and they know the dangers. And their duty. Those men came here precisely to do what you won't let them do, and none of us can understand why."

"I had no orders for *von Götzen*," Spicer-Simson said. "I was to take the *Kingani* and the *Hedwig von Wissmann*, but the Admiralty knew nothing of the *von Götzen*. If they had...."

"If they had," Hanschell interrupted, "they might not have sanctioned the expedition, and you would not have had your chance."

"A chance to hang myself for them one more time? They loose me on Africa and expect me to fend completely for myself and my command.

The locos were late, the oxen were late, we practically had to turn pirate to get down the Lualaba, and we arrived only to find that our supplies hadn't, and we had to go all the way back to Fungurume for them. And when we do get here, not only am I forced to fight against our enemy, on whom we have such pitiful intelligence that we aren't even aware of their greatest weapon until we are under its guns, but I have to battle the bloody Belgians, too. I had to take what I needed, when Stinghlamber should have been glad to give me all the assistance I requested."

"You can't let petty pique with Major Stinghlamber control...."

"It is not petty! He tried to usurp my command!"

"I remind you he was here first in a protectorate of his own country...."

"Protectorate! Ha!"

"...being attacked by his own ally—you!—after having staved off German attack from sea and land and seeing his men killed for eighteen months. Facing von Lettow-Vorbeck and having precious few resources, what would *you* have done?"

"You saw what I did! Against all odds, I accomplished what they sent me to do. It was nearly impossible. Frankly, Hanschell, I nearly broke several times just getting here."

"I never noticed."

"I couldn't let anyone see. I had to keep them going because we never received proper support from the Admiralty. After all I've done for them, to have them treat me like this. Maybe it's because I succeeded. That's the crux of it, isn't it? I succeeded. They can't deny that. I succeeded beyond their expectations, certainly, but perhaps even beyond their comprehension. And look how they reward me: denying me the equipment to fully vanquish the enemy in my hour of glory and promising a flogging when I can't because the rules are constantly shifting in the enemy's direction and favor. I need proper equipment!" Spicer-Simson pounded his hand with his fist, but the droop of his shoulders belied the force of the exclamation.

"Yes, Commander. You've proved yourself to them and to the rest of us. But the *von Götzen* still remains to be sorted out."

"I'll think of something," Spicer-Simson said, continuing to twist Jung's ring. He looked at Hanschell, a haunted look in his eyes. "I'll think of something."

11

But it didn't matter what Spicer-Simson said to Hanschell because his actions—or inaction—spoke louder than words to the men. The *von*

Götzen came and went several more times during the next few weeks, and still the commander did nothing, and the men's disappointment began to fester into discontent.

The commander's stonewalling was hard on all of them but especially so on Dudley, whose early success had left him hungering for more action. Day by day, he'd grown increasingly agitated, particularly on the occasions that the *von Götzen* was spotted steaming down the coast, and as his frustration mounted, it was only a matter of time before his highstrung temperament finally came to a head.

It happened while the lieutenant was supervising a group of African workers in strengthening the northern breakwater. One of the workers apparently wasn't working hard enough to suit Dudley and then had the temerity to make a comment about Dudley's parentage, not realizing that Dudley overheard him. But he found out quickly when Dudley punched him in the face. The African suffered a broken nose and a split lip, while Dudley cut his finger on the man's teeth.

As it turned out, the African got the better of the exchange. He was a cannibal, and his sharpened teeth cut Dudley quite deeply. Although the African's nose and lip began to heal nicely, within days, Dudley's finger was an ugly, swollen, septic sausage, and the infection began to spread to his hand. In the end, Hanschell could do nothing to save the finger. After having Rupia chloroform Dudley into unconsciousness, the doctor injected him with morphine then amputated the diseased digit before the infection could take the lieutenant's whole hand.

Even before Dudley woke, Hanschell had burned the finger in the little oven that Fundi had fabricated. As he stirred it among the coals to make sure it was completely charred, he couldn't help but reflect on all the missing fingers associated with the expedition—Tait's, Jung's, Kasemann's, and now Dudley's. Was there some sort of metaphysical symbolism hidden there? If so, it certainly was beyond Hanschell's understanding. But while he couldn't attach particular significance to the missing digits, he could take small heart in the fact that, in sheer numbers, the English count was two fingers, while the Germans were three and a half.

By the time the finger was ash, Dudley was moaning on his cot, and Hanschell went to see how he was doing.

12

"He'll be all right," Hanschell told Spicer-Simson.

The commander had come by the hospital to check on Dudley's condition, but the doctor already had sent the lieutenant back to his own hut, and Spicer-Simson seemed relieved that he didn't have to face him. In fact, the incident seemed to have taken as much of a toll on the commander as it had on the lieutenant. He looked up at Hanschell, something alien in his eyes. With a shock, the doctor realized it was fear—well concealed but there nonetheless.

"I know the men are disappointed in me."

"They are," Hanschell said honestly but not harshly.

"I don't understand why. I've brought them glory and promotions and awards."

"You think that the formal rewards of rank and medals enough for any man," Hanschell said sadly. "That's because those are what *you* crave. But sometimes men need more than that. Or not more, exactly, but something wholly different."

"Faith in their leaders is all they need," the commander said. "I have brought them this far. They must have faith."

"Faith is fickle, Commander. What they need is to understand."

"Wainwright does, I think, but not Dudley. I tried to explain to him, but he's an impetuous fool. Just look how he's injured himself over nothing. They're all fools if they think we can take the *von Götzen*. She's just too much for what we have."

"She is," Hanschell agreed. "As were the *Kingani* and the *Hedwig*, technically, when you took them on. But you did, and did it superbly. And without losing a man."

"No men lost?" Spicer-Simson held up his hand, and Lieutenant Jung's ring gleamed in the lantern light. "You have no idea what it's like to see someone get blown to bits, and all their body parts just lying around like so much offal. Living, breathing, laughing human beings one second, and bloody offal the next. And you're the one who gave the order. They're dead because of something you said."

"They'd have done the same to you."

"Don't you think I am aware of that constantly? Every time I stand there and watch the *von Götzen* steam by, I am thinking of victory, Hanschell. I truly am. But you know what I actually see? I see the faces of my crew, and I realized that they had come so far and done so much, and they'd done it for me. And in return, I've given them victory and honor, and I've done it without a single casualty among them." He lowered his eyes and shook his head. "I just couldn't do it. I couldn't send them out against certain death. Not after the bloodbath I saw aboard the *Kingani* and again on the *Hedwig*. It was nothing but meat, Hanschell. Meat littering the deck. Meat in the hold. We're nothing but meat, and I bear the evidence of that day and night."

He pulled Jung's signet off his finger and held it up, staring at it.

"How can it be, Hanschell, that something so insignificant can bring such pride and such guilt?"

"I don't know, Commander. Perhaps that is the nature of war."

"I'm sick of killing," Spicer-Simson said. "Oh that I were a young firebrand like Dudley, ready to step into the jaws of death for one more glimpse of glory. I thought that was what I wanted. All those years of trying and never quite measuring up. And then all this. God, Hanschell, who could have imagined it? I have a kingdom here. My officers are my dukes, my ratings are my navy, and the Africans are my loyal and devoted subjects. And I have the spoils of my victories. Yet I have nothing because the damn Admiralty can't seem to comprehend the gravity of the situation, and I'm expected to turn my men into bloody offal because some officious bureaucrat can't distinguish his arse from a hole in the ground."

"You still have your destiny," Hanschell said.

"Destiny has nothing to do with it, Doctor. If I thought it destiny, I'd still be at that forgotten desk back in a broom closet in the Admiralty. But I'm beyond itching for immortality. I'll settle for a pension."

"Somehow, I doubt that, Commander," Hanschell replied with a wan smile. "You uncorked that bottle long ago, and now there is nothing that can take the demon out and leave the liquor tasting as sweet."

"You're wrong, Hanschell. How can I think of immortality when the taking of this ring has taught me the value of mortality. Maybe that's something you don't think much on. Your whole training centers on mortality and the consequences of accident, illness, and simply not living properly."

"You're telling me you fear for the safety of the men."

"Yes, I fear for my men. Far better for Dudley that he spend a finger in frustration in lieu of being slaughtered fruitlessly."

"And yourself?"

"Are you calling me a coward?" Spicer-Simson's eyes flashed with some of their old flame.

"No man in this expedition can call you that. We all saw your bravery and your dogged resolve. But now we see your inaction, and men tend to believe what's set before them in the moment is the whole, blinding them to what's past or what the future might bring. Those matters are for leaders like you to consider, and you must explain it to the men in a way they can understand."

"I don't need their understanding, only their support."

"They've given you that, Geoffrey. You know these men would do anything for you because they already have. Now it's time to give them something back."

"I am. Their lives. I wish I could give this back as easily."

Spicer-Simson suddenly cast Jung's ring across the floor of the hospital hut. It came to rest, glittering coldly, at the feet of the fetish of the commander that sat against the wall.

"They want something more complex and important than their lives," Hanschell said. "Something even more than victory and the glory of being undaunted heroes and legendary warriors. They want to be what they've earned time and again: to be trusted compatriots. That's what you've robbed them of. That's why they feel betrayed. By not going after *von Götzen*, you have violated their trust, which is worse than your petty squabbling with Stinghlamber and borderline insubordination with the Admiralty over the Belgians' authority."

Instead of answering, Spicer-Simson went over, picked up the ring, and slipped it back onto his finger. Then he just stood there for a silent moment, staring at the fetish.

"A fine figure, wouldn't you say, Doctor," he said at last, bitterness lacing his voice. "A fine figure, this Bwana Chifunga Tumbo."

Chapter XVII: Searching for *St. George*

SPICER-SIMSON MAY HAVE BEEN feeling low, but the Admiralty was quite pleased. They were, it was true, puzzled by the way his recent messages went on at some length about this *Graf von Götzen*, but Sir Henry hadn't heard of it otherwise, and none of the rest of the flag officers had either. Most were of the opinion that it couldn't be much of a ship and speculated that Spicer-Simson was simply a stickler for completeness when he aired his complaints about the inadequacy of his flotilla.

"Commander Spicer-Simson has done such a marvelous job in making the region safe enough for a land assault," Sir Henry said in a staff meeting on the matter, "that active cooperation with the Belgians has become a practical possibility. I believe the time is ripe for us to proceed."

The other military lions gathered in the room readily gave their assent, and word was sent immediately to General Tombeur, who, along with General Molitor, had spent much of the last few months gathering, equipping, and arming two Belgian brigades to make the assault around the northern end of Tanganyika. Combined South African, Rhodesian, and British forces would march in from the south and north, and the total might of the allied armies would crush von Lettow-Vorbeck's tough and resistant troops in an inescapable vise.

Matters at Kalemie, however, weren't in such a positive and ready state.

Wainwright and Dudley were in Hanschell's hut, looking as befuddled as they were angry.

"The Belgians say that le commandant à la jupe is a coward." Dudley said.

"I've heard them," Hanschell told him then moved around Wainwright to rearrange some of the medicines in the drug cabinet. Not that they needed it, but he had to have something to do with his hands and something to concentrate on so he wouldn't have to look at the two lieutenants. They wanted him to tell them what was wrong with Spicer-Simson. Why, all of a sudden, he seemed to have caved in upon himself. Why he'd let the *von Götzen* steam past Kalemie several more times without giving chase or firing a shot. Why....

The trouble was, even Hanschell, who should know best, wasn't entirely sure of Spicer-Simson's reasons in refusing to engage the enemy. Hanschell recognized the validity of the commander's point of view when Spicer-Simson said his ships were too small and too lightly armed to handle something as large and well armed as the *von Götzen*. And he understood Spicer-Simson's professed reluctance to endanger his men without a good chance at victory.

But Hanschell suspected that something deeper lay behind the commander's retreat. He thought of Spicer-Simson as an interesting case study in military command. He could lead when it suited him and when lives weren't directly at stake, but now that he'd had a taste of true battle, he'd seen firsthand what horror war could wreak on human frailty. Not that Hanschell thought the commander was, personally, a coward. Far from it. Twice, now, Spicer-Simson had stood amid a hail of enemy fire without flinching. He had lived with personal failure for so long, Hanschell didn't think that death following on certain success would deter him if further success was attainable. What Spicer-Simson couldn't live with was the thought of another failure and seeing his own men turned to scraps of bloody meat to no purpose except his own humiliation. He'd had enough of that in his lifetime. And so he'd let the *von Götzen* pass by unmolested.

"Does it matter what the Belgians say?" Hanschell turned at last to face the two. "Stinghlamber is an aristocratic snob and an ineffectual leader. Thanks to him, the Belgians lost the lake in the first place. What really matters is that you remember how much the commander has done. You've followed him across Africa believing him little more than a buffoon, until he gives you a taste of glory. And now, when he continues to exercise his judgment—and convincingly, I might add—you reduce him to coward."

"I know it's not fair, Doctor," Wainwright said. "But it's not only the Belgians, though God knows they treat us worse than ever."

"And that's the commander's fault," Dudley put in. "All he does is send ill-tempered dispatches to Stinghlamber, further turning them against us."

"The real problem is our own men," Wainwright said. "Morale has sunk faster than *Hedwig*, and Dudley and I are the ones left to explain. The commander barely shows his face, and we haven't taken the boats out at all this past week."

"It's only been eleven days since you sank the *Hedwig*," Hanschell reminded him. "I think you should have greater patience. You know that the commander has informed the Admiralty of the *von Götzen*. As soon as the *Vengeur* and the *Baron Dhanis* are ready, he'll have two warships capable of taking on *von Götzen*. I'm sure he won't hesitate, then."

At that moment, Spicer-Simson entered the hospital. Hanschell's heart sank as he saw the defeated sag of the commander's shoulders, and apparently the two lieutenants noticed, as well, for neither could meet his eye.

"Bad news," Spicer-Simson began without preamble. "I've sent the Admiralty another dispatch about the *von Götzen*, but they simply don't believe that she's a threat. It's all that damn Lee's fault. He failed to gain sufficient information about the entire German fleet, and when he did find out about the *von Götzen*, he refused to tell me. They say that we've gotten the *Kingani* and the *Hedwig*, and that's enough."

"They won't send us anything larger?" asked Wainwright.

"They're shipping in two Short seaplanes. They'll have bombs and dropping gear, but I have been informed privately that the planes and dropping gear are out of date. All the new aircraft are being sent to Europe and the Mediterranean. Nobody cares what happens on Tanganyika, and I can't seem to make them understand. They're about to send British and South African colonial forces up from the south and Belgian forces from the northwest, and they think that now we've gotten the *Kingani* and *Hedwig* out of the way, the troops will be safe from bombardment. But *von Götzen* packs more of a punch, and range, than *Kingani* and *Hedwig* put together. Not to mention its potential for troop deployment. It'll be slaughter."

"Isn't there anything we can do, sir?"

"I did get a concession from them," Spicer-Simson said, running a hand over his thinning, close-cropped hair. "The Belgians have a steamer over in Leopoldville named the *Comte de Flandres*. It's iron-hulled and large enough to mount a large gun. If I can have her broken down and get her here, the Admiralty promises to send a modern gun to mount on her. I'm leaving for the interior tomorrow to inspect her and have her dismantled and shipped to us."

"But that'll take months, sir," Wainwright said. "*Toutou*'s back in action, the *Vengeur*'s new engines are being installed, and the *Baron Dhanis* is nearly finished. Both ships will be ready in half the time it'll take to cart in another ship and build it. With what we have, we're more than a match for the *von Götzen*."

"The *Vengeur* is a crapulent piece of junk," Spicer-Simson said, his mouth twisting bitterly. "It's had so many holes poked in it that it's now more patch than hull. Try firing any of the guns mounted on it, and the thing will shake to pieces. And *Baron Dhanis* is a wallowing tub if I ever saw one. She'll have absolutely no maneuverability. Besides, the Belgians are dead set on arming her with that pitiful pair of ancient landlubber guns they've got on our bluff, the stupid bastards. Shit!" He slapped his walking stick against his boot. "Why am I plagued with fools for allies and boats that can't mount ordnance of decent size? The bloody gunboats can't take anything more than those miserable three-pounders. *Fifi* is overmatched by her gun, and none of them is what I need—something that can pierce *von Götzen* hull. That's why I must get that steamer. The Belgians are sending her up the Congo as far as Stanleyville so I can inspect her."

"When will you be leaving, sir?" Dudley asked.

"Tomorrow. I'm taking Tyrer with me."

"Tyrer, sir?" Wainwright asked. "Wouldn't Tubby Eastwood be better? He's proved quite proficient in dealing with the Belgians."

"That's why I want him here," Spicer-Simson said. "You'll need all the help you can get with the new Belgian commander."

"A new commander, sir?" Wainwright asked.

"The man replacing Stinghlamber. The Belgians are saying Stinghlamber's health is the reason for the change, but I believe they've finally realized what an incompetent he is. They're bringing in a Lieutenant Colonel Moulaert, some engineer who's been in Leopoldville for the last year or so. An engineer!" He snorted. "Probably very good at building more fortification but completely lacking in tactical originality. Imagine, putting an engineer in charge during a major maneuver." He snorted again then turned to Wainwright. "I'm leaving you in charge. I'll want a full briefing when I return, but your main duty will be to deal with the airplanes. There won't be any pilots, of course. They need all of those over Europe. But the Belgians have a couple of bush pilots, so I suppose we'll have to let them have the planes. Under our command, of course. I've arranged with Stinghlamber to have the planes harbored in that little bay off Tongwe. That's only twenty miles from here, and we've already got telephone lines there. You'll have to transport supplies and oversee the construction of a base for the Shorts and to have the planes put together and tested. When the dropping gear arrives, have it installed."

"What about the *von Götzen*, Commander?"

"You are to leave her alone. Do not pursue or attack her. In the unlikely event that she attacks first, you may defend yourself, but otherwise steer clear of her. I don't want you losing any of the few ships I'm likely to have remaining in a vain attempt."

"And the Belgians, sir?"

"Take a measure of this Moulaert's mettle. Otherwise, let them rot. I'm commander of the lake, and until that command is rescinded, the ships take their orders from me. Disregard any orders that this Moulaert might give you. I'll handle him when I return."

2

The next morning, February 21, Spicer-Simpson departed Kalemie for the interior of the Congo, leaving the camp in a blue funk. Wainwright tried to rally the men with the task of constructing and supplying the little seaplane base at Tongwe, and for a time, the expedition seemed to regain a little of its former élan. They even shouted ribald insults at the *von Götzen* when it was seen steaming down the lake, though the ship was too distant for the Germans to notice.

Meanwhile, although Spicer-Simson may have thought he'd managed to get away from the Belgian command on the lake, he actually ran right into it. On his second day on the Lukuga River, the small passenger steamer he was on moored against the bank at a small village about halfway between Kabalo and Kongola to take on wood for the furnace. Another steamer, bound upriver, was there, and Spicer-Simson learned that Lieutenant Colonel Moulaert was aboard.

"Go inform the colonel that I shall be paying a call momentarily," he told Tyrer, and the sublieutenant hurried off.

After adjusting his tunic and donning his sword, Spicer-Simson left the steamer and walked up the path along the bank. Just as he came abreast of the other steamer, a shadow fell across him.

"Why have you left the lake?" a booming voice demanded in French.

Blinking, Spicer-Simson looked up and saw a bulky silhouette in a pith helmet leaning over the rail above him.

His face a wooden mask, Spicer-Simson went aboard and over to the man.

"*Lieutenant* Colonel Moulaert, I presume," he said in his most affected drawl. "Certainly you must be aware of the division of command on Tanganyika, and my recollection is that that I do not have to explain my movements to you, even if you were presently in command ashore. Which, I might add, you won't be for another four days."

Moulaert stiffened, and his thick neck seemed to puff with distended veins as his face reddened. He looked like he wanted to strike Spicer-Simson, but perhaps the commander's own muscular physique halted him.

"In the interest of solidarity, however, Lieutenant Colonel, I will inform you that my presence here in no way implies that I have left the lake. Do you see my men? Do you see my boats? Of course not. They are on Tanganyika, recovering from two decisive victories, retaking what your countrymen lost. I am here in the line of duty, reconnoitering the *Comte de Flandres* for possible use as part of my flotilla."

"The *Comte de Flandres*?" It was Moulaert's turn for sarcasm. "Impossible. She is a scow, completely unfit for warfare."

"I take it you have extensive experience with naval war ships, then, *Lieutenant* Colonel."

"Not really, but...."

"I have, in my time, commanded destroyers and other men-of-war," Spicer-Simson said. "You should leave the expert appraisal to me and stick to your engineering duties. The fort at Lukuga certainly could use some renovation."

"Speaking as an engineer, then," Moulaert said, "the *Comte de Flandres* is old and will be very difficult to salvage without losing significant portions that are held together by little more than rust."

"I still intend to inspect her."

"After that," Moulaert said dismissively, "you'd do better to take a look at the *St. George*."

"And what, pray tell, is that?"

"She belongs to the British consul in Leopoldville. That makes her one of your own, already, doesn't it, Commander?"

"I'll take it under advisement," Spicer-Simson said. "Now, I hear the boarding whistle." He turned, but Moulaert stopped him.

"One thing before you go, Commander. As commander at Lukuga, I will need to know of your plans and movements in the future. There is a land assault coming, and I will require your ships for support."

There was no mistaking Moulaert's tone—he wouldn't take no for an answer, even from Spicer-Simson. But he did not yet know the British commander, and Spicer-Simson failed to acknowledge the authority implicit in Moulaert's words.

"I have known of the impending land assault for months," Spicer-Simson retorted. "In fact, I have been consulted regularly during its planning. And may I remind you that I, and no one else, made it possible."

With that, he turned and left the steamer, Tyrer hurrying in his wake.

"What now, Commander?" Tyrer asked.

"Why, on to Stanleyville," Spicer-Simson replied blandly. "You don't think I'm going to take the word of some asinine land crab where watercraft are concerned, do you?"

When they reached Stanleyville, however, and inspected the *Comte de Flandres*, Spicer-Simson had to admit—to himself rather than Tyrer—that Moulaert's assessment of the steamer was correct. She was a rusted bucket of bolts, and though the dismantling wouldn't have been terribly difficult, the rebuilding would take more effort than creating a similar ship from scratch.

"It seems that the *Comte* has seen better days," was his only comment. "Just as well, I suppose. *Graf von Götzen* is our dragon, so perhaps we need a *St. George* to slay it."

They found the telegraph office, and Spicer-Simson sent a wire to the Admiralty inquiring if he could commandeer the vessel. The reply from the Colonial Office came a couple of days later, but it had to be forwarded to Leopoldville because Spicer-Simson had departed for there immediately after sending his telegram. By the time Spicer-Simson received the reply, he had finished going over the *St. George* from stem to stern.

"A fine vessel," he pronounced to *St. George*'s captain.

"Thank you, Commander," the captain responded. "She'll do ten knots." He was puzzled at why Spicer-Simson was here at all, but his confusion cleared up with the commander's next statement.

"I will need her on Tanganyika at the earliest opportunity. Please convey her to the shipyard and begin having her dismantled and prepared to ship to Kalemie."

"Are you certain, sir?" the captain asked, equally astounded and upset at the sudden loss of his command.

"I'm sure you will receive official word promptly," Spicer-Simson replied. "It will look good on your record if you helped matters by advancing the schedule."

"I'm not sure...," the captain began.

"You will be," Spicer-Simson said tartly. "Come, Tyrer. Let's visit the telegraph office. I'm certain our orders will have arrived by now."

Sure enough, the reply, was waiting.

"*St. George* entirely unsuitable," read the telegram. "Dismantling and rebuilding too costly. Suggest you find alternative."

That was the end of the message, but not the end of the affair.

"Why are landlubbers such fools, Tyrer?" Spicer-Simson complained. To the telegraph operator, he said, "Take a message: *St. George* inspected and perfectly suitable. Stop. Dismantling already begun. Stop. The fate of the invasion depends on it. Stop. Send confirmation immediately. Stop. Signed, Commander G. B. Spicer-Simson, R.N. Do you have it?"

"Yes, sir," the telegraph operator replied, and he read the message back.

Spicer-Simson turned to Tyrer. "Things will move along, now, don't you think, Sublieutenant?"

"Yes, sir. What do we do now? Go back to Kalemie?"

"Kalemie?" Spicer-Simson snorted. "And have to put up with that prick Moulaert?" He shook his head. "I have no intention of returning any sooner than I have to. Besides, I wish to supervise the dismantling of the *St. George*. That won't begin until tomorrow, though." He looked thoughtful. "I suppose we should get a drink, don't you think? I wonder if the hotel bar has any vermouth."

3

Word from Spicer-Simson finally reached the expedition in the middle of March.

"In Leopoldville," the message stated. "Arranging for a *St. George* to slay our Fafnir. She is being dismantled and shipped to Kalemie. Keep on the alert for Moulaert." The missive ended with a cryptic reference to a short journey Spicer-Simson had to make that would delay his return by a few weeks, but that if the *St. George* should arrive before him, they were to begin construction immediately.

The news seemed distant and remote and did nothing to cheer the officers or men as more weeks passed. Wainwright could do little but keep the camp running at its current low ebb, though several more appearances of the *von Götzen* only sapped morale further.

Even Hanschell, who had been willing to give Spicer-Simson the benefit of the doubt when the commander departed, was beginning to flag. The longer they were left leaderless and the more hopeless things became, the more he realized that Spicer-Simson had abdicated responsibility while using distance and intermediary subordinates to maintain authority. It was ethically worse than the doctor's own situation of having responsibility without authority because that was not a situation of the doctor's own making. Both qualities, he knew, had to work in tandem, or any enterprise was doomed.

Nevertheless, things in camp went smoothly if dully. The only real challenge came from Colonel Moulaert, who visited a few days after Spicer-Simson had departed. Knowing that the commander was gone, he had tried to pull rank on Wainwright and bully him into submission, but the lieutenant held firm, though to what ground he wasn't sure. He would say only that his commanding officer had left him with explicit instructions that he intended to carry out to the letter. But he had to concede that Moulaert had brought compelling news. Allied high command was stepping up its plans for the assault on German East Africa.

Wainwright wasn't surprised. He'd already received a dispatch, addressed to Spicer-Simson, from London via British headquarters in Nairobi, to that effect. The invasion would take place simultaneously on several fronts and, in fact, already had begun. British and South African troops were fighting their way north toward Bismarckburg at the southern end of Tanganyika and would arrive there within six weeks. Meanwhile, British forces near Kilimanjaro were on the move, and as they pressed in from the northeast, Belgians forces under General Tombeur and General Molitor would round the northern end of the lake. The Germans would be caught in the middle, hemmed in on all sides by Allied troops except to the west, where Tanganyika formed a barrier nearly as impenetrable as a mountain range.

Except for the *Graf von Götzen*. But after all, Spicer-Simson was there with his gunboats to halt any German resistance on the lake.

Wasn't he?

4

The Allied high command may have thought Spicer-Simson was where he was supposed to be, but the British and Belgians at Tanganyika knew he wasn't, though they weren't sure exactly where he was. But even if he wasn't where he was supposed to be, the war progressed on without him. And so did the work at Kalemie as *Vengeur*'s new engines were installed and the *Baron Dhanis* neared completion.

For the expedition, a brief flurry of excitement began on April 1, when the four Short 827 seaplanes arrived. All four had spent much of the previous year in Zanzibar serving with the RNAS No. 8 Squadron, gun-spotting for the monitor *Severn* and reconnoitering during British shore landings along the African east coast. When the call came that they were needed on Tanganyika, they were disassembled, packed into crates, and shipped by rail to Lukuga, and from there to Tongwe. Each of the biplanes was just over thirty-five feet long, had a wingspan of nearly fifty-four feet, and was powered by a one-hundred-and-fifty-horsepower Sunbeam engine appropriately named the Nubian. With a maximum speed of sixty-one miles per hour and a range of three and a half hours, the biplanes could carry several fifty-pound bombs in addition to a Lewis gun.

Two of the planes were assembled during the first two weeks of April by the Belgian pilots, who had been running mail up and down the Congo, and a couple of lorry mechanics. Someone should have warned the pilots that anyone associated with the Royal Naval African Expedition was prone to unusual turns of events. As soon as the two planes were built, the pilots were anxious to test them. They were set out on the water, and after the pilots revved the engines a few times, they took off.

The Shorts were built to fly at altitudes of up to about 2,000 feet, but Tanganyika, standing at more than 2,500 feet, was pretty rarified atmosphere for the low-horsepower Sunbeam Nubians, which had to fly higher still to gain any height over the lake. In addition, the round-bottomed pontoons were ill-designed for rough water, and the wind coursing through Tongwe Bay tended to be somewhat erratic, making any landing and take-off problematic.

One of the Shorts crashed on take-off, depositing its pilot and gunner in the mangrove trees lining the shore. Luckily, neither was seriously hurt. The second plane actually got into the air, but the pilot brought it in for a skewed landing on the light chop, irreparably damaging one of the pontoons.

Apparently the Belgian pilots were either rusty or unused to handling seaplanes, or maybe they'd been trained at the same camp as the Belgian gunners, who couldn't seem to hit even a stationary target.

The other two planes were assembled as quickly as possible, and both pilots, feeling lucky to be alive after their initial attempts, took greater care when they lofted in their maiden flight to visit Kalemie.

Dudley rushed out of the communications hut and over to where Wainwright and Cross stood with Jadot on the bluff, discussing some additions to the harbor.

"They just called from Tongwe," Dudley said breathlessly. "The planes are on their way. They'll be here in twenty minutes."

"Let's go out to the parade ground, then, and greet them," Wainwright said, feeling excited himself.

Nearly the entire expedition turned out, accompanied by the usual crowd of several hundred Africans. When the former began watching expectantly down the coast, the latter did likewise, little realizing the shock that they were about to receive. They knew that Bwana Chifunga Tumbo was lord over the mountains and the waters, and now they were about to discover he also was master of the skies.

The wait didn't take long.

At first, nothing could been seen because of the low-hanging clouds, but a steady twin drone began to infuse the air over the parade ground. The British talked excitedly among themselves, but their good mood didn't rub off on the Africans, who stood spell-bound, eyes wide with fear, as the sound grew.

The two seaplanes suddenly shot into view out of the clouds. The pilots spotted the parade ground, banked, and began flying in circles around the camp, waggling their wings. The British cheered and threw their caps into the air, while the Africans fell to their knees. Some raised their arms and wailed, while many hid their faces in the dirt. Abruptly, the airmen made a sudden dive toward the parade ground, pulling out just a hundred feet up. Even the British ducked, but pandemonium broke loose among the Africans. Terrified, most raced off into the surrounding bush, while others simply fell where they were, covering their heads and trembling. Several screaming mothers grabbed their children and ran for their huts.

Bwana Chifunga Tumbo's name was muttered on many lips. The last battle had been fought in the heavens, and now Bwana Chifunga Tumbo had brought huge, roaring birds of prey to carry his warriors to the heavens for the next fray. Once again, Spicer-Simson had proved his godhead, even if he wasn't present to enjoy the adulation.

Following the spectacular display of airmanship, the pilots circled twice more then headed for the water, landed, and taxied into Kalemie. The British rushed toward the path down the bluff, but for once, they weren't trailed by tribesmen.

As the men reached the harbor, the Shorts taxied up to the shore. The expedition members gathered around the planes, ogling. Few, even, of the English had been this close to an aircraft before. Each plane had its machine gun installed, but the bombs and dropping gear, like the commander, were lost in transit somewhere on the Dark Continent, so the planes were, for the time being, fancy toys instead of weapons of war.

Even so, the activity surrounding the planes reenergized the men, and their excitement reached its height a few days later when the bombs and dropping gear arrived. The lorry mechanics installed the gear, and though it was obsolete, the mechanics made it workable. Wainwright had a crude raft constructed, and one afternoon *Fifi* towed the raft out on the lake so the pilots could make a practice run. The British crowded aboard every available craft and followed *Fifi* out onto the lake to witness the event. The day was gorgeous, and the men jostled and chattered like schoolboys on a lark. Africans lined the bluff to watch, wondering what was going to happen.

After *Fifi* let loose of the towline and moved off a safe distance, the Shorts came in low over the water. At what seemed like an auspicious moment, the gunners each released a single bomb, then the pilots banked and rose as the gunners leaned out of their cockpits to watch the results. The first bomb missed the raft, sending a fountain of water over the makeshift craft, but the second hit the target. As the raft disintegrated into matchwood, the men let loose an uproarious cheer.

"Take that, *von Götzen*!" shouted Lamont.

Then everyone went home and had a boisterous evening meal at which the victories over the *Kingani* and *Hedwig* were relived at great length for the benefit of the Short crews.

But the next day, a somber mood settled over the camp once more. With the base for the planes completed, there seemed to be nothing else to do but brood. The Shorts could help sink the *von Götzen* but simply sat idle at their base because of Spicer-Simson's absence. And the complete lack of communication from the commander only made matters worse.

5

"Where is your commander?" Moulaert demanded. "Has he not returned?" It was April 28, and the new Belgian commander was worried. He was getting pressure from General Tombeur, and he knew his command remained incomplete without the flotilla at his disposal.

Wainwright was worried, too. Spicer-Simson had been gone for more than two months, and before today, there had been only the single cryptic message.

"The commander is inland," Wainwright temporized. "He's gone to bring back a ship large enough to sink the *Graf von Götzen*."

"Still not here?"

Moulaert couldn't believe it. What kind of commander was this Spicer-Simson? The man he'd met on the river seemed a pompous ass, but when Moulaert had arrived at Lukuga and heard about the expedition's extraordinary exploits, he'd begun to alter his opinion. Any man who could accomplish what Spicer-Simson had must be a first-class officer. Maybe he'd been hasty in judging the man too quickly. But now, after all this time had passed, what was he to think? It smacked of dereliction of duty. Spicer-Simson didn't need to supervise the dismantling of the *St. George*; he should be here.

Moulaert realized that Spicer-Simson might be staying away just to make him look like a fool. And Moulaert couldn't help but think the British commander might be right. Moulaert had barged in and unthinkingly fired the opening salvo of hostilities against the commander, who had had remarkable success in a very short time. And then he'd demanded complete accountability from Spicer-Simson. That couldn't have sat well with the British commander, and knowing that the friction that had developed between him and Spicer-Simson was his fault was a secret guilt that gnawed at Moulaert's conscience. But damn it! The man didn't have to pout did he? Not when a war was going on and success was just a short distance off?

"He has not said when he will return?"

Wainwright hesitated. Just that morning, he'd received a second message from Spicer-Simson. The commander was in Boma, the city at the mouth of the Congo River.

"Boma!" Dudley had fumed, when he heard. "That's sixteen hundred miles away! What the hell's he doing in Boma?"

"Visiting the British consul," Wainwright said, wanting to curse, himself.

"Shit!" Dudley spat, then shook his head. "I can't believe it."

Wainwright could, though. All too well. But he hesitated to say anything to Moulaert. Instead, he told the colonel that Spicer-Simson was still in the Congo arranging supplies and would return to Kalemie within a few weeks, praying all the while that his prediction would come true.

"I hope you're right," Moulaert said, sensing that Wainwright was hedging but realizing he'd get nothing out of the lieutenant. "General Tombeur has personally ordered me to coordinate with Commander Spicer-Simson. I want to put our differences behind us in view of our countries' joint advance on the Germans, but if he has not returned before the Allied advance, I *will* take command. You must understand, Lieutenant: We will need everything we've got to conquer the Germans, and that means your boats, with or without your commander."

Yes, thought Wainwright. With or without our commander, we will be forced back into the war though he's ordered us out of action. We need him, and I haven't the foggiest notion of where he is or how to find him or even if he really will come back.

6

Spicer-Simson did come back, on May 12, after being gone from Kalemie for eleven weeks. His African worshipers were ecstatic to see him again, but most of the men of the expedition remained disillusioned, believing that total victory over the Germans could have been accomplished long before now.

If Spicer-Simson noticed their chilly reaction to his return, he ignored it. But there was something different about him, Hanschell realized. He retained his pompous arrogance, but now it was tinged with a bitter turn. The next day, after morning parade, the commander called together his executive officers for a staff meeting.

"The *St. George* was nearly dismantled when I last left Leopoldville," he told them brusquely. "The engines and fittings already are being shipped up the Lualaba to Kabalo. They should arrive in Kalemie within another month, with the hull plating and frame soon to follow. Gentlemen, we should have a working vessel of sufficient size in a couple of months after that."

Wainwright refrained from arguing about time and opportunities lost because he had good news of his own. The *Vengeur*'s new engines had been tested, and the fifteen-hundred-ton *Baron Dhanis* waited only the mounting of the guns that Spicer-Simson had traded to Stinghlamber.

"Even before the *St. George* arrives, we'll have enough might to go up against the *von Götzen*, sir," Wainwright said, feeling his blood stir.

But the commander's bitter mood turned as foul as a tempest on Tanganyika when Wainwright informed him of the advancing front, the impending unified assault, and, most of all, Moulaert's repeated visits and their purpose.

"That goddamn Belgian pig!" Spicer-Simson said. "None of this would be possible without me, and he dares make threats against my command? Take a letter, Wainwright."

The commander fumed in silence while the lieutenant sat at the table and found paper and pen.

"Moulaert," Spicer-Simson dictated. "How dare you attempt to subordinate my officers and men, make threats against my command, and attempt to take unilateral action without my knowledge or consent? I remind you that it is I, not you, who have command of all Allied vessels on Lake Tanganyika by orders of our joint high commands. If that alone doesn't give you pause, recall that I have accomplished more here in six months than you and your predecessor have managed in two years. Henceforth, attend to your own affairs, safe ashore in your fortress, and let me deal with the Germans as I have successfully done in the past."

He stopped and, Wainwright could have sworn, took his first breath since beginning his dictation.

"Is that all, sir?" he asked, writing down the last of Spicer-Simson's tirade.

"No, but it will do for now. Make a fair copy for me to sign, and have a messenger deliver it immediately. I won't stand for that bastard's interference."

"What about the *Vengeur*?"

"I'll go reclaim her tomorrow and see what Moulaert has to say about it."

"It won't be easy, Commander. We can't man her ourselves. We'll have to use a Belgian crew, and they're not likely to take orders from us."

"The hell they won't!" Spicer-Simson snapped. "Goor will see to it."

"We might have a problem there, sir," Dudley ventured.

"If there is a problem, Lieutenant, let's hear it."

"It's Goor, sir. He's gone. They promoted Lieutenant Wautier, and he's been ignoring us and doing what Moulaert says."

"Well, they'd better remember who is commander of the naval forces on Lake Tanganyika," Spicer-Simson fumed. "I'll clap them in irons and have them court-martialed if they don't!"

Wainwright didn't have a chance to deliver Spicer-Simson's message because, no sooner than he left the commander's hut, Lieutenant Schelde arrived bearing orders for Spicer-Simson to attend Moulaert at Lukuga. Spicer-Simson read the note, then carelessly let it fall to the ground.

"Tell your new commander that if he wants to speak to me, he's more than welcome to come here," the commander said with clear disdain. "And you can deliver this as well." Spicer-Simson snatched the note out of Wainwright's hand and thrust it at Schelde, who took it as if it might be poison.

Schelde reluctantly returned to Lukuga with the message, which he handed to Moulaert, who turned red-faced as he read it.

"Who the hell does he think he is?" the lieutenant colonel ground out. "I'm the regional commander. How dare he refuse me?"

Schelde refrained from pointing out that the agreement between the British and their own government provided equal rank for Spicer-Simson and Moulaert. He didn't know his new commander well, but the iron-jawed scowl on Moulaert's face was easy to read.

"Gather the other officers," Moulaert ordered. "We're going to pay this upstart a visit."

Spicer-Simson was holding court on the parade ground when Moulaert entered the scene, trailing the other Belgian officers. The commander was blessing a child, the son of a local headman, and he pointedly ignored the Belgians until they were standing right in front of him.

"Commander Spicer-Simson," Moulaert demanded.

The commander glanced up as if noticing the Belgians for the first time. "Yes, Moulaert," he said. "I'll be with you in a moment."

Although angry at the British commander, Moulaert had come fully prepared to be as civil as possible, but Spicer-Simson's arrogant dismissal in favor of a native—and a child at that—left the Belgian officer sputtering. Spicer-Simson finished blessing the boy, shooed him off with a pat on the head, then calmly turned his attention to his new guest.

"What can I do for you?"

"You can obey orders, like a good officer," Moulaert said sharply.

"I'm sorry," Spicer-Simson frowned. He gestured to Tom, who quickly handed over the commander's cigarette holder, fresh cigarette in place. Spicer-Simson lit the cigarette, blew a puff into the air, and switched his focus from Moulaert to Wainwright. "Lieutenant, have we received any fresh orders from the Admiralty?"

"No, sir," Wainwright said.

"Well, Moulaert," Spicer-Simson said returning his gaze to the Belgian. "There you have it. No orders to obey...like a good officer."

"You insolent...." Moulaert unconsciously took a step forward as he spoke, but the rest of his sentence was washed away by a threatening mutter that rose from the massed Africans. Captain Dender put a restraining hand on Moulaert's arm, which Moulaert shook off with a glare at Dender. Dender stepped back, and while Moulaert didn't, he stopped and glanced around as a thousand pair of Ba-HoloHolo eyes appraised the tenderness of his flesh.

"Careful, Moulaert," Spicer-Simson said, a tight smile curling the corners of his mouth. "I suggest you save the hostilities for the Germans."

"Yes," Moulaert replied. Though anger remained on his face, he backed up a step then said sarcastically, "A pleasure to see you again, Commander."

"Since you've come so far, perhaps you'd like to take mess with us. Our cook has become expert at adapting the local supplies to our European tastes."

"Another time," Moulaert said. "I must see to my own command."

"Of course," Spicer-Simson said. "Another time."

With a wave, Moulaert gathered his officers, and the small group squeezed their way through the still-hostile dark crowd surrounding them. To their collective relief, they made it to the edge of the camp and were swallowed by the concealing bush.

Spicer-Simson watched them depart only as far as the edge of the parade ground, then he turned to his worshipers and waved to another mother, who hurried forward and handed him a wriggling boy. The youngster's eyes lit on the commander's tattoos, and he stopped squirming.

"You like that, do you?" the commander beamed. He flexed the muscles of his arm, sending the snakes there writhing. The boy laughed, and the commander joined in. The gathered Ba-HoloHolo murmured in approval.

7

The next morning, a Belgian runner brought a message from Moulaert. Spicer-Simson didn't tell his officers what Moulaert wrote, but it couldn't have been good news because the commander's foul mood deepened, as was soon evident to everyone in camp. About 11 o'clock, Spicer-Simson stalked off toward the Belgian fortress.

"Very good of you to come," Moulaert said when Spicer-Simson entered his office. He wanted to smooth over relations after the tension the day before. "May I get you something to drink?"

"Let's get on to business, Lieutenant Colonel," Spicer-Simson said abruptly. "I have urgent matters to attend to."

"Yes, well, General Tombeur has been worried that we have not taken any further action against the Germans for nearly three months. The *von Götzen*...."

"Perhaps the Belgian high command had better confer with the British military authorities, then," Spicer-Simson said gratingly. "According to the Admiralty, I was sent to take the *Kingani* and *Hedwig von Wissmann*, which I have done."

"But the *von Götzen* is still here," Moulaert said.

"The Admiralty is unconcerned about it; thus, so am I. The *von Götzen* will remain until a joint naval operation can dispose of her."

"That's exactly what I wanted to speak with you about, Commander," Moulaert said. "General Tombeur is preparing to move immediately on the

German-held territories. He wants one of your boats to take the commander of the 5th Battalion to our fort at Uvira. Then you are to return to provide naval support when your seaplanes bomb the *von Götzen*."

"You seem to have drawn up all my orders for me," Spicer-Simson commented dryly.

"I think it would be best if we work together to utilize your flotilla to the fullest capacity."

Spicer-Simson gave a derisive snort.

"You are an engineer, are you not, Lieutenant Colonel?"

"Yes, I was trained in the Academy of...."

"Where you learned your trade isn't important. I simply want you to tell me if you would allow me to design a bridge upon whose safety pedestrians and vehicles depend."

"You are not a trained engineer?"

"No."

"Then, of course not."

"For the same reason, Lieutenant Colonel, there would be no purpose in me allowing you to make decisions regarding naval tactics. There is a reason I am senior naval officer on Tanganyika—I am highly trained for it and have spent a lifetime honing my skills. And I certainly don't need an amateur, no matter how well intentioned, making a mess of matters he knows nothing about."

As Spicer-Simson said this, Moulaert felt rage growing in his breast, but he managed to say levelly enough, "Commander Spicer-Simson, surely you know that both our high commands want to see action. General Tombeur insists that we control the lake whether or not you have a hand in it."

"And they shall have it," Spicer-Simson replied, unflustered at Moulaert's attempt to bully him into submission. "I am currently in the process of drawing up a plan of operations."

"And you will share that with me when you are finished?"

"I'm afraid that won't be possible as this is a British military operation. I will forward it on to the Admiralty, who will, of course, disclose any pertinent information to your government. Whether or not they see fit to tell you is their concern, not mine."

"I see." And Moulaert thought he did see. His inadvertent opening salvo on Spicer-Simson back on the river had launched a conflict that would not easily be resolved—if ever—and he simply was taking a series of broadsides in return. It would be up to him to end the distention so that he could follow General Tombeur's orders as fully and as expediently as possible.

"It would seem," Moulaert said, "that our disagreement hinges on the division of authority sanctioned by our governments. Perhaps it might be best if we go over them to make certain they...."

"It would be a noble gesture, I'm sure," Spicer-Simson interrupted. "But not a very practical one. My orders were given verbally, so there is no fair-copy verse to read from to ascertain that your version is a variant."

"Very well, Commander. The lake remains yours for now, but I'm afraid I will have to retain control of the *Vengeur* and the *Baron Dhanis*."

"In direct contravention of orders?"

"I will bear the consequences."

"Indeed," Spicer-Simson said. "But no matter. *Vengeur* is a leaky tub worthless for combat, and the *Baron Dhanis* hasn't yet got ordnance. You may have them to piddle about on the water. I will soon have the *St. George*. In any case, I already have my instructions, and they do not include Uvira or your 5th Battalion commander. Now if you have nothing else to discuss, General Edwards has asked for my tactical advice."

He stood and departed, his swagger stick making sharp taps on the floor, leaving Moulaert dumbfounded that, in reclaiming his ships, he'd finally won the most protracted battle he'd yet engaged in since he'd taken command at Lukuga. But even if he had, Spicer-Simson had fired a crippling final shot by refusing to acknowledge the Belgian commander's authority or lend his support. Moulaert would have to proceed without the British flotilla.

8

The next morning, Spicer-Simson sent a message to Lieutenant Wautier demanding to see any and all orders that Moulaert had given regarding the Belgian portion of the Allied flotilla.

Wautier didn't so much refuse as pass the buck, replying to Spicer-Simson that he was forwarding the commander's request to Moulaert. Moulaert took four days to respond with a non-response that, as commander of the Belgian army and possessions, including all craft flying the Belgian flag, he needed to know why Spicer-Simson wanted the information.

The commander refused to explain, so matters between Kalemie and Lukuga, already sullen, came to a complete impasse. Wainwright aired his resentment of the situation to Hanschell during one of his regular office visits, and the doctor once more argued that the Allied flotilla was inadequate against the *von Götzen*.

"Stop trying to defend him," Wainwright said impatiently. "We've heard that blarney already. "We could easily have mounted a larger gun on the *Vengeur*. It may have been sluggish, but it and *Dix Tonne* still could be used in conjunction with the other boats to trap the *von Götzen* and sink it. If the commander hadn't thrown them away." The bitter scorn in his voice was hard for Hanschell to bear.

"But the gunboats' six-pounders will be useless...."

Wainwright cut him off. "You wouldn't say that if you'd seen what they did to the other two ships. *Mimi*'s couldn't penetrate the *Hedwig*'s hull, either, but it did considerable damage to anything on the deck and kept the German's occupied until *Fifi* could make the killing stroke. We could do the same with the *von Götzen*. And the gunboats are agile enough that there's little likelihood they'd be hit by the *von Götzen*'s guns. They would be more in danger from small arms fire, but their own guns could help scour the decks of that."

"It would be a hard-won victory. He's concerned for the safety of the men."

"More likely its his own skin he's worried about."

"But you saw how he stood there during the battles, with gunfire all around."

"I can't attempt to explain that. I just know we're stalled more thoroughly than we've ever been. Back on the trail, just before we crossed the Mitumba Mountains, he said something to me that I can't help but think of. He warned me not to let the men grow lax and cause the expedition to lose momentum, but now that's exactly what's happened to him—he's lost his momentum."

"He'll get it back."

"I'm not so sure. An endeavor such as this is like a steam engine. The thing might be in perfect working order, but until its boiler is heated and builds up enough pressure, it's just an inert mass. Maybe he could get things moving again, but frankly, I think he's lost the fire in his belly."

"Don't you think that if we forgave him his peccadilloes and eccentricities and more openly encouraged him, he might get it back?"

Wainwright gave Hanschell a sad look and shook his head.

"He doesn't deserve such a friend as you, Doctor. The man's a villain."

"He's not a villain," Hanschell insisted. "Maybe a dastard, but that's a different thing."

"Not to the men, it isn't." Wainwright stood. "I must get back to work. It seems I've become a harbor master instead of a warrior, and I've got a harbor to attend to."

The lieutenant left, but his mood of despair and frustration remained in the hospital hut. At last, Hanschell could brood on it no longer, and steeling himself, he went to confront Spicer-Simson once more.

He found the commander furiously scribbling a dispatch. He couldn't tell who it was addressed to, but that didn't really matter. Spicer-Simson was broadcasting the things like a wireless to anyone within range, from the Admiralty to upper-echelon British command in Africa to civilian authorities.

"What now, Doctor?" Spicer-Simson asked in an exasperated tone. "I have work to do."

"Yes, you do, but you're not doing it."

"What is that supposed to mean?" the commander demanded, bristling.

"For a time," Hanschell said, "I thought your obstinacy a good thing—a sign of dogged determination or even a higher nobility. Now I see it as a weakness—something mean and small."

"How dare you say that? Don't you know I can have you punished for insubordination?"

"Yes," Hanschell said. "You could, but you won't, because you know I'm right. It's gotten in the way of everything, and none of us understand why."

"Once more, then, Doctor," came the haughty reply. "The *von Götzen* is far beyond our capabilities...."

"I might have accepted that argument in the past, but no more. You went traipsing off into the jungle after a mythical hero to battle your dragon and, in the process, threw away the *Vengeur*, the *Baron Dhanis*, and *Dix Tonne*. Their firepower combined with the boats you already have could have defeated the *von Götzen*. Look how the navy cornered the *Königsberg* in the Rufiji, and she was a true battleship. And don't try to prick my conscience with the possibility that some of the men might be hurt or killed. Think of how many men in the land assault might suffer similar fates if you don't do something about the *von Götzen*."

"I don't have to justify my actions to you."

"I think you do."

"Get out!"

"Not until you tell me why you've let your personal quest for supremacy overcome your duty."

"The Admiralty. The goddamn Admiralty and their high-handed lack of regard. I did what they sent me to do, and I'll not do one whit more until they give me what I want."

And what would that be? Hanschell wondered. And would it ever be enough. For now he sensed a deeper truth to Spicer-Simson's refusal to engage the *von Götzen*. It had as much to do with a form of retaliation against the Admiralty as anything. It was similar to his acrimonious debate with the Belgians, only Spicer-Simson couldn't verbally attack the Admiralty as he could the Belgians. So he was simply dragging his heels, unwilling to risk his hard-won victories for leaders who had let him flounder, despite the fact that they had been all too ready to punish him in the past when he'd made errors. But although Spicer-Simson had made mistakes, he'd never lost faith in the service or let it down, and now, when he had performed for them beyond expectations, the Admiralty had not only let him down but virtually ignored his predicament and his very reasonable requests for more support.

"It's true that you've accomplished what the Admiralty sent you to do —clear the lake of the *Kingani* and *Hedwig von Wissmann*—and it will be simple to rest on your laurels and wait out the end of the war. But if you do, how will that affect your credibility with the Admiralty? All your hard-won victories won't be worth a farthing. Besides, there's too much at stake for personal pique to stand in the way. Your king and country need you."

"Good God, man!" Spicer-Simson retorted. "I'm not fighting for king and country! I'm fighting for my life!"

9

Apparently, nothing Hanschell said to Spicer-Simson had any effect. If anything, it further fueled the furious exchange of charges and counter charges of incompetency and illegitimacy with Moulaert, each letter chased by a telegram or report to British command in Nairobi, to General Edwards, or to the Admiralty.

Wainwright managed to catch the commander during a lull in his epistolary war, and he suggested once more that they go after the *von Götzen*.

"Request denied," Spicer-Simson said. "I have more important administrative matters to attend to."

"Administrative matters, sir?" Wainwright asked, the disappointment he felt twisting into malcontent.

"Yes. Moulaert swings some weight with General Tombeur, and he's been spreading lies about me to the Admiralty. I suspect they're going to force me to relinquish command of our boats."

"But they can't do that!" Wainwright said. "Not after all we've done."

"Can and probably will," Spicer-Simson said. "Time and again the Belgians have proved how incompetent they are. They lost the lake with steamers and we took it back with motorboats. But they're jealous, and they'll let that jealousy destroy all we've gained."

Although Wainwright could not get Spicer-Simson to agree to attack the *von Götzen*, he managed to get the commander's concession to send the Shorts, armed with their bombs, against the ship. The biplanes made several bombing runs at Kigoma during the next week, but to little effect aside from the destruction of a dock and some shore next to it. The pilots were not particularly intimidated by the ship's 105s and 88s, as impressive as they were, but a 37 millimeter anti-aircraft gun kept them at bay. They did report, however, the presence of a smaller ship being assembled in the Kigoma shipyard.

Perhaps Spicer-Simson's attitude about the *von Götzen* would have altered had he had but an inkling of the truth. For more than a month, Captain Zimmer had been reluctant to send out the ship. Even though it was finally ready for combat, all Zimmer's best captains and crews had been captured or killed, and he wasn't about to trust his last warship to less-experienced men. And now, the situation was even more grave.

The German high command knew that the lake already was lost and that if victory were to be snatched from the combined Allied assault, it would have to be won on land. To give his troops every possible advantage, Colonel von Lettow-Vorbeck had ordered Zimmer to remove the 105s and 88s from the *von Götzen* and transfer them inland to fortify German positions against the massing Allied assault. The anti-aircraft gun that intimidated the Short pilots was the only real gun aboard the ship. What the pilots thought were the big guns actually were wooden dummies mounted to fool the Allies into thinking the ship was still formidably armed.

But Spicer-Simson was too involved in his acrimonious debate with Moulaert to discover the fact that *Mimi* and *Toutou* alone could have captured the once-supreme *Graf von Götzen*.

Chapter XVIII: Bismarckburg

AT THE BEGINNING OF THE last week of May, Moulaert, at his wit's end, put in an appearance at Kalemie to plead with Spicer-Simson to throw the Tanganyika navy into the offensive at the north end of the lake.

"Request denied," Spicer-Simson said curtly.

"On what grounds?" Moulaert demanded.

"On the grounds that there are no grounds," the commander retorted. "Only water. I've returned your ships, and that's all you'll get from me. Just be thankful my gunners trained your incompetent men how to shoot straight."

"I think you will do well to heed General Tombeur's orders."

"His Majesty's Navy does not take orders from King Leopold or his subjects." Spicer-Simson blandly lit a cigarette and inserted it into his holder. "In any case, you obviously are not far enough up the chain of command to have been informed of the true state of affairs, so let me be the first to tell you. Brigadier General Northey is advancing on the southern front with British and Rhodesian forces, and I have instructions to lend him support. We are going to take part in the assault on Bismarckburg." He smoothed his tunic with an affected display of self-importance. "That's where the war in Africa will be decided. After all, you Belgian chaps can't really be expected to have any consequence on the outcome, even if you do manage to make it through Rwanda. And I very much doubt you'll accomplish that much." His upper lip wrinkled as if he smelled a bad odor.

Unable to offer a counter argument as strong as a direct command from Spicer-Simson's superiors, Moulaert left, vowing to himself that before the month was out he'd have General Tombeur confer with General Smuts regarding return of all authority on the lake—including Spicer-Simson's flotilla—to Belgian command.

True to his word, on May 25, Spicer-Simson left Kalemie, taking the entire expedition with him aboard *Fifi*, *Mimi*, and *Toutou*. For the first time in months, the men, normally jaded by their commander's volatile character and mercurial actions, seemed to want to trust that Spicer-Simson was again the inspired, if autocratic, leader they once thought him to be. They still had their hard-won flotilla, damn it, and even the *Graf von Götzen* would fear an encounter. In fact, most of the officers and men prayed that the *von Götzen* would be on hand to lend support to the German fortress at Bismarckburg. Then the commander would be compelled to take it on.

The promise of battle was not farfetched. Bismarckburg—present-day Kasanga—lay on Tanganyika's east coast about thirty miles from the lake's southern end. The German fortress there was well armed against both land and sea, with battlements overlooking the harbor. But the Germans were facing a formidable foe in the form of Murray's Column, commanded by Lieutenant Colonel Ronald Ernest "Kaffir" Murray. To a man, Murray's Column was filled with a rough and rugged bunch who'd fight an unnecessary order or the regular army as readily as they would the enemy. Most of them came from two companies of British South African Police from Rhodesia, and the rest were Rhodesian settlers. But though the settlers were volunteers, many of them had seen duty in the 1st Rhodesian Regiment, and they were no less tough than the BSAP soldiers they marched with.

Kaffir Murray was gathering his men at Kitute, a Northern Rhodesian village on the south tip of the lake. He wasn't, at present, a happy man. While crossing the Kalambo River on his way to Bismarckburg, his troops had been harried by the Germans, and he had wounded and dying men in the building he was using as a hospital. Worse, four messages from General Northey had arrived the day after he had, and all had ordered him to stay away from Bismarckburg.

But having come this far, Murray wasn't about to back down now. He knew Northey wouldn't complain about him disobeying orders if he took the fort and captured the Germans in it.

Murray had been in Kitute for the better part of a week, amassing supplies and ammunition for the assault, when Spicer-Simson's flotilla steamed thankfully into the tiny harbor on May 26. Two days of bad weather had put a damper on everyone's spirits. Half the trip had been spent sheltering in coves, riding out squalls. The men in *Mimi* and *Toutou* were especially miserable, having no cabin at all.

Although the men wanted to believe the commander had regained his former strength and élan and that more glory awaited, few illusions remained after the arduous trip. They could pretend all they wanted through the miserable hours as they were tossed by the rough seas, but by the time they reached Kitute, none did.

The boats caused quite a stir among Murray's men, and not half an hour after they tied up, Murray himself showed up on the dock. Spicer-Simson greeted him and invited him aboard *Fifi*. The Rhodesian commander was obviously impressed with the flotilla and the guns it mounted, though he gazed with a decided lack of sympathy on the commander's tattoos and unusual attire.

"Where the blazes did you come from?" he demanded heartily, his heavy, dark brows unknitting from their usual frown. "I didn't think the Allies had more than a native canoe to their name on the lake."

"You mean to say you haven't heard of our flotilla?" Spicer-Simson asked, surprised. "We transported those two gunboats," he gestured toward *Mimi* and *Toutou*, "up from Cape Town last fall, captured this ship from the Germans the day after Christmas, and sank another one in February."

"News to me," Murray said. "What are you doing here?"

"Here?" Spicer-Simson seemed at a loss. "We're here to help take Bismarckburg."

"Haven't received orders about you," Murray said. "Don't see you can help much, though. Land-based assault." He went forward toward *Fifi*'s 12-pounder. "What's your range with this?" He patted the barrel.

"Three thousand yards," the commander answered. "Perhaps we could shell the fort."

"You'd have to get within range of their guns." Murray shook his head. "I really don't think there is much for you to do here. You might go back up the lake and help the Belgians at the northern end."

"Lieutenant Colonel Moulaert assures me that his troops have that area well in hand," Spicer-Simson said. "And my orders from the Admiralty put me here, so I'm afraid you're stuck with me."

"Is that so? Well, maybe there is something you might do, after all. The German harbor is full of boats. As soon as they realize the fort is certain to fall, they'll try to escape by water. But now we have you to stop them."

"You want me to guard the harbor and prevent the Germans from escaping?"

"Most useful thing you can do for me."

"When will you make your assault?"

"We'll be leaving here in two or three days. Say four more for all my men and supplies to reach Bismarckburg, and then...." He shrugged. "Only the battle will tell how long after that. I think we can take it inside of a week." He peered hard at Spicer-Simson. "You just make sure those Germans don't escape."

2

The expedition remained in Kitute until a few days after Murray's column pulled out. On the morning of June 5, Spicer-Simson told Wainwright to prepare the boats and crews for a reconnaissance of Bismarckburg. When the flotilla, bobbing in the waves, came to a halt a safe distance outside the mouth of the harbor, the fort was easily visible, its white, crenellated walls standing starkly on top of a low hill. A small European settlement nestled in the shadow of the fort's walls, and between it and the harbor spread the huts of a sizable African village. Ten or twelve large canoes floated in the harbor, and scattered among them were five 100-ton dhows, their lateen rigging furled.

While Spicer-Simson had news that Murray's Column had just taken up positions outside the fort, none could be seen from the lake. Nor were any German troops visible. Even the African village lay in deadly stillness under the sun.

Wainwright and Dudley, who were commanding *Mimi* and *Toutou*, came aboard *Fifi* and approached Spicer-Simson as he stood in his ship's bows, staring through binoculars at the harbor and the fort.

"Looks pretty quiet, sir," Wainwright said, raising his own binoculars. "Where is everybody?"

"Preparing for battle, no doubt," the commander said absently.

"The dhows look to be unarmed," Wainwright said. "I do believe we could run in there and deal smartly with them right now." He glanced at Dudley.

"You could get blown to tinder," Spicer-Simson said, lowering his field glasses and staring at Wainwright. "Bismarckburg is at least as well armed as Lukuga, and even *Graf von Götzen* gives that a wide berth."

"But think of it, sir," Wainwright pressed. "The Belgians never once managed to hit any of the German ships, and *Mimi* and *Toutou* are much faster. Certainly too fast for them to range us before we sink every one of those boats. We'd be in and out without a scratch, sir."

"Won't you let us give it a go, Commander?" Dudley added.

"Gentlemen," Spicer-Simson said, voice tinged with acid. "If the fort's commandant, Lieutenant Hasslacher, is worth a single grain of salt, and I would suppose he is, he'll have his gunners trained to drop a shell anywhere they want in the entire harbor at an instant's notice. I don't care how fast a rat runs around in a box, half a dozen men standing around with pistols are sure to hit him. I've said it before, but you force me to say it once again. I'll not risk my craft on a futile gesture of bravado. Besides," he shrugged his round shoulders and lifted the binoculars to his eyes. "I have my orders. We are to stand by and prevent the Germans from escaping, not attack. And from what I can see, absolutely nothing is going on." He lowered the binoculars again. "I don't know why that blasted Murray had to send us out here before he was fully prepared. He's all bivouacked in his cozy tent, yet I do believe he expects us to live aboard our boats until he's good and ready for us. But I think not. We'll return to Kitute, and when Murray actually does attack Bismarckburg, we'll come back and take the dhows in hand." He turned and called to Tait. "Bring her up to speed and make for Kitute!" Facing Wainwright and Dudley, he said, "Back to your boats, gentlemen, or you may have to swim for it."

Dudley looked as if he wanted to speak, but Wainwright immediately sensed the futility of argument. Jabbing the other lieutenant on the shoulder, he said, "Let's go, Dudley," he muttered. "We'll have our chance."

He hopped over the rail to *Mimi*'s deck, and Dudley went to *Toutou*, and the flotilla headed back to Kitute.

And stayed there. The crew anchored the boats and lived ashore in some of the facilities that Murray's troops had abandoned when they marched to Bismarckburg. The quarters were considerably better than what they had in Kalemie—real houses instead of huts—but the men were so disgusted at the way things had turned out that none of them seemed to appreciate the fact. Morale declined to a new low, and with it went the last dregs of discipline. Hanschell soon found himself powerless to keep the men out of the local bars and brothels, and even the officers no longer seemed to care.

As the next few days passed, there was much speculation on Spicer-Simson's course of action, most daring to christen it non-action. "Surely we'll mobilize" and "As soon as the battle starts, we'll be Murray's left flank on the water" were a couple of the more generous statements bandied about, but in truth, the battle already had begun.

3

If battle it could rightly be called. Murray's men were probably some of the meanest troops in all of Africa, but they were hampered by a nearly fatal flaw—chivalry. Or perhaps it was just sheer audacity.

Bang! went the gun of one of Murray's men when he spied a German or askari scurrying about on the fort's walls. A few minutes later, another shot barked. And later another. One at a time and in unhurried tempo. In short, these rugged, battle-hardened Rhodesians considered it bad manners for more than one man to fire at a time. Some of the squads even drew lots to determine firing order. And while there were several machine guns in the column's armaments, no one even thought of taking such unfair advantage over the enemy. And so it went for several days, the Rhodesians shooting like sportsmen—and occasionally actually hitting someone—while the Germans settled in to wait. No wonder it appeared from the lake as if nothing was happening.

Lieutenant Hasslacher wasn't particularly concerned about a bunch of pot-shots taking flakes off his walls. Cannon might have worried him, but Murray's Column seemed not to have any big guns. Actually, he was puzzled that the attackers went about their task in such a leisurely fashion, but that would give Colonel von Lettow-Vorbeck more opportunity to squash the Allies elsewhere while the southern front was stalled outside his gates. The fort could wait. It had its own wells, and it was well stocked with food and other necessities as well as munitions. If a siege was called for, Bismarckburg could sit it out for a long, long time.

Then, on the morning of June 7, after the battle had puttered along for three days, one of Hasslacher's men came into his office and announced that a British askari had approached under a white flag and demanded that the fort surrender.

"Who sent you?" Hasslacher asked after he'd gone to the gate, which was a heavy wooden slab studded with iron and pierced with loopholes.

"Bwana Murray, himself, send me. He say for you to give up and end your war."

"Go back to your bwana," Hasslacher said with a laugh, "and tell him that anytime *he* wants to surrender, he's welcome to use your white flag."

With that Hasslacher shut the gate and returned to his office.

The askari went back to Murray's tent to deliver Hasslacher's rejoinder. On his way, he passed by a BSAP squad enjoying the day under the veranda of a commandeered house in the European settlement. They were led by Dr. J. M. Harold, a former international rugby player and all Irish.

"You," Harold called out to the askari. "What did he say?"

"Him say we surrender," the askari called back over his shoulder as he hurried on and disappeared behind the next house.

"What did he say?" asked another BSAP.

"I think he said they've surrendered,'" Harold answered.

"What?" blinked a second BSAP. "Already? We've only been here a few days."

"I suppose Murray's Column's put the fear of God into 'em," a third commented.

"We'll, boys," Harold said. "What say we move in and be the first at the gate?"

His suggestion was answered with a hearty round of "Good shows" and "Jolly goods," and a few minutes later, Harold and half a dozen of his men strolled up to the gate.

"Open up in there!" Harold said loudly to the eyes staring unbelievingly at him through the loopholes.

The same guard who had fetched Hasslacher for the askari went to his commander's office again, this time with the news that a bunch of crazy Rhodesians were pounding at the front gate and demanding it be opened.

"Why didn't you shoot them?" Hasslacher asked.

"They weren't shooting at us," the corporal explained. "They just walked right up."

Hasslacher went to the gate and stared through a loophole at the Rhodesians. They were armed, but to a man their weapons were slung over their shoulders, and all appeared to be far more casual than open hostilities warranted. Hasslacher opened the gate and stared at the leader of the Rhodesians.

"What can I do for you, gentlemen?" he asked.

"Oh, just coming in," said a nonchalant Dr. Harold. He started to take a step toward the threshold, but Hasslacher stopped him.

"What do you mean: Just coming in?"

"Well, you've surrendered."

"I've done nothing of the sort," snorted Hasslacher. "In fact, you are now my prisoners."

Silence followed for several seconds as Harold thought back more carefully over what the departing askari had said. But before he could finish the recollection, he leapt to a conclusion.

"Like hell we are!" he shouted. "You're ours! Grab him!"

Harold's Rhodesians, almost as startled as Hasslacher himself, snatched the German's arms and clothing and dragged him kicking and yelling out of the gateway and across the open ground in front of the fort.

They were fifty feet away before the Germans at the gate and on the walls realized what had happened and started firing. The hail of bullets tore up the ground all around, and three of the BSAPs were hit before they reached a gully deep enough to shelter them from the fire from the walls. One took a bullet through the cheek, the second a bullet in the thigh, and the third was mortally wounded in the groin.

They stayed in the gully for most of the day, pinned down by the German fire despite returned volleys from their own comrades, who now thought it sporting, at least for the moment, to shoot all at once. Without water, desiccating under the hot sun, all they could do was duck and listen to the pitiful moans of the dying man. At last, Hasslacher had a suggestion.

"Your gambit was courageous, Doctor," he told Harold. "But we are all now in a hopeless situation. We must end this before we all are killed or die of thirst."

"Agreed," Harold replied, never having intended any sort of courageous gambit in the first place. "What do you suggest?"

"We both will call a cease fire, and when the shooting has stopped, I will return to the fort, and you will return to your own lines."

"That sounds like a fair arrangement," Harold said. "Shall we?"

Hasslacher pulled his handkerchief from his pocket and held it gingerly above the lip of the gully. Within a few minutes, all firing from both forces stopped, and Hasslacher, followed by Harold, slowly rose into full view.

"Cease fire!" Hasslacher yelled to the fort, and Harold gave a blast on his trusty rugby whistle, which he carried through thick and thin.

Unarmed German troops rushed from the fort carrying pitchers of lemonade, which they served to everyone. While the BSAPs and Hasslacher quenched their thirst, British stretcher bearers came and collected the wounded. Captain R. W. M. Langham, a Rhodesian officer who came out with the stretcher bearers, accepted a glass of lemonade and toasted Harold and Hasslacher.

"I don't imagine," he said, "that many actions have been terminated by a whistle like a Football Association cup tie."

And then, lemonade drained, there was nothing left to do but for Hasslacher and his men to return to the fort and Harold and his own men to retreat to the house they'd been in when, what seemed an eternity ago, the askari had passed on his way back to Murray.

As Harold lay down to sleep, he vowed that things would go differently in the morning.

4

That same evening, Geoffrey Spicer-Simson informed his men that Murray's troops would be attacking the fort on the morrow, and the boats would need to be on hand to prevent the Germans from escaping. He estimated that would probably happen after nightfall, but he warned his men that they might have to spend one or more nights aboard the boats. Whether or not he realized it, most would have been happy to have remained at Bismarckburg for all of the last three days. Wainwright and Dudley, especially, rankled at their shore beds in Kitute so far from the action at Bismarckburg.

Then Spicer-Simson led his flotilla out of the tiny harbor at Kitute. When the boats reached Bismarckburg about mid-morning, Wainwright, in *Mimi*, was in advance of the others and so was the first to take in the scene.

"What that, sir?" Mollison asked, pointing at the fort—specifically at the flagpole sticking above the fort and the flag flying from it.

Wainwright stared and was about to comment when Flynn burst out with, "Oh, hell! It's the Union Jack!"

It *was* the British flag, flapping in the tepid breeze, and in that instant, Wainwright realized the unmistakable and cruel truth: Not a single dhow or canoe remained in the harbor. He lifted his binoculars and could see troops in BSAP uniforms lounging around the docks on the waterfront.

He glanced back across the fifty feet of water that separated *Mimi* from *Fifi*, trying to make out the commander's face, but Spicer-Simson was half turned away, and only the tensed set of his round shoulders betrayed his inner workings. That and the droop of cigarette holder normally clenched at a jaunty angle between the commander's teeth.

"I shouldn't like to be the commander," Mollison commented dryly.

Nor I, Wainwright thought.

Then he saw Spicer-Simson give a subdued signal to his wheelman, and *Fifi* headed toward the harbor. Wainwright waved for Mollison to follow. In a few minutes, the ships were at the docks, and the last hundred yards weren't pleasant as jeers and laughter floated across the water to meet them.

After *Fifi* and the other boats were tied up, Spicer-Simson stepped onto the dock, accompanied by Wainwright, Dudley, and Hanschell.

"Where the hell were you navy chaps?" a young Rhodesian lieutenant called out. He was slouching lazily against a piling. "The Germans sneaked out, and you let them get away by sea. We had the fort surrounded."

"You don't look like Colonel Murray," Spicer-Simson sniffed, adjusting his cap with its insignia. "Don't you soldiers know how to salute a superior officer?"

"Why, yes, sir." The lieutenant raised a desultory hand almost to the bill of his cap then let it fall. He didn't bother to stand upright.

"I want to speak with Murray," the commander said. "Take me to him."

"If it's Murray you want, he's straight up there." The lieutenant's backsides remained cemented to the piling, and his desultory hand made another feeble rise in the general direction of the fort. "You can't miss it. Just follow your nose."

Spicer-Simson flushed, and he looked as if he was about to say something more to the Rhodesian, but instead, he turned to his own men.

"Wainwright, stay here. Dudley and Hanschell, if you will accompany me, please." He started up the street toward the fort at a brisk pace.

After more than six months in the commander's presence, Hanschell had become so accustomed to Spicer-Simson's personal garb that he no longer noticed it. And today, the commander was dressed no differently than usual in his blue shirt, gold-braided cap, and of course, his leather skirt. But while time had inured Hanschell to Spicer-Simson's clothing, the rugged Rhodesians had no such advantage. Nor did they know the arduousness of dragging two boats across half of Africa or of the battles with the *Kingani* and the *Hedwig von Wissmann*. They only knew that the navy had showed up at Bismarckburg too late, and that its commander cut a ridiculous figure.

"Chase me, Charlie!" said one wag, and laughter broke out, telegraphing to any soldier within earshot that something was afoot. They flocked to see what it was.

"Ooh-la-la!" sang another as soon as he caught a glimpse of Spicer-Simson, and his mate gave a big smacking sound and said, "Kiss me, Gertie!"

Whistles, laughter, and more ribald comments followed them up the street, all the way to the gates of the fort. Hanschell could see that Dudley was extremely distressed by the Rhodesians' sarcasm, but he knew the young lieutenant would survive. After all, he could go home knowing he wasn't the target of the jest. It was the commander Hanschell worried about.

Through it all, Spicer-Simson faced straight ahead, jaw stiffly set. But by the time the docks were behind them, a lifeless pallor had replaced the flush that had suffused him at the Rhodesian lieutenant's insolence. Hanschell had to admire the commander's resolute march, but if his footsteps didn't falter, surely his heart must.

A sergeant met them at the gates, and under his glare, the catcalls died out, and most of the curious and derisive BSAPs wandered off to see what the Germans had left behind in the European settlement.

"Colonel Murray is this way, sir," the sergeant said following a sharp salute. He led them to an anteroom and said to Spicer-Simson "He'll see you alone, sir."

"It's all right, gentlemen," Spicer-Simson told Hanschell and Dudley. "Wait for me here." His voice was steady, but Hanschell could see panic lurking in the commander's gray eyes. Failure, the doctor read there. One more chance, and one more failure. Maybe the worst.

And then Spicer-Simson disappeared into the heart of the fortress, and all Hanschell and Dudley could do was wait. The sergeant brought them each a cup of tea, but Hanschell had little taste for it, and it soon grew tepid on the small table beside his chair. He noticed that Dudley wasn't drinking his, either.

"You mustn't blame the commander," Hanschell said to the lieutenant.

"And why not?" Dudley's voice was harsh and his face was still red from enduring the soldiers' ridicule. "He's a cowardly fool, and he's made us all look like cowardly fools."

"You think that now with all these idiots around. But you won't later, when you've had a chance to remember all he accomplished."

"He didn't do it without us," Dudley retorted fiercely.

"Nor you without him," was the doctor's quiet reply.

After that, they waited in silence, Dudley stewing in his anger, Hanschell wishing he had some drug that would take away everyone's pain.

At last, the commander emerged, face livid and taut. His hands extended outward, as if he was feeling his way through a dense fog or a darkened room. Hanschell later heard from Doctor Harold that Spicer-

Simson made no attempt to excuse his failure and even admitted that he'd miscalculated. But Kaffir Murray cared little for men who did not fight and even less for the man who'd thwarted his complete and total victory by sleeping while the enemy escaped out the back door. He told the commander to shut his mouth then verbally keelhauled Spicer-Simson's intelligence, drew and quartered his military competence, and decapitated his courage. Finally, he told him to submit his miserable self forthwith to voluntary confinement pending an investigation. And that was an order!

If Hanschell knew nothing of the details at the moment, he could guess their substance. Right now, though, there was nothing he could do but take the devastated commander's arm and follow the sergeant, who lead them through the fort to the officers' residential wing. There they left Spicer-Simson in his temporary quarters: a small, white room with a window that opened on a scenic view of the harbor and Tanganyika beyond.

"He'll be all right here, sir," the sergeant said, helping them seat the commander in a chair beside a table. "I'll make sure he's comfortable."

It was impossible to tell if Spicer-Simson was comfortable or not. He just sat there, not stirring, not speaking, only staring blankly through the window with his gray eyes lost in some realm where thought would be torment.

"What's wrong with him, Doctor?" Dudley asked as they quietly left the room.

"He's taxed to the limit, Lieutenant," Hanschell replied, shutting the door. "He's given what he could, and he's used up."

"If you'll send your quartermaster up, I'll see to it that all your men are billeted nearby," the sergeant said. "They'll have to put up with the ruffians outside, but I'm afraid there's not much help for that."

"You don't feel like they do, Sergeant?" Hanschell asked.

"I'm just a sergeant, but I've been with Colonel Murray a long time, and I hear things those blokes in the tents don't. The colonel had me check on your commander's exploits, and while I can't say I understand what happened here, I know your expedition's been through a lot."

"Thank you, Sergeant. We'll send up our paymaster, Lieutenant Eastwood. He serves as our quartermaster."

"Thank you, sir. I'll be looking for him."

Then there was nothing for Hanschell and Dudley to do but go back through the streets of Bismarckburg to the docks in the harbor, enduring more insults. They remained as silent as Spicer-Simson had been when he emerged from Murray's office, but Dudley broke it once after a particularly offensive insult floated down the street after them.

"By God!" he growled. "The commander's had it in the neck! I'd like to...." But there were no words for it because words were inadequate.

When they reached the docks, they told Wainwright what had transpired, most of the men standing around listening with growing anger.

"Those bastards!" Flynn spat. "What the hell do they know about the commander or what we've done?"

"Somebody ought to go up there and teach 'em a lesson," Lamont said, and his words were followed by assent from many of the men. But Wainwright calmed them down.

"No use in fighting a battle we can't win, boys," he said. "What we need to do right now is support the commander. From what Doctor Hanschell says, he needs us more than ever."

Calling one of the ratings to follow him, Wainwright went to *Fifi*, rummaged in Spicer-Simson's trunk, and came up with a bottle of vermouth.

"Take this up to the commander," he ordered. "He could probably use it right now. And take a bottle of water and a couple of glasses, too."

The rating hurried off with the bottle, and Wainwright stepped back onto the dock and, with Hanschell and Dudley, stared up at the fort that held their commander.

Silently, Wainwright shook his head.

Chapter XIX: The Duration

"What's the news, Doctor?"

It was the day after Murray had torn into Spicer-Simson, and the officers were clustered down at the docks, feeling despondent as they looked up at the white fort on the hill from which Hanschell had just come.

"Much the same," Hanschell said. "He's spending most of his time lying on his bed. I can tell he's been writing something, but all I ever see of it are crumples of paper that he burns."

"Did you tell him what I propose?"

"I did. He says you have permission to send the Shorts after the *von Götzen*, but not the boats."

"I just don't know what to make of it," Wainwright sighed. "I know he's not a coward. Look at all he did. Why couldn't he just have gone ahead after *von Götzen* and Bismarckburg?"

"He took what he could take without risk."

"Without risk?" Wainwright snorted. "I was there, Doctor. The lead flew all around us, and he didn't seem to give a damn about his own skin."

"Maybe it isn't his own skin he's worried about."

"You said before that he was worried about us." A short laugh escaped Wainwright's lips. "I don't believe it. I've known the commander to be concerned about the safety of his boats, but his men?"

"No, not you, either. Spicer-Simson was sent here to eliminate *Kingani* and *Hedwig von Wissmann*. He took the necessary risks to do that, but perhaps that was all he was prepared to take. He did what the Admiralty asked, and when he requested sufficient resources to do more, they ignored him. You know as well as I that they've steadfastly refused to consider the *von Götzen* a credible threat. But you've seen her, where they haven't, and you know she's a powerful adversary."

"That still doesn't answer for that fiasco in letting the Germans escape."

"Perhaps not, but I did learn from Murray's sergeant that Murray didn't expect the Germans to vacate the fort so quickly. It seems that there was some fluke the day before when some of Murray's men captured the fort's commander. It was only temporary, and they had to surrender him, but apparently the incident spooked the Germans into leaving without putting up more of a fight. Had the fluke not happened, the battle probably would have gone on days longer, in which case, the commander's timing was quite sufficient."

"I guess we'll never really know," Wainwright said. "But it's a hell of an end to things."

"I'd rather remember the commander for his spectacular successes," Hanschell said quietly, "not damn him for refusing to futilely throw his men to certain death or for matters beyond his control."

"I'm not sure you're right, Doctor, but I bow to your experience. You know the commander better than any of us."

"I might once have thought that true, but no more. But I do know this much: He's a beaten man."

"He may be," Wainwright said. "But the rest of us aren't. If you'll excuse me, I think I'll have the Shorts throw a few bombs at the *von Götzen*."

2

On June 11, the Shorts did lob several bombs at the German behemoth, but the nervous pilots remained intimidated by *von Götzen*'s anti-aircraft gun and were afraid to come in close enough to ensure a strike. They reported that one of the bombs made a direct hit, but in reality, though the bombs did some minor damage to the nearly empty waterfront, all missed the ship.

The day after that, in the same packet that brought the message from the pilots, a telegram came for Spicer-Simson. He ordered that the men assemble for full-dress parade, and he emerged from his room for the first time in days and went down to the harbor. He was pale and haggard, and he was wearing trousers. He also was without his sword, but he did have his swagger stick and cigarette holder, though the latter did not jut at its former jaunty angle.

"Gentlemen," he said without preamble when they were arrayed at attention. "I have received another congratulatory message from the Admiralty. They commend us for our support in helping Colonel Murray take Bismarckburg." He looked blankly around at the men, not seeming to notice their disgusted expressions and sullen shame. "That will be all."

Without another word or gesture, Spicer-Simson turned and left the docks.

Wainwright dismissed the men, and they watched as their commander trudged up the street toward the fort, swagger stick hanging listlessly from his hand.

Three days later, Colonel Murray deployed his forces up the lake's eastern shore, leaving Spicer-Simson in command at Bismarckburg.

"Pockets of German resistance will remain, even after we pass," the colonel told Spicer-Simson. "I expect your boats to lend support where needed."

"They will," Spicer-Simson promised.

"They'd better," Murray snapped, his dark brows beetling. "One more failure, and I'll see you cashiered."

As soon as Murray left, Spicer-Simson retired to his room. And remained there, seemingly oblivious to anything outside its whitewashed walls. Hanschell visited him twice a day, but of the other officers, only Eastwood came to see him, and even he quit coming after a few days.

"He just sits there, Doctor," Eastwood said. "What can we do?"

"We'd better do something," Wainwright said. "Murray's men need us to mop up after him. He's due to meet up with Tombeur and our Kenyan forces any day for the final push on Tanga. Murray won't like it if there are Germans left at his rear to harry his troops. They're asking for support."

"Let's go up," Hanschell suggested. "You'd be better at explaining all this to him."

"Send Dudley out," Spicer-Simson said in a distant voice when Wainwright and Hanschell relayed the situation. "He can take *Fifi* and *Mimi*. Why don't you go with them, Doctor. Your services might be needed. There's certainly nothing you can do here."

"What about me, sir?" Wainwright asked, dejected at not being given another chance at command.

"I want you to go back to Kalemie. The *St. George* will have arrived by now, and I'll need someone I can trust to see she's put together properly. And you can send the Shorts out again."

"But they reported they hit the *von Götzen*," Wainwright said.

"Best check, Wainwright. Be safer than sorry. Take Cross and Lamont with you—they're damn fine engineers."

So, the next day, the two boats left, with Hanschell, Wainwright, Cross, and Lamont aboard as passengers, bound for the northern reaches of the lake, leaving Spicer-Simson with only a skeleton staff, a lot of empty time, and the four white walls of his self-enforced confinement.

The pieces of the *St. George* were at Kalemie, but the past month had wrought more significant changes at the harbor. The recently arrived

Belgian political officer had, in Spicer-Simson's absence, begun drawing up a list of port regulations. There was little that Wainwright could do about that, so he simply sent a report to Spicer-Simson, then he and the two engineers began seeing to the rebuilding of the *St. George*. None of them knew why they bothered, even though they learned that the Short pilots had erred in believing the *von Götzen* sunk and that it was still afloat.

Wainwright sent the planes out several more times, but to no avail. Apparently the pilots were too leery of the anti-aircraft gun and the machine gun and small arms fire from the harbor, and though they dropped a few more bombs into the harbor, none touched the *von Götzen*.

3

Captain Zimmer had more to worry about than a couple of biplanes showing up every few days to tear up more of his unused docks. By mid July, the railway leading to Kigoma had been seized by Belgian forces, in part under Colonel Moulaert's command, completely cutting off the port. Colonel von Lettow-Vorbeck could spare nothing to retake the line, for Allied forces had his own troops under attack from all directions. At last a messenger from von Lettow-Vorbeck got through the Belgian lines, and Zimmer found himself facing the inevitable. Belgian troops were as close as Gottorp, only sixty miles away, were advancing steadily, and would be in Kigoma in a matter of days. Zimmer was to do what he could to keep war materiels out of Allied hands and then escape south down the lake.

Reluctantly, on July 26, Zimmer boarded the *Graf von Götzen* for the last time and ordered her to steam out just past the harbor mouth, where the water was a little deeper but before it shelved off into Tanganyika's unplumbed depths. There, he ordered the engineers to shut down the engines and, when they were cool, grease all the machinery. While that was happening, he had the anti-aircraft gun dismantled and loaded, along with all its ammunition, aboard the waiting *Wami*.

Thank goodness the bigger guns had been removed and transferred to the infantry months ago, he thought as he sent *Wami* back into the harbor.

Wami returned, and eventually the preparations aboard the *von Götzen* were complete.

"Don't worry," Zimmer said to the lieutenant in charge of *Wami*. "We'll refloat her after we've driven the Allies out of East Africa. Never doubt that for a moment." Then he waved to the three engineers who were still aboard, and the men disappeared below decks. They weren't gone long, and when they came back onto the deck, everyone hurried aboard *Wami*, which chugged a safe distance away from the *Graf von Götzen*.

At first, nothing seemed different about the ship that, even with dummy wooden guns, had terrorized Lake Tanganyika. But after a few minutes, a definite settling was noticeable. As the settling grew more pronounced, the men on *Wami* removed their hats. At last, with a great hiss of air, the scuttled ship sank out of sight. *Wami* turned and headed for the harbor.

4

For the rest of the day, the lieutenant in command of *Wami* ferried the last of Kigoma's contingent thirty miles down the coast to the mouth of the Mlagarasi River. From there, it would be but a relatively short hike inland to meet up with one of the contingents of von Lettow-Vorbeck's forces.

Zimmer went with the last group before sundown, and the lieutenant made several more trips the following day without incident. But as luck would have it, on his final trip, one of the British gunboats appeared from behind a promontory. By now, the Germans knew all too well what had happened to the *Kingani* and *Hedwig von Wissmann*, and the lieutenant was filled with fear, realizing that he could never out-run the British gunboat. Nor dare he fight. With no arms aboard save a few rifles and pistols, *Wami* was ill-equipped for battle, and through his binoculars, the lieutenant could see the dark snout of a gun mounted on the British boat's bow—one of those three-pounders, no doubt, that had taken the *Kingani*.

With no other option, the lieutenant turned and ran straight for shore. They were still some miles north of the Mlagarasi, but better to have to walk an extra day than spend the remainder of the war in a prison camp.

No sooner had the *Wami* begun its flight, than the other boat gave chase. Had the lieutenant realized the truth, however, he might have tried to bull his way past, for the other craft was not *Mimi* or *Toutou* but only lowly *Vedette*, a Lewis gun, not a three-pounder, on its bow.

The lieutenant had kept *Wami* hugging the shoreline as closely as possible, and it didn't take him long to run his boat aground on the rocky shingle. Within minutes, his men, carrying their meager equipment, were scrambling up the shallow bluff. As a final act, the lieutenant raked coals from the small boiler's furnace and spilled them into *Wami*'s wooden hull. The boat ignited, and as smoke trailed over the water, the lieutenant followed his men up the embankment and vanished into the bush without ever knowing he'd been spooked rather than attacked.

The Belgians weren't exactly sure why *Wami* had fled from them. All they'd planned to do was follow at a safe distance to see what the Germans were up to. But as soon as *Wami* went up in flames, they turned and headed down the coast to find *Fifi*, which they'd seen patrolling in the vicinity earlier in the morning.

Dudley still commanded *Fifi*. After nearly six weeks, Spicer-Simson remained in his room in Bismarckburg, refusing to do more than issue vague orders to assist in mopping up operations as required. When the Belgians puttered up and shouted that the Germans had abandoned *Wami* on the shore and set it afire, Dudley had Tait bring around the helm and follow the Belgian motorboat to the scene.

Only a dim pall of smoke marked the spot when they arrived, most of the hull already having burned to the waterline.

"Drop a few shells on her," Dudley instructed Waterhouse. "It's not much, I know, but we've nothing better to do."

Waterhouse complied, and in a few minutes, nothing remained of *Wami* but a few scraps of wood floating on the water.

Curious why the *Wami* would dare venture so far south of Kigoma without the protection of *Graf von Götzen*, Dudley told Tait to make for the German harbor. He had no intention of disobeying the commander's orders not to engage the German warship, but it wouldn't hurt to gain a little intelligence.

As it turned out, there was precious little intelligence to be gained. As *Fifi* steamed past the harbor mouth, there was no sign at all of the *von Götzen*. Puzzled as to what that could mean, Dudley made for Kalemie, where he could confer with Wainwright and send a wire to the commander to let him know that the *von Götzen* was not in Kigoma and might be loose anywhere on the lake.

5

The news seemed to put a tiny spark of vitality back into Spicer-Simson's eyes, especially when he learned that the construction of the *St. George* was going well. He wired Wainwright to send out the Shorts to see what they could see. The pilots didn't find the *von Götzen*, but they reported that there seemed to be no activity at all at Kigoma, not even construction on the *Adjutant*, which lay half-built in the shipyard.

Wainwright told them to make a second run, increasing the radius of their search. This time, they came back with the news that they'd spotted the hulk of the *von Götzen* sunk in thirty feet of water just north of Kigoma.

"They must have scuttled her," Spicer-Simson mused when Wainwright sent him the news. "Our forces were getting too close."

He sent a dispatch off to Nairobi, stating that the seaplanes had identified the sunken *von Götzen*, and that the Allies now had total control of Lake Tanganyika.

But the news made little impact since General Tombeur and the combined Allied forces were, that very day, occupying Kigoma. The Germans were all but defeated, though von Lettow-Vorbeck had held on far more tenaciously than anyone expected.

Even so, Spicer-Simson's private war with the Belgians had not yet reached a cease fire. He was not languishing in his room all those hours and days and weeks. He was composing and sending lengthy and rambling dispatches detailing Belgian usurpation of his authority as senior naval officer, their flagrant disregard of agreements with the British government, their incompetency in war, and most of all, their conniving imperialism on the lake, to the detriment of the British Empire. Foremost among these latter were the unreasonable port regulations they'd instituted at Kalemie. It was *his* harbor, damn it, engineered and built by *his* expedition in the face of obstinate obstruction from Stinghlamber. Kalemie, by God, was a British harbor, and their taking it was tantamount to international brigandry.

These messages flowed steadily through the local telegraph office, and Spicer-Simson had Seaman Tasker stationed there to personally handle his correspondence and any other war news.

Around the middle of August, the commander began to appear outside on the fort's ramparts and occasionally down in the harbor. He was largely ignored by the men, who couldn't forgive him the gaffe at Bismarckburg, but Spicer-Simson didn't respond to their snubbery—didn't even seem to notice them or anyone else. He shambled about like a ghost seeking some faint whisper of life at which to grasp. Frequently, he scanned Tanganyika through his binoculars, and the times Hanschell approached, he could hear the commander muttering under his breath.

"I suppose our part is nearly finished," Hanschell ventured on one of his visits.

"Our role may not yet be played out," Spicer-Simson replied arcanely.

We've taken the lake, and the Germans are done for. What more is there?"

"I'm expecting a dispatch," the commander said, raising his binoculars and aiming them toward the lake. "Any day, now."

A dispatch *did* arrive on August 23, but if Spicer-Simson was truly expecting one, this wasn't it. Nor was it one he wanted. Worse, it was addressed to Colonel Moulaert. It was from Jan Smuts, commander in chief of the Allied offensive against German East Africa, and it came in on the common wire, so the British received it at the same time as the Belgians.

The offensive had gone well, Smuts said, largely because "German armed vessels on the lake had been bombed and destroyed by seaplanes."

Seaman Tasker was in concord with his fellows in his contempt of Spicer-Simson's lapses, but this insult to the entire expedition brought a flush of emotion that outweighed his disdain, and he hurried over to the commander's quarters.

"A lotta gall they have, sir," he said, handing the message to Spicer-Simson. "After all we've done."

The commander read the dispatch, his face growing pale.

"I need Doctor Hanschell," he told Tasker. "Please send for him immediately."

6

When Hanschell arrived two days later, he found Spicer-Simson once again had taken to his bed. He pulled a chair over, sat by his stricken leader, and stared at his slack jaw and sunken eyes that now had the color of old pewter instead of bright steel, noting the unhealthy pallor that had crept in beneath the deep tan.

"What am I to do?" Spicer-Simson's voice, once a brash bray, was but a faint wheeze. "Now they don't even believe me at all. I'm ruined."

He gestured toward the dispatch from Smuts where it lay on the bedside table. Hanschell read it, feeling his own ire rise. But he suppressed the emotion. There would be time enough for that later if necessary. Right now, he had a patient to attend.

"It doesn't matter what Smuts thinks," he told Spicer-Simson. "What matters is that your work is done. It's time to go home."

"Back to England?" A flicker of hope dawned in the commander's eyes, only to be quenched by a cloud of depression. "But what will they think?"

"They'll think you are what you are," Hanschell said. "The hero who, through valiant effort and against tremendous odds, took Lake Tanganyika."

"But how shall it be accomplished? I'm the commander here...."

"And I'm your doctor. You've been ill. Africa is no place for an Englishman, and you've been here too long. Oh, Tanganyika is healthy enough, and we've got a fine sea breeze, but you've been traipsing around in the Congo looking for the *St. George*. I shouldn't doubt you've picked up some nasty tropical bugs that I can't cure on the spot. England's the place for that."

"Yes, please. I need to go home. I can't face the scrutiny any more."

"The scrutiny?" Hanschell was perplexed.

"I feel like a bug stuck under your microscope, ready for discovery and cataloging as some exotic species."

"I assure you, Geoffrey, that I never...."

"Oh, I know it's not you. You're the only one I can call friend. It's everyone else. Like the men—all watching me, often without looking at me. But I can't blame them, either. I am, after all, their leader, and it is only right and necessary that they watch me. But there were the Belgians, constantly probing for a weakness with which to thwart my plans by impeding my physical needs—the precious few, that is, that actually were being met by the Admiralty—in direct contravention of all their agreements when I formally consented to lead the expedition."

"The Admiralty let you down."

"Yes, yes. And it was their scrutiny that mattered most—what they were seeing and what they refused to see. I didn't want to make a wrong decision and then be shown for a fool. I've had enough of that from them. In no manner would I endure that one more time. Better to endure censure for lack of initiative. But I reached their goal for them and executed what they required post-haste and with no casualties. Even better, I captured one of the enemy ships and sunk another and, most likely, kept the third at bay even if I didn't attack it. But I have the evidence of the Admiralty's own dispatches making light of the *von Götzen* and the remaining German defenses on the lake. If it comes to a confrontation with them over this, they're as much to blame for ignoring my repeated requests for more resources. And it's not as if the Admiralty wasn't desperate to have a leader for this insane venture. Honestly, Hanschell, I don't know why I allowed myself to become involved. It's all been so terribly consuming."

He looked up at Hanschell, eyes so bright that the doctor thought that he really must have come down with fever. But Hanschell could do no more than think this before the commander hurried on.

"Do you know what was the worst? It was the Africans. Always the Africans, always there, always watching. Oh, it's not that they're Africans, if that's what you're thinking. It's just that there were so *many* of them. I drew them like iron filings to a magnet. You can't imagine what it's like being a god, Hanschell. It's damnably hard. You can't make a mistake, can't falter. Do you know what men do to gods who fall? And these were not just *any* men. They were Ba-HoloHolo. If we failed in the slightest way against the *von Götzen*, they would have torn us to pieces. Not just me—all of us."

Spicer-Simson hung his head.

"I simply couldn't take it any more, Hanschell. I thought it was the success that I wanted, but when I got it, it broke me."

"Don't worry," Hanschell reassured him. "I'll take care of matters immediately. Who will you leave in charge? Wainwright?"

"Not Wainwright," Spicer-Simson said, looking at the doctor. "He's a good man and performed invaluable service for the expedition, but he's an engineer, not a seaman. I'll recommend that Dudley take over until my replacement arrives."

"Are you sure, Commander? Dudley is a bit of a hothead."

"A quick temperament is more suited to a line officer," Spicer-Simson responded. "Besides, Dudley is young and has more opportunity for advancement than Wainwright. I know only too well what it's like to be an old lieutenant, which is probably all Wainwright could ever be even if he is a clever innovator. He'll most likely return to his farm, anyway, when the fighting's done. I want to give Dudley a chance to advance if he chooses to stay in the Royal Navy."

He sighed and closed his eyes, and after a long pause, Hanschell thought he must have fallen asleep. But then he opened his eyes and said, "Would you take it down for me, Hanschell?"

Hanschell wrote out Spicer-Simson's dispatch, which included his recommendations concerning his officers and, to his surprise, a belated declaration formally relinquishing Spicer-Simson's command over the Belgian ships. Then Hanschell composed a dispatch of his own and, that afternoon, sent both to General Edwards.

"Commander Spicer-Simson is ill," the doctor's message read. "He is suffering from acute nervous disability and a variety of tropical diseases. In my professional opinion, he should be invalided immediately and returned to England."

The reply came rather too quickly and was worded too eagerly, but in short, it agreed that Commander Spicer-Simson's job was finished and he was ordered home forthwith. His recommendation regarding Dudley was approved until the Admiralty could send out a replacement.

A good thing, too, Hanschell thought, carefully folding the telegram and putting it in his breast pocket. Spicer-Simson already had left Bismarckburg and was in Kitute, preparing to depart to visit General Edwards before traveling to England.

Something that Spicer-Simson said during their final conversation was still fresh in Hanschell's mind, though.

"We could never have done it without you, Doctor," the commander said with a wan smile. "You kept us healthy and fit."

Wainwright, who'd come to Bismarckburg to see Spicer-Simson off, received his own accolades regarding the way he'd managed to move the expedition along despite every obstacle placed it its path.

Then the two men watched as Spicer-Simson boarded the motorboat to be ferried to Kitute.

"I want you to know something, Hanschell," Wainwright said after Spicer-Simson was out of sight. "The commander was wrong about the lack of illness. Or, not wrong, exactly, because you certainly did your job. I don't think I shall ever forget the sight of those government rest houses all ablaze." He chuckled, then sobered as he went on. "But he was off the mark. He should have praised your for your faith in him. That was what gave him credibility at the first, and without that, this expedition would never have reached the Mitumba Mountains, much less gotten over them to take *Kingani* and *Hedwig*."

"And what a shame that would have been," Hanschell replied with a smile. "Look at all the Belgian money we would have lost."

Both men laughed.

7

"A frightful place, sir," Spicer-Simson told Sir Henry Jackson, the British First Sea Lord. "If I never see the Congo again, it will be too soon. What was going on at Tanganyika was nothing compared to there. Barbarism at its worst. And the sickness was frightful. I was there six weeks trying to commandeer the *St. George* from the British consul, and while that may not seem long, it was enough for me to contract a mélange of tropical diseases. The good Dr. Hanschell diagnosed malaria and dysentery, and there were a few even he couldn't identify. I'm still a little shaky."

He *was* still a little shaky, but the sea voyage home had done him a world of good, and he'd lost the listless bewilderment that had inhabited his mind those last two months in Africa. In fact, he barely remembered them at all.

"We had reports through the Belgians that something was wrong." Sir Henry leaned back in his leather-upholstered chair and tented his fingers. "What was it, they said, Sir David?"

"Something about you refusing to attack the *Graf von Götzen*," Admiral Gamble replied. "Nothing to it, of course, eh, Spicer-Simson?"

The three men were in Sir Henry's office, where Spicer-Simson had received his original orders what seemed a lifetime ago. The red leather chairs were grouped around the fireplace, but with the summer weather pleasant outside, the hearth was cold and bare.

"Well, Admiral, of course, at first, I couldn't attack the *von Götzen*. She simply was too large and well armed for my gunboats. That's why I went for the *St. George*. And then we refloated the *Alexandre Delcommune* and renamed her *Vengeur*, and with her added to the flotilla, we could have sunk the *von Götzen*. But by then the Germans knew we were on the lake looking for her, and they refused to come out and face us."

"Didn't your seaplanes eventually sink her?" asked Sir David. "We had some such word from Jan Smuts."

"General Smuts is a very capable soldier," Spicer-Simson said discreetly, "but on this matter ill-informed. Actually, the Germans knew we were closing in, and they were afraid we'd take *von Götzen* as we had *Kingani* and turn it against them. They scuttled her with us just minutes away."

"You tell an interesting tale," said Sir Henry. "And the public has taken it to heart. The press has, of course, followed your exploits since the taking of the *Kingani*, and some of your men have been talking to the reporters."

"I'll take them in hand at once, sir," Spicer-Simson said, not mentioning that he'd given several interviews himself.

"No need. Our men are having a terrible time on the Continent, though we are making headway, and a bit of good news and public heroics might be just the thing to brighten up the folks here at home."

A discreet knock sounded at the door, and Sir Henry called out, "Come in."

Sir Henry's adjutant entered.

"We're ready, Sir Henry."

"Very good. Whitley, see to it that Commander Spicer-Simson finds his way to the courtyard." He turned to Spicer-Simson, who stood, formally straight. "Would that all our naval actions went as well as yours, Commander."

"Thank you, sir." Spicer-Simson saluted and held the pose until Sir Henry nodded.

"Be off with you, now, Commander. Sir David and I will join you shortly."

"Sir." Spicer-Simson spun on his heel and strode purposefully from the room, followed by the adjutant.

"What do you think, David?" Sir Henry asked as the door shut behind them.

"I'm not certain," said Sir David. "He's a curious sort of chap. The Belgians seem to believe he was a lot more trouble than he was worth.

And I hold with my initial assessment that he doesn't seem to be real leadership material."

"But he did take Tanganyika. That much is clear."

"Yes, he did all we asked him to do, if not any more than that. But the reports are conflicting, and I'll want a closer look at all the dispatches."

"Naturally," said Sir Henry, standing and smoothing his uniform. "But right now, I think it's time we went out and gave the commander his due."

Spicer-Simson had joined several other naval officers in the courtyard outside naval headquarters. The award ceremony was relatively brief. There was a war on, after all, and admirals are in short supply and high demand. When Sir Henry got to Spicer-Simson, he began by repeating the message King George had sent immediately following the capture of *Kingani*, congratulating the success of His Majesty's remotest expedition. Spicer-Simson was promoted to full captain, and the Distinguished Service Order was pinned on his tunic. And despite his turmoils with the Belgian command at Lukuga, King Leopold had named him Commander of the Crown of Belgium and awarded him the Belgian Croix de Guerre with three palms.

That night, Geoffrey Spicer-Simson celebrated with Amy in high style at one of London's swankiest clubs.

"My dear," Spicer-Simson told his wife over a glass of wine while they waited for their meal. "Your fashions were ravishing. The African's were quite taken with my skirt."

"You were in all the papers," she said proudly. "It's so terribly exciting."

During the meal, the commander—the captain—was asked several times to autograph this newspaper or that magazine account of the expedition. His favorite headline was "Nelson Touch on African Lake," but Amy was partial to the title "The Jules Verne Expedition." Neither mentioned that "The Strangest Story of the War" probably was the most accurate.

"You're a hero, dear," Amy told him that night after they returned home and Toutou was dancing around their feet, trying to gain their attention, while Mimi lounged regally on the ottoman. "A real hero."

"I am, aren't I?" He seemed as amazed as she was proud.

"I always knew it was so."

"I did, too," he said, and he sounded like he meant it.

"But I'm so glad you're home. I missed you terribly, and I was frightfully worried that something might happen to you."

"To tell the truth, my dear, I'm glad to be home, myself. The success of the whole affair was on my shoulders, and I don't mind telling you that it was an awful ordeal."

"Well, let's think no more on it tonight," she said, tugging his arm and giving him a sly wink. "We have business to attend to upstairs."

"Indeed."

They went up the stairs and soon shut off the lights.

If Spicer-Simson had known of the ordeal to come, though, the night might not have been so blissful.

8

Despite other pressing matters, it didn't take Admiral Gamble long to examine the record, voluminous though it was for the relatively short duration of a year. It didn't take long, because he was curious. And suspicious.

Gamble already was aware of the friction between Spicer-Simson and both Stinghlamber and Moulaert, having seen many of the dispatches Spicer-Simson sent during his time on the lake. But now, as Gamble reread them, he realized that he had merely glossed over them in his initial reading, automatically giving Spicer-Simson the benefit of the doubt. After all, the man had accomplished the nearly impossible feat of dragging the boats intact across half a forbidding continent and, on top of that, quickly taking the two ships he had been sent after.

Spicer-Simson's successes lent him credibility and were featured in four official Admiralty dispatches. They were to his credit—and much of the reason behind his promotion and awards. Thus, Gamble had backed Spicer-Simson at every juncture, particularly with delegation of naval authority on the lake. Even the counterarguments of the Belgian high command coming on the heels of Stinghlamber and Moulaert's complaints had not altered Gamble's opinion—or the opinion of anyone else on the Admiralty staff. Everyone had come to believe, based on the flurry of messages that had come in since the expedition had reached the lake, that the Belgian military authorities in Africa were a difficult and incompetent lot whose behavior often bordered on the aberrant. Spicer-Simson was simply a determined officer trying to deal expediently with the business of winning the war.

But the further Gamble read, the more he saw that most of Spicer-Simson's dispatches and reports seemed a confused jumble. Confused, that is, until he requested and received reports from John R. Lee, Major

Stinghlamber, Colonel Moulaert, and others, at which point, a clearer picture began to emerge.

According to Lee, Spicer-Simson had seized command of the expedition through a web of underhanded misconduct and treacherous misinformation, sending Lee off on wild goose chases then reporting him drunk and truant. Stinghlamber wrote at length about the commander's arrogance, his forcible seizure of the Belgian vessels, and his unorthodox uniform. As he read about the leather skirt, Sir David raised an eyebrow. And, Moulaert told how Spicer-Simson had literally vanished for nearly three months, leaving his command in complete limbo at a critical juncture in the assault on German East Africa.

Then there was that scathing message that came in over regular channels, spotted by an alert information officer on Sir David's staff. It was from Brigadier General Northey, the British commander on the southern front, who had just recently received a complete report from Lieutenant Colonel Ronald Murray, the man who had lead the successful assault on Bismarckburg. The details of Spicer-Simson's dereliction in failing to prevent the Germans from escaping were stated in no uncertain terms.

Most damning of all, however, was the tenor of Spicer-Simson's later dispatches, which more and more took on a tone of rambling madness as events on the lake progressed—or deteriorated—following the sinking of the *Hedwig von Wissmann*.

Cautioning his staff to let no word of the reports find their way into the press, Sir David met in closed session with Sir Henry. It was a long meeting.

"Good, God," Sir Henry sighed as Sir David finished laying out the facts of the case. "I'm dumbfounded. We knew it was desperate venture, but had we known that Spicer-Simson was such a foolhardy character, I don't think we would have sent him."

"There was no one else," Sir David reminded him.

"Even so, it appears that our man not only flagrantly disobeyed his orders and sent us false information," said Sir Henry, "but he has filled his official communications with malicious and libelous lies against the Belgians and subjects of our own crown. What a damn embarrassment."

"At the very least," Gamble replied. "Before taking a step that severed cooperation before the command of the lake was assured, he should have written to Colonel Moulaert, received a reply, and transmitted the correspondence to the Admiralty. His tactless behavior might have contributed to a serious disaster."

"Can we even be certain that he performed his duty at all?" Sir Henry asked stroking his chin. "That dispatch from Smuts says that the German boats were all sunk by the seaplanes. Could that be true?"

"I don't think so. The facts damn the man in many cases, but he did take the *Kingani* and sink the *Hedwig von Wissmann*. His own men corroborate the story, and the Belgians witnessed it all as well. And word has come in that Captain Zimmer did scuttle the *Graf von Götzen*, as Spicer-Simson stated."

"Lucky for him, or our new captain would find himself a rating. I'm sorry, now, that we gave him that promotion. I'd just as soon clap him in irons."

"I think we should tread cautiously," Sir David said. "There are several major considerations that warrant in Spicer-Simson's favor."

"I don't see what."

"You and I personally ordered him to Lake Tanganyika to attack and sink the *Kingani* and the *Hedwig von Wissmann*. He performed those duties most expediently, even capturing the *Kingani* and adding it to our own fleet. You do recall that it was the first German ship to be captured in conflict. And he took away our first German ensign after his second victory."

"Sheer luck, if you ask me," groused Sir Henry. "Is there a second reason?"

"Dr. Hanschell's official report. He diagnosed Spicer-Simson with a variety of tropical diseases that, he says, rendered the man incapable of command during the expedition's final months on the lake."

"Medical men," Sir Henry snorted. "Let them heal, but keep them out of war. Anything else?"

"General Edwards, who was far closer to the action than we are and knows the Belgians there, backs Spicer-Simson's account, up to a point."

"Up to a point!" Sir Henry waved his hand rapidly in the air for a moment then calmed down. "Well, Edwards is a good man and an able leader. He wouldn't say what he believed to be untrue." Sir Henry peered closely at Gamble. "But that doesn't make it true."

"That's so, Sir Henry, but there is one final consideration. Perhaps the most convincing of all."

"It had better be to counter dereliction of duty."

"Judge for yourself." Sir David drew from his briefcase a handful of newspapers and magazines and laid them on Sir Henry's desk. "Commander Spicer-Simson's Exploits on Lake Tanganyika" and "The Hero of the Gunboats," read the headlines on the top two, and the rest had titles as sensationalistic. Sir Henry stopped at the one that proclaimed, "Nelson Touch on African Lake."

"Are we to be bullied by the popular press?"

"They state the truth, so far as it goes, and we dare not say further. Spicer-Simson is a popular hero, and we've publicly acknowledged it by promoting him and awarding him the Distinguished Service Order. Even the Belgians have abetted with their bloody titular commission and Croix de Guerre." He held out his hands in a helpless gesture. "I don't see what we can do."

"We can put the bloody fool in his place and keep him there," Sir Henry fumed. He called in his adjutant and told him to have Captain Spicer-Simson brought immediately.

Spicer-Simson, in full-dress uniform, arrived within the hour. He was feeling buoyant. Accolades and fame were his current lot in life, and he was ready for whatever the Admiralty might now offer in the way of an assignment. Perhaps a real cruiser....

He thought he was ready.

The first sign that things weren't going right was that Sir Henry didn't let him stand at ease but made him hold his rigid posture.

"We've been reviewing the facts of your case, and we are sorely troubled," Sir David began. Then he went on to roughly detail Spicer-Simson's peccadilloes and failures in the course of the Royal Naval African Expedition. "We are all too aware that your tactless behavior might have contributed to a serious disaster."

"I consider you a rascal, if not an outright rogue," said Sir Henry. "If I had my say, you'd be on your way to the brig right now. But Sir David has provided arguments in your favor."

"You are to return to desk duty immediately," Sir David said, "where you shall remain for the duration. Sir Henry's adjutant has your assignment. Pick it up from him on your way out. Dismissed."

Captain Spicer-Simson saluted as sharply as he could manage considering the weakness in his knees. He turned and left the favor of the Admiralty.

In half an hour, he arrived at his new assignment. Or old. The desk he'd now be commanding was the exact same desk where he'd been sitting eighteen months earlier when Sir David approached the major of the Royal Marines with the proposition that he lead the Royal Naval African Expedition. He sat down, looked at the dusty desk, then at the neighboring cubbies. The sallow lieutenant was gone, but the major was still there, looking back.

"I've been reading about your exploits, Captain," the major said. "Exciting stuff. Jolly good show. Congratulations on your success and your promotion."

"Thank you, Major," Spicer-Simson said, taking out his handkerchief and wiping the dust from the desk. He didn't feel particularly well, but he managed to flash a weak grin.

9

"You're looking fit, old man," Hanschell said. And he meant it, though he'd expected to find Spicer-Simson dejected following his reassignment to his old desk in the intelligence office.

"Fit," repeated Spicer-Simson, looking down his nose at the doctor. "Yes, I suppose that covers it. I must say, that if I have to sit in that musty office every day, these certainly do help." He patted his chestful of ribbons.

The two men were sharing lunch in a popular restaurant near the building that housed Spicer-Simson's office. It was nearly a year after the expedition had returned, but Hanschell only recently had made it home. In the expedition's last weeks, he'd suffered a minor wound and a bout of fever that kept him in an African hospital for many months.

"Congratulations on those," the doctor said. "And your promotion. I understand that you received some sort of financial reward, too."

"Prize money," Spicer-Simson corrected. "For the capture of the *Kingani* and a handful of dhows."

Hanschell remembered quite clearly that all the dhows had escaped, but he said nothing as his former commander continued.

"And, of course, there was the blood money for enemy casualties," Spicer-Simson snorted. "I had to take them to court over that. Can you believe it? I save the entire African theater of war from disaster and then have to sue for my rightful due. But the court found my cause was just."

"Good for you," Hanschell responded, though his feelings were ambivalent. He could see the heroic in his former commander, but there was something pitiful, as well, and he wasn't certain he would ever entirely be able to separate the two.

A lad of about twelve took that moment to shyly approach their table, carrying a magazine in his hand.

"Pardon me, sir. Might you be the gentleman in this story?" The boy laid the magazine on their table, turned to "The Jules Verne Expedition."

"Quite right, lad," Spicer-Simson replied. "One of my favorite renditions, that."

"My mum says you might sign it for me."

"She does, does she? Which one is your mum?"

"There," the boy pointed. "With the other two ladies. Mrs. Whitburn and Miss Legee."

"The one dressed in the yellow frock?"

"That's her."

The woman smiled and gave a small wave, and Spicer-Simson flashed her his toothy grin.

"I will indeed sign your magazine," Spicer-Simson said, and he did so with a flourish.

"Was it hard?" the boy asked.

"Hard? You mean taking the boats to Africa?"

"No, sir. Killing the Germans. My dad is in France, and his letters say the war is frightfully hard on body and soul."

"It is, lad. It is hard. We must do it because it is the right thing to do, but that makes the job no more pleasant."

"My mum says she hopes there will never be another war, that we've already lost too many of our own. Do you think there will be, sir?"

"I don't know. Maybe war is mankind's nature. Be we should always strive for peace."

"I hope I don't have to go to war. The books and magazines make it sound heroic, but I don't think I'd like killing people. Not even Germans."

"There's the rub," Spicer-Simson said. "To be the hero, you have to kill."

"My dad's killed some Germans. Maybe he'll be a hero like you."

"He already is, lad. Now you run along back to your mum."

"Yes, sir. And thank you."

The boy hurried back to his own table, and Spicer-Simson turned to Hanschell, apparently dismissing the boy from his mind.

"I've been keeping up with my men," he said. "Being an intelligence officer has its advantages."

"Oh?" Hanschell said with curiosity. "What's become of them?"

"All have returned to duty, of course, but they've been flung to the winds and are in theaters of war as far as the Admiralty could scatter them. Wainwright's in...."

Spicer-Simson began a detailed accounting of where the men now were, but Hanschell listened with only half his mind. The other half was dwelling on the final part of the exchange with the boy, the part about killing to be a hero, for he noticed that Spicer-Simson, as he talked, was twisting compulsively, if absently, at the signet ring that he'd cut from the hand of Lieutenant Jung.

Acknowledgements

I want to thank:

Philip Montgomery of the Woodson Research Center at Rice University (Lee Pecht, director) for researching the Short biplanes and especially for giving the book its first test read and making some very needed suggestions.

Researchers at the London Library who corresponded with me and gave me early aid.

Chuck Thurmon, for the excellent cover.

All the family and friends who listened to me rant about the story for the last few years, and especially the ones who read drafts of the book: Melissa Kean, Steve Kean, Maggie Kean, Robert Trojanowski, Kitty Trojanowski, Linda Del Angel, and Paul Estrada.

And, of course, my wife, Julie, and my daughters, Sydney and Mariko.

Sources

I would be completely remiss if I did not cite my initial source: "Bwana Chifunga Tumbo," by Doug Hansen, which was a two-page spread in the second issue of *Forbidden Knowledge*, a sarcastic underground comic book published in 1978. It isn't exactly an accurate rendition—Hansen's conflates the two sea battles and ends before Bismarckburg and disgrace, and apparently, he didn't know the difference between Old World and New World monkeys and apes, for his depiction of Josephine includes a tail. Even so, it is the one version that most succinctly captures the crazy tenor of Spicer-Simson's expedition, not to mention the man himself.

As soon as I read Hansen's story back in 1978, I realized that this had to be real as well as incredible, and I lazily began doing some research on the subject. The more I found out, the more I was drawn in, and finally, I knew I had to write about it.

The one sure thing about history is that you're following those who've come before. And that is doubly true of writing history. *Lord of the Loincloth* can be called cumulative history because it is pulled together from a number of sources. Call it an editorial collage, if you will.

The better the source, the more I used—and abused—it. A bibliography can be found at the end of this chapter, but a few sources deserve special thanks: *The Phantom Flotilla*, by Peter Shankland; *The Battle for the Bundu*, by Charles Miller; *Military Operations: East Africa, Vol. I*, compiled by Lieutenant Colonel Charles Hordern; "Gunboats on Lake Tanganyika," by Peter Kemp; "The Great War in Africa, 1914–1918," by Byron Farwell; "The Operations on Lake Tanganyika in 1915," by Geoffrey B. Spicer-Simson; and "Transporting a Navy Through the Jungles of Africa in War Time," by Frank J. Magee. Without these sources to plunder, my effort would have been impossible.

Part of the need to portray historical characters with any accuracy lies in the use of direct quotes. So, to preserve the verisimilitude of the characters and to maintain the language they used, I have lifted from my sources a number of quotes they attribute to specific individuals and incorporated them into my narrative. Almost all of Spicer-Simson's tall tales, for example, were lifted directly (or with minor editorial changes) from Shankland's book. Since he relied on three eyewitnesses to write his account—the principal being Doctor Hanschell—I can only suppose that his renditions of Spicer-Simson's yarns are relatively accurate, and thus I've used them to a great extent to show the man in words as close to his own as possible.

Throughout, I've intentionally left out footnotes, not wanting to break the flow of the narrative, though doing so makes me feel a little like a minor Spicer-Simson, usurping credit for the work of others. But I am deeply indebted to those historians who previously have written about Spicer-Simson's expedition. Because I've left out footnotes, I'm resorting to the following clunky method to cite quotations that are direct or substantially paraphrased from my source material.

Chapter I
p. 26, direct quote: "In any case, it is both the duty and the tradition of the Royal Navy to engage the enemy wherever there is enough water to float a ship." (Shankland, p. 20; Miller, p. 197)

p. 31, paraphrase: One of Hanschell's favorites concerned an experience Spicer-Simson claimed to have had while on the Yangtze. He said he'd discovered an unknown channel—unknown to outsiders, that is. The locals knew it quite well since a great number of their kin had drowned in its rapids and whirlpools. (Shankland, p. 107)

p. 31, direct quote: "The assent was so perilous that the Chinese built a temple beside it in a grove of fir trees, and the pilots and junk men go there to pray before attempting it. I like to think that I was able to help the poor devils. I expect they still remember me." (Shankland, p. 107)

Chapter II
p. 50, direct quote: "And how do you think I got the command?" (Shankland, p. 12)

p. 50 direct quote: "Simply by eavesdropping! Admiral Gamble came in here to offer it to my colleague, a major from the Royal Marines. I listened on tenterhooks, for I saw at once what a chance it would be for me. When, much to my relief, the major turned it down, I stepped up and offered to go in his place!" (Shankland, p. 12)

p. 50, direct quote: "An amazing situation, isn't it?" (Shankland, p. 13)

p. 50, direct quote: "It would be your responsibility to get every man out to the lake alive and well and safely back again." and "Yes, I'll go," (Shankland, p. 13–14)

p. 51, direct quote: "You're going to Lake Tanganyika." and "Amy Spicer-Simson phoned me this afternoon, and we had a long talk about it." (Shankland, p. 74)

p. 56, direct quote: "As they are not convicts, you had better find proper names for them." (Farwell, p. 222)

p. 60, direct quote: "Dear boy." (Shankland, p. 44)

p. 60, direct quote: ...known to the barmaids of London's West End as "the Piccadilly Johnny with the glass eye." (Shankland, p. 44)

p. 65, direct quote: "I understand that you are joining the *navy*." (Shankland, p. 31; Farwell, p. 223)

p. 66, direct quote: "The matter is very urgent." (Farwell, p. 223)

p. 66, direct quote: "I am aware that it is urgent, thank you. Now go do as I say. Sit on a park bench and draw up a list of your requirements. And remember, you are in the *navy*. "Good luck." (Shankland, p. 31; Farwell, p. 223)

p. 66, direct quote: "Don't worry, Doctor. Everything will turn out all right. Just do the best you can." (Shankland, p. 31)

Chapter III
p. 74, direct quote: "We don't need more than a little quinine, do we? We can buy some in Cape Town." (Shankland, p. 35; Farwell, p. 224)

p. 74, direct quote: "The bar's open—come and have a drink!" (Farwell, p. 224)

p. 74, direct quote and paraphrase: "Much more than that, Doctor. At least in the tropics. It's the best thing for those sorts of climes. Whiskey, too, for that matter." The commander indicated Hanschell's own glass. "A lot of men will drink beer and wine there, but those are headachy drinks. Vermouth is the best, and really such a pleasant, wholesome drink, too. In fact, the idea of vermouth alone is attractive, for it is made from the dried flowers of chamomile, to which the later pressings of the grape have been added. One has only to smell dried chamomile flowers to find that their fragrance is that of hay meadows in an English June." (Dolby, from the chapter "Sherry and Bitters")

p. 80, direct quote: "The most dangerous animal I have ever had to face was a rhinoceros. The charges of the lion and the elephant have caused the deaths of many hunters, but the rhinoceros is the only animal who rushes upon a man without the least warning or provocation. This one, I remember—it was in wild, broken country near the banks of the Gambia River. The first I heard of him was a violent snort. I recognized its significance at once. It meant that he was suspicious but hadn't yet got wind of me. Quickly and silently—one learns to move silently when one's life depends on it—I moved to one side of the track where the undergrowth was thickest. Gently, I parted the bushes with my hands, and there he was in full view, not fifty yards away, his head raised, listening. Soon, reassured, he began feeding again, tearing up the ground with his huge horn, unearthing roots which he seized with thick, prehensile lips and crushed with his teeth. Then I noticed the birds. They were fluttering round him, alighting on his back and picking out insects from the thick folds of his skin. It was those birds which nearly cost me my life." (Shankland, p. 37-38)

p. 81, direct quote with a minor variation: "Suddenly, one let out a screech, and they all flew away. The rhino threw up his head, breathing heavily with rage, and then he scented me! In an instant, with a sound like thunder, he was pounding straight for me! You know I have my ammunition specially made for me, and my finely tempered bullet went right through the animal from stem to stern, piercing lungs, heart, spleen, and liver on the way. I was really sorry that I had to kill him, for these animals are becoming increasingly rare, but of course I had to in self-defence. His horn was exceptionally long—about twenty-seven inches, as I recollect. Anyway, it was so remarkable that I presented it to the Natural History Museum." (Shankland, p. 37–38)

p. 82, paraphrase: "I came on the spoor of a waterbuck just after noon," Spicer-Simson said. "I knew he was out there in the bush, not fifteen minutes ahead of me. I tracked him as silently as any native and came upon him feeding, completely oblivious to my approach." And "Unfortunately, he caught wind of me before I could get a clear shot, and he charged me through the bush," Spicer-Simson went on, ignorant of the side conversation between Hyde and the doctor. "Not as frightening as a rhino, I assure you, but just as deadly with those sharp antlers. My trusty rifle spoke, and he went down in a heap. In a moment I had him on my shoulders, and after an hour or two, I reached camp. We feasted that night, I can tell you." (Shankland, p. 39)

p. 82, direct quote: "....about the size of a pony" (Shankland, p. 39)

p. 85, direct quote: "Stars are my line of work." and "Is that so?" and "I certainly wouldn't know it from what you've been telling us." (Shankland, p. 41)

p. 85, direct quote: "I am a navigating officer." (Shankland, p. 41)
    p. 72, direct quote: "He'd make a damned poor navigating officer." (Shankland, p. 41; Farwell, p. 225)

p. 85, direct quote: "Furthermore, this is no man's expedition, it is the Royal *Naval* African Expedition, and I am in command." (Shankland, p. 39)

p. 85, paraphrase: All his doubts of the previous evening vanished. No matter what the men thought of him, Spicer-Simson was their commander, and he would never let them forget that fact. (Shankland, p. 39)

p. 86, paraphrase: Hanschell had not met Lee, who had left for Africa weeks before Spicer-Simson invited Hanschell to join, but the men Lee had chosen— Wainwright and Eastwood among them—seemed to be forthright, and they had nothing but good words to say about the big-game hunter. And their trust of Lee had automatically spilled over to the expedition members who did not know him. Hanschell was anxious to meet the man who had originated the plan for the expedition and even now was paving the way for it across a bleak and dangerous landscape. (Shankland, p. 39)

p. 86, direct quote: "One meets Lee's type all over Africa. They do a bit of shooting, prospecting, contracting for native labor, and so on, but really they're little more than tramps." (Shankland, p. 39)

p. 86, paraphrase: "Tosh, Doctor. Lee is a johnny-come-lately. The Admiralty had me studying the feasibility of a similar expedition long before Lee showed up. That's why they chose me to lead." (Shankland, p. 39)

p. 86, paraphrase: Hanschell, remembering that Spicer-Simson originally told him that he'd overheard the plan when Admiral Gamble offered leadership of the expedition to the major of the Royal Marines, started to say something then thought better of it. But he did not think better of Spicer-Simson for usurping Lee's role in creating the expedition. (Shankland, p. 39)

p. 87, direct quote: I can only think that the hand of God is over this expedition." and "I have no information on that, but of one thing I am sure—Spicer-Simson's hand will be over it." (Shankland, p. 40)

p. 89, paraphrase: "Of course none of *us* have been doctors...." (Shankland, p. 50)

p. 92, the dialogue is a direct quote and the rest is paraphrased: "No smoking here!" he ordered loudly. A couple of the passengers who had lit cigarettes went to the side and tossed them into the ocean. "Why ever not?" Spicer-Simson asked with exaggerated surprise. His cigarette holder was clenched in his teeth, but at the moment it was empty, and none of the other expedition officers or ratings were smoking. "Because of the danger of igniting petrol fumes," the captain answered. "Well, I'm damned. What nonsense!" Spicer-Simson snorted. "We're far out of reach of any vapors up here." "No smoking here!" the captain yelled. "Those are my orders!" Spicer-Simson looked taken aback at the captain's vehemence, and he suddenly flushed crimson. But he said nothing in reply, giving only a tight-lipped smile that drew his upper lip away from his teeth and a shrug that seemed to say it was all a small matter and beneath him. (Shankland, p. 42–43)

p. 92, direct quote: "You know what these merchant service fellows are like." and "Actually, I could have ordered the captain off the deck there and then. As a commander of the Royal Navy on the active list in time of war, I can order any merchant skipper to turn his ship over to me." (Shankland, p. 43)

p. 92, paraphrase: Cross was giving his derisive sneer. (Shankland, p. 43)

p. 93, direct quote: "Bloody liar!" (Shankland, p. 49)

Chapter IV

p. 111, direct quote: "Lieutenant Cross of the Royal Seahorse Engineers!" (Shankland, p. 49)

p. 112, paraphrase: Don't you know that horse thieves are hanged in this country?" (Shankland, p. 49)

p. 112, direct quote: "Captain Cross!" and "Telegram for Captain Cross!" (Shankland, p. 49)

p. 112, direct quote: "I can readily understand that an engineer lieutenant would want to be thought an army captain, but as he is now serving under a Royal Naval commander and in a Royal Naval expedition, I, Commander Spicer-Simson of the Royal Navy must order that in future, Engineer *Lieutenant* Cross will bear that in mind and keep his army preferences to himself until he has left the Royal Navy." (Shankland, p. 50)

p. 133, direct quote: "It doesn't really matter about the mountains. I've served at sea for so many years that I'm not likely to find any of these landsmen's problems too difficult." (Shankland, p. 51)

Chapter V

p. 162, direct quote: "And last, I want to remind you that we have another enemy more deadly than the Germans. By that, I mean tropical diseases. Very special precautions are necessary which will be explained to you by Surgeon Hanschell. Every man in this expedition, in matters of health and hygiene, will unquestioningly follow his instructions. Each of you will report daily to the surgeon for a dose of quinine. All drinking water will be boiled. Fly whisks will be issued today, and it will be the duty of every man to whisk flies off his neighbor, irrespective of rank. You, being untrained, can't tell the difference between household flies and the dreaded tsetse flies which carry sleeping sickness, so all flies are to be treated as dangerous. All ailments, all cuts, scratches, and abrasions will be reported to the surgeon. We cannot burden ourselves with useless invalids. If any man falls sick, he will be given a supply of food and water and left with one of the natives to get back to Elizabethville under his own power. Am I clear?" (Shankland, p. 61)

p. 170, direct quote: "It's my own design. Had my wife sew it up for me before I left." (Shankland, p. 142; Farwell, p. 230)

p. 170, direct quote: "I told them six inches wasn't enough beam! We need twelve inches for this terrain."(Shankland, p. 62-63)

Chapter VI

p. 193, direct quote: 'Lee's deliberate purpose in recommending that route to the Admiralty had been to lead to expedition astray.' (Shankland, p. 66)

Chapter VII

p. 211, direct quote: "I'm not surprised you got lost. You're not a seaman, are you?" (Shankland, p. 87)

p. 213, direct quote: "You don't have to go." and "It's up to you entirely." (Shankland, p. 74)

p. 221, direct quote: "We saw hundreds of them up by the Gambia River. They were a regular pest, raiding the plantations and digging up the ground nuts. We had to beat the bush to get rid of them. Of course, from my river craft, I shot more of them than anybody else. I got tired of firing at them with a rifle—only one round at a time, you know—so I changed over to my double-barreled shotgun. I bagged so many with buckshot that I had a letter of commendation from the governor. I preserved the pelts of the finest specimens in Cooper's Sheep Dip—there's lots of it out in the Gambia, you know—and I had a fur coat with a little cap to match made of them for my wife. It was the envy of all the other women! None of them could get one like it." (Shankland, p. 76)

p. 224, direct quote: "It's easily seen that none of you were trained in sail. When I was a midshipman on the *Voltage*, I would stand on the quarterdeck with a rifle and shatter a bottle, six times out of six, that was swinging from the weather yardarm of a topsail yard." (Shankland, p. 77)

p. 226, direct quote: "Why has nobody heard my shots?" and "Doesn't anybody keep watch?" (Shankland, p. 77)

p. 228, direct quote: "That dirty stiff walking about the camp over there." and "You don't deserve much credit for keeping a thing like that alive." (Shankland, p. 79)

p. 231, direct quote: "Just the same with buffalo. You've got to face up to them. It's only when they lower their heads to charge that they expose the vital spot." (Shankland, p. 78)

Chapter VIII

p. 238, direct quote: "You think we're here to protect you from the Congo?" and "No, Commander, we are here to protect the Congo from you!" and "I don't think the Congo is in much danger from us." and "We're nearly all amateurs, you know, except for the commander." and "Exactly! "That's precisely the point. You English have a genius for amateurism. That's what makes you so dangerous. It's almost always obvious what professionals are going to do, but who but amateurs could have dreamed up an expedition like this!" and "You appear to think better of our prospects than most of your colleagues." and "I think two dozen English amateurs with guns in their hands are capable of any folly—or of any heroism—and no government in its right senses would allow them to wander about unwatched. We'll be very relieved to get you out of the Katanga again, I can assure you. You British nearly took it over once, with fewer men than this!" and "You can't really believe that we have any designs against our own allies." and "Perhaps not. "I can only say that you English amateurs are unpredictable and have been quite good in the past at taking other people's colonies." and "And when we heard that the good doctor was busy burning down government property, the vice governor general thought that it was time someone should investigate." (Shankland, p. 89-90)

p. 246, direct quote: "Merde! He's at it again! Amateur! Amateur! You want to burn up all the Congo?" (Shankland, p. 93)

p. 246, direct quote: "The ground is sterilized, now." and "You may camp here if you like, sir." (Shankland, p. 92)

p. 248, direct quote: "Thank God England's got a navy," (Shankland, p. 92)

p. 248, direct quote: "All hand-picked men, recruited only from the cannibal tribes." and "They're fiercer. And better nourished." (Shankland, p. 93)

Chapter IX

p. 262, direct quote: "Have not the services of Mr. Lee been dispensed with, and has he not been so informed?" (Shankland, p. 100)

Chapter X

p. 285, direct quote: "The river has been well worth reporting as most of the previous reports are erroneous." (Shankland, p. 108)

p. 285, paraphrase: "At present," the message wound up, "there are no steamers prepared to make the journey upstream until the rains come, and I have come down with a touch of fever, making my return to Bukama impossible at this time. Tyrer and I will proceed on to Lukuga." (Shankland, p. 108)

p. 292, paraphrase: "Not really. Too many people and too much splashing. They're really very sneaky creatures and not nearly as brave as they are fearsome to look at. They'd rather creep up on a solitary man and drag him into the water with as little fuss as possible." (Shankland, p. 114-115)

p. 301, paraphrase: "I'm quite astonished at your home and wife," Hanschell said to Mauritzen. "It seems as if a touch of European elegance lies in the heart of the Dark Continent." (Shankland, p. 121-122)

p. 301-302, direct quote: "I don't know why you should be surprised, old man. European women have a peculiar propensity for being European under all circumstances. I remember how they were in Peking during the Boxer Rising, when the country was overrun by armed bands of fanatics shouting 'Death to the foreigners!' and looting, burning, and murdering. Of course, things should never have been allowed to come to such a pass. The situation had been badly handled. By the time I arrived on the scene, the nine hundred European residents, including two-hundred-and-fifty women and children, had taken refuge in the British Legation, and an international force was trying to get through to them. We had to go up the Pei-ho River to Tien-tsin and, from there, fight our way to Peking, ninety miles farther on. The ancient Chinese city was surrounded by a wall thirty feet high. Although I had been gravely wounded, I easily outdistanced my men and was the first to scale the wall. Fighting off the Boxers guarding the Shaow-men Gate, I threw it open from the inside, and the bluejackets rushed in. When we reached the legation by the sluice-gate entrance, I had the biggest surprise of my life. We were on a green, sunny lawn where beautifully dressed women were strolling up and down with parasols. They were half-starved, and they had been living for the past two months in the constant expectation of being massacred, but they looked as if they were at an English garden party. The contrast with the filth and horror surrounding them on all sides could not have been greater. I was reminded of this when we entered the Mauritzen's barge." and "There was, however, one important difference, sir." and "Mrs. Mauritzen shows not the least desire to be rescued." (Shankland, p. 121-122)

p. 304, direct quote: "...disgusting! Go get some clothes on! Where do you think you are, back in Donegal?" (Shankland, p. 123)

p. 314, direct quote: "But I'm told that no prisoners escape then. They can't get through the wall of mosquitoes that surround it!" (Shankland, p. 126)

Chapter XI

p. 327, direct quote: "There was plenty of game before the war, but we've been short of meat, and now there's little left to shoot." and "What a pity. The hunting of big game is a passion with me. It's given me some of the most exciting moments of my life. I once had a remarkable day's shooting in the Galapagos Islands—the place, you know, where Darwin did his work. They're teeming with wild pigs and goats, and especially with wild cattle. In most countries, you know, it's considered too dangerous to hunt these animals except on horseback, but although there were no horses, I was determined to try my luck. It seemed a most uninviting place, all scrub and cinders, but it was a perfect day, and when we had climbed to a thousand feet or so, we came to a beautiful plateau with hills and woodlands—lots of cover for the animals—and good pasture in the valleys. I always took my chief petty officer with me; the best gun bearer I ever had. The wild cattle were really wild—not used to being hunted. You had to kill, not wound. The first to fall, I remember, was a huge black bull with wicked-looking horns. I dropped him at forty yards, but before I could go up and examine him, two more thundered down on us. I bagged them with a left and right. 'There's another coming up astern, sir,' my CPO told me, putting a loaded rifle into my hand. I turned and saw that the one I had thought dead was charging us, not ten yards distant, head down, tail in the air, foam flying from his blood-stained nostrils. I gave him a ball between the eyes, and he fell dead at my feet." (Shankland, p. 133-134)

p. 327, paraphrase: Spicer-Simson paused to stuff another forkful of mutton into his mouth, not noticing the slightly horrified expression on the women's faces. Hanschell, remembering the last "bull" the commander had bagged at pointblank range while it was staked and tied like a lamb at the slaughter, wondered if it would eventually become a wild, dangerous, thundering behemoth in some future retelling. (Shankland, p. 134)

p. 328-329, direct quote: "As soon as war was declared, I was ordered to take command of a flotilla of destroyers based at Harwich. Night after night, we put to sea, hoping to intercept the German forces that from time to time made lightning raids on our vital communications with France. It was not very spectacular, you know—days and nights of cold and wet discomfort, but always with the possibility that something might happen. I wonder if any of you know, or can imagine, what it is like to be on a night patrol off the enemy coast in winter? There is little protection on the bridge of a destroyer, and one feels damnably alone, surrounded by impenetrable darkness. There is no light but the dim glow of the binnacle, the gleam of the bow wave, and the receding wake. One is almost mesmerized by the hiss of the water and the even throbbing of the engines—then suddenly a far-off searchlight stabs the darkness and goes out again. The dense smoke rolling from the funnels makes a blacker cloud against the leaden surface of the sea. Two hours before dawn. Action stations! I always close up the guns' crews and have everything in readiness for the first gray light of morning, for who knows what it will disclose? Of course, one has done it a hundred times, but it is impossible to relax, for one day it will happen, as it happened to me, that the first streak of dawn discloses an enemy vessel. Then it's a question of getting off the first salvo—a matter of seconds between destroying and being destroyed! But this time, the enemy wasn't another destroyer, but a cruiser! Our first salvo was right on the target—but what could our two-inch guns do against the six-inch guns of our huge antagonist? Columns of water sprang up all around us—she had opened fire and straddled us! The descending columns cascaded across our decks. I take no credit for what happened next, for we really had no alternative. It was impossible to retreat—we would have been blown out of the water long before we were out of range. I rang for full ahead and, zigzagging wildly, closed the enemy! At only 1,000 yards range, I loosed off all my torpedoes, and one of them struck her amidships! Then I turned and made off under the protection of a smokescreen. It was purely a matter of self-defense, you see. I had little hope, perhaps little intention, of sinking the cruiser. One's actions in such cases are purely instinctive, but we heard later that after a long tow by her escorting destroyers, she capsized and sank just off the Jade Estuary." (Shankland, p. 134-136)

p. 329, paraphrase: Hanschell, knowing that the only torpedoing Spicer-Simson had taken part in had been on the receiving end when the *Niger* had been sunk while anchored, kept a straight face that masked his pity. (Shankland, p. 136)

p. 333, direct quote: "I designed it myself, and my wife made it for me. (Shankland, p. 142; Farwell, p. 230)

p. 334, direct quote: "...two little cruiser automobiles." (Shankland, p. 139)

p. 342, direct quote: "The storm was ferocious. It broke over our camp in a hurricane of wind accompanied by ear-splitting bursts of thunder and vivid lightning, which illuminated the country for miles around. The lake itself became a raging sea, enormous breakers rolling up and crashing on the shore, uprooting trees, and demolishing native huts...." (Magee, p. 349)

p. 351, direct quote: "Of course it's an ingenious idea! All my life I've had ingenious ideas; the difficulty is to get people to accept them. I remember in the Channel Maneuvers of 1905—it was to test the efficiency of submarines—we had been told that the exercise was to be as realistic as possible. I had the idea of passing a line to another destroyer and dragging for periscopes. The submariners complained that I nearly sank one of their boats, but that doesn't matter. The point is that the idea worked! Of course, in peacetime, there's little scope in the Home Fleet for an officer with ingenious ideas—that's why I went to China. It was the result, as it happened, of another exercise, this time to test the defenses of Portsmouth Harbor. The destroyer I had the honor to command was part of the attacking force. We approached under cover of night and it was, I thought, a wonderful opportunity to distinguish myself. I went full ahead and made straight for the harbor entrance! Unfortunately, the coxswain was dazzled for a moment by searchlights, and instead of crashing into the harbor boom as I had intended, I piled my ship up on the beach at Southsea. I was court-martialed for hazarding my ship—and what was the result? I was able to demonstrate that mine had been the most dashing attack of all, and that I had got nearer to our objective than any other captain. So I was exonerated; even highly commended! But it so happened that there wasn't another destroyer available for me at the time, so I was given command of a gunboat on the Yangtze River. I was tired of fleet maneuvers, anyway. I wanted some real action, and that was the place for it! You will find that the best officers of the Royal Navy, the men to whom the country turns in its hour of need, all had their training on the China Station, and in gunboats." (Shankland, p. 145)

Chapter XII

p. 358, direct quote: "When I was on the Gambia, even the governor acknowledged the superiority of my rank to his." and "I understood that the governor of a colony took precedence over everybody." and "Not in this case. The governor of Gambia, before he joined the colonial service, had been a major in the army. A major does not forfeit his rank on resigning; therefore, he was still a major, and as such, had to acknowledge the seniority of my rank of lieutenant-commander. We served together on the Boundary Commission. But these Belgians don't even know the difference between a commandant and a commander." (Shankland, p. 148)

p. 384, direct quote: "What's the matter with you? Aren't you a signalman? Aren't you qualified?" and "Yes, sir. It's only your semaphore I can't read, sir." (Shankland, p. 155)

p. 385, direct quote: "The Belgians, for certain reasons, are not greatly esteemed in the Congo, but these people have learnt that we English are a different sort of white man, (Spicer-Simson, p. 755)

p. 387, direct quote: "What's this? A whorehouse? Take all that down and burn it." (Shankland, p. 155; Farwell, p. 232)

p. 387, direct quote: "It's not fair! The Admiralty never intended that I should go afloat. I was supposed to command the base, to be responsible for discipline, for naval routine, for arranging supplies. Lee and Hope were supposed to command the launches. Now, owing to their disgraceful defection, I have to go out in the launches and take all the risks of a naval action, because there isn't another worthy seaman among you!" (Shankland, p. 157)

Chapter XIII

p. 393, direct quote: "Chief Waterhouse, dismiss the men. All hands clean into fighting rig." (Shankland, p. 158; Farwell, p. 233)

p. 393, direct quote: "Man the launches for immediate action." (Shankland, p. 158)

p. 394, direct quote: "Nonsense! You're much too valuable to risk afloat!" (Shankland, p. 158)

p. 394-395, direct quote: "It will be all right. I've felt all along that the hand of God is over this expedition." and "Why shouldn't it be equally over the Germans?" and "You'll see. We'll all get home safely, every one of us." (Shankland, p. 159)

p. 398, direct quote: "Die Engländer sind hier!" (Shankland, p. 165; Farwell, p. 235)

p. 402, direct quote: "Your gallant commander has rammed his prize!" (Shankland, p. 160-161; Farwell, p. 235)

- p. 408, direct quote: "He was quite unmoved. He stood there in full view of the enemy, his long cigarette holder in his mouth and his eyes glued to his binoculars. The only thing that shifted him was the collision when I rammed the prize. He fell flat, and when he got up, he was laughing!" (Shankland, p. 162)
- p. 408, direct quote: "And it wasn't easy, sir." and "Sometimes the gun was pointing down at the sea, sometimes up at the sky. And the commander worried me, sir. He was shouting all the time, and I couldn't hear a word he said, what with the wind and the spray and the roar of the motors—and him with that cigarette holder between his teeth. He must have thought I put up a very poor show." (Shankland, p. 163)
- p. 408, direct quote: "How many rounds did we fire?" and "Thirteen, sir." (Shankland, p. 163)
- p. 409, paraphrase: "Twelve hits out of thirteen, Chief. I call that wonderful shooting considering you were on your knees doing it." (Shankland, p. 163)
- p. 409, direct quote: "Certainly not. Petty Officer Flynn couldn't have possibly scored a hit without a qualified naval officer to spot for him and give him the ranges." and "Brum Waterhouse can have all the hits, sir. I'm happy so long as none of theirs hit me." (Shankland, p. 163)

Chapter XIV
- p. 422, direct quote: 'His Majesty the King desires to express his appreciation of the wonderful work carried out by his most remote expedition.' (Shankland, p. 167; Miller, p. 206; Magee, p. 359)
- p. 424, paraphrase: "...demoralizing Africa." (Shankland, p. 207; Farwell, p. 238)
- p. 429: "They can't be waterspouts. They're not moving." (Shankland, p. 174)
- p. 429, direct quote: "These are waterspouts. Very dangerous things! Only local atmospheric pressures prevent them from moving. At sea, one learns to give them a wide berth. I've seen much bigger ones off the China Coast— tremendous vortices of thousands of tons of water. I remember one incident as captain of a gunboat on the Yangtze River. I'm expert, you know, at speaking Chinese, and I had warned a large junk to alter course to avoid one of these waterspouts: They're enormously strong craft, hundreds of years old, some of them—three masts, sails of matting. The gear's so heavy it takes a dozen men to steer them. The Chinese believe that the air is full of devils, and that particularly nasty ones run up and down the waterspouts, and so at the head of the mainmast, they fix a long painted bamboo. If a devil sits on it, it breaks, and he falls into the sea. Inge-

nious idea, but that junk would have done better to listen to my advice. The waterspout broke over it—my little gunboat was nearly capsized by the huge wave set up by the falling water. When the dense cloud of spray had blown away, there was nothing left of the junk but a few shattered timbers and a little painted bamboo stick bobbing up and down in the sea." (Shankland, p. 174-175)

p. 441, direct quote: "Navyman god!" (Shankland, p. 206)

p. 441, direct quote: "Navyman god. You come see?" (Shankland, p. 206)

Chapter XVI

p. 459 direct quote: "Nothing for you, Doctor. We've no casualties, and there are no wounded survivors." (Shankland, p. 189)

p. 459 direct quote: "We'd never have done it if it hadn't been for Mr. Wainwright. And a pretty close thing it was, sir. It was our last two shells that got her!" (Shankland, p. 190)

p. 463 direct quote: "Commander Spicer-Simson, here is Senior Lieutenant Job Odebrecht, on your invitation, to join our mess." (Shankland, p. 192)

p. 471 direct quote: 'I doubt whether any one tactical operation of such miniature proportions has exercised so important an influence on enemy operations.' (Farwell, p. 242)

p. 473 direct quote: "HMS Fifi, 9 February, 1916." and "Fundi, RN" (Shankland, p. 201; Miller, p. 209)

p. 473, direct quote: "Her big guns don't scare anyone, sir. We wouldn't have been the ones to sink—we were on a winning run! But I suppose our luck will change now." (Shankland, p. 199)

Chapter XVII

p. 485, direct quote: "Why have you left the lake?" (Shankland, p. 204)

Chapter XVIII

p. 509, direct quote: "What can I do for you, gentlemen?" and "Oh, just coming in." and "What do you mean: Just coming in?" and "Well, you've surrendered." and "I've done nothing of the sort. In fact, you are now my prisoners." and "Like hell we are! You're ours! Grab him!" (Farwell, p. 245)

p. 510, direct quote: "I don't imagine that many actions have been terminated by a whistle like a Football Association cup tie." (Farwell, p. 246)

p. 510, direct quote: "Oh, hell! It's the Union Jack!" (Shankland, p. 211)

p. 511, direct quote: "Where the hell were you navy chaps? The Germans sneaked out, and you let them get away by sea. We had the fort surrounded." (Shankland, p. 211; Farwell, p. 247)

p. 511, direct quote: "...he's straight up there. You can't miss it. Just follow your nose." (Shankland, p. 212; Farwell, p. 247)

p. 512, direct quote: "Chase me, Charlie!" and "Ooh-la-la!" and "Kiss me, Gertie!" (Shankland, p. 212; Farwell, p. 247)

p. 513, direct quote: "By God! The commander's had it in the neck! I'd like to...." (Shankland, p. 213)

Chapter XIX

p. 529, direct quote: "At the very least. Before taking a step that severed cooperation before the command of the lake was assured, he should have written to Colonel Moulaert, received a reply, and transmitted the correspondence to the Admiralty. His tactless behavior might have contributed to a serious disaster." (Shankland, p. 218)

If I missed correctly attributing any passages or failed to mention a direct contribution, please forgive me, for such was not my intention.

Bibliography

Barnes, C. H. *Shorts Aircraft Since 1900*. (Putnam, 1967, p. 104–105.)
Beard, Peter H. *The End of the Game*. (Doubleday, 1977.)
Boyes, John. *King of the Wa-Kikuyu*. (St. Martin's Press, 1993.)
Burke, Fred. *Africa: Selected Readings*. (Houghton Mifflin Company, 1969.) Burton, Anthony. *Traction Engines: Two Centuries of Steam Power*. (Chartwell Books, Inc., 2000.)
Casey, Louis S. and John Batchelor. *The Illustrated History of Seaplanes and Flying Boats*. (Hamlyn Publishing Group/Phoebus Publishing Company/BPC Publishing Company Ltd., 1980, p. 19.)
De Weerd, H. A. "The Tanganyika Expedition." (*United States Naval Institute Proceedings*, October 1932, p. 1461.)
Dolbey, Robert V., Captain, RAMC. *Sketches of the East African Campaign*. (John Murray, 1918.) (Also available at Project Gutenberg: http://www.gutenberg.org/.)
Paul Elk Ltd. *The Conquest of the Air*. (Paul Elk Ltd., 1972, p. 127.)
Farwell, Byron. *The Great War in Africa: 1914–1918*. (Norton, 1986, p. 206–249.)
Forbath, Peter. *The River Congo*. (E.P. Dutton, 1979.)
Forester, C. S. *The African Queen*.
Hansen, Doug. "Bwana Chifunga Tumbo." (*Forbidden Knowledge* #2, Last Gasp Publishing Co., 1978, p. 20–21.)
Harvey, Donald Joseph. "World War I." (Source unknown.)

Hordern, Charles, Lieutenant-Colonel. *Military Operations: East Africa, Vol. I, August 1914–September 1916*. (His Majesty's Stationery Office, 1941, p. 192–194.)

Kemp, Peter. "On Lake Tanganyika." *The Marshall Cavendish Illustrated Encyclopedia of World War I*. Young, Peter, Brigadier, ed. (Marshall Cavendish, date unknown).

Magee, Frank. "Transporting a Navy Through the Jungles of Africa in War Time." (*National Geographic Magazine*, Vol. XLII, #4, October 1922, p. 331–362.) Marshall, S. L. A., Brigadier General USAR (Ret.). "The War in Africa." (*The* American Heritage History of World War I. Random House, 1966.)

Mbiti, John S. Introduction to African Religion. (Praeger Publishers, 1975.) McLynn, Frank. Hearts of Darkness: The European Exploration of Africa. (Carroll & Graf Publishers, Inc., 1992.)

Miller, Charles. *Battle for the Bundu: The First World War in East Africa*. (MacMillan, p. 194–211.)

Miller, Denis. *A Source Book of Traction Engines*. (Ward Lock Ltd., 1983.) Newbolt, Henry. *Naval Operations, Vol. IV*. (Longmans, Green and Co., 1928, p. 80–85.)

Page, Melvin E. *Africa and the First World War*. (St. Martin's Press, 1987, p. 186–207.)

Perrett, Bryan. "Spicer-Simson Pulls It Off." (*Gunboat! Small Ships at War*, Castle Books, 2000, p. 128–142.)

Shankland, Peter. The Phantom Flotilla: The Story of the Naval African Expedition, 1915–16. (Collins, 1968.)

Sibley, J. R., Tanganyikan guerrilla: East African campaign, 1914–1918 (Ballantine, 1971, Pan/Ballantine, 1973.)

Speake, Graham, ed. *Cultural Atlas of Africa*. (Andromeda Oxford, 1998.) Spicer-Simson, G. B. "The Functions of the Fleet." (Typescript of lecture given by Spicer-Simson to the Intelligence Division of the Naval Staff in Calais on 6/9/18, 2 pages.)

Spicer-Simpson. "The Operations on Lake Tanganyika in 1915." (*RUSI Journal*, Royal United Service Institution, Vol. LXXIX, #516, November 1934, p. 752–764.)

Spicer-Simson, G. "The Tanganyika Expedition." (*U.S. Naval Institute* Proceedings. September 1933, p. 1337.)

Stoecker, Helmuth. German Imperialism in Africa: From the Beginning until *the Second World War*. (C. Hurst & Co./Humanities Press International, 1986, p. 93–280.)

United States Naval Institute. "The Naval War on Tanganyika Lake." (*United States Naval Institute Proceedings*. May, 1933, p. 729–733.)

United States Naval Institute. "The Tanganyika Expedition." (*United States Naval Institute Proceedings*. September, 1933, p. 1337–1339.)

von Berchem, Beda "The Naval War on Tanganyika Lake." (*United States Naval Institute Proceedings*, May 1933, p. 729.)

Who's Who. "Spicer-Simson, Commander Geoffrey Basil." (*Who's Who, 1943*, p. 2925.)

Wood, W. J. Leaders and Battles: The Art of Military Leadership (Novato, CA: Presidio Press, 1984, p. 272–300.)

Young, Francis Brett. *Marching on Tanga*. (Alan Sutton, 1984.)

Also from Phosphene Publishing Co.
The Clay Guthrie Mysteries

THE DEAD DETECTIVE

Teetering on the edge of the gutter, ex-cop Clay Guthrie is offered a way out of his bitter isolation. All he has to do is locate a stolen sculpture. The task seems simple enough until Guthrie finds himself enmeshed in a series of surreal events that push him to the breaking point. His disturbingly dangerous employers threaten him with pain and death if he fails, and the mysterious old man who is their antagonist forces Guthrie to act on his behalf, warning that worse horrors will greet his success. The only way Guthrie can survive is to find the sculpture and help the old man destroy the terrible power that lives within it. But first, he must endure a series of trials that test his endurance and drive him into the core of his own corruption.

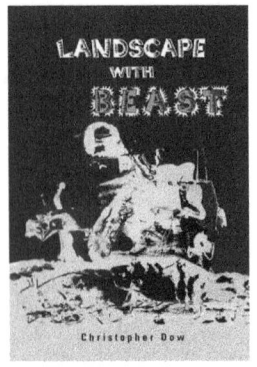

LANDSCAPE WITH BEAST

"Who better to send into a grave situation to find out what lies buried there than one who has known death? Such as you." With those words, mysterious old Tereba sends Guthrie on the trail of a missing artist. Having to deal with a witch from an ancient lineage and the ultimate hunter seeking the ultimate prey didn't bother him, but the doorway to another world was a different matter. Out there an unknowable predator waited, and it wanted nothing more than to lay waste to everything in its path. But Guthrie couldn't refuse. He knew that anythingTereba directed his way would be as interesting and important as it might be dangerous, and those were lures he couldn't resist. Besides, when he set a trap for his nemesis, the bait wasn't the only thing that disappeared into the unknown along with the artist. Now Guthrie's client had vanished, too.

THE TEXAS TROLL UNLIMITED

When a frightened railroad employee tells Clay Guthrie that a monster in a boxcar ate his co-worker, Guthrie finds himself drawn into a web of corrupt and warped ambition and wanton violence. Traveling to far West Texas in search of the monster, Guthrie and the trainman encounter an organization whose goal is the total destruction of social order and whose weapon is an abomination from the past. Waging a guerrilla war against their enemies beneath the harsh Texas sun, they quickly discover that the nights hold a mortal danger more terrible than their human enemies. With the fate of civilization in the balance, they must eliminate the humans who stand in their way before they can root out and confront a canny and clever inhuman foe.

DARKNESS INSATIABLE

Clay Guthrie is sent by his mysterious employer to track down a missing man, but finding the objet of his search in an unnatural place and in an impossible condition provides no easy answers. Far worse, he encounters a town the grip of an unknown, unseeable, and malevolent force that thrives on turmoil and destruction and has left the utter annihilation of three other towns in its wake. What will it take to learn the cause and remedy it before it's too late? And who—or what—will get in the way?

Phosphene Publishing Company

publishes books and DVDs relating to literature, history, the paranormal, film, spirituality, and the martial arts.

For other great titles, visit
phosphenepublishing.com

3.1 (11/24)